Fungal Biology

4th edition

Jim Deacon

Institute of Cell and Molecular Biology, University of Edinburgh, UK

Blackwell
Publishing

BLACKWELL PUBLISHING
350 Main Street, Malden, MA 02148-5020, USA
9600 Garsington Road, Oxford OX4 2DQ, UK
550 Swanston Street, Carlton, Victoria 3053, Australia

First edition published 1980
Second edition published 1984
Third edition published 1997
Fourth edition published 2006 by Blackwell Publishing Ltd

1   2006

Library of Congress Cataloging-in-Publication Data

Deacon, J.W.
  Fungal biology / J.W. Deacon.—4th ed.
      p. ; cm.
  Rev. ed. of: Modern mycology. 3rd ed. 1997.
  Includes bibliographical references and index.
  ISBN-13: 978-1-4051-3066-0 (pbk. : alk. paper)
  ISBN-10: 1-4051-3066-0 (pbk. : alk. paper)
  1. Mycology.   2. Fungi.
  [DNLM: 1. Fungi.   2. Mycology. QK 603 D278i 2006]   I. Deacon, J.W.
Modern mycology.   II. Title.

  QK603.D4 2006
  579.5—dc22

                                                2005004137

A catalogue record for this title is available from the British Library.

Set in 8/10.5pt Stone Serif
by Graphicraft Limited, Hong Kong
Printed and bound in England
by TJ International, Padstow, Cornwall

The publisher's policy is to use permanent paper from mills that operate a sustainable forestry policy, and which has been manufactured from pulp processed using acid-free and elementary chlorine-free practices. Furthermore, the publisher ensures that the text paper and cover board used have met acceptable environmental accreditation standards.

For further information on
Blackwell Publishing, visit our website:
www.blackwellpublishing.com

Contents

Preface

Fungal Biology (4th edition) is the successor to three previous editions of "Modern Mycology." The text has been fully updated and expanded to cover many new developments in fungal biology. Each of the 17 chapters is largely independent, with a clear theme and cross-referencing, so that the text can be used to focus on selected topics.

The early chapters deal with the unique structure and organization of fungi and fungus-like organisms, including modern experimental approaches in fungal biology, and the many ways in which fungi respond to environmental cues. These chapters also cover the diversity of fungi, and fungal products including immunosuppressants, antibiotics, and mycotoxins that contaminate food.

Recent developments in fungal genetics, molecular genetics, and genomics are discussed within the framework of a "biochemical and molecular toolbox," using in-depth examples such as the roles of virus-like double-stranded RNA for the control of chestnut blight, and the population dynamics of Dutch elm disease. Major sections of the text deal with the development of fungi as commercial biological control agents of plant pathogens and insect pests. In addition, one of the three new chapters deals with the symbiotic associations of fungi with plants and animals, and the biology of lichens. Plant pathogens and plant defense also are covered in depth, using selected examples of all the major pathosystems.

Two final chapters are devoted to the "moulds of man," covering the biology, pathogenicity, and virulence factors of the major fungal diseases of humans, and the antifungal drugs used to treat these conditions.

This text is designed to appeal to both undergraduates and postgraduates. The emphasis throughout is on the functional biology of fungi, with several examples from recent research, and many tables and illustrations. The text is supported by a comprehensive website (available via **www.blackwellpublishing.com/deacon**), with over 600 images, many in color, including "Special Focus Topics" and "Profiles of Significant or Interesting Fungi." My own images are identified, and can be used freely, without restriction. The website also has a large interactive (randomized) test bank of multiple-choice questions, designed to aid self-assessment and reinforcement of key learning outcomes.

I wish to thank many colleagues who have contributed to this book by providing images and resources. They include many of my doctorate students, and Nick Read's research group at the University of Edinburgh, who have been supportive throughout.

Jim Deacon
Edinburgh

Chapter 1

Introduction: the fungi and fungal activities

This chapter is divided into the following major sections:

- the place of fungi in the "Tree of Life" – setting the scene
- the characteristic features of fungi: defining the fungal kingdom
- the major activities of fungi as parasites, symbionts and saprotrophs
- fungi in biotechnology

Fungi are a unique group of organisms, different from all others in their behavior and cellular organization. Fungi also have an enormous range of activities – as pathogens of crop plants or humans, as decomposer organisms, as experimental "model organisms" for investigating genetics and cell biology, and as producers of many important metabolites. The uniqueness of fungi is a prominent feature of this book, which adopts a functional approach, focusing on topics of inherent interest and broad significance in fungal biology.

The uniqueness of fungi is reflected in the fact that they have the status of a **kingdom**, equivalent to the plant and animal kingdoms. So, fungi represent one of the three major evolutionary branches of multicellular organisms.

In terms of biodiversity, there are estimated to be at least 1.5 million different species of fungi, but only about 75,000 species (5% of the total) have been described to date. For comparison, there are estimated to be 4.9 million arthropod species and about 420,000 seed plants (Hawksworth 2001, 2002).

If the estimate of the number of fungal species is even remotely accurate then we still have much to learn, because even the fungi that we know about play many important roles. To set the scene, we can mention just a few examples:

- Fungi are the most important causes of **crop diseases**, responsible for billions of dollars worth of damage each year, and for periodic devastating disease epidemics.
- Fungi are the main **decomposers** and recyclers of organic matter, including the degradation of cellulose and wood by the specialized enzyme systems unique to fungi.
- Fungi produce some of the most toxic known metabolites, including the **carcinogenic aflatoxins** and other **mycotoxins** in human foods and animal feedstuffs.
- With the advance of the acquired immune deficiency syndrome (AIDS) and the increasing role of transplant surgery, fungi are becoming one of the most significant causes of death of **immunocompromised** and **immunosuppressed patients**. Fungal diseases that were once extremely rare are now commonplace in this sector of the population.
- Fungi have an enormous range of **biochemical activities** that are exploited commercially – notably the production of antibiotics (e.g. **penicillins**), **steroids** (for contraceptives), **ciclosporins** (used as immunosuppressants in transplant surgery), and enzymes for food processing and for the soft drinks industry.
- Fungi are **major sources of food**. They are used for bread-making, for mushroom production, in several traditional fermented foods, for the production of Quorn™ mycoprotein – now widely available in supermarkets and the only survivor of the many

"single-cell protein" ventures of the late 1900s – and, of course, for the production of alcoholic drinks.

- Fungi can be used as "cellular factories" for producing **heterologous** (foreign) **gene products**. The first genetically engineered vaccine approved for human use was produced by engineering the gene for hepatitis B surface antigen into the yeast (*Saccharomyces cerevisiae*) genome. In this way the antigen can be produced and exported from the cells, then purified from the growth medium.
- The **genome sequences** of several fungi have now been determined, and in several cases the genes of fungi are found to be homologous (equivalent) to the genes of humans. So, fungi can be used to investigate many fundamental cell-biological processes, including the control of cell division and differentiation relevant to biomedical research.
- Fungi are increasingly being used as commercial **biological control agents**, providing alternatives to chemical pesticides for combating insect pests, nematodes, and plant-pathogenic fungi.

The first part of this book (Chapters 1–9) deals with the growth, physiology, behavior, genetics, and molecular genetics of fungi, including the roles of fungi in biotechnology. This part also includes an overview of the main fungal groups (Chapter 2). The second part (Chapters 10–16) covers the many ecological activities of fungi – as decomposers of organic matter, as spoilage agents, as plant pathogens, plant symbionts, and as pathogens of humans. A final chapter is devoted to the ways of preventing and controlling fungal growth, because this presents a major challenge in modern *Fungal Biology*."

The place of fungi in the "tree of life" – setting the scene

The **Tree of Life Web Project** is a major collaborative internet-based endeavor (see Online resources at the end of this chapter). Its aim is ultimately to link all the main types of organism on Earth according to their natural phylogenetic relationships. The hope is that this will lead us closer to the very root of life on earth, which is currently estimated to be some 3.6–3.8 billion years ago (1 billion = 1000 million years; 10^9 years). However, fungi arrived much later on the scene. The oldest known fossil fungi date to the Ordovician era, between 460 and 455 million years ago – a time when the largest land plants are likely to have been bryophytes (liverworts and mosses). This accords remarkably well with recent phylogenetic analyses based on comparisons of gene sequences, discussed below.

Carl Woese of the University of Illinois at Urbana-Champaign, USA, has championed the use of molecular phylogenetics. The basis of this is to identify genes that are present in all living organisms and that have an essential role, so they are likely to be highly conserved, accumulating only small changes (mutations and back mutations) over large spans of evolutionary time. Comparisons of these sequences can then indicate the relationships between different organisms. There are limitations and uncertainties in this approach, because of the potential for lateral gene transfer between species and because there are known to be variable rates of gene evolution between different groups of organisms. However several highly conserved genes and gene families can be used to provide comparative data.

Most phylogenetic analyses are based primarily on the genes that code for the production of **ribosomal RNA**. Ribosomes are essential components of all living organisms because they are the sites of protein synthesis. They occur in large numbers in all cells, and they are composed of a mixture of RNA molecules (which have a structural role in the ribosome) and proteins. In **prokaryotes** (non-nucleate cells) the ribosomes contain three different size bands of ribosomal RNA (**rRNA**), defined by their sedimentation rates (S values, also known as Svedberg units) during centrifugation in a sucrose solution. These three rRNAs are termed 23S, 16S, and 5S. In **eukaryotes** (nucleate cells) there are also three rRNAs (28S, 18S, and 5.8S). The genes encoding all of these rRNAs are found in multiple copies in the genome, and the different rRNA genes can be used to resolve differences between organisms at different levels.

For most phylogenetic analyses the genes that code for 16S rRNA (of prokaryotes) and the equivalent 18S rRNA (of eukaryotes) are used. These **small subunit rDNAs** contain enough information to distinguish between organisms across the phylogenetic spectrum. Using this approach, several different phylogenetic trees have been generated, but many of them are essentially similar, and one example is shown in Fig. 1.1.

Several points arise from Fig. 1.1, both in general terms and specifically relating to fungi.

- Ribosomal DNA sequence analysis clearly demonstrates that there are three evolutionarily distinct groups of organisms, above the level of kingdom. These three groups – the **Bacteria**, **Archaea**, and **Eucarya** (eukaryotes) – are termed **domains** and the differences between them are matched by many differences in cellular structure and physiology.
- Beneath the level of domains, there is still uncertainty about the taxonomic ranks that should be assigned to organisms. Plants, animals, and fungi are almost universally regarded as separate **kingdoms**

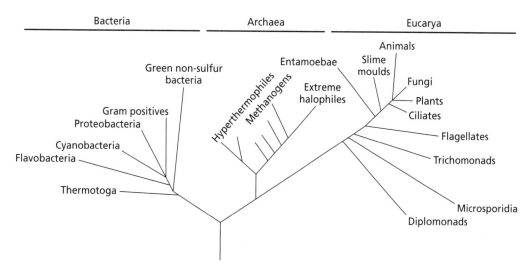

Fig. 1.1 A representation of the **Universal Phylogenetic Tree**, based on comparisons of the genes encoding small-subunit (16S or 18S) ribosomal RNA. The lengths of the lines linking organisms to their nearest branch point represent inferred evolutionary distances (rRNA gene sequence divergence). (Based on a diagram in Woese (2000) but showing only a few of the major groups of organisms.)

(Whittaker 1969). But, arguably, this status could also apply to the many "kingdoms" of bacteria, especially the enormous **Proteobacteria** kingdom which includes most Gram-negative bacteria. And, it could be argued that the many separate groups of unicellular eukaryotes (amoebae, slime moulds, flagellates, etc.) should also be regarded as kingdoms, based on their apparently long-term separation as judged by rDNA sequence divergence. However, many of these lower eukaryotes are still poorly studied, so they are often referred to collectively as "protists," pending further resolution of their relationships.

- The major multicellular organisms – the **animals**, **plants**, and **fungi** – form a cluster at the very top of the Eucarya Domain, so they are often termed the "**crown eukaryotes**". The interesting feature of these groups is that they seem to have diverged from one another at roughly the same time, and then underwent a major, rapid expansion and diversification. The time when this happened, **roughly half a billion years ago**, coincides with the period when the land surfaces were colonized by primitive plants such as bryophytes (mosses and liverworts) and when there were only three major continental land masses: (i) a land mass including present-day North America and Europe, located near the equator; (ii) part of modern Siberia, towards the north; (iii) a land mass consisting of present-day South America, Africa, Antarctica, India, and Australia in the southern hemisphere.

- Currently, the earliest **fossil evidence of fungi** dates to the Ordovician period, between 460 and

455 million years ago, but it is almost certain that aquatic fungi would have been present before that time, perhaps dating back to about 1 billion years ago. The Chytridiomycota are widely believed to be among the most ancient of the presently known fungi – not least because they have motile flagellate cells, indicating their dependence on free water. By contrast, in the Devonian period (417–354 million years ago) there is abundant evidence of fossil fungi associated with primitive land plants. For example, representatives of several major groups of fungi have been found in the Rhynie Chert deposits of Aberdeenshire, Scotland, representing the Devonian era. The early fossil fungi of the Rhynie deposits are very well preserved and, intriguingly, occur in close association with the underground organs of early land plants. These early terrestrial fungi, belonging to a newly defined group, the Glomeromycota (see Fig. 2.4), are remarkably similar to the arbuscular mycorrhizal fungi that colonize the roots of nearly 80% of present-day land plants (Fig. 1.2). So it seems that these fungi co-evolved with early land plants, and that their hyphae could have facilitated the uptake of mineral nutrients and water from soil, just as they do today (Lewis 1987; Chapter 13).

- Having made the case for a long-term association between fungi and land plants, we need to correct a widely held misconception: there is now **strong evidence that fungi are more closely related to animals than to plants** (Baldauf & Palmer 1993). The fungi evolved as an early branch from the animal

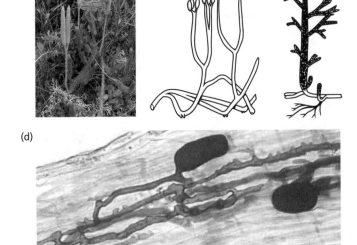

(a) (b) (c)

(d)

Fig. 1.2 (a) A present-day "club-moss," *Lycopodium*, which represents a primitive member of the ferns (pteridophytes), and (b,c) two fossil pteridophytes (*Asteroxylon mackiei*, and *Rhynia major*) from the Rhynie chert deposits. (d) Swollen vesicles of a present-day mycorrhizal fungus are remarkably similar to vesicles found in fossils from the Rhynie deposits (417–354 million years ago).

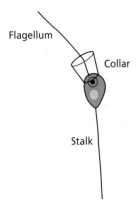

Fig. 1.3 *Codosiga gracilis*, a member of the choanoflagellates (organisms with a single flagellum and a collar), considered to be the common ancestors of both fungi and animals. (Based on a drawing from: http://microscope.mbl.edu/scripts/microscope.php?func=imgDetail&imageID=4575)

lineage, and both groups probably have a common origin in one of the simple unicellular eukaryotes. Currently it is believed that the most likely common ancestor of both the fungal and the animal kingdoms is a protozoan of the group termed **choanoflagellates**, also known as the collar-flagellates (Fig. 1.3). These resemble both the earliest branch of animals (the sponges) and the earliest branch of fungi (the chytrids). It is a humbling thought that humans should have evolved from something like this!

The characteristic features of fungi: defining the fungal kingdom

To begin this section we must make an important distinction between the **true fungi** and a range of **fungus-like organisms** that have traditionally been studied by mycologists, but are fundamentally different from fungi. Here we will focus on the true fungi, often termed the **Mycota** or **Eumycota**. We will discuss the fungus-like organisms in Chapter 2.

All true fungi have a range of features that clearly separate them from other organisms and that serve to define the fungal kingdom (**Mycota**). These features are outlined below:

• All fungi are **eukaryotic**. In other words, they have membrane-bound nuclei containing several chromosomes, and they have a range of membrane-bound cytoplasmic organelles (mitochondria, vacuoles, etc.). Other characterisitics, shared by all eukaryotes, include: cytoplasmic streaming, DNA that contains noncoding regions termed introns, membranes that typically contain sterols, and ribosomes of the 80S type in contrast to the 70S ribosomes of bacteria ("S" refers to Svedberg units, as mentioned earlier).

• Fungi typically grow as filaments, termed **hyphae** (singular: hypha), which extend only at their extreme tips. So, fungi exhibit **apical growth** in contrast to many other filamentous organisms (e.g. filamentous green algae) which grow by repeated cell divisions within a chain of cells (intercalary growth). Fungal hyphae branch repeatedly behind their tips, giving

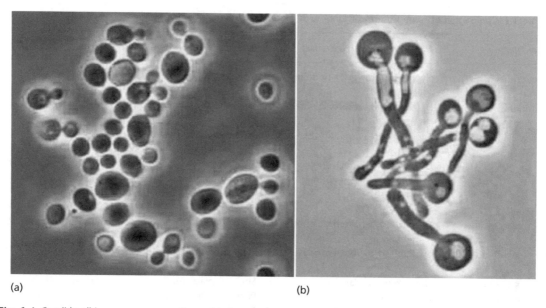

(a)   (b)

Fig. 1.4 *Candida albicans,* a common dimorphic fungus that grows on the mucosal membranes of humans. Normally it is found as a budding yeast (a), but the yeast cells can produce hyphae (b) for invasion of the tissues.

rise to a network termed a **mycelium**. However, some fungi grow as single-celled yeasts (e.g. *Saccharomyces cerevisiae*) which reproduce by budding, and some can switch between a yeast phase and a hyphal phase in response to environmental conditions. These **dimorphic fungi** (with two shapes) include several species that are serious pathogens of humans (Chapter 16). They often grow as yeast-like cells for proliferation in the body fluids but convert to hyphae for invasion of the tissues (Fig. 1.4).

- Fungi are **heterotrophs** (chemo-organotrophs). In other words, they need preformed organic compounds as energy sources and also as carbon skeletons for cellular synthesis. The cell wall prevents fungi from engulfing food by phagocytosis, so fungi absorb simple, soluble nutrients through the wall and cell membrane. In many cases this is achieved by secreting enzymes at the hyphal tips to degrade complex polymers and then absorbing the simple, soluble nutrients released by the depolymerase (polymer-degrading) enzymes.

- Fungi have a distinctive range of **wall components**, which typically including **chitin** and **glucans** (polymers of glucose with predominantly β-1,3 and β-1,6 linkages). Short lengths of cellulose (a β-1,4-linked polymer of glucose) have been detected in some fungal walls, especially in some of the primitive fungi. However fungi differ from plants because they do not have cellulose-rich cell walls.

- Fungi have a characteristic range of soluble carbohydrates and storage compounds, including **mannitol** and other sugar alcohols, **trehalose** (a disaccharide of glucose), and **glycogen**. These compounds are similar to those of some animals – notably the arthropods – but are different from those of plants.

- Fungi typically have **haploid nuclei** – an important difference from almost all other eukaryotes. However, fungal hyphae often have several nuclei within each hyphal compartment, and many budding yeasts are diploid. These differences in nuclear status and nuclear arrangements have important implications for fungal genetics (Chapter 9).

- Fungi reproduce by both sexual and asexual means, and typically produce **spores**. Fungal spores vary enormously in shape, size and other properties, related to their various roles in dispersal or dormant survival (Chapter 10).

In summary, we can define fungi by the following characteristic features (Table 1.1):

- eukaryotic
- typically grow as hyphae, with apical growth, but sometimes as yeasts
- heterotrophic – they depend on pre-formed organic nutrients

Table 1.1 Comparison of some features of fungi with those of animals and plants.

Character	Fungi (and chapter reference)	Animals	Plants
Growth habit	Hyphal tip growth or budding yeasts (3, 4)	Not hyphal	Multicellular tissues
Nutrition	Heterotrophic, absorb soluble nutrients (6, 11)	Heterotrophic, ingest food	Photosynthetic
Cell wall	Typically contains chitin (3)	Absent, but chitin is found in insect exoskeletons	Mainly cellulose
Nuclei	Usually haploid; nuclear membrane persists during division (9)	Typically diploid; the membrane breaks down during nuclear division	Diploid; the membrane breaks down during nuclear division
Histones	Histone 2B	Histone 2B	Plant histones
Microtubules	Sensitive to benzimidazoles and griseofulvin (17)	Sensitive to colchicine	Sensitive to colchicine
Lysine synthesis	Synthesized by AAA pathway (7)	Not synthesized, must be supplied	Synthesized by DAP pathway
Golgi cisternae	Unstacked, tubular (3)	Stacked, plate-like	Stacked, plate-like
Mitochondria	Plate- or disk-like cisternae (3)	Plate- or disk-like cisternae	Tubular cisternae
Translocated carbohydrates	Polyols (mannitol, arabitol, etc.), trehalose (7)	Trehalose in insects	Glucose, fructose, sucrose
Storage compounds	Glycogen, lipids, trehalose (7)	Glycogen, lipids, trehalose in some	Starch
Mitochondrial codon usage	UGA codes for tryptophan	UGA codes for tryptophan	UGA codes for chain termination
Membrane sterols	Ergosterol (7, 17)	Cholesterol	Sitosterol and other plant sterols

AAA, alpha-amino adipic acid pathway; DAP, diamino-pimelic acid pathway.

- typically have a haploid genome
- have walls composed primarily of chitin and glucans
- absorb soluble nutrients through the cell wall and plasma membrane
- produce spores.

The major activities of fungi: pathogens, symbionts, and saprotrophs

As we have already seen, all fungi require organic nutrients for their energy source and as carbon nutrients for cellular synthesis. But a broad distinction can be made according to how these nutrients are obtained: (i) by growing as a **parasite** (or a **pathogen** – a disease-causing agent) of another living organism; (ii) by growing as a **symbiont** in association with another organism; or (iii) by growing as a **saprotroph** (saprophyte) on nonliving materials. These topics are covered in detail in Chapters 11–14.

Fungal parasites of plants

A large number of fungi are adapted to grow as parasites of plants, obtaining some or all of their nutrients from the living tissues of their host. Many of these associations are quite specific because the fungus infects only one type of host, and sometimes it is so specific that the fungus cannot grow at all in laboratory culture – it is an obligate parasite that can grow only in the host tissues. Many examples of this are found among the rust fungi and powdery mildew fungi (Chapter 14), while other examples are found in the fungus-like downy mildews (Chapter 2), and the plasmodiophorids (Chapter 2). These host-specific fungi are termed **biotrophic** parasites (*bios* = life; *trophy* = feeding) because they feed from living host cells without killing them, often by producing special nutrient-absorbing structures to tap the host's reserves. At the other end of the spectrum are many common fungi that aggressively attack plant tissues. They are termed **necrotrophic** parasites (*necros* = death) because they kill the host tissues as part of the feeding process – for example by producing toxins or degradative enzymes. A common example is the fungus *Botryotinia fuckeliana* (more commonly known by its former name, *Botrytis cinerea*) which rapidly destroys soft fruits such as strawberries, raspberries, and grapes, covering the fruit surface with its gray sporing structures.

The fungal (or fungus-like) parasites of plants are enormously significant, accounting for more than 70% of all the major crop diseases, and for many devastating epidemics. To cite just a few examples:

• Potato blight caused by the fungus-like organism *Phytophthora infestans* destroyed the potato crops of Ireland in the 1840s, leading to the starvation of up to one million people, and large-scale emigration to the rest of Europe and the USA. Even today the control of *P. infestans* and its close relatives, the downy mildew fungi, accounts for about 15% of world fungicide sales. *The Advance of the Fungi* by E. C. Large (1940) provides a fascinating and highly readable account of potato blight and its legacy.
• Dutch elm disease, caused by *Ophiostoma novo-ulmi* and *O. ulmi* (Chapter 10), has destroyed most of the common elm (*Ulmus procera*) trees in Britain and Western Europe in the last 30 years, as it did in North America earlier in the 1900s. Similarly, chestnut blight caused by the fungus *Cryphonectria parasitica* (Chapter 9) has devastated the native American chestnut (*Castanea dentata*) population in the USA – an epidemic that can be traced to the first recorded diseased chestnut tree in the New York Zoological Garden in 1904 (Chapter 9). And, at the time of writing, a new species of *Phytophthora* (*P. ramorum*) is causing **sudden oak death** in southwestern USA and has already spread to several parts of Europe (Chapter 14).

Fungal symbionts of plants

Many fungi form symbiotic associations with plants, in which both of the partners are likely to benefit. The two most important examples are **lichens** and **mycorrhizas**. Lichens are intimate associations between two organisms – a photosynthetic partner (a green alga or a cyanobacterium) and a fungus – which together produce a thallus that can withstand some of the most inhospitable environments on Earth (Fig. 1.5). Typically, the fungus encases and protects the photosynthetic cells, and also absorbs mineral nutrients from trace levels in the environment, while the photosynthetic partner provides the fungus with carbon nutrients. There are about 13,500 lichen species across the globe, and they play essential roles

as pioneer colonizers of habitats where no other organisms can grow, including rock surfaces and unstable, arid mineral soils (Chapter 13).

Mycorrhizas are intimate associations between fungi and the roots or other underground organs of plants. There are many types of mycorrhizal fungi, which have evolved independently of one another and which serve different roles. In almost all cases these fungi depend on the plant for a supply of carbon nutrients, while the plants depend on the fungi for a supply of mineral nutrients (phosphorus, nitrogen) from the soil. As we will see in Chapter 13, phosphorus is often the critical limiting factor for plant growth, because soil phosphates rapidly form insoluble complexes with organic matter or with divalent cations (Ca^{2+}, Mg^{2+}) and cannot easily diffuse to the plant roots. Mycorrhizal fungi help to alleviate this problem by providing an extensive hyphal network for capturing mineral nutrients and transporting them back to the roots. However, some other mycorrhizal fungi serve a quite different role. Orchids and some nonphotosynthetic plants are absolutely dependent on fungi for all or part of the plant's life, because the plant feeds on sugars supplied by a soil fungus.

Lichens and mycorrhizas are not the only examples of symbiosis. In recent years many plants have been found to harbor symptomless **endophytic fungi** within the plant walls or intercellular spaces. These fungi apparently do no harm to the plants. Instead they can be beneficial because they help to activate plant defense genes and produce insect anti-feedant compounds such as the ergot alkaloids. But this is a double-edged sword, because the toxins can cause serious damage to grazing animals such as horses, cattle, and sheep (Chapter 11).

Fungal pathogens of humans

In contrast to the many fungal parasites of plants, there are only some 200 fungi that infect humans or other warm-blooded animals. In fact, humans have a high degree of innate immunity to fungi, with the exception of the dermatophytic fungi which commonly cause infections of the skin, nails, and hair. However, the situation changes drastically when the immune system is compromised, and this is becoming common in patients with AIDS, transplant patients whose immune system is purposefully suppressed, patients suffering from cancer or advanced diabetes, and patients undergoing prolonged corticosteroid therapy. In any of these circumstances there is a significant chance of infection from fungi that pose no serious threat to healthy people. For example, the widespread and extremely common airborne fungus *Aspergillus fumigatus* normally grows on composts and in soil, but it has become one

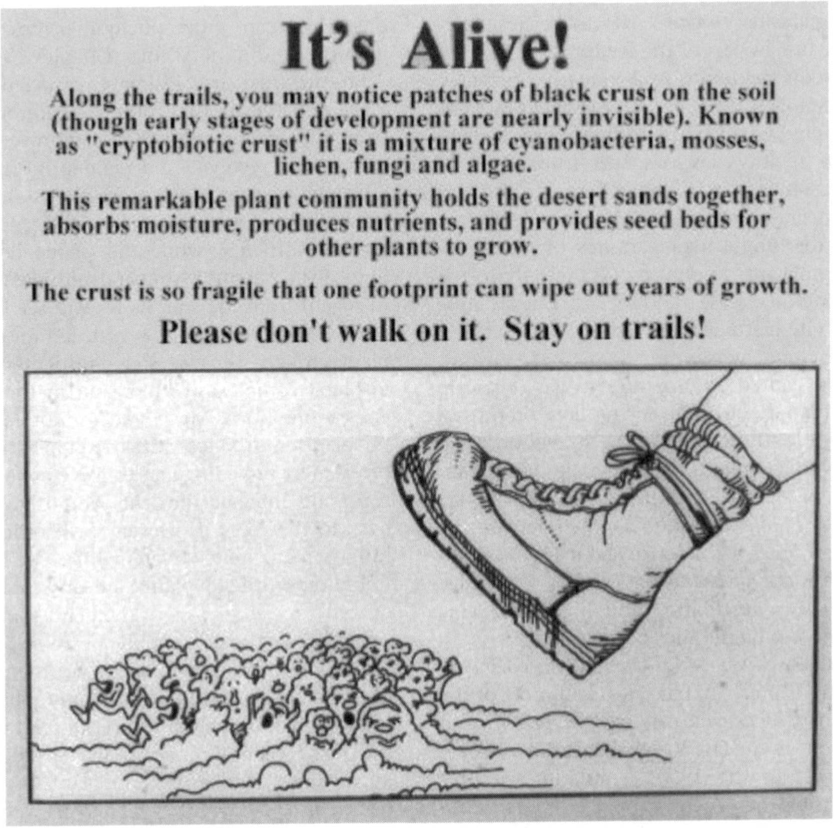

It's Alive!

Along the trails, you may notice patches of black crust on the soil (though early stages of development are nearly invisible). Known as "cryptobiotic crust" it is a mixture of cyanobacteria, mosses, lichen, fungi and algae.

This remarkable plant community holds the desert sands together, absorbs moisture, produces nutrients, and provides seed beds for other plants to grow.

The crust is so fragile that one footprint can wipe out years of growth.

Please don't walk on it. Stay on trails!

Fig. 1.5 A sign in Arches National Park, Utah, USA. *Get your boots off our microbes!*

of the most significant invasive fungi in deep surgical procedures, and the survival rate can be as low as 30%. Many other fungi that can grow at 37°C have spores that are small enough to enter the lungs and reach the alveoli. These fungi were virtually unknown until recently but are now extremely common causes of infection in immunodeficient patients. For example, the fungus-like organism *Pneumocystis jiroveci* (previously named *P. carinii*) commonly causes pneumonia in patients infected with the human immunodeficiency virus (HIV). The onset of this disease in patients with HIV is regarded as one of the "AIDS-defining" symptoms.

Only a handful of antifungal drugs are available to treat the major human mycoses, and most need to be administered at low doses over a prolonged time to avoid excessive toxicity. Many of these drugs are expensive so they offer little hope to the poorest people in the developing world. Chapter 16 is devoted to the mycoses of humans, and Chapter 17 deals with the drugs available to treat these conditions.

Fungal parasites as biological control agents

Fungi parasitize many types of host, including other fungi (**mycoparasites**, Chapter 12), insects (**entomopathogens**, Chapter 15), and nematodes (**nematophagous fungi**, Chapter 15). In the past, such fungi might have been regarded as curiosities, but now they are recognized as being significant population regulators of their hosts and as potential **biological control (biocontrol) agents** of major pests or plant pathogens. We discuss biocontrol at many points in this book, notably in Chapters 12 and 17.

Fungal saprotrophs

Saprotrophs (saprophytes) are organisms that feed on dead organic matter (*sapros* = death; *trophy* = feeding). Fungi play a major role in this respect because they produce a wide range of enzymes that degrade complex polymers such as starch, cellulose, proteins, chitin, aviation kerosene, keratin, and even the most complex lignified materials such as wood. In fact, there are few naturally occurring organic compounds that cannot be degraded by one fungus or another. One of the few exceptions is sporopollenin, the highly resistant polymer found in the walls of pollen grains.

Fungi are particularly important in the decomposition of cellulose, which represents about 40% of plant cell wall material and is the most abundant natural polymer on Earth. Grazing animals (ruminants) also consume significant amounts of cellulose, but this is broken down in the rumen (in effect, a large anaerobic fermentation vessel) and the rumen fungi are thought to play a significant role in the decomposition process. The breakdown of polymers by fungi is intimately linked to hyphal growth which provides both penetrating power and the coordinated release of extracellular enzymes and subsequent reabsorption of the enzymic breakdown products (Chapter 6). But different fungi are adept at degrading different types of polymer, so fungal saprotrophs often grow in complex, mixed communities reflecting their different enzymic capabilities (Chapter 11).

Although the decomposer fungi play vital roles in the recycling of major nutrients, they can also be significant spoilage agents. A well-known example is the dry-rot fungus, *Serpula lacrymans*, which is a major cause of timber decay in buildings (Chapter 5). Similarly the "sooty moulds" that commonly grow on kitchen and bathroom walls are extremely difficult to eradicate (Fig. 1.6). They utilize the soluble cellulose gels that are used as stabilizers in emulsion paints or as wall-paper pastes. These common fungi include species of *Alternaria*, *Cladosporium* and *Sydowia polyspora* (previously called *Aureobasidium pullulans*) which discolor the walls because of their darkly pigmented hyphae and spores. However, their natural habitat is the surface of leaves or the decaying stalk tissues of plants, and they occur in buildings only because they find similar conditions (and substrates) to those in their natural environment (Chapter 8). Public health authorities are now paying increasing attention to safety in the workplace, and particularly to the potential roles of fungi in "sick building syndrome," which has been linked (tenuously) to infant cot death. The conditions in underventilated buildings can certainly promote the growth of moulds, including *Stachybotrys chartarum*, another common sooty mould. But there is no definitive evidence to link these fungi to sick building syndrome.

Some saprotrophic fungi pose a serious threat to human and animal welfare by growing on stored food products and producing **mycotoxins**. These are a diverse range of fungal secondary metabolites, often found in improperly stored materials. For example, **aflatoxins** are commonly produced in groundnuts and cottonseed meal. They are among the most potent known carcinogens and are strongly implicated in hepatomas. Similarly, the toxins produced by several *Fusarium* species on grain crops are implicated in esophageal cancer in Africa, and in kidney carcinomas. The pathways leading to the production of these compounds are discussed in Chapter 7; the maintenance of safe storage conditions is covered in Chapter 8.

Fungi in biotechnology

Fungi have many traditional roles in biotechnology, but also some novel roles, and there is major scope for their future commercial development (Wainwright 1992). Some of these roles are outlined below.

Foods and food flavorings

In 1994 the total world production of edible mushrooms was estimated to be over 5 million tonnes, with a value of US $14 billion. Much of the mushroom-growing industry is based on strains of the common cultivated mushroom *Agaricus bisporus* (or *A. brunnescens*) discussed in Chapters 5 and 11. But *Lentinula edodes* (the Shiitake mushroom, which is grown on logs; Fig. 1.7), *Volvariella volvacea* (the padi straw mushroom, which is grown on rice straw), and *Pleurotus ostreatus* (the oyster mushroom, Fig. 1.8) are traditionally grown in Japan and southeast Asia, and are now widely available in western supermarkets.

Fig. 1.6 Part of a bathroom ceiling where the paint has flaked away, revealing extensive growth and sporulation of sooty moulds.

(a)

(b)

Fig. 1.7 (a,b) Commercial culture of the shiitake mushroom, *Lentinula edodes*, on inoculated logs. (Courtesy of Robert L. Anderson (photographer) and USDA Forest Service; www.forestryimages.org)

Fungi are used to produce several traditional foods and beverages, including alcoholic drinks (ethanol from the yeast *Saccharomyces cerevisiae*) and bread, where the yeast produces CO_2 for raising the dough. *Penicillium roqueforti* is used in the later stages of production of the blue-veined cheeses such as Stilton and Roquefort, to which it imparts a characteristic flavor. *P. camemberti* is used to produce the soft cheeses such as Camembert and bries; it grows on the cheese surface, forming a "crust," and produces proteases which progressively degrade the cheese to give the soft consistency. Less well known but equally significant is the role of fungi in the fermentation of traditional foods around the world. For example, *Rhizopus oligosporus* is used to convert cooked soybean "grits" to a nutritious staple food, called **tempeh** (Fig. 1.8). This involves only a short (24–36 hour) incubation time, during which the fungus degrades some of the fat and also degrades a trypsin inhibitor in soybeans, so that the naturally high protein content of this crop is more readily available in the diet, and a "flatulence factor" is broken down during this process. The food termed **gari** is part of the staple diet in southern Nigeria; it is produced from the high-yielding root crop, cassava, perhaps better known in its processed form, tapioca. Raw cassava contains a toxic cyanogenic glycoside termed linamarin, which is removed during a prolonged and largely uncontrolled fermentation in village communities. Much of this process involves bacteria, but the fungus *Galactomyces geotrichum* (asexual stage: *Geotrichum candidum*) gives the product its desired flavor. Details of the production of several traditional Asian fermented foods can be found in Nout & Aidoo (2002).

A major development in recent years has been the introduction of an entirely new type of food, termed **Quorn™ mycoprotein** (Fig. 1.9). This is produced commercially by growing a fungus (*Fusarium venanatum*) in large fermentation vessels, then harvesting the fungal hyphae and processing them into meat-like chunks and various oven-ready meals. *Quorn* (as it is now called) is widely available in British and European supermarkets. It has an almost ideal nutritional profile, with a high protein content, low fat content, and absence of cholesterol (Table 1.2). The production of Quorn is discussed in detail in Chapter 4.

Fungal metabolites

Metabolites can be grouped into two broad categories (Chapter 7):

- **Primary metabolites:** the intermediates or end products of the common metabolic pathways of all organisms (sugars, amino acids, organic acids, glycerol, etc.) and which are essential for the normal cellular functions of fungi.

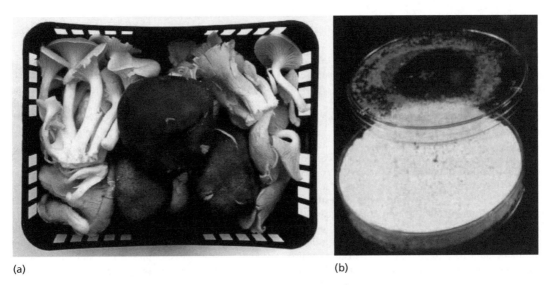

(a)   (b)

Fig. 1.8 (a) A tray of exotic mushrooms from a supermarket in the UK, including shiitake (centre) and the oyster mushroom, *Pleuotus ostreatus*, reputed to be an aphrodisiac. (b) An attempt to produce a homemade cake of tempeh, which tasted only marginally better than it looks.

Table 1.2 Nutritional composition of Quorn™ mycoprotein, compared with traditional protein sources. (Data from Trinci 1992.)

	Units	Quorn	Cheddar cheese	Raw chicken	Raw lean beef	Fresh cod
Protein	g 100g⁻¹	12.2	26.0	20.5	20.3	17.4
Dietary fibre	g 100 g⁻¹	5.1	0	0	0	0
Total fats	g 100 g⁻¹	2.9	33.5	4.3	4.6	0.7
Fat ratio	Polyunsaturated: saturated	2.5	0.2	0.5	0.1	2.2
Cholesterol	mg 100g⁻¹	0	70	69	59	50
Energy	kJ 100g⁻¹	334	1697	506	514	318

(a)   (b)

Fig. 1.9 Quorn: (a) the package and (b) one of several products: "Fillets in a Mediterranean marinade with tomato, red wine, and herbs."

- **Secondary metabolites:** a diverse range of compounds formed by specific pathways of particular organisms; they are not essential for growth, although they can confer an advantage to the organisms that produce them (e.g. antibiotics, fungal toxins, etc.).

Several metabolites of both groups are produced commercially from fungal cultures (Turner 1971; Turner & Aldridge 1983). One of the best examples of a fungal primary metabolite is **citric acid**, with an estimated global production of 900,000 tons in the year 2000 (Ruijter *et al.* 2002). Citric acid produced on this vast scale is the mainstay of the soft drinks industry (lemonade, etc.) because it has a tart taste and also enhances flavor, reduces sweetness, and has antioxidant and preservative qualities. Specially selected, overproducing strains of *Aspergillus niger* are used for the commercial production of citric acid, but several other conditions are necessary – the cultures must contain high levels of readily metabolizable sugars (up to 20% or more) and the concentration of either phosphate or nitrogen must be kept low, to limit the amount of fungal growth. In these conditions 80% or more of the sugar supplied to the cultures is converted into citric acid, which is then exported from the cells and accumulates in the culture medium. The effect of this is to lower the pH of the culture medium to 3.0 or less, which the fungus tolerates well. This secretion of the acid is a crucial feature, because fungal cells tightly regulate their internal pH. Recent evidence indicates that cells of *A. niger* maintain their intracellular pH at 7.7 when the cells are exposed to external pH levels ranging from 1.5 to 6.

Other organic acids are produced commercially by fungal fermentations. **Gluconic acid** (estimated annual global production of 50,000–100,000 tons) is used mainly as a food additive, and is produced by specific strains of *A. niger*, grown at normal pH. This acid is produced by the direct oxidation of glucose, catalyzed by the enzyme glucose oxidase. **Itaconic acid** (global production 70,000–80,000 tons) is produced by *Aspergillus terreus* and is used as a co-polymer in the manufacture of paints, adhesives, etc.

In some respects the production of citric acid and itaconic acid is similar to the **production of ethanol** by *Saccharomyces* spp. – the basis of the alcoholic drinks industry. Both types of product accumulate in the culture medium when **growth is restricted** by some factor but when the biochemical machinery continues to operate, like the engine of a car taken out of gear. For example, ethanol accumulates as a metabolic end-product when yeast is grown in a sugar-rich medium favoring metabolism, but in **anaerobic conditions** that limit cell growth.

In contrast to the bulk metabolites mentioned above, a vast range of **secondary metabolites** are produced by fungi, and they include several high-value products with pharmaceutical applications. A small selection of these is shown in Table 1.3. The best-known examples are the **penicillins** – a group of structurally related β-lactam antibiotics that are synthesized naturally from small peptides. As explained in Chapter 7, the naturally occurring penicillins such as penicillin G (produced by *Penicillium chrysogenum*) have a relatively narrow spectrum of activity. But a wide range of other penicillins can be produced by chemical modification of the natural penicillins. All modern penicillins are semisynthetic compounds; they are obtained initially from fermentation cultures but are then structurally modified for specific desirable properties. Schmidt (2002) reviewed the manufacture and therapeutic aspects of β-lactam antibiotics, including the **cephalosporins** which are structurally related to the penicillins. Remarkably, despite their age (the penicillins were first produced commercially in the late 1940s),

Table 1.3 Some valuable secondary metabolites produced commercially from fungi.

Metabolite	Fungal source	Application
Penicillins	*Penicillium chrysogenum*	Antibacterial
Cephalosporins	*Acremonium chrysogenum*	Antibacterial
Griseofulvin	*Penicillium griseofulvum*	Antifungal
Fusidin	*Fusidium coccineum*	Antibacterial
Ciclosporins	*Tolypocladium* spp.	Immunosuppressants
Zearalenone	*Gibberella zeae*	Cattle growth promoter
Gibberellins	*Gibberella fujikuroi*	Plant hormone
Ergot alkaloids and related compounds	*Claviceps purpurea* and related fungi	Many effects including: antimigraine, vasoconstriction, vasodilation, antihypertension, anti-Parkinson, psychiatric disorders

the β-lactam antibiotics still share 50% of the world market for systemic antibiotics, with sales in 1998 worth about US $4 billion for penicillins and about US $7 billion for the more recently developed cephalosporins.

Several non-β-lactam antibiotics are also produced by fungi. They include **griseofulvin** (from the fungus *P. griseofulvum*) which has been used for several years to treat dermatophyte infections of the skin, nails and hair of humans, although recently it has been replaced by less toxic drugs (Chapter 17). **Fusidic acid** (from various fungi) has been used to control staphylococci that have become resistant to penicillin, and there is renewed interest in a range of other natural fungal products for treating the systemic fungal infections of humans (Chapter 17). **Ciclosporins** from various fungi (but principally from species of *Tolypocladium*) are used as immunosuppressants to prevent organ rejection in transplant surgery. In fact, 17 different fungal taxa are reported to produce ciclosporins. Another powerful immunosuppressant is the antibiotic **gliotoxin** (from *Trichoderma virens*), which is better known for its role in biological control of plant pathogenic fungi (Chapter 12). The production and use of these immunosuppressants was reviewed by Kürnsteiner *et al.* (2002). As a final example, the **ergot alkaloids** and related toxins of the ergot fungus, *Claviceps purpurea* (Chapter 14), have many important pharmacological applications (Keller & Tudzynski 2002). The four-membered ring structure of the D-lysergic acid derivatives of ergot alkaloids mimic the ring structures of neurotransmitters (dopamine, epinephrine (adrenaline), and serotonin: Fig. 1.10). However, at present many of the ergot derivatives are too nonspecific in their modes of action to meet their true potential in treating human disorders.

Even these few examples raise fascinating questions about the roles of fungal secondary metabolites. What functions do they serve in fungi and what competitive advantage do they confer? In recent years many of the genes encoding the secondary biosynthetic pathways have been identified and sequenced. This should lead both to an understanding of their roles and to the potential construction of transgenic strains that overproduce valuable metabolites.

Some of the polysaccharides of fungi have potential commercial value. **Pullulan** is an α-1,4-glucan (polymer of glucose) produced as an extracellular sheath by *Sydowia polyspora* (formerly *Aureobasidium pullulans*), one of the sooty moulds. This polymer is used in Japan to make a film-wrap for foods. A potential new market could develop from the discovery that fungal wall polymers or their partial breakdown products can be powerful elicitors of plant defense responses (Chapter 14) so they might be used to "immunize" plants. For example, the β-glucan fractions from walls of the yeast *S. cerevisiae* have this effect. So too does **chitosan**, the de-acetylated form of chitin in fungal cell walls (Chapters 3 & 7). At present, chitosan is used on a large scale in Japan for clarifying sewage, but the source of this chitosan is crustacean shells. Fungi are an alternative, easily renewable source of this and other polymers.

Enzymes and enzymic conversions

Saprotrophic fungi and some plant-pathogenic fungi produce a range of extracellular enzymes with important commercial roles (Table 1.4). The **pectic enzymes** of fungi are used to clarify fruit juices, a fungal **amylase** is used to convert starch to maltose during bread-making, and a fungal rennet is used to coagulate milk for cheese-making. A single fungus, *Aspergillus niger*, accounts for almost 95% of the commercial production of these and other bulk enzymes from fungi, although specific strains of the fungus have been selected for particular purposes. The methanol-utilizing yeasts (*Candida lipolytica*, *Hansenula polymorpha*, and *Pichia pastoris*) have potential commercial value because they produce large amounts of **alcohol oxidase**, which could be used as a bleaching agent in detergents. The wood-rotting fungus *Phanerochaete chrysosporium* is extremely active in degrading lignin; it has the

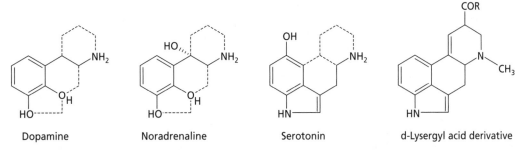

Fig. 1.10 Structural similarities between three neurotransmitters (dopamine, noradrenaline, and serotonin) and the D-lysergic acid derivatives of ergot alkaloids.

Table 1.4 Some fungal enzymes produced commercially. (Based on Wainwright 1992.)

Enzyme	Fungal source	Application
α-Amylase	*Aspergillus niger, A. oryzae*	Starch conversions
Amyloglucosidase	*A. niger*	Starch syrups, dextrose foods
Pullulanase	*Aureobasidium pullulans*	Debranching of starch
Glucose aerohydrogenase	*A. niger*	Production of gluconic acid
Proteases (acid, neutral, alkaline)	*Aspergillus* spp. etc.	Breakdown of proteins (baking, brewing, etc.)
Invertase	Yeasts	Sucrose conversions
Pectinases	*Aspergillus, Rhizopus*	Clarifying fruit juices
Rennet	*Mucor* spp.	Milk coagulation
Glucose isomerase	*Mucor, Aspergillus*	High fructose syrups
Lipases	*Mucor, Aspergillus, Penicillium*	Dairy industry, detergents
Hemicellulase	*A. niger*	Baking, gums
Glucose oxidase	*A. niger*	Food processing

potential to be developed for delignification of agricultural wastes and byproducts of the wood-pulping industry, so that the cellulose in these materials could be used as a cheap substrate for production of fuel alcohol by yeasts (Chapter 11).

In addition to these examples of "bulk" enzymes, fungi have many internal enzymes and enzymic pathways that can be exploited for the bioconversion of compounds such as pharmaceuticals. For example, fungi are used for the bioconversion of steroids, because fungal enzymes perform highly specific dehydrogenations, hydroxylations and other modifications of the complex aromatic ring systems of steroids. Precursor steroids are fed to a fungus, held at low nutrient level either in culture or attached to an inert bed, so that the steroid is absorbed, transformed and then released into the culture medium from which it can be retrieved.

Heterologous gene products

Genetic engineering of fungi, particularly *Saccharomyces cerevisiae*, has developed to the stage where the cells can be used as factories to produce pharmaceutical products, by the introduction of foreign (heterologous) genes, as we already noted for the hepatitis B vaccine. There are several advantages in using yeast to synthesize such products. *S. cerevisiae* is already grown on a large industrial scale, so companies are familiar with its culture. It is a **GRAS** organism, i.e. "generally regarded as safe." Its genome was the first to be sequenced, and its genetics and molecular genetics are well-researched (Chapter 9). Furthermore, yeast has a well-characterized secretory system for exporting gene products into a culture medium. Examples of heterolog-

ous gene products that have been produced experimentally from yeast include **epidermal growth factor** (involved in wound healing), **atrial natriuretic factor** (for management of hypertension), **interferons** (with antiviral and antitumor activity), and **α-1-antitrypsin** (for potential relief from emphysema). There are, however, disadvantages in using *S. cerevisiae*. In particular, this fungus is genetically quite different from other fungi and other eukaryotes, including its use of different codons for some amino acids, so it does not always correctly read the introduced genes. For this reason attention has switched to some other fungi, such as the fission yeast *Schizosaccharomyces pombe* and the filamentous fungus *Emericella* (*Aspergillus*) *nidulans*, for both of which the genomes have now been sequenced.

Online resources

Forestry Images. http://www.forestryimages.org. [Many high-quality images of fungi, diseases, forestry practices, etc.]

Fungal Biology. http://www.helios.bto.ed.ac.uk/bto/FungalBiology/ [The website for this book.]

Tree of Life Web Project. http://tolweb.org/tree?group=life. [A major source of information on fungal systematics and phylogeny.]

General texts

Alexopoulos, C.J., Mims, C.W. & Blackwell, M. (1996) *Introductory Mycology*, 4th edn. John Wiley, New York.

Carlile, M.J., Watkinson, S.C. & Gooday, G.W. (2001) *The Fungi*, 2nd edn. Academic Press, London.

Jennings, D.H. & Lysek, G. (1999) *Fungal Biology: understanding the fungal lifestyle*, 2nd edn. Bios, Oxford.

Kendrick, B. (2001) *The Fifth Kingdom*, 3rd edn. Mycologue Publications, Sidney, Canada.

Turner, W.B. (1971) *Fungal Metabolites*. Academic Press, London.

Turner, W.B. & Aldridge, D.C. (1983) *Fungal Metabolites. II*. Academic Press, London.

Wainwright, M. (1992) *An Introduction to Fungal Biotechnology*. Wiley, Chichester.

Webster, J. (1980) *Introduction to Fungi*, 2nd edn. Cambridge University Press, Cambridge.

Cited references

Baldauf, S.L. & Palmer, J.D. (1993) Animals and fungi are each other's closest relatives: congruent evidence from multiple proteins. *Proceedings of the National Academy of Sciences, USA* **90**, 11558–11562.

Hawksworth, D.L. (2001) The magnitude of fungal diversity: the 1.5 million species estimate revisited. *Mycological Research* **105**, 1422–1432.

Hawksworth, D.L. (2002) Mycological Research News. *Mycological Research* **106**, 514.

Keller, U. & Tudzynski, P. (2002) Ergot alkaloids. In: *The Mycota X. Industrial Applications* (H.D. Osiewicz, ed.), pp. 157–181. Springer-Verlag, Berlin.

Kürnsteiner, H., Zinner, M. & Kück, U. (2002) Immunosuppressants. In: *The Mycota X. Industrial Applications* (H.D. Osiewicz, ed.), pp. 129–155. Springer-Verlag, Berlin.

Large, E.C. (1940) *The Advance of the Fungi*. Henry Holt, New York.

Lewis, D.H. (1987) Evolutionary aspects of mutualistic associations between fungi and photosynthetic organisms. In: *Evolutionary Biology of the Fungi* (eds Rayner, A.D.M., Brasier, C.M. & Moore, D.), pp. 161–178. Cambridge University Press, Cambridge.

Nout, M.J.R. & Aidoo, K.E. (2002) Asian fungal fermented food. In: *The Mycota X. Industrial Applications* (H.D. Osiewicz, ed.), pp. 23–47. Springer-Verlag, Berlin.

Ruijter, G.J.G., Kubicek, C.P. & Visser, J. (2002) Production of organic acids by fungi. In: *The Mycota X. Industrial Applications* (H.D. Osiewicz, ed.), pp. 213–230. Springer-Verlag, Berlin.

Schmidt, F.R. (2002) Beta-lactam antibiotics: aspects of manufacture and therapy. In: *The Mycota X. Industrial Applications* (H.D. Osiewicz, ed.), pp. 69–91. Springer-Verlag, Berlin.

Trinci, A.P.J. (1992) Myco-protein: a twenty-year overnight success story. *Mycological Research* **96**, 1–13.

Woese, C.R. (2000) Interpreting the universal phylogenetic tree. *Proceedings of the National Academy of Sciences, USA* **97**, 8392–8396.

Chapter 2

The diversity of fungi and fungus-like organisms

This chapter is divided into the following major sections:

- overview of the fungi and fungus-like organisms
- the true fungi (Kingdom Mycota): Chytridiomycota, Glomeromycota, Zygomycota, Ascomycota, Basidiomycota, mitosporic fungi
- the cellulose-walled fungus-like organisms (Kingdom Straminipila)
- other fungus-like organisms: slime moulds, cellular slime moulds (acrasids and dictyostelids), and plasmodiophorids

In this chapter we focus on the major groups of fungi and fungus-like organisms, covering the whole span of fungal diversity in its broadest sense. We will use selected examples to illustrate key features of the fungal groups, and their biological significance. There will be some surprises in store. For example, we will see that some of the most devastating plant pathogens are not fungi at all, but belong to an entirely separate kingdom. We will see that some of the organisms once considered to be among the most "primitive" – the microsporidia, trichomonads, and diplomonads (see Fig. 1.1) – are derived from fungi by the loss of features such as mitochondria, which they once possessed. We will also see how the development of molecular methods for determining the relationships between organisms has enhanced our understanding of fungi in many respects, but there is still no consensus on the best way to construct phylogenetic trees. In the words of Patterson & Sogin (Tree of Life Web Project, see Online Resources): "The consequence . . . has been to demolish the model of the 1990s, but not to replace it with something better."

Overview of the fungi and fungus-like organisms

Box 2.1 shows all the fungi and fungus-like organisms that are currently considered to be fungi in the broadest sense. The vast majority are true fungi, sometimes

Box 2.1 The several types of organism that constitute the fungi in a broad sense.

Kingdom: Fungi (Mycota)
Probably derived from a choanoflagellate ancestor
    Phylum **Chytridiomycota**
    Phylum **Zygomycota**
    Phylum **Glomeromycota**
    Phylum **Ascomycota**
    Phylum **Basidiomycota**

Kingdom: Straminipila
Probably derived from the protist group containing golden-brown algae, diatoms, etc.
    Phylum **Oomycota**
    Phylum **Hyphochytridiomycota**
    Phylum **Labyrinthulomycota**

Fungus-like organisms of uncertain affinity
    Phylum **Myxomycota** (plasmodial slime moulds)
    Phylum **Plasmodiophoromycota** (plasmodiophorids)
    Phylum **Dictyosteliomycota** (dictyostelid slime moulds)
    Phylum **Acrasiomycota** (acrasid slime moulds)

termed Eumycota (*eu* = "true"). Until recently, all the true fungi were assigned to four phyla – Chytridiomycota, Zygomycota, Ascomycota, and Basidiomycota. But in 2001, a fifth phylum was erected – the Glomeromycota (arbuscular mycorrhizal fungi and their relatives). These had previously been included in the Zygomycota. It will be recalled from Chapter 1 that the Glomeromycota were associated with the earliest land plants (Schuessler *et al.* 2001) and are still associated with the vast majority of plants today.

Within the **Kingdom Fungi**, the Ascomycota and the Basidiomycota have many features in common, pointing clearly to a common ancestry. The phylum Chytridiomycota has traditionally been characterized on the basis of motile cells with a single posterior flagellum. This phylum was redefined recently, based on sequence analysis of the nuclear genes encoding small subunit (SSU) ribosomal RNA (18S rDNA). This has revealed that some nonmotile fungi, previously assigned to the Zygomycota, are closely related to the Chytridiomycota and must be reassigned. An example is the fungus *Basidiobolus ranarum* (Chapter 4), now transferred to the Chytridiomycota. The status of Zygomycota (as currently defined after excluding the Glomeromycota) is still unclear. Some of its members may need to be separated into new groups.

Nevertheless, the current view is that all organisms within the Kingdom Fungi constitute a monophyletic group (all derived from a common ancestor), sharing several features with animals (see Table 1.1). Gene sequence analyses provide the basis for a natural phylogeny, especially when data for the SSU rDNA are supported by sequence analysis of other gene families, such as the tubulin and actin genes.

The **Kingdom Straminipila** (straminipiles, or stramenopiles) is now universally recognized as being distinct from the true fungi. It consists of one large and extremely important phylum, the Oomycota, and two small phyla, the Hyphochytridiomycota (with about 25 species) and Labyrinthulomycota (with about 40 species). The Phylum Oomycota is remarkable in many ways. It includes some of the most devastating plant pathogens, including *Phytophthora infestans* (potato blight), *Phytophthora ramorum* (sudden oak death in California), *Phytophthora cinnamomi* (the scourge of large tracts of *Eucalyptus* forest in Australia), and many other important plant pathogens, including *Pythium* and *Aphanomyces* spp. But perhaps most remarkable of all is the fact that Oomycota have evolved a lifestyle that resembles that of the true fungi in almost every respect. We discuss this group in detail later in this chapter and at several points in this book.

The **fungus-like organisms of uncertain affinity** include four types of organism: the acrasid cellular slime moulds, the dictyostelid cellular slime moulds, the plasmodial slime moulds (Myxomycota), and the plasmodiophorids. For most of their life these organisms lack cell walls, and they grow as either a naked protoplasmic mass or as amoeboid cells, converting to a walled form at the onset of sporulation. There is no evidence that they are related to fungi, but they have traditionally been studied by mycologists, and they have several interesting features, which are discussed towards the end of this chapter.

Against this background, we now consider the individual phyla in more detail.

The true fungi (Kingdom Mycota)

Chitridiomycota

The Chytridiomycota, commonly termed chytrids, number about 1000 species (Barr 1990) and are considered to be the earliest branch of the true fungi, dating back to about 1 billion years ago. They have cell walls composed mainly of **chitin** and **glucans** (polymers of glucose) and many other features typical of fungi (see Table 1.1). But they are unique in one respect, because they are the only true fungi that produce **motile, flagellate zoospores**. Typically, the zoospore has a single, posterior whiplash flagellum, but some of the chytrids that grow in the rumen of animals have several flagella (Chapter 8), and some other chytrids (e.g. *Basidiobolus ranarum*, recently transferred to the Chytridiomycota based on SSU rDNA analysis), have no flagella. This provides a good example of the value of DNA sequencing in determining the true phylogenetic relationships of organisms.

Ecology and significance

Most chytrids are small, inconspicuous organisms that grow as single cells or primitively branched chains of cells on organic materials in moist soils or aquatic environments. They are considered to play significant roles as primary colonizers and degraders of organic matter in these environments. Two common examples are *Rhizophlyctis rosea* (a strongly cellulolytic fungus that is common in natural soils) and *Allomyces arbuscula*, both shown in Fig. 2.1.

Often, the Chytridiomycota are anchored to their substrates by narrow, tapering **rhizoids**, which function like hyphae in secreting enzymes and absorbing nutrients. Sometimes the rhizoidal system is extensive, or the **thallus** (the "body" of the fungus) resembles a string of beads, with inflated cells arising at intervals along a rhizoidal network. An example of this is the fungus *Catenaria anguillulae*, shown in Fig. 2.2. A few chytrids grow as obligate intracellular parasites of plants (e.g.

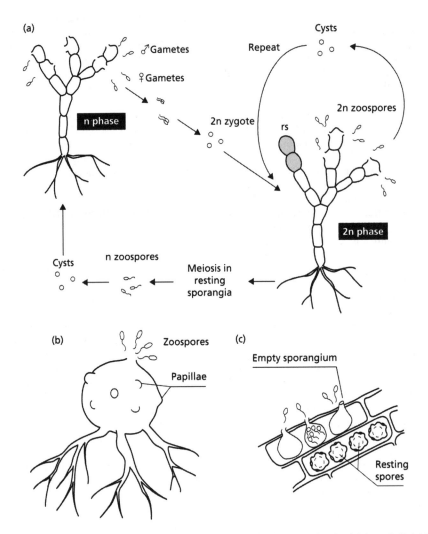

Fig. 2.1 Chytridiomycota. (a) Life cycle of *Allomyces*, which alternates between haploid (n) and diploid (2n) generations. The haploid thallus produces male and female gametangia that release motile gametes. These fuse in pairs and encyst to produce 2n zygotes, which germinate to produce a 2n thallus. The 2n sporangia release zoospores for recycling of the 2n phase. Thick-walled resting sporangia (rs) are formed in adverse conditions; after meiosis these germinate to release haploid zoospores. (b) *Rhizophlyctis rosea*, a common cellulolytic fungus in soil. It grows as a single large cell, up to 200 μm diameter, with tapering rhizoids. At maturity, the large, inflated cell converts into a sporangium, where the cytoplasm is cleaved around the individual haploid nuclei, and large numbers of zoospores are released through the exit papillae. (c) *Olpidium brassicae* grows as naked protoplasts in root cells of cabbages. At maturity, the protoplasts convert to sporangia, which release zoospores into the soil. These spores encyst on a host root, germinate, and release a protoplast into the host. Thick-walled resting spores are produced in adverse conditions and can persist in soil for many years (see Fig. 2.3).

Olpidium brassicae; Figs. 2.1c, Fig. 2.3), of small animals (e.g. *Coelomomyces*; see Fig. 15.5), of algae, or of fungal spores. But very few chytrids can be considered to be economically important – the most notable example is *Synchytrium endobioticum*, which causes potato wart disease, where the potato tubers develop unsightly galls that render them unmarketable.

Having said this, the significance of chytrids lies mainly in their fascinating biology. Anybody who has watched a chytrid zoospore crawling like an amoeba along the body of a nematode, searching for the best site to encyst, and then winding in its flagellum, encysting and penetrating the host will never forget the experience (Fig. 2.2). All these events can be

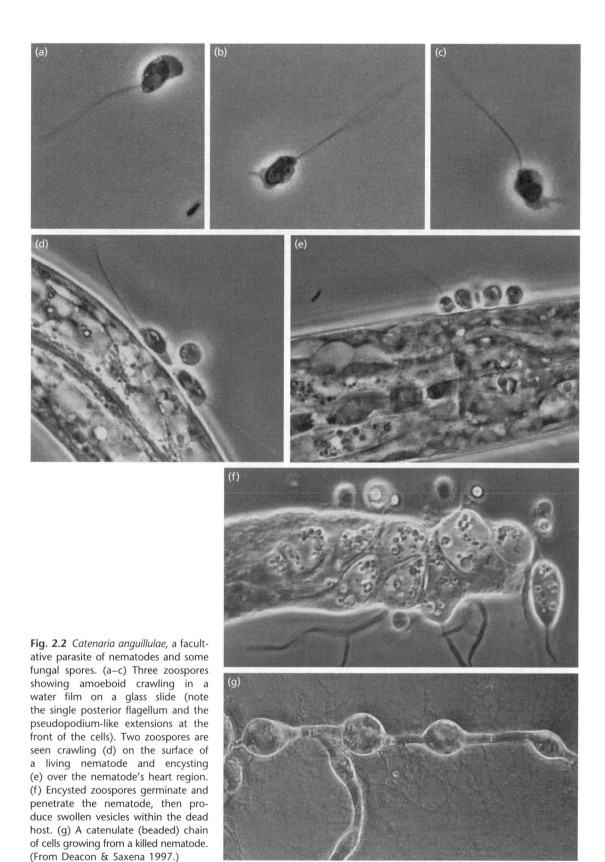

Fig. 2.2 *Catenaria anguillulae*, a facultative parasite of nematodes and some fungal spores. (a–c) Three zoospores showing amoeboid crawling in a water film on a glass slide (note the single posterior flagellum and the pseudopodium-like extensions at the front of the cells). Two zoospores are seen crawling (d) on the surface of a living nematode and encysting (e) over the nematode's heart region. (f) Encysted zoospores germinate and penetrate the nematode, then produce swollen vesicles within the dead host. (g) A catenulate (beaded) chain of cells growing from a killed nematode. (From Deacon & Saxena 1997.)

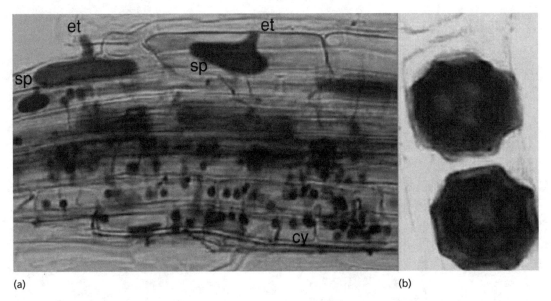

(a)   (b)

Fig. 2.3 *Olpidium brassicae*, a biotrophic (obligate) parasite commonly found in cabbage roots. The root cell contents were destroyed by treatment with hot KOH, then rinsed, acidified, and stained with trypan blue to reveal fungal structures within the roots. (a) Two sporangia (sp) about 30 µm long, with exit tubes (et) and many germinating zoospore cysts (cy). (b) Two thick-walled resting spores of *O. brassicae*, about 25 µm diameter, within a root cell.

followed in simple glass chambers on microscope slides (Deacon & Saxena 1997).

The Chytridiomycota are difficult to isolate by standard methods such as dilution plating of soil onto agar plates. But they can be found easily by suspending small "bait" particles on the surface of natural waters or in dishes of flooded soil. In these conditions, chytrid zoospores accumulate on baits such as cellulose, chitin, keratin, insect exoskeleton, or pollen grains. Then they encyst and produce rhizoids for anchorage to the substrate. There is strong experimental evidence that chytrid zoospores accumulate *selectively* on different types of bait (Mitchell & Deacon 1986). For example, when pieces of cellulose or purified crab-shell chitin were added to zoospore suspensions, the zoospores were seen to encounter the baits at random, but then changed their swimming pattern, making frequent random turns, and often encysted within 3–5 minutes. Zoospores of *Allomyces arbuscula* and *A. javanicus* accumulated and encysted on both cellulose and chitin, whereas zoospores of *Chytridium confervae* encysted preferentially on chitin, and zoospores of *Rhizophlyctis rosea* accumulated and encysted only on cellulose. The most likely explanation is that the zoospores have surface-located receptors that recognize different structural polymers – a phenomenon well known in zoospores of the fungus-like Oomycota (Chapter 10).

Taxonomy and relationships

Currently, the Phylum Chytridiomycota is subdivided into five orders (**Blastocladiales, Chytridiales, Monoblepharidales, Neocallimastigales,** and **Spizellomycetales**) based largely on ultrastructural features of the zoospores, which seem to indicate conserved patterns of evolution in the different lineages. For example, the zoospores of all members of the Order Blastocladiales (including *Catenaria anguillulae*) have a conspicuous nuclear cap that surrounds the nucleus and is filled with ribosomes (see Fig. 10.12), but different arrangements of the organelles are found in the other chytrid orders. Many of these differences relate to the arrangement of microtubular elements associated with anchorage of the flagellum, and the arrangements of mitochondria and lipid storage reserves that are essential for zoospore motility. We return to these points in Chapter 10, where we discuss zoospore ultrastructure. We should also note that the Order Neocallimastigales is unique in being obligately anaerobic. These organisms have recently been shown to have a **hydrogenosome**, equivalent to a mitochondrion, for generating energy (Chapter 8). But there is still doubt about whether the Neocallimastigales is a natural phylogenetic grouping. SSU rDNA analysis of a wider range of chytrids should help to clarify their relationships.

Reproduction and other features of Chytridiomycota

Most true fungi have a haploid genome but some species of *Allomyces* (Fig. 2.1) can alternate between haploid and diploid generations. This is also true for some species of *Blastocladiella*. There are different patterns of the life cycle within *Allomyces* spp. but one of these patterns, exemplified by *A. arbusculus* and *A. macrogynus* (Fig. 2.1), involves a predominantly diploid phase. Sporangia are produced on the diploid colonies, and cytoplasmic cleavage within the sporangium leads to the release of diploid zoospores. These encyst and then germinate to produce further diploid colonies. This process can continue as long as the environmental conditions are suitable, but at the onset of unfavorable conditions the fungus produces thick-walled resting sporangia. Meiosis then occurs within the resting sporangia, leading to the release of haploid zoospores, which encyst and then germinate to produce haploid colonies. These colonies produce male and female gametangia, which release male and female haploid gametes. The gametes of opposite mating type fuse to form a diploid zygote, which encysts and then germinates to repeat the diploid phase of the life cycle. In other *Allomyces* spp. there is no separate gametophyte generation; instead this is probably represented by a cyst, which germinates to produce a further asexual diploid colony.

One of the most intriguing features of chytrid zoospores is their ability to undergo prolonged amoeboid crawling, by pseudopodium-like extensions of the cell. This occurs both on glass surfaces and on the surfaces of potential hosts such as nematodes – a searching behavior for locating suitable sites for encystment (Fig. 2.2). Then the zoospores round-up and wind-in their flagellum by rotating most or all of the cell contents (Deacon & Saxena 1997).

Many details of the Chytridiomycota can be found in Fuller & Jaworski (1987).

Glomeromycota

As noted in Chapter 1 (see Fig. 1.2), fungal fossils resembling the common and economically important **arbuscular mycorrhizal (AM) fungi** were associated with plants in the Rhynie chert deposits of the Devonian era about 400 million years ago (mya). They probably played a vital role in the establishment of the land flora. Recent detailed analysis of these fungi, based on SSU rDNA sequences, has shown convincingly that they are distinct from all other major fungal groups, so they have been assigned to a new phylum, the **Glomeromycota** (Schuessler *et al.* 2001). The relationships between this group and other fungi are still unclear, but Fig. 2.4 shows that there is very

strong bootstrap support for linking all of these AM fungi.

The arbuscular mycorrhizal fungi are considered in detail in Chapter 13, so we will discuss them only briefly here. They share several characteristic and quite remarkable features:

- They are found growing within the roots of the vast majority of plants, and yet they cannot be grown independently, in the absence of a plant.
- When they penetrate the roots they grow predominantly between the root cortical cells and often (but not always) produce large, swollen vesicles which are believed to function as food storage reserves.
- The AM fungal hyphae penetrate individual root cortical cells to form intricately branched **arbuscules** (tree-like branching systems). But they do not kill the root cells, and instead they establish an intimate feeding relationship, which seems to benefit both partners. Fig. 2.5 shows the extent of this type of relationship, in roots cleared of plant protoplasm, then stained with trypan blue.

Recently discovered fossil hyphae and spores from Wisconsin, USA, date back to the Ordovician, 460–455 mya (Redecker *et al.* 2000) and therefore pre-date the vascular plants (i.e. plants with water-conducting tissues). Examples of these early fossil fungi are shown in Fig. 2.6. They are remarkably similar to the AM fungal spores, although it is emphasized that they were not associated with plants. Nevertheless, they can provide calibration points in phylogenetic analyses, as shown in Fig. 2.7.

One final point – and perhaps the most remarkable – was the discovery of a novel type of symbiosis, first reported in 1996 and still known from only a few natural sites in Germany. In this symbiosis, AM fungi are attached to plant roots and grow up to the soil surface, where the AM fungus produces transparent bladders. These bladders engulf and internalize the cells of a cyanobacterium, *Nostoc punctiforme*, so that the fungus obtains sugars from the cyanobacterial partner. The resulting organism is called *Geosiphon pyriforme*, a member of the *Geosiphonaceae* (Glomeromycota) shown in Fig. 2.4. We return to this in Chapter 13.

Zygomycota

Five major features serve to characterize the phylum Zygomycota:

1 cell walls composed of a mixture of **chitin, chitosan** (a poorly- or non-acetylated form of chitin) and **polyglucuronic acid;**

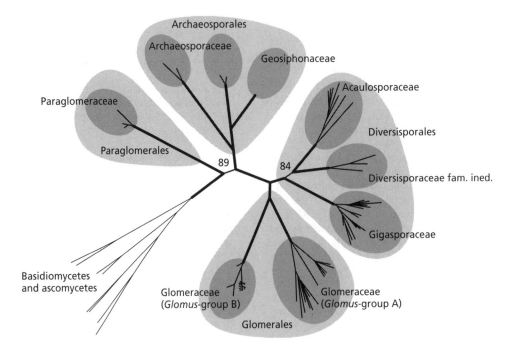

Fig. 2.4 Proposed generalized taxonomic structure of the AM fungi and related fungi (Glomeromycota). Thick lines delineate "bootstrap" values (indicating relatedness between the main branches) above 95%. Lower values (89% and 84%) are shown on two of the branches. (Reproduced by courtesy of Schuessler *et al.* 2001, and the British Mycological Society.)

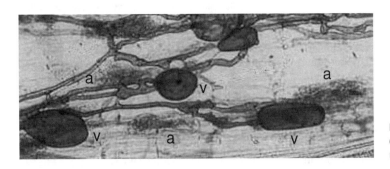

Fig. 2.5 Vesicles (v) and arbuscules (a) of present-day arbuscular mycorrhizal fungi in clover roots.

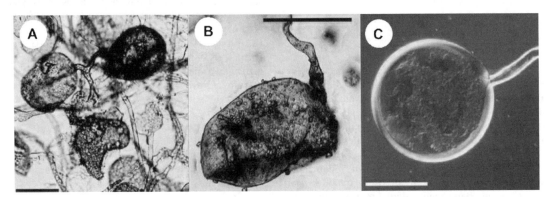

Fig. 2.6 (A,B) Fossil hyphae and spores from the Ordovician, about 460 mya, compared with a spore (C) of a present-day *Glomus* species (an arbuscular mycorrhizal fungus). All scale bars = 50 μm. (Images courtesy of Dirk Redecker; see Redecker *et al.* 2000.)

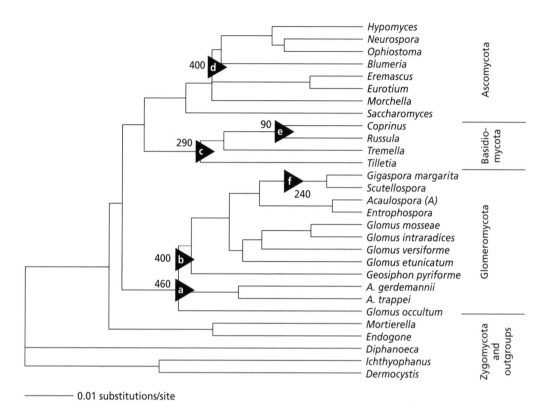

———— 0.01 substitutions/site

Fig. 2.7 A phylogenetic tree based on small subunit ribosomal RNA gene sequences. Triangle a indicates fossil spores of the Ordovician (460 mya). Triangle b indicates fossil spores of presumed arbuscular mycorrhizal (AM) fungi from the Rhynie chert (400 mya). Triangle c indicates a fossil clamp connection (Chapter 4) – the earliest evidence of Basidiomycota (290 mya). Triangle d indicates Ascomycota from the Rhynie chert (400 mya). Triangle e indicates a gilled mushroom in amber (90 mya). Triangle f indicates AM fungi of the *Gigaspora* type in the Triassic (about 240 mya). As reference points, the "outgroups" are *Diphanoeca* (a choanoflagellate from which fungi are thought to have arisen), and two early divergent forms of animals, *Ichthyophanus* and *Dermocystis*. (Courtesy of Redecker *et al.* 2000.)

2 hyphae that typically **lack cross walls**, so all the nuclei are contained within a common cytoplasm (a coenocytic mycelium);

3 the production of a thick-walled resting spore – the **zygospore** – which if formed by a sexual process involving the fusion of two gametangia;

4 the production of **asexual spores** by cytoplasmic cleavage within a sporangium;

5 a haploid genome.

Ecology and significance

The Zygomycota contains a diverse range of species, with different types of behaviour, but for convenience we start with the most familiar examples in the Order **Mucorales**. This includes several common species of the genera *Mucor*, *Rhizopus*, *Phycomyces* and

Thermomucor (a thermophilic fungus that can grow up to about 60°C). All these fungi grow mainly as saprotrophs in soil, on animal dung, on composts in the case of thermophilic species, or on various other substrates such as over-ripe fruits. They grow rapidly, often covering an agar plate in 24–36 hours, and they are among the commonest fungi found on soil dilution plates. They have been termed "sugar fungi", to denote the fact that they often depend on simple, soluble carbon nutrients. However, some fruit-rotting species such as *Rhizopus stolonifer* produce pectic enzymes that can rapidly liquefy apples and soft fruits (see Fig. 14.6). In addition, several species of *Mortierella* degrade chitin by producing chitinases (as also do several chytrids). Chitin is a common and important polymer, because it is a major component of fungal walls and of the exoskeletons of insects and other invertebrates.

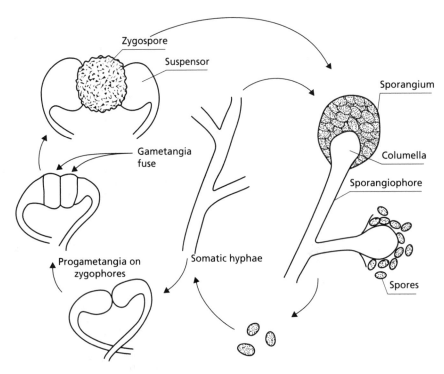

Fig. 2.8 General life cycle of Mucorales. Asexual reproduction occurs by the production of sporangia which release haploid spores. Sexual reproduction occurs by the fusion of gametangia at the tips of zygophores, leading to a diploid zygospore that undergoes meiosis to release haploid spores.

A generalized life cycle of these fungi is shown in Fig. 2.8. After a phase of rapid somatic (vegetative) growth, they produce erect aerial branches (**sporangiophores**) behind the colony margin, then a thin-walled **sporangium** is formed at the tip and the sporangial contents are cleaved around the individual nuclei to produce many uninucleate spores (**sporangiospores**). These spores are darkly pigmented, with melanin in their walls. They are released and usually wind-dispersed by breakdown of the thin sporangial wall, but in some species they are water-dispersed. The tip of the sporangiophore projects into the sporangium as a **columella**, which remains after the spores have been shed (Fig. 2.8). Sporangiospores can germinate rapidly in suitable conditions, repeating the asexual cycle.

Within the Zygomycota there are several variations in the ways that sporangia and sporangiospores are produced. Several species produce sporangia at the tips of long, unbranched sporangiophores (Fig. 2.9a) but the thermophilic species often produce branched sporangiophores (Fig. 2.10b). *Thamnidium elegans* is a psychrophilic (cold-loving) species which can grow on meat in cold-storage, and at one stage was patented for tenderizing steak. It produces a large sporangium at the

tip of a long sporangiophore, but the sporangiophore also produces complex branching structures that terminate in many few-spored **sporangioles** (Fig. 2.9b).

Other variations are found in different Orders and Families of the Zygomycota (Fig. 2.11). In this respect it is important to recognize that the spore-bearing structures of fungi have functional significance – their role is to maximize the chances of dispersing spores to sites where the fungus can establish new colonies. We cover this topic in detail in Chapter 10.

Many members of the Mucorales also have a sexual stage. In appropriate culture conditions aerial branches, termed **zygophores**, grow towards one another and swell at their tips to produce **progametangia** (Fig. 2.8). Then the progametangia develop a complete septum, which cuts off each **gametangium** from the subtending hypha. The gametangia then fuse to form a single cell that develops into a **zygospore** – a thick-walled, ornamented resting spore. Coupled with this, the terminal parts of the zygophores can swell, to produce bulbous suspensors (Figs. 2.8; 2.12). Zygospores exhibit a period of constitutive dormancy, but eventually germinate to produce a sporangium, releasing haploid spores. Many species in the Mucorales are heterothallic (outcrossing) and develop zygospores only when

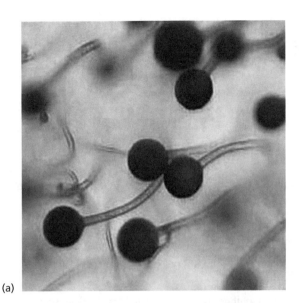

(a)

(b)

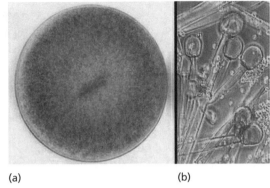

(a) (b)

Fig 2.10 (a) A 2-day-old colony of *Thermomucor pusillus* (previously called *Mucor pusillus*), a thermophilic species common in composts. (b) The characteristic branched sporangiophores of this fungus; the spores have been shed but the sporangiophores and columellae remain, like drumsticks.

strains of opposite mating type (designated plus and minus) are opposed on agar plates. Nuclear fusion and meiosis usually occur within the zygospore. This whole process is orchestrated by volatile hormones termed **trisporic acids**, discussed in Chapter 5.

In addition to the Order Mucorales, the Zygomycota contains a wide range of fungi that parasitize other small organisms, including nematodes, other small invertebrates, and other fungi. For example, the genus *Piptocephalis* (Fig. 2.11c), in the **Order Zoopagales**, contains several species that parasitize other Zygomycota and occasionally other fungi. *Piptocephalis* is one of several genera that produce few-spored sporangia, termed **merosporangia**, with linearly arranged spores. In nature the *Piptocephalis* spp. probably grow as obligate parasites on other fungi, especially other Zygomycota, although they can be grown in pure culture if supplied with the necessary vitamins. They penetrate the host hyphae without killing them, by producing specialized, nutrient-absorbing **haustoria**, and in this way they tap the host's nutrients to support their own development. This type of interaction seems to be governed by specific recognition factors (lectins), discussed in Chapter 12 (see Fig. 12.10).

Fig. 2.9 (a) The characteristically unbranched sporangiophores of most *Mucor* species. The sporangia are still intact and contain many darkly pigmented spores with melanized walls. (b) *Thamnidium elegans*, showing one of several variations in the production of sporangiospores by

Zygomycota. This fungus produces typical sporangia on long sporangiophores, but complex branching structures arise from the sporangiophore and terminate in small, few-spored **sporangioles**.

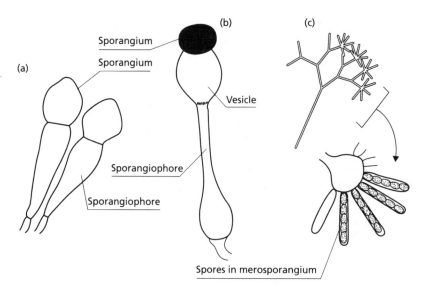

Fig. 2.11 (a) Sporangia of insect-pathogenic members of the Entomophthorales (e.g. *Entomophthora, Pandora* spp.) are released and function like spores. (b) *Pilobolus*, a common fungus on herbivore dung, has a sporangium containing many spores. It is mounted on a vesicle, which ruptures at maturity to shoot the sporangium onto surrounding vegetation (Chapter 10). (c) *Piptocephalis* is a parasite of other Zygomycota. It produces branched sporangiophores with **merosporangia** (each containing a few linearly arranged spores) at the branch tips.

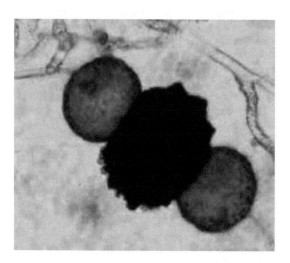

Fig. 2.12 A mature, thick-walled, warty zygospore (sexual spore) of *Rhizopus sexualis*, produced by the fusion of two gametangia. The bulbous swellings on either side are termed suspensors.

The **Order Entomophthorales** is notable because it includes many fungi that parasitize insects, and that can cause spectacular population crashes of their insect hosts (discussed in Chapter 15). Many are host-specialized, typically infecting only a narrow range of insects, but others have relatively broad host ranges. Techniques have been developed to mass produce some of these fungi in solid or liquid culture systems, raising the prospect of using them as practical biocontrol agents of insect pests. The Entomophthorales are notable for the way in which they are dispersed. As shown in Fig. 2.11a, the hyphae emerging from a dead insect produce sporangiophores, each bearing a large sporangium at its tip. The tip of the sporangiophore has a **columella**, which bulges into the spore and is double-walled (one wall being the tip of the columella, and the other being the wall of the sporangium). The progressive build-up of turgor pressure causes the sporangium to be shot suddenly from the columella, projecting the sporangium to a distance of about 4 cm (Webster 1980). (Technically, the sporangium in this case is a single spore, which is not enclosed in a sporangium wall, so it is termed a conidium, like the many types of conidia produced by the Ascomycota – see later.) The sporangium (conidium) is sticky and must adhere to an insect in order to infect. If it does not land on an insect it will germinate to produce a secondary conidium, which is similarly dispersed, and this process will be repeated until the conidium lands on a suitable host or until the nutrient reserves are exhausted. Chapter 15 provides a much fuller account of these and other insect-pathogenic fungi.

Some other fungi of the Entomophthorales display a dispersal strategy similar to that described above. One example is *Conidiobolus coronatus*, a common saprotrophic fungus in soil. Another example is *Basidiobolus ranarum*, which is found in the gut, the faeces or the intestines of insect-eating reptiles or amphibians (e.g. frogs, from which the species name *ranarum* is derived). This fungus has recently been transferred to the Chytridiomycota, based on SSU rDNA sequencing. A third fungus, *Pilobolus* (Fig. 2.11b) in the Order Mucorales, shows a parallel mode of dispersal. This fungus is common in animal dung and produces a large sporangium which is shot free from the dung, but the sporangium subsequently breaks down to release many individual spores. We return to this topic when we discuss spore dispersal, in Chapter 10.

Zygomycota as pathogens of humans

Several common members of the Mucorales, especially *Rhizopus arrhizus*, *Absidia corymbifera* and *Rhizomucor pusillus* (a thermophilic species) can cause serious, life-threatening infections of humans. Similar infections can occasionally be caused by *Conidiobolus* (Entomophthorales) and *Basidiobolus ranarum*. Collectively, these diseases are termed **zygomycosis** (infections caused by members of the Zygomycota). In the vast majority of cases the infection is associated with predisposing factors. For example, diabetics with ketoacidosis are among the most vulnerable to infection, which often develops in the rhino-cerebral area (the nasal passages, progressing into the brain). But many other predisposing factors can be involved, including severe malnourishment in children, severe burns, cancer, lymphoma, and immunodeficiency or immunosuppression.

These fungal infections can develop rapidly by spread through the arteries and invasion of the surrounding tissues. They often require surgical excision of the affected areas. Even so, the prognosis is poor – the overall mortality rate is about 50% and can rise to 85% in patients with the rhino-cerebral form of the disease. The zygomycoses fall clearly into the category of opportunistic infections, because they are associated with impaired host defenses. They are among the most difficult infections to control because of their rapid, invasive growth.

Ascomycota

The phylum Ascomycota contains about 75% of all the fungi that have been described to date. It is not only the most important phylum, but also the most diverse, and many of the relationships within this group have yet to be resolved by modern molecular methods. The one feature that characterizes all members of this phylum is the **ascus** – a cell in which two compatible haploid nuclei of different mating types come together and fuse to form a diploid nucleus, followed by meiosis to produce haploid sexual spores, termed **ascospores**. In many species the meiotic division is followed by a single round of mitosis, leading to the production of eight ascospores within each ascus (Fig. 2.13). In the more advanced members of the group, many asci are produced within a fruiting body, termed an **ascocarp**. This can take various forms – a flask-shaped **perithecium** with a pore at its tip, a cup-shaped **apothecium**, a closed structure that breaks down at maturity, termed a **cleistothecium**, or a **pseudothecium** which is usually embedded in a pad of tissue, termed a **stroma**. In other cases, the asci are produced singly and are not enclosed in a fruiting body – for example, this is true for the budding yeast *Saccharomyces cerevisiae* (which produces only 4 ascospores in a naked ascus) and the fission yeast *Schizosaccharomyces octosporus* (eight-spored asci). Several of these points are illustrated in Figs. 2.13–2.15.

The vast majority of Ascomycota also produce asexual spores by mitosis. These **mitospores** (as opposed to meiospores, derived by meiosis) are extremely common and function in dispersal. In fact, a substantial number of Ascomycota are known only in the form of their mitosporic (asexual) stages. Their sexual stages remain to be discovered and in some cases might have been abandoned, but SSU rDNA sequence analysis can be used to link them to the Ascomycota. To give just one example, the extremely common fungus, *Aspergillus fumigatus*, has never been shown to produce a sexual stage. This fungus is becoming increasingly common as a major cause of death in hospital patients undergoing deep surgical procedures, and it also causes the serious respiratory disease termed **aspergillosis** (Chapters 8 and 16). In cases such as this it is important to know the sources of genetic variation – is there a sexual stage that we have yet to discover?

Like many members of the Ascomycota, *N. crassa*, has a sexual stage, leading to the production of asci containing ascospores. *N. crassa* is heterothallic, requiring strains of different mating types (termed **A** and **a**) for sexual development. The female sex organ, termed an **ascogonium**, is often a coiled, multinucleate hypha with a receptive trichogyne. It is fertilized either by contact with a spermatium (a small uninucleate cell that cannot germinate) or by a conidium. Then the ascogonium produces an **ascogenous hypha** that will eventually give rise to the asci. The process by which this happens is shown in Fig. 2.16.

(a)

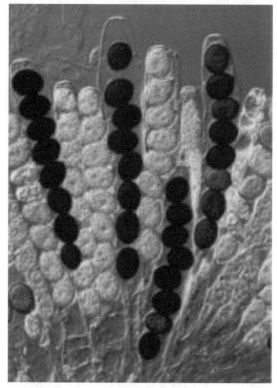

(b)

Fig. 2.13 (a) A developing **perithecium** of *Sordaria macrospora* which, at maturity, will have a pore at its tip to release the ascospores. (b) Several asci in different stages of development within the perithecium; each ascus contains eight ascospores, about 20 µm diameter. (Courtesy of N. Read; from Read, N.D. & Lord, K.M. 1991 *Experimental Mycology* 15, 132–139.)

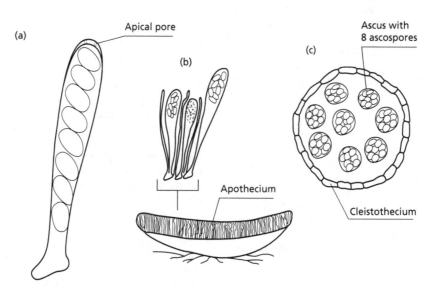

Fig. 2.14 Diagrammatic representation of some fruiting structures of Ascomycota. (a) A single ascus containing eight ascospores which, at maturity, will be shot forcibly through the narrow apical pore. (b) A cup-shaped **apothecium** containing many asci interspersed with sterile "packing" hyphae (paraphyses). The paraphyses help to laterally constrain the developing asci so that they project beyond the surface of the cup to release their spores. (c) A **cleistothecium** (closed ascocarp) such as that of *Eurotium*, the sexual stage of some *Aspergillus* spp. The asci, each with eight ascospores, are released when the cleistothecium wall breaks down at maturity to release the ascospores.

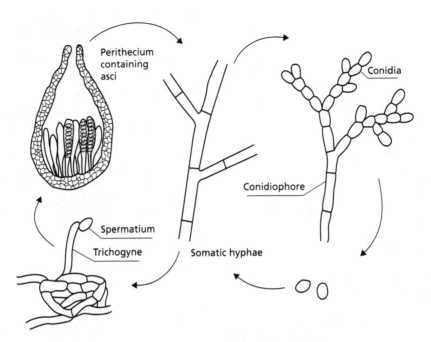

Fig. 2.15 Stages in the life cycle of *Neurospora crassa* and many similar fungi in the Ascomycota.

Fig. 2.16 Early stages in the production of asci from an ascogenous hypha. (a) The ascogenous hypha grows for some distance then curls back at its tip to form a crosier. (b) Synchronous nuclear division occurs as the crosier bends back upon itself. (c) Septa are laid down so that the penultimate cell (located at the top) has one nucleus of each mating type. This cell will extend and will give rise to the first ascus, when the nuclei fuse and then undergo meiosis and one round of mitosis, to produce eight ascospores. Meanwhile, the tip cell of the crosier fuses with the ascogenous hypha, and the nucleus is transferred. (d) A new branch develops from the fused cell, and this branch forms a further crosier, to repeat the whole process. Thus, eventually, a large number of ascus initials are produced, which will give rise to a large number of asci. The final stage in this process is the development of a fruitbody tissue surrounding the asci. The haploid ascospores are either **A** or **a** mating type, and eventually germinate to produce haploid colonies, to complete the life cycle.

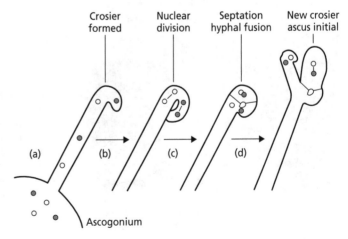

Taxonomy and relationships

The Ascomycota is considered to be a monophyletic group (all members sharing a common ancestry) that dates back to the coal age, at least 300 million years ago. This group is closely related to the Basidiomycota (the phylum containing mushrooms, etc.) as a sister group that shares similar features.

Currently, the Ascomycota is divided into three subgroups, although the status of some of the early-diverging members is still unclear:

1 **Archaeascomycetes**, including the fission yeast, *Schizosaccharomyces pombe*, and some primitive plant pathogens such as *Taphrina* spp. (Fig. 2.17).
2 **Hemiascomycetes** – the true yeasts such as *Saccharomyces cerevisiae*, used to produce alcohol and for bread-making. These fungi do not produce ascocarps.
3 **Euascomycetes**, comprising the majority of species that produce hyphae and ascocarps.

In addition to the main distinguishing feature, the ascus, the Ascomycota are characterized by **cell walls composed primarily of chitin and glucans**, by **hyphae that have cross walls** (septa; singular septum) at regular intervals but with a central pore that allows the passage of nuclei and other cellular organelles, and by the production of **non-motile asexual spores** (mitospores, termed **conidia**) that are produced in various ways, but never by cytoplasmic cleavage in a sporangium.

The life cycle of Ascomycota

We will use *Neurospora crassa* as an example to illustrate significant features of the life cycle of Ascomycota (Fig. 2.15). *Neurospora* grows as a branched network of hyphae with perforated septa, and produces aerial hyphae that terminate in branched chains of conidia (asexual spores). These are formed by a bud-like process involving swelling and repeated septation. When mature, the conidia develop a pink coloration and are easily removed from the hyphae by air currents. This cycle of sporulation, dispersal and spore germination occurs repeatedly when substrates are readily available.

There are several variations in sexual reproductive systems of the Ascomycota, which we need mention only briefly. For example, *N. crassa* and *N. sitophila* are heterothallic (out-crossing) but some species (*Neurospora tetrasperma*, *Podospora anserina*) produce only 4 binucleate ascospores in each ascus, and are homothallic. A single ascospore can give rise a colony that will produce the sexual stage.

Ecology and significance

Because the Ascomycota is a very large and important phylum, many aspects of its biology are covered in the later chapters of this book. The group as a whole includes many economically important plant pathogens, such as the powdery mildew fungi of many crop plants (Chapter 14), the vascular wilt fungi (Chapter 14), *Ophiostoma ulmi* and *O. novo-ulmi*, which cause the devastating Dutch elm disease that swept repeatedly across Europe and the USA in the last century (Chapter 10), the equally devastating chestnut blight disease in the USA, caused by *Cryphonectria parasitica* (Chapter 9) and several toxigenic fungi, such as *Claviceps purpurea* (ergot of cereals; Chapter 7).

The Ascomycota also includes several pathogens of humans, domesticated animals and livestock. For example, the ubiquitous dermatophytic (ringworm) fungi are estimated to infect about half of the total human population, especially in the tropics and subtropics, but also in developed countries, where infections such as "athlete's foot" and "nail fungus" are common. The ascomycetous yeast, *Candida albicans*, is a common commensal organism in the gut and on other mucosal membranes of humans, causing irritation to people who wear dentures and to women during menstruation or pregnancy. And, some ascomycetous fungi (*Blastomyces dermatitidis*, *Histoplasma capsulatum*) can cause life-threatening

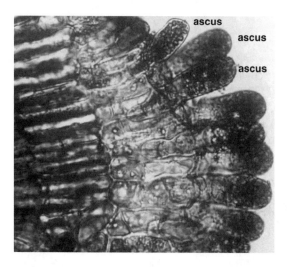

Fig. 2.17 *Taphrina populina*, a plant-parasitic member of the Archaeascomycota which causes yellow blister of poplar leaves (*Populus* spp.). The fungus grows as a mycelium in the plant leaves, causing leaf distortion, and it produces naked asci which project from the surface of the leaves. Within the asci the ascospores often bud, so that the asci become filled with a mixture of ascospores and budding yeast cells (particularly evident in the top ascus). (See further images on the website for this book (see Online resources).)

diseases of humans whose immune system is deficient or impaired. The "Moulds of Man" are covered in Chapter 16.

In a different context, some of the Ascomycota form mycorrhizal associations with forest trees (e.g. *Tuber* spp – the truffles) and an estimated 96% of the 13,500 known species of lichens have Ascomycota as the fungal partner. Very few of these lichenized fungi grow in a free-living state. In terms of numbers, the ascomycetous lichen fungi probably represent the largest single component of biodiversity in the entire fungal kingdom (Chapter 13). Many ascomycetous yeasts (*Saccharomyces*, *Candida oleophila*, etc.) commonly grow as saprotrophs on leaf and fruit surfaces, and some are now being marketed as biological control agents to prevent fruit rots (Chapters 11, 12). Many other ascomycetous fungi degrade cellulose and other structural polymers; e.g. *Chaetomium* in soil and composts, *Xylaria* and *Hypoxylon* as agents of wood decay, *Sordaria* and *Ascobolus* in herbivore dung, and *Lulworthia* on wood in estuarine environments (Chapter 11).

Basidiomycota

According to Kirk *et al.* (2001), the Basidiomycota contains about 30,000 described species, which is 37% of the described species of true fungi. Although the most familiar examples of this group are the mushrooms and toadstools, there is an enormous diversity of species, including basidiomycetous yeasts, many important plant pathogens, and some serious human pathogens. The one single feature that characterizes the group is the **basidium** in which meiosis occurs, leading to the production of sexual spores (**basidiospores**) that usually are produced **externally** on short stalks termed sterigmata (Fig. 2.19).

Several other features are found in the Basidiomycota. The hyphae often have a complex **dolipore septum** (Chapter 3) that prevents nuclei from moving between different hyphal compartments. The cell wall typically is composed of chitin and glucans (but chitin and mannans in the yeast forms). The nuclei typically are haploid, but throughout most of the life of a colony each hyphal compartment contains two nuclei, representing two different mating compatibility groups. A mycelium of this type is termed a **dikaryon**. The significance of this will be explained shortly.

Taxonomy and relationships

The Basidiomycota is a monophyletic group (all of its members having a common ancestry) and is a sister group to the Ascomycota. In other words, both of these groups have a common ancestor. Three major sub-groups are recognized within the Phylum Basidiomycota:

1 **Urediniomycetes**, including the **rust fungi** (Uredinales) which are economically significant plant pathogens of many crops and wild plants.
2 **Ustilaginomycetes**, including the **smut fungi** (Ustilaginales) some of which again are significant plant pathogens, and gain their name from their black, sooty spores.
3 **Hymenomycetes**, including mushrooms, puffballs, and jelly fungi.

However, the phylogenetic relationships between and within these three groups are still unclear.

Significant features in the life-cycle of Basidiomycota

We will begin by considering the generalized life-cycle of a typical mushroom (Fig. 2.18). Basidiospores, each containing a single haploid nucleus, germinate and grow into hyphal colonies that have a single nucleus in each hyphal compartment. This phase is termed a **monokaryon** (meaning that it has one nuclear type in each hyphal compartment). The monokaryon can produce small **oidia** that either act as fertilizing elements or they can germinate to produce further monokaryotic colonies. The next stage of development occurs when the hyphae of two monokaryons of different mating compatibility groups fuse with one another, or when an oidium of one mating compatibility group attracts a hypha of a different mating compatibility group. This causes a strong "homing response" but the chemical attractants causing it have not been identified. Once the two compatible strains have fused by the process termed **plasmogamy**, all subsequent growth occurs by **dikaryotic** hyphae (with two nuclei – one of each compatibility group – in each hyphal compartment).

The fungus can grow for many weeks, months or even years in this form, producing an extensive network of dikaryotic hyphae, but in response to environmental signals (Chapter 5) it will produce a mushroom or other type of fruitbody. All the tissues of the mushroom are composed of dikaryotic hyphae, but at a relatively late stage in development of the mushroom the gills become lined with (dikaryotic) basidia. Within each basidium the two nuclei fuse to form a diploid nucleus (**karyogamy**) which subsequently undergoes meiosis. Sterigmata then develop from the top of each basidium, and the four haploid nuclei migrate into the developing basidiospores (Fig. 2.19). Sometimes only two spores are produced, instead of four – a common example being the cultivated mushroom, *Agaricus bisporus*. This is because, after meiosis in the basidium, 2 nuclei enter each spore, and the germinating

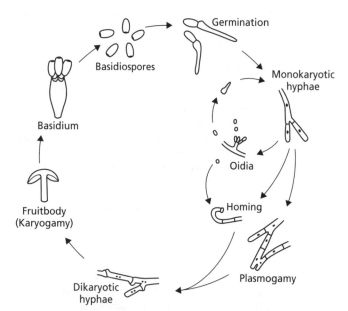

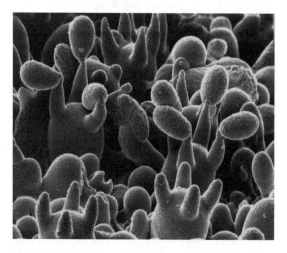

Fig. 2.18 A generalized life cycle of a mushroom; see text for details. (Courtesy of Maria Chamberlain.)

Fig. 2.19 Scanning electron micrograph of the basidia of a mushroom, *Coprinus cinereus*. Note the inflated basidia, each with four sterigmata, and the basidiospores, some of which have already been released, or collapsed during preparation of the specimen. (Courtesy of Dr Chris Jeffree.)

basidiospores produce a dikaryotic mycelium that is homothallic (self-fertile). However, the vast majority of mushroom-producing fungi are heterothallic – about 90%. We will discuss the compatibility systems of these fungi in Chapter 5.

The way in which the basidiospores are released is not fully understood, but just before their release a small droplet, termed the hylar droplet, is seen at the base of the spore, then the spore is "popped" from the sterigma into the space between the gills, where it falls vertically, free from the fruitbody. After a period of dormancy the haploid basidiospores germinate to produce monokaryotic hyphae, which repeat the whole cycle.

One of the most intriguing aspects of this life cycle is the manner in which the dikaryon develops. This is shown diagrammatically in Fig. 2.20. The dikaryotic tip cell extends and synthesizes protoplasm until it attains a critical cytoplasmic volume. Then a small backwardly-projecting branch arises, and the two nuclei divide synchronously – one of the nuclei dividing along the axis of the hypha, and the other dividing so that one of the daughter nuclei enters the branch. The branch is then cut off by the development of a septum (cross-wall) and a septum also develops in the main axis of the hypha. The effect of this is to create a tip cell with two nuclei of different compatibility types, a penultimate cell with a single nucleus, and a branch with a single nucleus, isolated by septa. The branch then fuses with the penultimate cell, by dissolution of the wall, and the nucleus migrates into the penultimate cell, restoring the dikaryotic condition. This process occurs every time the nuclei divide. The resulting "bumps" on the hyphae are easily seen by light microscopy, and are termed **clamp connections** (Fig. 2.20). However, not all members of the Basidiomycota produce clamp connections, and yet they still grow as dikaryons, suggesting that other regulatory mechanisms are involved.

If we refer back to the development of asci in Fig. 2.16, we see a remarkable similarity in the behavior

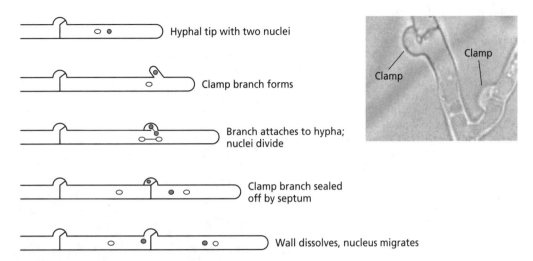

Hyphal tip with two nuclei

Clamp branch forms

Branch attaches to hypha; nuclei divide

Clamp branch sealed off by septum

Wall dissolves, nucleus migrates

Clamp

Clamp

Fig. 2.20 The role of clamp connections in maintaining a regular dikaryon.

of clamp connections (Fig. 2.20) and the development of asci in the Ascomycota. In both cases the nuclear distribution is maintained by a hypha growing back upon itself and fusing with the cell behind it. And in both cases there is a dikaryotic phase, although it occurs only temporarily in the ascus initials. These lines of evidence reinforce the view that the Ascomycota and Basidiomycota are closely related.

Ecology and significance

The Phylum Basidiomycota exhibits the most extreme variation of any group of fungi. Some of the yeasts on plant surfaces are basidiomycetous, although they are far outnumbered by the ascomycetous yeasts. A common example is the salmon-colored yeast, *Sporobolomyces roseus*, which grows on senescent leaf surfaces. It divides by budding but also has a sexual stage, in which single yeast cells produce a sterigma and release a basidiospore, apparently by the same mechanism as in mushrooms. *S. roseus* is a common inhabitant of leaf surfaces, and the spore population in the air can be high enough to cause respiratory allergies. But a few basidiomycetous yeasts are much more significant and can be life-threatening pathogens of humans, especially of people with impaired immune defenses. The most notable examples are *Blastomyces dermatitidis*, *Cryptococcus neoformans* and *Histoplasma capsulatum*, all of which can produce sexual stages when strains of different mating types are crossed in laboratory conditions, but the sexual stages are seldom found in nature. The sources of inoculum that initiate infection in the lungs are common environmental samples, such as birds' droppings and even the bark of eucalypt

trees (Chapter 16). The important lesson from this is that even the apparently insignificant saprotrophic fungi of plant surfaces and other materials can have a significant impact on human health.

Some of the most economically important plant pathogens are members of the Basidiomycota. They include the many species and genera of **rust fungi** (Uredinales) that are specialized to infect particular types of plant. One of the best-known examples is *Puccinia graminis*, which causes black stem rust of wheat. These fungi can have complex life cycles, with up to four different types of spore being produced at different stages of the life cycle (Chapter 14). Another important group of pathogens, the **smut fungi** (Ustilaginales), are unusual because they grow only as yeasts in culture but as mycelia in their hosts. Again, these are discussed in Chapter 14.

The most familiar fungi in the Basidiomycota belong to the class Hymenomycetes, and include almost all the larger fungi, such as the mushrooms (commonly known as toadstools), woody brackets, puffballs, earth-stars, splash cups, and jelly-like structures. Most of these "higher" Basidiomycota degrade polymers such as cellulose, hemicelluloses and lignin. They are found in composts (e.g. *Coprinus*), leaf litter (e.g. *Mycena*), the thatch of dead leaf sheaths in old grasslands (e.g. *Marasmius oreades*, which produces "fairy rings"; Chapter 11) and in woody substrates, where they play a major role in wood decay (Chapter 11). Some of these fungi cause major tree diseases; e.g. *Armillaria* spp. in hardwood stands (Chapter 5) and *Heterobasidion annosum* in conifers (Chapter 12).

Many other mushroom-producing fungi form mycorrhizal associations with forest trees, although this applies mainly to the cooler regions of the world, because most tropical and subtropical trees have arbuscular-mycorrhizal fungi. Common examples of mycorrhizal fungi found in the cooler regions include species of *Amanita*, *Russula*, *Cortinarius*, *Boletus*, *Hebeloma*, *Lactarius*, etc. They play major roles in facilitating the uptake of mineral nutrients from soil. But, whereas the arbuscular mycorrhizal fungi of most land plants seem to be involved primarily in the uptake of phosphates from soil (Chapter 1), the ectomycorrhizal fungi of forest trees (excluding the tropics and subtropics) seem to be most significant in degrading soil proteins and providing the trees with nitrogen (Chapter 13).

Several members of the Basidiomycota are grown commercially as mushroom crops (e.g. *Agaricus bisporus*, *Volvariella volvacea*, *Lentinula edodes*), while others are "cultivated" by insects to provide a food source for the insect colonies. Two classic examples of this are seen in the "fungus gardens" of the leaf-cutting attine ants (*Atta* species) and in a sub-family of termites that lack cellulolytic bacteria in their guts. In both cases, the insects obtain their main food source by bringing pieces of leaf or other material into the nests and inoculating this with spores of basidiomycetous fungi. The insect colony then feeds on the fungal hyphae, and weeds-out any contaminant fungi. Cladistic analysis has traced the origin of these mutualistic associations to about 50–60 million years ago (Chapter 13).

Other members of the Basidiomycota, such as *Amanita phalloides* – "the destroying angel" – produce deadly toxins that bind to the cytoskeletal proteins in cells. These toxins have become important tools in cell biology for investigating cell dynamics (Chapter 3).

The diversity of fruiting bodies (basidiocarps) produced by Basidiomycota

As a means of illustrating the considerable diversity of form and function of basidiocarps, a range of examples are shown, with annotations, in Figs 2.21–2.32. All these examples can be downloaded in color, on the accompanying website for the book (see **Online Resources**).

Cyathus striatus

Fig. 2.21 Basidiocarps of one of the bird's nest fungi (genus *Cyathus*), about 1 cm diameter. These fungi are commonly seen in soil enriched with wood chippings. The small gray-brown cups open at maturity by rupture of the thin upper membrane (see arrows), revealing a cluster of egg-like **peridioles**. The basidiocarp acts as a splash cup. When raindrops hit the peridioles these are flung out of the cup, but each periodiole is attached to a thin **funicular cord**, with a sticky **hapteron** at its base, and this attaches to any object that it strikes, such as a leaf blade.

(a)

(b)

Fig 2.22 (a,b) Earth stars (*Geastrum* spp.), about 6 cm diameter, growing in leaf litter beneath birch trees. These basidiocarps are like puffballs, but are initially enclosed in thick outer wall layers. These layers split and curl back, helping to raise the basidiocarp above the leaf litter. Basidiospores develop and mature within the closed fruit-body, which then develops a pore at its tip. The spores are "puffed" through this pore when raindrops hit the papery casing of the "spore chamber."

Fig. 2.23 Puffballs (*Lycoperdon* spp.) in an immature white stage (a) and mature brown stage (b), about 3 cm diameter. These are similar to the earthstars shown in Fig. 2.22, but frequently are attached to woody substrates. Puffballs produce basidiospores from basidia that develop within the fruitbody. The spores are not shot from the sterigmata but drop off and accumulate within chambers inside the developing fruitbody. Mature puffballs have a thin, papery outer casing with a pore at the top. Spores are puffed from the pore by raindrops.

Fig. 2.24 Two characteristic wood-rotting fungi of Scottish birch woods. (a) *Fomes fomentarius* (the hoof fungus), about 20 cm diameter. This hard, woody, perennial, bracket fungus is very common. The hyphae growing within the wood cause a brown rot. The underside of the basidiocarp consists of minute pores through which the basidiospores fall and are then wind-dispersed. (b) A bracket-shaped polypore of the fungus *Piptoporus betulinus*, the "razor strop" fungus.

Fig. 2.25 (a) An evanescent bracket of *Polyporus squamosus* (Dryad's saddle), about 25 cm diameter, growing on a senescent poplar tree. The spores are released through pores on the underside, but the bracket senesces and decays within a few months and is usually replaced annually. (b) A large conical-shaped fruitbody of *Phaeolus schweinitzii*, parasitic on the roots of conifers. The cap is about 15 cm diameter, cream-colored at the margin but dark brown in the center. In this fresh specimen the toadstool is covered with a downy felt on the upper surface.

Fig. 2.26 (a) A basidiocarp of *Boletus scaber* (10 cm diameter, mauve-brown in color), growing as a mycorrhizal associate of birch trees. The underside (b) shows that basidiospores are produced in pores, not on gills, giving rise to the common name for these fungi – the polypores.

Fig. 2.27 Two examples of **jelly fungi**, with a moist, jelly-like appearance. (a) Ear-shaped basidiocarps of *Auricularia auricula*, commonly seen as a saprotroph on the dead branches of elder trees (*Sambucus* spp.). The dark brown fruitbody (about 6 cm) can dry completely and shrivel, but swells and resumes its shape after rains. (b) *Tremella mesenterica* (about 5 cm diameter) growing as a saprotroph on a dead tree branch. Jelly fungi have several distinctive structural and ultrastructural features, including basidia that are deeply divided like tuning forks or with four long sterigmata, or with septate basidia. Some have a haploid yeast-like budding phase, and some can bud conidia from the basidiospores.

(a)

(b)

(a)

(b)

Fig. 2.29 (a,b) Thin, leathery brackets of the polypore fungus *Coriolus versicolor* (about 5–7 cm diameter), which often occurs abundantly on logs or fallen hardwood trees. The brackets grow at their margins and often show concentric zones of different colors – hence the species name, *versicolor*.

Fig. 2.28 (a) A further example of a jelly fungus: a forked basidiocarp of *Calocera viscosa* (about 5 cm high, color orange-yellow), which grows from underground decaying stump tissues of trees. (b) A basidiocarp of the fungus *Lactarius torminosus* (color salmon-pink), frequently seen as a mycorrhizal associate of birch trees. This poisonous fungus is commonly known as the woolly milk cap because of its woolly appearance. The specimen is about 10 cm diameter.

(a)　　　　　　　　　　　　　　　　(b)

Fig. 2.30 Two examples of edible gilled fungi. (a) A typical cluster of basidiocarps of *Pleurotus ostreatus*, the "oyster fungus" (each about 10 cm diameter) growing on a log. The gills are pure white, and the upper surface is mauve-brown. (b) The orange-yellow colored chanterelle, *Cantharellus cibarius*, which typically grows on mossy banks beneath birch trees.

(a)　　　　　　　　　　　　(b)　　　　　　　　　　　　(c)

Fig. 2.31 The "ink caps" (*Coprinus* spp.) and related fungi, commonly found on animal dung or in composts. Typically, the basidiocarps develop rapidly but last for only a short time. The cap containing the gills then senesces (deliquesces) from the base upwards, dripping an inky fluid containing the black basidiospores. Mycological enthusiasts have been known to write with this material, but that was before computers! (a) *Coprinus comatus* (known as shaggy cap, or lawyer's wig), about 15 cm tall at maturity. (b) Similar basidiocarps 2–3 days after maturity and showing deliquescence. (c) A related genus, *Psathyrella*, growing on an agar plate. *Coprinus* and related species are among the few mushroom-producing fungi that can be grown easily in agar culture. They have been used extensively in genetical and developmental studies. (Images courtesy of Maria Chamberlain.)

Fig. 2.32 *Amanita muscaria*, the "fly agaric" (about 12 cm diameter). The cap is bright red, with white scales. The stalk (stipe) bears a ring (annulus) just below the cap, and at the base of the stalk is a cup-shaped volva (not clearly visible, but see Fig. 13.6). This species is a common mycorrhizal associate, especially of birch (*Betula* spp.) and pines (*Pinus* spp.). It is extremely poisonous.

Mitosporic fungi

Finally, among the true fungi, we consider the large number of ascomycetous fungi that produce conidia, but whose sexual stages are absent, rare or unknown. In the past, these fungi have variously been termed "Deuteromycota" (Deuteromycotina) or "Fungi Imperfecti" but the more appropriate term is **mitosporic fungi**, indicating that the spores are produced only by mitotic nuclear division. The majority of these fungi are likely to be assigned to genera of the Ascomycota by gene sequence comparisons. But at present, the genera and species with no known sexual stage are given provisional generic names, termed **"form genera."** When a sexual stage is discovered the fungus must be renamed, and described according to the features of its sexual reproductive stage – the so-called "perfect state." (Other terms used include **anamorph** for the asexual stage, and **teleomorph** for the sexual stage)

Some very common and important fungi are known only by their asexual stages. Examples include fungi of the form-genera *Alternaria* (although one species has a sexual stage, termed *Lewia infectoria*), *Aspergillus* (although two species have a sexual stage: *Fenellia flavipes* for *Aspergillus flavipes*, and *Emericella nidulans* for *Aspergillus nidulans*), *Cladosporium*, *Humicola*, *Penicillium*, *Phialophora*, etc.

The interesting question is: why have so many fungi largely or completely abandoned sexual reproduction? The answer might be that they have developed alternative means of genetic recombination, such as a "parasexual cycle" that has been described in (mitosporic) genera such as *Penicillium* and *Aspergillus* (Chapter 9).

The methods of conidial production

The mitosporic fungi, like the asexual stages of known Ascomycota, produce conidia in various ways, but never by cytoplasmic cleavage in a sporangium. Some of the many methods of conidium production are shown in Fig. 2.33. It has been proposed that the main types of conidium development can be either thallic (essentially by a fragmentation process involving septation, as in the fungus *Geotrichum candidum*; Fig. 2.33g) or blastic, involving swelling of the conidium initials before they are separated by septa (e.g. *Cladosporium* (Fig. 2.33a) and *Alternaria* (Fig. 2.33b)).

Most types of conidium development would fall into the "blastic" type. For example, the conidia of *Aureobasidium pullulans*, now named by its sexual (teleomorph) stage, *Sydowia polyspora*, produce conidia by budding at or near the septa (Fig. 2.33h). The conidia of *Humicola* spp. are produced by a ballooning of the tip of a conidiophore branch (Fig. 2.33d). The conidia of many fungi are produced by extruding spores from flask-shaped cells termed **phialides** (*Aspergillus*, *Penicillium*, *Phialophora*; Figs. 2.33c,e,f). Other variations are seen in the ways that conidiophores are arranged. For example, *Penicillium* spp. typically have brush-like conidiophores, terminating in phialides, and the successive conidia accumulate in chains (Fig. 2.33e). *Aspergillus* spp. typically have a conidiophore with phialides arranged on a swollen vesicle, and again they produce chains of conidia (Fig. 2.33c). The conidiophores can be aggregated into a stalk termed a **synnema** or **coremium** (e.g. *Pesotum*, Fig. 2.33i) or the conidia can arise from a pad of tissue (an **acervulus** as in *Gloeosporium*, Fig. 2.33k) or they can be produced in a flask-shaped **pycnidium** (e.g. *Phomopsis*, Fig. 2.33j).

Ecology and significance

As will be clear from the comments above, mitosporic fungi are extremely common and produce abundant conidia which are dispersed by wind, rain or animals. Many of these fungi are common saprotrophs, with important roles in nutrient recycling. Several have conidia that are large enough 20–40 μm to impact in the nostrils, causing hay-fever symptoms (e.g. *Alternaria*). Others have conidia that are small enough (about 4 μm diameter) to escape impaction in the upper respiratory tract, and reach the alveoli where they can cause acute allergic responses (Chapter 10) or in some cases (e.g. *Aspergillus fumigatus*) establish long-term

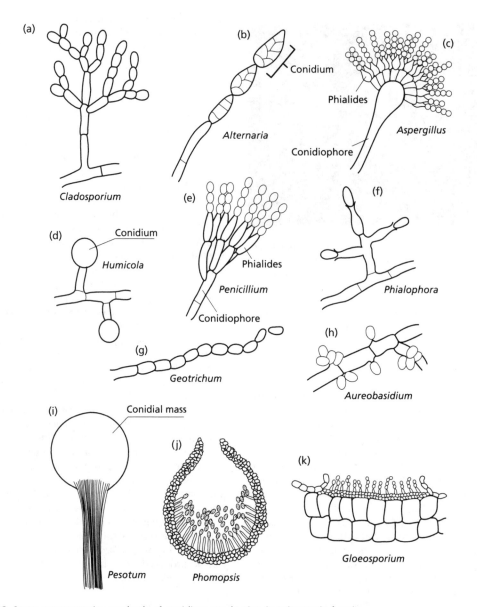

Fig. 2.33 Some representative methods of conidium production in mitosporic fungi.

foci of infection in the lungs. Mitosporic fungi such as *Penicillium*, *Aspergillus* and *Fusarium* spp. can also be major spoilage agents of stored food products (Chapter 8), and produce several dangerous mycotoxins (e.g. the aflatoxins of *Aspergillus flavus*, discussed in Chapter 7). Several mitosporic fungi are major plant pathogens, including the vascular wilt fungi *Verticillium dahliae* and *Fusarium oxysporum* (Chapter 14). Others parasitize insects – for example, the many strains of *Metarhizium* and *Beauveria* that have potential for biological control of insects (Chapter 15).

Mitosporic fungi have many other beneficial roles and are exploited commercially for their vast range of useful metabolites (antibiotics, enzymes, immunosuppressants, organic acids, etc.).

The fungus-like organisms

In addition to the true fungi, several fungus-like organisms have traditionally been studied by mycologists. This legacy dates back a long time and it is defensible, so long as we recognize that the fungus-like organisms are **not true fungi**. The main fungus-like organisms that can be considered as fungi in a broader sense (Box 2.1) are:

- The **Oomycota**, which belong to the Kingdom **Straminipila** (stramenipiles; Dick 2001) and are closely related to the golden-brown algae and diatoms.
- The acrasid and dictyostelid slime moulds (**Acrasiomycota** and **Dictyosteliomycota**).
- The plasmodial slime moulds (**Myxomycota**).
- The plasmodiophorids (**Plasmodiophoromycota**).

The Oomycota

The Oomycota are the most economically important fungus-like organisms and play extremely significant roles in many environmental processes. They are particularly important as plant pathogens, and cause some of the most devastating plant diseases, such as **potato blight** (caused by *Phytophthora infestans*), **sudden oak death** (caused by *Phytophthora ramorum*), many seedling and **"decline" diseases** caused by *Pythium* spp., and several **downy mildew diseases**. These diseases are treated in detail in Chapter 14. Many other aspects of the Oomycota are covered in Chapters 3 and 10.

One of the most remarkable features of Oomycota is that they behave exactly like the true fungi. Their hyphae exhibit apical growth and penetrate plant cells by producing cell-wall-degrading enzymes. They also use similar infection strategies to those of fungi. So, by a remarkable degree of convergent evolution the Oomycota have developed a lifestyle equivalent to that of fungi (Latijnhouwers *et al.* 2003). Nevertheless, they are wholly unrelated to the true fungi. Apart from their lack of photosynthetic pigments, they have many plant-like features, including:

- cell walls composed primarily of glucans, including cellulose-like polymers;
- diploid nuclei, in contrast to the haploid nuclei of most fungi;
- cell membranes composed of plant sterols, in contrast to ergosterol, the characteristic fungal sterol;
- energy storage compounds similar to those of plants, in contrast to the polyols (sugar alcohols) and trehalose found in most fungi;
- a range of ultrastructural features more closely related to those of plants than of fungi, including stacked, plate-like golgi cisternae (the fungi have tubular golgi cisternae), and tubular mitochondrial cisternae (the fungi have plate-like or disc-like mito-chondrial cisternae;
- different sensitivities to a range of antifungal agents.

The characteristic life-cycle features of Oomycota

Figure 2.34 illustrates the major life-cycle features of Oomycota, represented by the plant-pathogen, *Phytophthora infestans*, and water moulds such as *Saprolegnia* spp.

Typically, the somatic (vegetative) stages are **broad, fast-growing hyphae** with **diploid nuclei**. The hyphae lack cross-walls (septa) except when they produce complete, unperforated septa to isolate the reproductive structures.

Asexual reproduction involves the production of a **multinucleate sporangium**, which is separated by a septum. The sporangial contents are then cleaved around the individual nuclei, resulting in the production of diploid **zoospores**, each with an anteriorly directed tinsel-type flagellum and a trailing whiplash flagellum. The zoospores are released when the tip of the sporangium breaks down, but in some species (e.g. *Phytophthora infestans*) the whole sporangium can be released and functions as a dispersal spore. Depending on environmental conditions (especially temperature) this can germinate by producing a hypha or it can undergo cytoplasmic cleavage and release zoospores. The important **downy mildew pathogens** of several crop plants release sporangia that are wind-dispersed and germinate by a hypha to infect their hosts.

Sexual reproduction typically involves the production of a male sex organ (**antheridium**) and a female sex organ (**oogonium**) that may contain one or several eggs. Meiosis occurs within these sex organs, and fertilization is achieved by the transfer of a single haploid nucleus through a fertilisation tube to each haploid egg (Fig. 2.34b). In some *Phytophthora* species (e.g. *P. cactorum*) the antheridia are attached laterally to the oogonium. But in others (e.g. *P. infestans*) the oogonial hypha grows through the antheridium and then swells to form the oogonium. In any case, fertilization leads to the development of one or more thick-walled, diploid **oospores**. These usually have a constitutive dormant period before they germinate to produce either diploid hyphae or a sporangium that releases diploid zoospores.

There are many variations on these basic themes. For example, some of the Oomycota are homothallic (self-fertile) and some heterothallic. Some species (e.g. *Pythium oligandrum*) characteristically develop oospores parthenogenetically, without a sexual process. In some of the water moulds (Saprolegniales) the zoospores released from the sporangia encyst immediately and then germinate to produce a secondary zoospore, which can swim for many hours. In *Pythium* spp. the sporangia do not release their zoospores "directly" by dissolution of the tip of the sporangium. Instead, the sporangium produces a short exit tube and the tip of this breaks down to form a membrane-bound vesicle

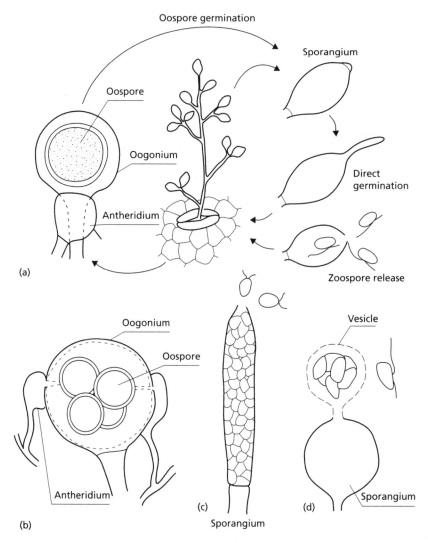

Fig. 2.34 Characteristic features of Oomycota. (a) Stages in the infection cycle of *Phytophthora infestans* (potato blight). Strains of opposite mating types in the host tissues produce antheridia and oogonia. Fertilization leads to the production of thick-walled resting spores (oospores). After a dormant period, the oospores germinate by a hyphal outgrowth, which produces lemon-shaped, wind-dispersed sporangia. These initiate infection either by germinating as a hyphal outgrowth (in warm conditions) or by cytoplasmic cleavage to release motile biflagellate zoospores in cool, moist conditions. Infected leaves produce sporangiophores through the stomatal openings, leading to spread of infection to other plants. (b) Sexual reproduction by *Saprolegnia*: several oospores are formed in each oogonium. (c) Asexual reproduction in *Saprolegnia*: the sporangium develops at the tip of a hypha, then the protoplasm cleaves to release several primary zoospores (each with two anterior flagella). These encyst rapidly and the cysts release secondary zoospores (laterally biflagellate) for dispersal to new sites. (d) Asexual reproduction in *Pythium*. At maturity the sporangium discharges its contents into a thin, membraneous vesicle. The laterally biflagellate zoospores are cleaved within this vesicle which then breaks down to release the zoospores.

(Fig. 2.34d). The contents of the sporangium are extruded into this vesicle, where the zoospores differentiate and then are released by rupture of the vesicle.

Classification, ecology, and significance

There are estimated to be 500–800 species in the Oomycota, classified in 5 or 6 Orders. However, some of these Orders contain only a few organisms, which will be mentioned where appropriate in later chapters. Here we will focus on the two major Orders, **Saprolegniales** and **Peronosporales**.

The **Saprolegniales**, commonly known as water moulds, are extremely common in freshwater habitats and in some brackish, estuarine environments. They can often be baited by immersing hemp seeds in natural waters, but it is difficult to free the cultures from bacteria. Several species of *Achlya* and *Saprolegnia* are common saprotrophs in freshwater habitats, and contribute to the recycling of organic matter, but *S. diclina* is an important pathogen of salmonid fish. The related *Aphanomyces* spp. are perhaps best-known as aggressive root pathogens (e.g. *A. euteiches* on pea and beans). But *A. astaci* is a pathogen of crayfish and has virtually eliminated the native European crayfish (the disease being known as crayfish plague). The introduced American crayfish, however, is resistant to the disease and has now replaced the European crayfish in almost all sites.

The Order **Peronosporales** is by far the largest and most economically important group. It includes many serious plant pathogens. For example, many *Pythium* spp. cause seedling diseases and attack the feeder roots of almost all crop plants (Chapter 14), but a few *Pythium* spp. parasitize other fungi and have potential for biocontrol of plant diseases (Chapter 12). The Peronosporales also includes all the *Phytophthora* spp., causing diseases such as potato blight (*P. infestans*). One of the most aggressive plant pathogens is *Phytophthora cinnamomi*, with an extensive host range. It causes root rot of pines, eucalypts, fruit trees, and many other plants in warmer regions of the globe, and can invade natural vegetation with devastating consequences. A similarly aggressive species, *Phytophthora ramorum*, causes sudden oak death in California (Chapter 14) and is currently spreading across Europe, posing a threat to many native European trees. Also in this group are the important **downy mildew** pathogens (e.g. *Bremia lactucae* on lettuce, *Plasmopara viticola* on grape vines) which are obligate biotrophic parasites (Chapter 14).

The remaining Orders include **Leptomitales** (a small group of aquatic fungi, such as *Leptomitis lacteus* which is common in sewage-polluted waters). This group has several interesting and peculiar features, such as fermentative rather than respiratory metabolism

(Chapter 8), a requirement for sulfur-containing amino acids because they cannot use inorganic sulfur sources, and hyphae that are constricted at intervals, like chains of sausages. The walls of some species contain significant amounts of chitin (which is unusual for Oomycota) and the hyphal constrictions are often plugged by cellulin granules, composed mostly of chitin.

The Order **Lagenidales** is a small group of common symptomless parasites of plant roots (e.g. *Lagena radicicola*; Macfarlane 1970) or parasites of algae, fungi, or invertebrates (e.g. *Lagenidium giganteum* on nematodes, mosquito larvae, etc.). Often these cannot be grown in culture in the absence of a host.

The cellular slime moulds

The cellular slime moulds comprise two groups of organisms – the **dictyostelid** cellular slime moulds and the **acrasid** cellular slime moulds. These two groups are biologically similar to one another, but the dictyostelids seem to be a monophyletic group, whereas the acrasids seem to be more disparate. Both types of organism grow and divide as haploid, unicellular amoebae, which engulf bacteria and other food particles by phagocytosis. They are commonly found in moist organic-rich soil, leaf litter, animal dung, and similar types of substrate. When nutrients become depleted, the amoebae undergo a complex developmental sequence in which many thousands of amoebae aggregate and eventually produce a stalked fruiting body (Fig. 2.35). This process has been studied intensively in the dictyostelid slime mould, *Dictyostelium discoideum*, which has become a model system of cell communication and differentiation.

At the onset of starvation, a few amoebae act as an aggregation center and produce periodic pulses of cyclic AMP (cAMP). This causes other amoebae to join the aggregate by chemotaxis, and the amoebae then undergo streaming along defined tracks so that they aggregate as a mound. Then the cells in the mound differentiate into two sorts – the pre-stalk cells and the pre-spore cells, each in its defined position where it will later become a stalk cell or a spore. At this stage the mound topples over, and the resulting slug (**grex**) migrates to the surface of the substrate where it differentiates into a fruiting body (**sorocarp**) with a stalk and a sporing head. The walled spores are subsequently released and wind-dispersed.

A sexual cycle has also been described, leading to the production of thick-walled, diploid **macrocysts**, which undergo meiosis followed by repeated mitotic divisions. Then the cytoplasm is cleaved and many haploid amoebae are released to repeat the cycle. Detailed coverage of the dictyostelids can be found in Raper (1984).

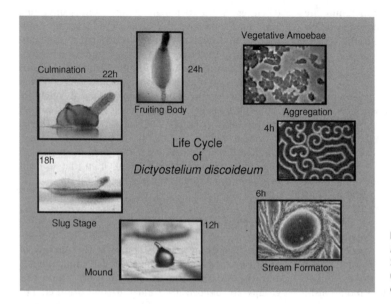

Fig. 2.35 The life cycle of *Dictyostelium discoideum*. (Courtesy of Florian Siegert ©. From: http://www.zi.biologie. unimuenchen.de/zoologie/dicty/ dicty.html)

(a)   (b)   (c)

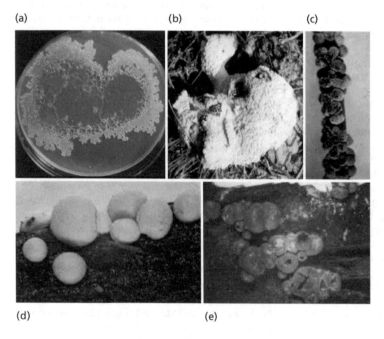

(d)   (e)

Fig. 2.36 Myxomycota. (a) A plasmodium of *Physarum polycephalum* growing on an agar plate. Note how the plasmodium, which was inoculated in the center of the plate, has migrated towards the margin, where the protoplasm has amassed. (b) A decaying bracket fungal fruitbody (about 15 cm diameter) completely covered by cream-colored sporing structures of the myxomycete *Fuligo septica*. (c) Mature sporangia of the slime mould *Physarum cinereum* on a grass blade about 4 mm diameter; many of the sporangia have opened to release the dark gray spores. (d) Immature, pink sporangia of *Lycogala* on the surface of rotting wood; each sporangium is about 1 cm diameter. (e) Similar sporangia 5 days later, when the sporangia had matured, turned gray, and consisted of a thin papery sac that ruptured to release the spores.

Myxomycota – the plasmodial slime moulds

The plasmodial slime moulds are wall-less organisms that consist of a multinucleate network of protoplasm (the **plasmodium**; Fig. 2.36) exhibiting rapid, rhythmic surges of protoplasmic streaming. These organisms are commonly seen in the autumn on moist rotting wood, on senescing fungal fruitbodies and on similar organic substrata where bacteria are abundant. They engulf bacteria and other food particles by phagocytosis. The plasmodium typically develops within a substrate but then migrates to the surface in response to nutrient depletion and the whole plasmodium converts into fruiting structures (**sporangia**) containing

(a)

(b)

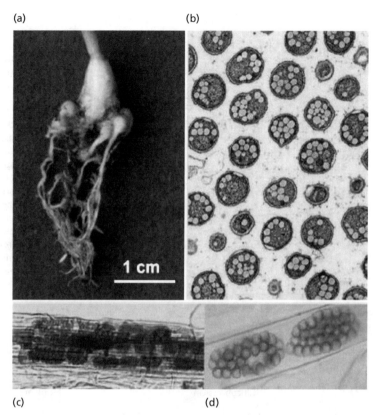

1 cm

Fig. 2.37 Plasmodiophorids. (a) Clubroot symptoms caused by *Plasmodiophora brassicae* on a cruciferous seedling: note the grossly deformed root system. (b) Large numbers of thick-walled resting spores of *P. brassicae* in clubroot tissue. (c,d) Clusters of thick-walled resting spores of *Polymyxa graminis* at different magnifications in the roots of grasses. ((a,b) Courtesy of J.P. Braselton, Ohio University; see http://oak.cats.ohiou.edu/~braselto/plasmos/)

(c)

(d)

many haploid spores that are wind-dispersed. These spores germinate to produce either amoeboid cells termed **myxamoebae** or flagellate swarmers, which fuse in pairs, and the resulting diploid cell grows into a plasmodium. *Physarum polycephalum* is perhaps the best-known example. It has been studied intensively by morphogeneticists because it can be maintained on defined (bacterium-free) media. Full details of the Myxomycota can be found in Martin & Alexopoulos (1969).

Plasmodiophorids

Plasmodiophorids (Fig. 2.37) are obligate intracellular parasites of plants, algae, or fungi. They grow only in a host organism, and cannot be grown in laboratory culture. As a group, the plasmodiophorids show no obvious relationship to fungi or to other fungus-like organisms, so their phylogenetic status remains unclear (Down *et al.* 2002). The most important organism of this group is *Plamodiophora brassicae*, which causes the damaging clubroot disease of cruciferous crops (the cabbage family). In this disease the roots are severely deformed, with proliferation of cell divi-

sion caused by the production of phytohormones, and this can lead to serious yield losses. A related organism, *Spongospora subterranea*, causes powdery scab of potato tubers. In addition to these, several common plasmodiophorids, such as *Polymyxa* spp., grow as symptomless parasites in the roots of many plants, causing little or no damage, but they can act as vectors of some economically important plant viruses (Chapter 10).

Significant stages in the life cycle of *P. brassicae* are shown in Fig. 2.38, although there is still doubt about the details of some stages. The thick-walled resting spores can persist in soil for many years, and only a low proportion of them will germinate in any one year. So, once this fungus is established in a site it is almost impossible to eradicate. The resting spores eventually germinate to release a motile, biflagellate zoospore, which locates a host root by chemotaxis and then encysts in a defined orientation. The cyst protoplast is then injected into a root epidermal cell or a root hair by means of a bullet-like structure. The protoplast then grows into a small **primary plasmodium** within the infected cell, and converts into a sporangium, which releases zoospores when the root cell dies or when an exit tube is formed. At this stage, the zoospores are

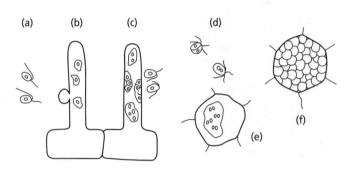

Fig. 2.38 Stages in the life cycle of *Plasmodiophora brassicae*, which causes clubroot disease of cruciferous crops. (a) Uninucleate **primary zoospores** are released from germinating resting spores in soil. (b) The zoospores encyst on a root hair or root epidermal cell and inject a protoplast into the host cell. (c) The protoplasts grow into small **primary plasmodia** and then convert to sporangia, which release zoospores into the soil when the root cell dies, or when an exit tube is formed. (d) The zoospores fuse in pairs. (e) The fused zoospores infect root cortical cells and develop into **secondary plasmodia**. (f) At maturity the secondary plasmodia covert into **resting spores** which often completely fill the cortical cells. These spores are finally released into soil when the roots decay. Eventually, the resting spores germinate to release haploid zoospores which repeat the infection cycle.

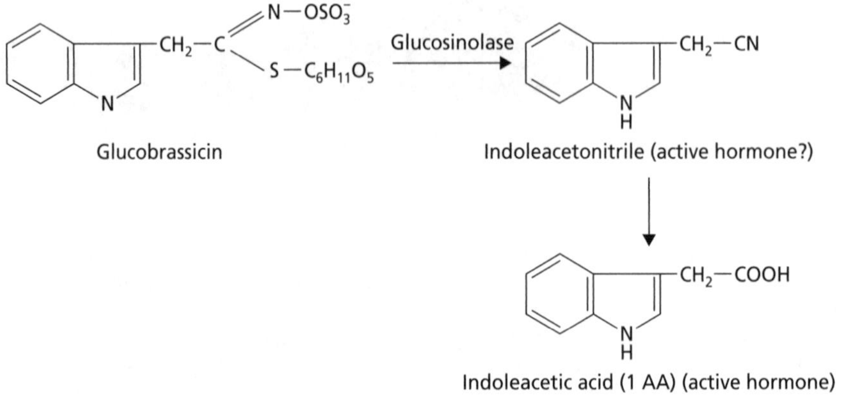

Glucobrassicin

$\xrightarrow{\text{Glucosinolase}}$

Indoleacetonitrile (active hormone?)

Indoleacetic acid (1 AA) (active hormone)

Fig. 2.39 Proposed conversion of glucobrassicin to the plant hormone indoleacetic acid by *Plasmodiophora brassicae*.

believed to fuse in pairs, and the resulting secondary zoospores infect root cortical cells, where they develop into large, **secondary plasmodia** that will eventually produce thick-walled **resting spores**. There is evidence that meiosis occurs just before the development of resting spores, suggesting that the primary zoospores are haploid, and therefore that the secondary zoospores and zoosporangia are diploid.

The main damage caused by *Plasmodiophora* lies in the fact that the root responds to infection of the cortex by undergoing rapid cell expansion (**hypertrophy**) and cell division (**hyperplasia**), leading to the development of large galls. The plant nutrients diverted to these galls severely reduce the shoot growth and yield of crops such as cauliflower, while the galls themselves

make root crops unmarketable. The proposed explanation for gall development is that glucobrassicin, a characteristic compound in cruciferous plants, is converted to the plant hormone indolylacetic acid (IAA) by the action of an enzyme, glucosinolase (Fig. 2.39).

As a group, the plasmodiophorids are difficult to study because they cannot be grown in laboratory culture, and so many basic aspects of their biology remain unclear. They show no obvious relationship to other fungus-like organisms, so they seem to represent a basal lineage of the early eukaryotic organisms. They are very common, because when the roots of almost any grass plant are cleared of protoplasm and stained with trypan blue the roots are frequently seen to contain resting spores of *Polymyxa graminis* (Fig. 2.37) or similar organisms such as *Ligniera* spp. They do not cause deformation of the roots, and even *Plasmodiophora brassicae* seems able to grow in the roots of many cruciferous plants, including the common weed *Capsella bursa-pastoris* (shepherd's purse), without causing galls, perhaps because these plants do not contain high levels of glucobrassicin or do not respond to infection by producing active hormones. Comprehensive accounts of the plasmodiophorids can be found in Karling (1968) and Buczacki (1983).

Online resources

Fungal Biology. http://www.helios.bto.ed.ac.uk/bto/FungalBiology/ [The website for this book.]

Plasmodiophorid Home Page. http://oak.cats.ohiou.edu/~braselto/plasmos/

Slime moulds. Superb images from George Barron (also including many fungi). http://www.uoguelph.ca/~gbarron/myxoinde.htm

Tree of Life Web Project. Major source of information on fungal systematics and phylogeny. http://tolweb.org/tree?group=life

Zoosporic Fungi online. http://www.botany.uga.edu/zoosporicfungi/

Reference texts

Buczacki, S.T. (1983) *Zoosporic Plant Pathogens: a modern perspective*. Academic Press, London.

Dick, M.W. (2001) *Straminipolous Fungi*. Kluwer Academic Publishers, New York.

Fuller, M.F. & Jaworski, A., eds (1987) *Zoosporic Fungi in Teaching and Research*. Southeastern Publishing Corporation, Athens, GA.

Hawksworth, D.L., Kirk, P.M., Sutton, B.C. & Pegler, D.N. (1996) *Ainsworth and Bisby's Dictionary of the Fungi*, 8th edn. CAB International, Wallingford, Oxon.

Karling, J.S. (1968) *The Plasmodiophorales*. Hafner Publishing Co., New York.

Kirk, P.M., Cannon, P.F., David, J.C. & Stalpers, J. (2001) *Ainsworth and Bisby's Dictionary of the Fungi*, 9th edn. CAB International, Wallingford, Oxon.

Martin, G.W. & Alexopoulos, C.J. (1969) *The Myxomycota*. Iowa University Press, Iowa City, IA.

Raper, K.B. (1984) *The Dictyostelids*. Princeton University Press, Princeton, NJ.

Cited references

Barr, D.J.S. (1990) Phylum Chytridiomycota. In: *Handbook of Protoctista* (Margulis, L., Corliss, J.O., Melkman, M. & Chapman, D.J., eds), pp. 454–466. Jones & Bartlett, Boston.

Buczacki, S.T. (1983) *Zoosporic Plant Pathogens: a modern perspective*. Academic Press, London.

Deacon, J.W. & Saxena, G. (1997) Orientated attachment and cyst germination in *Catenaria anguillulae*, a facultative endoparasite of nematodes. *Mycological Research* **101**, 513–522.

Dick, M.W. (2001) *Straminipolous Fungi*. Kluwer Academic Publishers, New York.

Down, G.J., Grenville, L.J. & Clarkson, J.M. (2002) Phylogenetic analysis of *Spongospora* and implications for the taxonomic status of the plasmodiophorids. *Mycological Research* **106**, 1060–1065.

Fuller, M.F. & Jaworski, A., eds (1987) *Zoosporic Fungi in Teaching and Research*. Southeastern Publishing Corporation, Athens, GA.

Karling, J.S. (1968) *The Plasmodiophorales*. Hafner Publishing Co., New York.

Kirk, P.M., Cannon, P.F., David, J.C. & Stalpers, J. (2001) *Ainsworth and Bisby's Dictionary of the Fungi*, 9th edn. CAB International, Wallingford, Oxon.

Latijnhouwers, M., de Wit, P.J.G.M. & Govers, F. (2003) Oomycetes and fungi: similar weaponry to attack plants. *Trends in Microbiology* **11**, 462–469.

Macfarlane, I. (1970) *Lagena radicicola* and *Rhizophydium graminis*, two common and neglected fungi. *Transactions of the British Mycological Society* **55**, 113–116.

Martin, G.W. & Alexopoulos, C.J. (1969) *The Myxomycota*. Iowa University Press, Iowa City, IA.

Mitchell, R.T., & Deacon, J.W. (1986) Selective accumulation of zoospores of Chytridiomycetes and Oomycetes on cellulose and chitin. *Transactions of the British Mycological Society* **86**, 219–223.

Raper, K.B. (1984) *The Dictyostelids*. Princeton University Press, Princeton, NJ.

Read, N.D. & Lord, K.M. (1991) *Experimental Mycology* **15**, 132–139.

Redecker, D., Kodner, R. & Graham, L.E. (2000) Glomalean fungi from the Ordovician. *Science* **289**, 1920–1921.

Schuessler, A., Schwarzott, D. & Walker, C. (2001) A new fungal phylum, the Glomeromycota: phylogeny and evolution. *Mycological Research* **105**, 1413–1421.

Webster, J. (1980) *Introduction to Fungi*, 2nd edn. Cambridge University Press, Cambridge.

Chapter 3

Fungal structure and ultrastructure

This chapter is divided into the following major sections:

- the structure of a fungal hypha
- fungal ultrastructure
- the hypha as part of a colony
- the structure of yeasts
- fungal walls and wall components
- septa
- the fungal nucleus
- cytoplasmic organelles
- the fungal cytoskeleton and molecular motors

Fungi have many unique features in terms of their structure, cellular components, and cellular organization. These features are intimately linked to the mechanisms of fungal growth and therefore to the many activities of fungi as decomposer organisms, plant pathogens, and pathogens of humans. In this chapter we consider the main structural and ultrastructural features of fungi, and some of the powerful techniques that enable us to view the dynamics of subcellular components in living fungal hyphae, giving an insight into the way that fungi grow.

Overview: the structure of a fungal hypha

The hypha is essentially a tube with a rigid wall, containing a moving slug of protoplasm (Fig. 3.1). It is of indeterminate length but often has a fairly constant diameter, ranging from 2 μm to 30 μm or more (usually 5–10 μm), depending on the species and growth conditions. Hyphae grow only at their tips, where there is a tapered region termed the **extension zone**; this can be up to 30 μm long in the fastest-growing hyphae such as *Neurospora crassa* which can extend at up to 40 μm per minute. Behind the growing tip, the hypha ages progressively and in the oldest regions it may break down by autolysis or be broken down by the enzymes of other organisms (heterolysis). While the tip is growing, the protoplasm moves continuously from the older regions of the hypha towards the tip. So, a fungal hypha continuously extends at one end and continuously ages at the other end, drawing the protoplasm forward as it grows.

The hyphae of most fungi have cross walls (**septa**; singular **septum**) at fairly regular intervals, but septa are absent from hyphae of most Oomycota and Zygomycota, except where they occur as complete walls to isolate old or reproductive regions. Nevertheless, the functional distinction between septate and aseptate fungi is not as great as might be thought, because septa have pores through which the cytoplasm and even the nuclei can migrate. Strictly speaking, therefore, septate hyphae do not consist of cells but of interconnected compartments, like the compartments of a train. As we will see later, this enables cellular components to move either forwards or backwards along dedicated pathways, so that a hypha can function as an integrated unit.

All fungal hyphae are surrounded by a wall of complex organization, described later. It is thin at the apex (about 50 nm in *Neurospora crassa*) but thickens to about 125 nm at 250 μm behind the tip. The plasma membrane lies close to the wall and seems to be firmly attached to it because hyphae are difficult to plasmolyse.

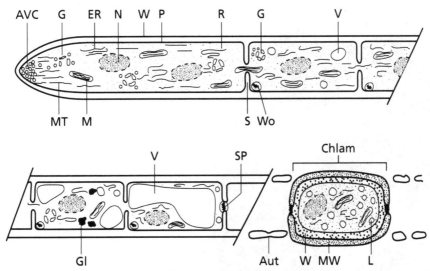

Fig. 3.1 Diagrammatic representation of a fungal hypha, showing progressive aging and vacuolation behind the hyphal tip. In the oldest regions, the walls may break down by autolysis or the mycelial nutrients may accumulate in chlamydospores (thick-walled resting spores that serve in dormant survival). Aut = autolysis; AVC = apical vesicle cluster; Chlam = chlamydospore; ER = endoplasmic reticulum; G = Golgi/Golgi equivalent; Gl = glycogen; L = lipid; M = mitochondria; MT = microtubules; MW = melanized wall; N = nucleus; P = plasmalemma; R = ribosomes; S = septum; SP = septal plug; V = vacuole; W = wall; Wo = Woronin body.

Fungal ultrastructure

Transmission electron microscopy of thin sections of fungal hyphae has been one of the most important tools for understanding the behavior of fungal hyphae, and for elucidating the mechanisms of apical growth. The initial studies on fungal ultrastructure relied on **chemical fixation** methods: the hyphae were "fixed" by immersion in aldehydes such as glutaraldehyde, then post-treated with electrondense substances (osmium tetroxide and uranyl acetate) to give maximum contrast. The superb images of Oomycota shown in Figs 3.2 and 3.3 are widely acknowledged to be among the best that have ever been produced by this technique. But a newer technique termed **freeze substitution** was developed in the late 1970s, and this provides even greater resolution (Howard & Aist 1979) so it has now become the standard (Fig. 3.4). In this technique, the living hyphae are plunged into liquid propane (−190°C) causing almost instantaneous preservation of the hyphae, then transferred to cold (−80°C) acetone with osmium tetroxide and uranyl acetate, and slowly brought back to room temperature.

Comparison of the electron micrographs obtained by these two methods should be made with caution, for at least two reasons:

1 The chemically fixed hyphae belong to a species of *Pythium* (Oomycota), which is not a true fungus, and

the specimen was stained to give maximum resolution of the internal organelles, so the wall is poorly stained.

2 The hyphae prepared by freeze substitution represent a true fungus, *Athelia* (*Sclerotium*) *rolfsii*, and some of the organelles of fungi are different from those of the Oomycota, as discussed later in this chapter.

However, it is notable that the degree of preservation of the hyphae is much better in the freeze-substituted material. In particular, the plasma membrane has a smooth profile compared with the many indentations seen with the chemical fixation method, and the subcellular organelles are more clearly defined.

The zonation of organelles in the apical compartment

All actively growing fungal hyphae – including hyphae of the fungus-like Oomycota – show a clearly defined polarity in the arrangement of organelles from the hyphal tip back towards the base of the apical compartment. The extreme hyphal tip (Figs 3.2, 3.4) contains a large accumulation of membrane-bound vesicles, but no other major organelles. These vesicles show differences in electron density, suggesting that they have different contents. Most of the vesicles are thought to be derived from **golgi bodies** (or the func-

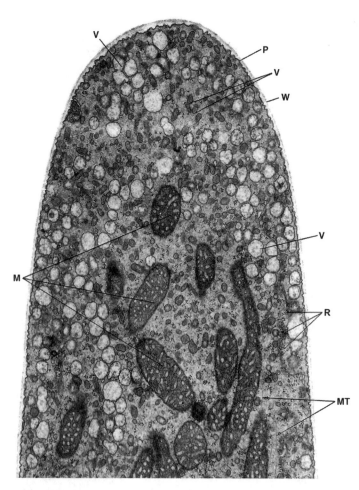

Fig. 3.2 Electron micrograph of the hyphal tip of *Pythium aphanidermatum* (Oomycota) prepared by chemical fixation. The apex contains a cluster of Golgi-derived vesicles (V) of at least two types: large with electronlucent contents and small with electrondense contents. Mitochondria (M) are abundant in the subapical zone but are absent from the extreme tip. Other components include: the hyphal wall (W), plasma membrane (P) which has an indented appearance due to chemical fixation, microtubules (MT) which are poorly resolved, and clusters of ribosomes (R). (Courtesy of C. Bracker; from Grove & Bracker 1970.)

tional equivalent of golgi bodies in fungi) further back in the apical compartment, and to be transported to the apex where they contribute to apical growth. In support of this view, cytochemical staining of some of the vesicles has shown that they contain **chitin synthase**, the enzyme responsible for producing the chitin component of fungal walls.

The collection of vesicles at the hyphal tip is termed the **apical vesicle cluster** (AVC). At high magnification in the true fungi the center of the AVC seems to be a vesicle-free core consisting of a network of **actin microfilaments**, with microtubules running through this and sometimes extending up to the extreme tip of the hypha. Early light microscopists had recognized the existence of the AVC as a phase-dark body in the hyphal tips of septate fungi, when viewed by phase-contrast microscopy. It was termed the **Spitzenkörper** ("apical body"), and it was noted to disappear when growth was stopped by applying a mild shock to the hyphal tip, and to reappear when growth restarted. Moreover, the Spitzenkörper was seen to shift towards

the side of a hyphal tip when hyphae changed their growth direction, and more recently the Spitzenkörper has been shown to split prior to the formation of a hyphal branch, so that two tips are produced from the original one. All this evidence points to a central role of the Spitzenkörper in fungal tip growth – a body that functions as the equivalent of a "tip-growth organelle."

Behind the extreme tip is a short zone with few or no major organelles. Then typically there is a zone rich in **mitochondria** (Figs 3.2, 3.4). Through the actions of **ATPase**, this zone almost certainly generates the proton-motive force (H$^+$ gradient across the cell membrane) that drives nutrient uptake at the hyphal tip, as explained in Chapter 6 (Fig. 6.3). The mitochondria of fungi have plate-like cisternae similar to those of animals and plants (see Fig. 3.12), whereas the mitochondria of Oomycota have tubular cisternae (Figs 3.2 and 3.3). Further back along the apical compartment there is increasing development of a system of **branched tubular vacuoles** (Fig. 3.4). These extend

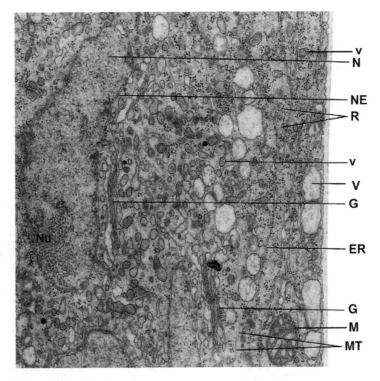

Fig. 3.3 Part of a subapical region of a hypha of *Pythium aphanidermatum* showing: endoplasmic reticulum (ER); Golgi bodies (G) consisting of a stack of pancake-like cisternae, typical of animals and Oomycota, but distinct from those of fungi (see Fig. 3.4); mitochondrion (M); microtubules (MT); nucleus (N); nuclear envelope (NE); nucleolus (Nu); ribosomes (R); large and small vesicles (V and v). (Courtesy of C. Bracker; from Grove & Bracker 1970.)

through the septa into the hyphal compartments further back, providing a transport system for the movement of metabolites, including phosphates which often can be in short supply.

The distribution of **nuclei** varies between and within fungal groups. The aseptate fungi (e.g. Zygomycota) and fungus-like organisms (e.g. Oomycota) contain many nuclei within a common cytoplasm, so these fungi are coenocytic. Many septate fungi (e.g. Ascomycota and mitosporic fungi) also have several nuclei in the apical compartment, but sometimes only one or two nuclei in compartments behind the apex. However, nuclei can squeeze through the septal pores, so the concept of a single nucleus determining the functions of a fixed volume of cytoplasm does not really apply to most fungi. Instead, we should regard many fungi as having several nuclei sharing a common pool of cytoplasm. This has important implications for the genetic fluidity of fungi, discussed in Chapter 9.

Many members of the Basidiomycota (mushrooms, toadstools, etc.) have a regular arrangement of one nucleus in each compartment of the monokaryon, but two nuclei (of different mating compatibility groups) in each compartment of the dikaryon (see Fig. 2.20). This arrangement is ensured by a special type of septum, termed the **dolipore septum** (see later), which has only a narrow pore, too small to allow nuclei to pass through.

The hypha as part of a colony

Fungal colonies typically develop from a single germinating spore, which produces a **germ tube** (a young hypha) that grows and branches behind the tip. As the original hypha and the first-order branches grow, they produce further branches behind their tips. These branches diverge from one another until, eventually, the colony develops a characteristic circular outline (Fig. 3.5). Then, in the older parts of a colony where nutrients have been depleted, many fungi produce narrow hyphal branches that grow towards one another instead of diverging, and fuse by tip-to-tip contact, involving localized breakdown of their walls (Fig. 3.6). This process of **hyphal anastomosis** creates a network for the pooling and remobilization of protoplasm to produce chlamydospores or other, larger differentiated structures (Chapter 5).

The phenomena seen in Figs 3.5 and 3.6 are remarkable for several reasons:

- In nutrient-rich conditions, the hyphae and branches at the colony margin always diverge from one another, positioning themselves so that they grow in the spaces between existing hyphae. These positioning mechanisms are very precise, but we have no firm evidence of their causes – possibly local gradients of carbon dioxide, or other growth metabolites?

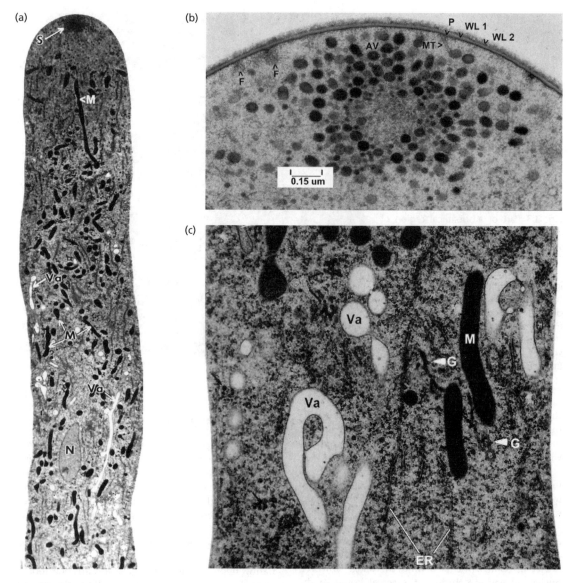

Fig. 3.4 Electron micrographs of hyphae of *Athelia (Sclerotium) rolfsii* prepared by freeze substitution. (a) Part of the apical compartment, about 7 μm diameter, showing the distribution of major organelles: Spitzenkörper (S); mitochondria (M); tubular vacuoles (Va); nucleus (N). (b) The Spitzenkörper at the extreme tip of the hypha consists of a cluster of apical vesicles (AV) with contents of varying electron density, surrounding a vesicle-free zone; a microtubule (MT) extends to the plasma membrane (P). Next to the plasma membrane are bundles of F-actin (filasomes, F). The wall consists of at least two layers of different electron density (WL 1, WL 2). (c) Subapical region of the hypha, showing mitochondria, tubular vacuoles, endoplasmic reticulum (ER), and Golgi bodies (G) (Courtesy of R. Roberson; see Roberson & Fuller 1988, 1990.)

- Nutrient-poor conditions near the center of a colony create exactly the opposite effect, in which hyphae grow towards one another and fuse at the points of contact. As seen in Fig. 3.6, this again involves precise orientation of hyphal tip growth, because the "homing response" is extremely accurate and always leads to tip-to-tip fusion. We have no knowledge of the factors involved.
- Hyphal anastomosis only occurs between members (colonies) of the same species, and seldom, if ever, between unrelated species. Even within a species, strains that are genetically different can fuse, but this

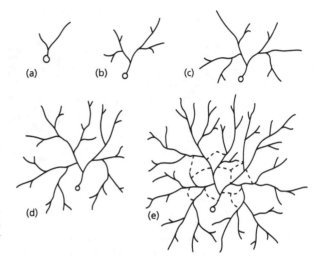

Fig. 3.5 (a–e) Stages in the development of a fungal colony from a germinating spore. The broken lines in (e) represent narrow anastomosing hyphae near the center of the colony.

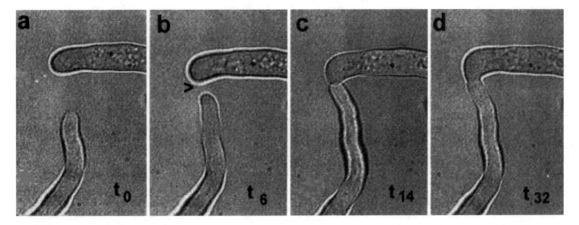

Fig. 3.6 Videotaped sequence of anastomosis of two hyphae of *Rhizoctonia solani*. The times shown are minutes after the start of video recording. The upper hypha had stopped growing at t_0 but began to produce a branch (arrowhead) at t_6 in response to the approaching hyphal tip. The hyphal tips met at t_{14}. Dissolution of the tip walls and complete hyphal fusion was achieved at t_{32}. (From McCabe *et al.* 1999.)

usually leads to rapid death of the fused hyphal compartments – a reaction governed by heterokaryon incompatibility (*het*) loci (see Figs 9.6, 9.7).

• The hyphae of Oomycota seldom if ever exhibit "vegetative" anastomosis, but they do undergo hyphal fusions when they produce sexual hyphae of opposite mating types; the development of these is controlled by sex hormones (see Fig. 5.22).

The structure of yeasts

Several fungi grow as budding, uninucleate yeasts, rather than as hyphae. Common examples include the bread yeast, *Saccharomyces cerevisiae* (Ascomycota, Fig. 3.7), *Cryptococcus* spp. (Ascomycota, Fig. 3.7),

and *Sporobolomyces roseus* (Basidiomycota). Some other fungi are characteristically **dimorphic** (with two shapes): they switch between a hyphal and a yeast form in response to changes in environmental conditions (Chapter 5). A good example is the genus *Candida* (Ascomycota), including the common species *Candida albicans*, which can be a significant pathogen of humans (Chapter 16). Therefore, yeasts are not fundamentally different from hyphal fungi; they merely represent a different growth form, adapted to spread in nutrient-rich water films and similar types of environment.

As shown in Fig. 3.8, each yeast cell has a single nucleus and a typical range of cytoplasmic organelles, including a large, conspicuous vacuole and usually several phase-bright lipid bodies. The vacuole and the lipid bodies are often the only structures clearly visible

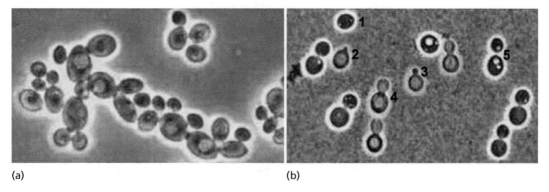

(a)   (b)

Fig. 3.7 (a) The common budding yeast, *Saccharomyces cerevisiae*, viewed by phase-contrast microscopy. (b) *Crypto-coccus albidus*, a yeast that produces a rigid polysaccharide capsule over the cell surface, seen by mounting the cells in a suspension of India ink particles. Different stages in the budding process are labeled from 1 (nonbudding cell) to 5 (cell separation). Note the presence of bright lipid bodies in the plane of focus of some cells (e.g. 5) and the presence of a large central vacuole in the plane of focus of some other cells (e.g. 2, 3, 4).

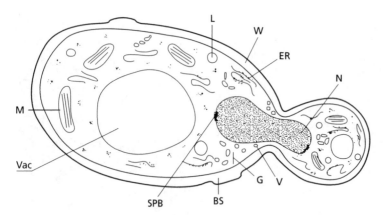

Fig. 3.8 Diagrammatic representation of a budding yeast, *Saccharomyces cere-visiae*, about 5 μm diameter. BS = bud scar; ER = endoplasmic reticulum; G = Golgi; L = lipid body; M = mitochon-drion; N = nucleus; SPB = spindle-pole body; V = vesicle; Vac = large central vacuole; W = wall.

by light microscopy. Most yeasts divide by budding from one or more locations on the cell surface. During this process a small outgrowth appears at the bud site, then the bud progressively elongates before expanding into a rounded form, by synthesis of new wall com-ponents over the whole of the cell surface. When the bud has nearly reached its final size, the nucleus of the mother cell migrates towards the bud site and divides, so that one nucleus remains in the mother cell and a second nucleus enters the daughter cell.

The final separation of the two cells is achieved by the development of a septum. In *Saccharomyces* this occurs when a ring of chitin is produced at the "neck" site, and this ring of chitin expands inwards until it becomes a complete chitin plate between the mother and daughter cell. Then other wall materials are deposited on each side of this chitin plate, and the cells separate by enzymic cleavage of the wall between the chitin plate and the daughter cell. This process leaves a **bud scar** on the mother cell and a **birth scar** on the

daughter cell. These scars are inconspicuous by normal light microscopy, but the chitin plate on the mother cell can be seen clearly if yeasts are treated with a fluores-cent brightener such as Cellufluor (calcofluor), which binds to chitin and fluoresces blue when the cell is observed under a fluorescence microscope.

By using fluorescent dyes that bind to chitin, we can count the number of bud scars on the cell surface. This reveals that *Saccharomyces cerevisiae* is a **multipolar budding** yeast – it always buds from a different point on the cell surface, never from a previous bud site, whereas some other yeasts exhibit **bipolar budding** – the buds always arise at the same positions, often at the poles of the cell. In theory, the cells of bipolar species are immortal, with no limit to the number of times they can bud, whereas the cells of multipolar species would eventually run out of potential bud sites – calculated to be up to 100, although only about 40 bud scars have been seen on a single cell. However, in practise this distinction is unimportant because in exponentially growing cultures (where one cell produces two in a unit of time, two produce four in the next unit of time, and so on) one half of all the cells will be first-time mothers, one quarter will be second-time mothers, and an infinitessimally small number will be very old mothers. In fact, the typical chitin content of a yeast population seldom exceeds more than 1–2% of the total wall material, and almost all of this chitin is found in the bud scars, not in the rest of the yeast cell wall. We return to this topic in Chapter 4, because some important advances have been made in understanding growth polarity, based on the use of cell-cycle mutants in the laboratory of the Nobel Laureate, Sir Paul Nurse.

The taxonomy of yeasts is complicated by the lack of obvious morphological features, so in recent years it has relied heavily on molecular approaches. There are probably many hundreds, if not thousands, of yeast species that have yet to be clearly demarcated and described. Recent evidence from the ascomycetous budding yeasts suggests that these are a monophyletic group – all with a common ancestor – and they are not simply reduced forms derived from the mycelial Ascomycota. They are also distinct from the "fission yeasts" such as *Schizosaccharomyces* species, which do not bud but instead form filaments that fragment by septation into brick-like cells (arthrospores, or arthroconidia).

Fungal walls and wall components

In recent years it has become clear that fungal walls serve many important roles, quite apart from the obvious role of providing a structural barrier. For example, the way in which a fungus grows – whether as cylindrical hyphae or as yeasts – is determined by the wall components and the ways in which these are assembled and bonded to one another. The wall is also the interface between a fungus and its environment: it

protects against osmotic lysis, it acts as a molecular sieve regulating the passage of large molecules through the wall pore space, and if the wall contains pigments such as melanin it can protect the cells against ultraviolet radiation or the lytic enzymes of other organisms. In addition to these points, the wall can have several physiological roles. It can contain binding sites for enzymes, because many disaccharides (e.g. sucrose and cellobiose) and small peptides need to be degraded to monomers before they can pass through the cell membrane, and this is typically achieved by the actions of wall-bound enzymes (Chapter 6). The wall also can have surface components that mediate the interactions of fungi with other organisms, including plant and animal hosts. All these features require a detailed understanding of wall structure and architecture.

The major wall components

The primary approach to investigating the wall composition of fungi is to disrupt fungal cells and purify the walls by using detergents and other mild chemical treatments, then use acids, alkalis and enzymes to degrade the walls sequentially. Although relatively few fungi have been analysed in detail, these treatments show that fungal walls are predominantly composed of polysaccharides, with lesser amounts of proteins and other components. The major wall components can be categorized into two major types: (i) the **structural (fibrillar) polymers** that consist predominantly of straight-chain molecules, providing structural rigidity, and (ii) the **matrix components** that cross-link the fibrils and that coat and embed the structural polymers.

The main wall polysaccharides differ between the major fungal groups, as shown in Table 3.1. The Chytridiomycota, Ascomycota, and Basidiomycota typically have chitin and glucans (polymers of glucose) as their major wall polysaccharides. Chitin consists of long, straight chains of β-1,4 linked *N*-acetylglucosamine

Table 3.1 The major components of fungal walls.

Taxonomic group	Structural (fibrillar) components	Matrix components
Chytridiomycota	Chitin Glucan	Glucan ?
Zygomycota	Chitin Chitosan	Polyglucuronic acid Mannoproteins
Ascomycota	Chitin β-(1→3), β-(1→6)-glucan	Mannoproteins α(1→3)-glucan
Basidiomycota	Chitin β-(1→3), β-(1→6)-glucan	Mannoproteins α(1→3)-glucan
Oomycota (not true fungi)	β-(1→3), β-(1→6)-glucan Cellulose	Glucan

residues (Chapter 7), whereas the fungal glucans are branched polymers, consisting mainly of β-1,3-linked backbones with short β-1,6-linked side chains. The Zygomycota typically have a mixture of chitin and chitosan (a poorly acetylated or nonacetylated form of chitin; Chapter 7), polymers of uronic acids such as glucuronic acid, and mannoproteins. The Oomycota (which are not true fungi) have little chitin, and instead they have a mixture of a cellulose-like β-1,4-linked glucan and other glucans.

Having made these points, it is important to recognize that the wall composition of a fungus is not fixed, but can change substantially at different stages of the life cycle. This is also true for many **dimorphic fungi**, which can grow as either hyphae or budding yeast-like cells (Chapter 5).

Wall architecture

The major wall components of fungi can be thought of as the bricks and mortar, but it was only in 1970 that we began really to understand the architecture of the fungal wall. For this, Hunsley & Burnett developed an elegently simple technique which they termed **enzymatic dissection**. They mechanically disrupted fungal hyphae so that only the walls remained, and then treated the walls with combinations and sequences of different enzymes, coupling this with electron microscopy to detect any changes that the enzyme treatments had caused. If, for example, the surface appearance of the wall changed after treatment with a particular enzyme "X," then the substrate of "X" is likely to be the outermost wall component. So, by using various sequences and combinations of enzymes it was possible to strip away the major wall components and to see their relationships to one another. Three fungi were used in this work, but we will take *Neurospora crassa* as an example to illustrate the essential features (Fig. 3.9).

In mature regions of hyphae the wall of *Neurospora* was shown to have at least four concentric zones; these are shown as separate layers in Fig. 3.9, but in reality they grade into one another. The outermost zone consists of amorphous glucans with predominantly β-1,3 and β-1,6 linkages, which are degraded by the enzyme laminarinase. Beneath this is a network of glycoprotein embedded in a protein matrix. Then there is a more or less discrete layer of protein, and then an innermost region of chitin microfibrils embedded in protein. The total wall thickness in this case is about 125 nm. But the wall at the growing tip is thinner (c. 50 nm) and simpler, consisting of an inner zone of chitin embedded in protein and an outer layer of mainly protein. So it is clear that the wall becomes stronger and more complex behind the extending

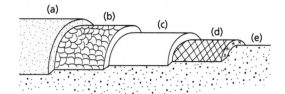

Fig. 3.9 Diagram to illustrate the wall architecture in a "mature" (subapical) region of a hypha of *Neurospora crassa* as evidenced by sequential enzymatic digestion. (a) Outermost layer of amorphous β-1,3-glucans and β-1,6-glucans. (b) Glycoprotein reticulum embedded in protein. (c) A more or less discrete protein layer. (d) Chitin microfibrils embedded in protein. (e) Plasma membrane. (Based on Hunsley & Burnett 1970.)

hyphal tip, as further materials are added or as further bonding occurs between the components.

Neurospora seems to have an unusually complex wall architecture, because a glycoprotein network has not been seen in some other fungi. Nevertheless, the general pattern of wall architecture of hyphae is fairly consistent: the main, straight-chain microfibillar components (chitin, or cellulose in the Oomycota) are found predominantly in the inner region of the wall, and they are overlaid by nonfibrillar or "matrix" components (e.g. other glucans, proteins, and mannans) in the outer region. However, there is substantial bonding between the various components, serving to strengthen the wall behind the apex. In particular, some of the glucans are covalently bonded to chitin, and the glucans are bonded together by their side chains. We return to this topic in Chapter 4, when we consider the mechanisms of apical growth.

The extrahyphal matrix

In addition to the main structural components of the wall, some yeasts can have a discrete polysaccharide capsule, and both hyphae and yeasts can be surrounded by a more or less diffuse layer of polysaccharide or glycoprotein, easily removed by washing or mild chemical treatment. These extracellular matrix materials can have important roles in the interactions of fungi with other organisms. For example, the yeast *Cryptococcus neoformans* is a significant pathogen of humans; its polysaccharide capsule masks the antigenic components of the cell wall so that the fungus is not engulfed by phagocytes and can proliferate in the tissues (Casadevall 1995). In a different context, the fungus *Piptocephalis virginiana* (Zygomycota; see Chapter 12) parasitizes other Zygomycota such as *Mucor* spp. on agar media, but does not parasitize them in liquid culture

Fig. 3.10 Diagrammatic representation of a simple septum of *Neurospora crassa*. The septum develops as an ingrowing ring from the lateral wall of the hypha. Associated with this is a modification of the lateral wall (LW), including a proliferation of the glycoprotein reticulum (GR). G = glucan layer; GR/P = glycoprotein reticulum embedded in protein; P = protein; C = chitin. (Based on Trinci A.P.J. (1978) *Science Progress* 65, 75–99.)

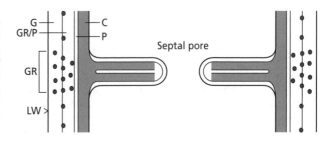

where the host fungi produce an extracellular polysaccharide (Chapter 12). In fact, the production of an extracellular matrix is often influenced by growth conditions. We noted in Chapter 1 that pullulan is produced commercially from *Aureobasidium pullulans*; in this case its synthesis is favored by an abundant sugar supply in nitrogen-limiting growth conditions (Seviour *et al.* 1984). Some of the other gel-like materials around fungi also have potential commerical roles.

Septa

Septa (cross-walls) are found at fairly regular intervals along the length of most hyphae in the Ascomycota, Basidiomycota, and mitosporic fungi, although the septa are perforated to allow continuity of the protoplasm. If part of the hypha is damaged, then a Woronin body or a plug of coagulated protoplasm rapidly seals the septal pore to localize the damage. Then the hypha can regrow from a newly formed tip behind the damaged compartment, or in some cases a new tip can grow into the damaged compartment (see Fig. 12.12). Clearly, these "damage limitation" responses help to conserve the integrity of the hypha, and might often be necessary when hyphae are eaten by insects, or are attacked by parasitic fungi (mycoparasites), or when hyphae of different vegetative compatibility groups attempt to anastomose (Chapter 9). However, there are several fungi and fungus-like organisms that do not produce septa in the normal vegetative hyphae – for example the Zygomycota (*Mucor*, etc.) and the fungus-like Oomycota (*Pythium*, *Phytophthora*, etc.). These **aseptate** fungi are more vulnerable to damage.

Septa might help to provide structural support to hyphae, especially in conditions of water stress. But one of their main roles seems to be to enable differentiation. By blocking the septal pores, a fungal hypha is transformed from a continuous series of compartments to a number of independent cells or regions that can undergo separate development (Chapter 5).

Septa can be seen by normal light microscopy, but electron microscopy reveals several different types of

septum. The Ascomycota and mitosporic fungi have a **simple septum** (Fig. 3.10) with a relatively large central pore, ranging from 0.05 to 0.5 μm diameter, which allows the passage of cytoplasmic organelles and even nuclei. The development of these septa occurs remarkably quickly, usually being completed within a few minutes. They develop as an ingrowing ring from the lateral walls of the hypha, and this is associated with localized modifications of the lateral walls themselves, including a localized proliferation of a glycoprotein reticulum in the walls of *Neurospora crassa*.

The Basidiomycota also have simple septa when they are growing as monokaryons (with one nucleus in each cell). But they often have a more complex **dolipore septum** (Fig. 3.11) when strains of different mating compatibility groups fuse to form a dikaryon, with two nuclei in each compartment. The dolipore septum has a narrow central channel (about 100–150 nm diameter) bounded by two flanges of amorphous glucan. On either side of this septum are bracket-shaped membraneous structures termed **parenthosomes**, which have pores to allow cytoplasmic continuity but which prevent the passage of major organelles. Thus, the Basidiomycota often have a more regular arrangement of nuclei compared with other fungi. But we will see in Chapter 5 that the septa are selectively degraded when the Basidiomycota begin to form a

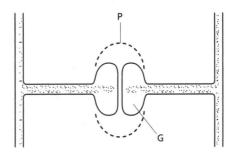

Fig. 3.11 The dolipore septum, found in many members of the Basidiomycota. Large deposits of glucan (G) line the narrow central pore, and specialized perforated membranes termed parenthosomes (P) prevent major organelles from passing through the septal pore.

complex fruiting body, such as a mushroom. This enables the mass redistribution of materials necessary to fuel the development of the fruiting structure.

From even this brief account, it is clear that septa play several important roles in fungal biology. They can isolate compartments or they can allow the free passage of organelles through the septal pores, and they can be degraded to allow the mass translocation of nutrients and cytoplasmic components to sites of future development.

The fungal nucleus

Fungal nuclei are usually small (1–2 μm diameter) but exceptionally can be up to 20–25 μm diameter (e.g. *Basidiobolus ranarum*, Chytridiomycota; Chapter 4). They are surrounded by a double nuclear membrane with pores, as in all eukaryotes. However, fungi are notable for several peculiar features of their nuclei and nuclear division (Heath 1978). First, the nuclear membrane and the nucleolus remain intact during most stages of mitosis, whereas in most other organisms the nuclear membrane breaks down at an early stage during nuclear division. A possible reason for this is that the retention of a nuclear membrane might help to prevent dispersion of the nuclear contents in hyphal compartments that contain several nuclei and rapidly streaming protoplasm. Second, in fungi there is no clear metaphase plate; instead the chromosomes seem to be randomly dispersed, and at anaphase the daughter chromatids pull apart along two tracks, on spindle fibres of different lengths. A third point of difference is that fungi have various types of **spindle-pole bodies** (also called **microtubule-organizing centers**). They are responsible for microtubule assembly during nuclear division. In Ascomycota and mitosporic fungi the spindle-pole bodies are disc-shaped, whereas in Basidiomycota they are often composed of two globular ends connected by a bridge. However, the significance of these differences is unclear. All fungi need a microtubule-organizing center to ensure that the chromosomes separate correctly during nuclear division.

The vast majority of fungi are **haploid** with chromosome numbers ranging from about 6 to 20. For example there are six chromosomes in *Schizophyllum commune* (Basidiomycota), seven in *Neurospora crassa* (Acomycota), eight in *Emericella (Aspergillus) nidulans* (Ascomycota), 16 in *Saccharomyces cerevisiae* (Ascomycota), and 20 in *Ustilago maydis* (Basidiomycota). A few fungi are naturally diploid (e.g. the yeast *Candida albicans*, and members of the Oomycota). And some fungi can alternate between haploid and diploid generations – e.g. *S. cerevisiae*, and *Allomyces* spp. (Chytridiomycota) – or exist as polyploid series (*Allomyces* and several *Phytophthora* spp.).

Many aspects of fungal genetics and fungal genomes are discussed in Chapter 9, but here we need to address one of the most remarkable features: **fungi are the only major group of eukaryotic organisms that are haploid**. Almost all other eukaryotes are diploid, including plants, animals, Oomycota, and the many single-celled organisms loosely termed "protists." The likely reason for the haploid nature of fungi is that it confers a specific advantage. On the one hand, as we have seen, fungal hyphae usually contain several nuclei in each hyphal compartment, and there is continuity of protoplasmic flow between the compartments. If a recessive mutation occurs in one of the nuclei, then the fungus will still behave as wild-type, but the recessive mutation will be retained, thereby storing potential variability – the hallmark of diploid organisms. On the other hand, a mutated nucleus can be exposed to selection pressure periodically, when the fungus produces uninucleate spores or when a branch arises and a single "founder" nucleus enters it. So, in at least some respects, haploid fungi with multinucleate compartments have many of the advantages associated with both a haploid and a diploid lifestyle. We return to this subject in Chapter 9.

Cytoplasmic organelles

Many of the cellular components of fungi are functionally similar to those of other eukaryotes, but differ in some important respects. The major differences are discussed below.

The plasma membrane

As in all eukaryotes, the fungal plasma membrane consists of a phospholipid bilayer with associated transmembrane proteins, many of which are involved directly or indirectly in nutrient uptake. The membrane also can anchor some enzymes. In fact, the two main wall-synthetic enzymes, chitin synthase and glucan synthase, are **integral membrane proteins**; they become anchored in the membrane in such a way that they produce polysaccharide chains from the outer membrane face (Chapter 4). The plasma membrane has a third important role, in relaying signals from the external environment to the cell interior – the process termed **signal transduction**, discussed in Chapter 4.

The fungal plasma membrane is unique in one important respect – it typically contains **ergosterol** as the main membrane sterol, in contrast to animals, which have **cholesterol**, and plants which have phytosterols such as **β-sitosterol**. The Oomycota also have plant-like sterols. But some plant-pathogenic fungi such as *Pythium* and *Phytophthora* spp. are unable to synthesize sterols from nonsterol precursors and instead need to be supplied with sterols from the host. Ergosterol is the primary target of several fungicides that are used to control plant-pathogenic fungi. Ergosterol is also a primary target of several antifungal drugs that are used to treat human mycoses (Chapter 17).

The fungal secretory system: golgi, endoplasmic reticulum, and vesicles

Fungi have a **secretory system**, consisting of the endoplasmic reticulum (ER), the Golgi apparatus (or Golgi equivalent), and membrane-bound vesicles. Proteins destined for export from the cell are synthesized on ribosomes attached to the ER, then enter the ER lumen and are transported to the Golgi. During their progressive transport through the Golgi cisternae, proteins undergo various modifications, including partial cleavage and reassembly, folding into a tertiary structure, and the addition of sugar chains (glycosylation). Then the proteins, or glycoproteins, are packaged into vesicles which bud from the maturing face of the Golgi, and are transported to the plasma membrane for secretion. This intricate postal system sorts and delivers proteins to specific destinations, including the enzymes (pectinases, cellulases, proteases, etc.) that are destined for export to degrade polymers in the surrounding environment.

The secretory system is involved in at least some aspects of fungal tip growth, because glycoproteins are only produced in the Golgi and are transported in vesicles to the sites of wall growth (Chapter 4). Additionally, the Golgi is important for the commercial production, and release from the cell, of foreign (heterologous) gene products (Chapter 9). The ability of a protein to enter the ER is determined by a signal sequence at the N-terminus, which is subsequently removed. Without this sequence the protein will remain in the cell.

Once again, the fungal secretory system shows some unique properties. The typical Golgi apparatus of animals, plants, and Oomycota consists of a stack of membraneous cisternae (Fig. 3.12a). But fungi have a **Golgi equivalent** which is ultrastructually quite different from a typical Golgi – it consists of sausage-shaped strings, beads, and loops (Fig. 3.12b).

Chitosomes

Most of the vesicles seen in hyphal tips have not been chemically characterized, but some of the smaller ones (microvesicles) resemble particles that have been purified *in vitro* and are termed **chitosomes**. These particles were initially discovered by homogenizing hyphae,

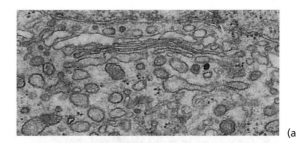

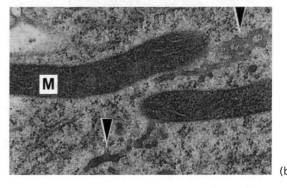

Fig. 3.12 (a) Magnification of part of Fig. 2.3, showing the typical Golgi body found in plants and animals. (b) Magnification of part of Fig. 2.4, showing mitochondria (M) with plate-like cisternae typical of fungi, and the "Golgi equivalent" consisting of sausage-shaped rings, beads, and loops (arrowheads).

removing the wall material and major membranous organelles by centrifugation, and then subjecting the supernatant to sucrose density centrifugation so that its components separated out as bands at different depths in the tubes. When one of these bands was examined by electron microscopy it was found to contain chitosomes – small spheroidal bodies, 40–70 nm diameter, each surrounded by a "shell" about 7 nm thick (Fig. 3.13). When the band containing these particles was incubated in the presence of an activator (a protease enzyme) and UDP-N-acetylglucosamine (the sugar-nucleotide from which chitin is made), each particle was seen to produce a coil of chitin inside it, then the particles ruptured to release ribbon-like chitin microfibrils, each composed of several chitin chains. If the supernatant was first treated with digitonin (a saponin which solubilizes sterol-containing membranes) and then centrifuged in a density gradient, the chitosomes were no longer found; instead, the addition of activator and UDP-N-acetylglucosamine resulted in the production of chitin in a different band corresponding to a lighter fraction of the homogenate. But subsequent removal of the digitonin caused the chitosomes to reappear, indicating that they

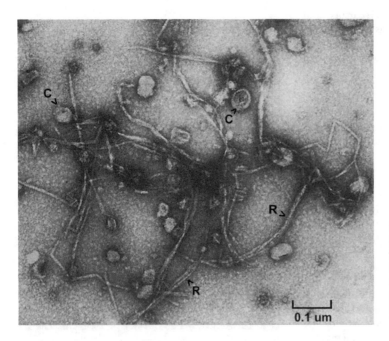

Fig. 3.13 Ribbon-like aggregates of chitin microfibrils (R) produced *in vitro* from chitosomes (C) isolated from *Mucor rouxii*, when incubated with substrate (*N*-acetylglucosamine) and a proteolytic activator. (Courtesy of C.E. Bracker; from Bartnicki-Garcia *et al.* 1978.)

are **self-assembling aggregates of the enzyme chitin synthase**, each particle containing sufficient enzyme molecules to produce, after proteolytic activation, a chitin microfibril composed of several chitin chains. So, this elegant collaborative study by biochemists and electron microscopists led to the identification of one of the types of vesicle often seen in the hyphal apex. One of the questions still to be resolved is how these particles reach the hyphal tip after they have been assembled further back in the hyphae. One suggested mechanism is that they are packaged into membrane-bound **multivesicular bodies** such as those sometimes seen in electron micrographs (Fig. 3.20).

Vacuoles

The vacuolar system of fungi has several functions, including the storage and recycling of cellular metabolites. For example the vacuoles of several fungi, including mycorrhizal species (Chapter 13), accumulate phosphates in the form of polyphosphate. Vacuoles also seem to be major sites for storage of calcium which can be released into the cytoplasm as part of the intracellular signalling system (Chapter 5). Vacuoles contain proteases for breaking down cellular proteins and recycling of the amino acids, and vacuoles also have a role in the regulation of cellular pH. All these important physiological roles are in addition to the potential role of vacuoles in cell expansion and (possibly) in driving the protoplasm forwards as hyphae elongate at the tips.

Vacuoles often are seen as conspicuous, rounded structures in the older regions of hyphae, but recent work has shown that there is also a tubular vacuolar system extending into the tip cells. It is an extremely dynamic system, consisting of narrow tubules which can dilate and contract, as inflated elements travel along them in a peristaltic manner (Fig. 3.14). This was demonstrated by studying the movement of a fluorescent compound, carboxyfluorescein (CF), in living hyphae of a range of fungi. The hyphae were treated with a nonfluorescent precursor, carboxyfluorescein diacetate (CFDA), which is lipid-soluble so it enters the plasma membrane and is then rapidly absorbed from the cytosol into the vacuoles. There the acetate groups are cleaved by esterases to yield CF, which is membrane-impermeable so it remains in the vacuolar system. Rees *et al.* (1994) showed that the tubular vacuoles which contain CF form a more or less continuous system which passes through even the dolipore septa of Basidiomycota, transporting materials backwards and forwards in the hyphae. This has important implications for the bidirectional translocation of materials within single hyphae (Chapter 7), and perhaps especially for mycorrhizal systems (Chapter 13).

Endocytosis and vesicle trafficking

Studies on plant cells, animal cells, and budding yeast suggest that the uptake of substances through the

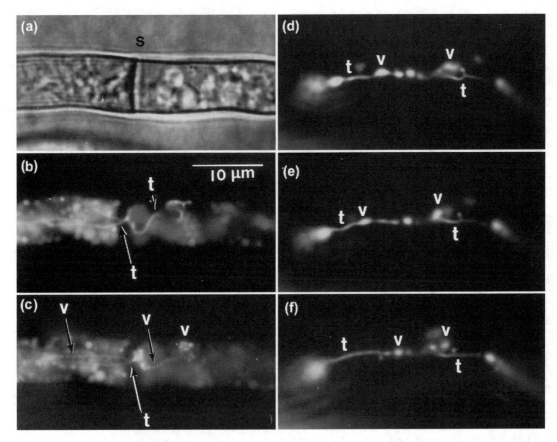

Fig. 3.14 The dynamic tubular vacuolar system of living fungal hyphae, treated with a fluorescent dye that becomes localized in the vacuoles. (a) Hypha of *Penicillium expansum* seen by normal bright-field microscopy, showing the presence of a septum (S). (b) The same hypha viewed by fluorescence microscopy, showing a narrow tubule (t) passing through the septum and then branching. (c) The same hypha showing dilated vacuoles (V) after they have passed through the septum. (d–f) Hypha of *Aspergillus niger* photographed at 8-second intervals, showing a succession of peristalsis-like movements causing the vacuolar dilations to travel along the hypha from left to right. (Courtesy of B. Rees, V. Shepherd and A. Ashford; from Rees *et al.* 1994.)

cell membrane might be achieved by an active process of **endocytosis** – the converse of exocytosis, which is widely recognized to occur when vesicles fuse with the plasma membrane and release their contents into the surrounding environment (Figs 3.2, 3.4). However, there is still some debate about whether fungal hyphae have an endocytotic system, and (if they do so) whether it is equivalent to the endosomal system commonly reported in other cell types.

The evidence for endocytosis in fungal hyphae is based on the use of amphiphilic steryl dyes such as FM4-64 and FM1-43, which are reported to be marker dyes for endocytosis in many different organisms. These vital dyes can be taken up by cells and are then visual-

ized by laser scanning confocal microscopy. Figure 3.15 shows confocal images of *Neurospora crassa* treated with two of these dyes – FM4-64 which strongly labels the plasma membrane and Spitzenkörper, and FM1-43 which labels the mitochondria but only weakly labels the Spitzenkörper. Essentially similar labeling patterns with FM4-64 were found when hyphae of eight different fungi were tested, including representatives of Ascomycota, Basidiomycota, mitosporic fungi, and Zygomycota (Fig. 3.16). The reasons for the different labeling patterns of these dyes are still unclear, but the strong labeling of the plasma membrane by FM4-64 is consistent with reports that this dye becomes intercalated between the outer and inner layers of the cell

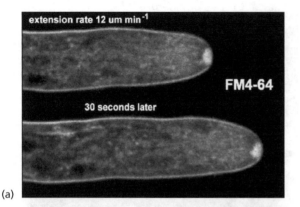

(a)

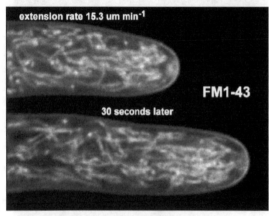

(b)

Fig. 3.15 (a,b) Hyphal tips of *Neurospora crassa* treated with two steryl dyes to detect putative endocytosis. Both steryl dyes, FM4-64 and FM1-43, became internalized, and hyphae treated with both dyes continued to grow during confocal laser imaging, as shown by the images taken at 30-second intervals. (Courtesy of N.D. Read; from Fisher-Parton *et al.* 2000.)

membrane. It then seems to be internalized by an active process, because it cannot passively cross the inner leaf of the plasma membrane. Similar hyphae treated with a metabolic inhibitor (sodium azide) showed no evidence of dye uptake.

In several instances (Fig. 3.17) a satellite Spitzenkörper was seen to originate at some distance behind the hyphal tip and it moved towards the tip, ultimately fusing with the main Spitzenkörper. Although none of these lines of evidence provides definitive proof of the existence of an endosomal system in filamentous fungi, the evidence is at least compatible with an endosomal system in which vesicle trafficking between different organelles provides a mechanism for recycling of membranes and their contents between different subcellular compartments.

On this basis, a hypothetical model of the organization of a vesicle-trafficking network, including the possible involvement of endosomes, has been proposed (Fig. 3.18). This is, however, an area of controversy because Torralba & Heath (2002) did not find evidence of endocytosis in *Neurospora crassa* hyphae. Studies with mutants may be needed to resolve this issue. Many proteins associated with endocytosis in *Saccharomyces cerevisiae* have homologs in the *Neurospora* genome sequence and could be candidates for mutational studies to test whether filamentous fungi have an equivalent endosomal system (Read & Kalkman 2003).

The cytoskeleton and molecular motors

The cytoskeleton plays a major role in the internal organization of eukaryotic cells, providing a dynamic structural framework for transporting organelles, for cytoplasmic streaming, and for chromosome separation during cell division. The three main elements of the cytoskeleton are: (i) **microtubules**, consisting of polymers of tubulin proteins, (ii) **microfilaments**, consisting of the contractile protein actin, and (iii) **intermediate filaments** which provide tensile strength. The distribution of the cytoskeletal components can be visualized in living hyphae by using fluorescent vital dyes (Fig. 3.19). These elements are considered to play a major role in coordinating hyphal tip growth (Heath 1994).

In electron micrographs of hyphae, the microtubules are seen as long, straight tubules, about 25 nm diameter, occurring either singly or as parallel arrays. They are seen mainly in the peripheral regions of the hypha but can also extend up to the membrane at the extreme tip (Fig. 3.4). Microtubules are also seen to be closely associated with membrane-bound organelles (e.g. Fig. 3.20), indicating that they could provide a tramline system for organellar movement. Consistent with this, the benzimidazole fungicides that have been used widely to control plant pathogens (Chapter 17) exert their antifungal action by binding to fungal microtubules, and this causes hyphae to stop growing. The benzimidazole compounds similarly block nuclear division by binding to the spindle microtubules.

Microtubules are dynamic cellular components. They depolymerize in response to treatments such as cold shock, and conversely they can be stabilized by compounds such as **taxol**, the toxin from yew trees (which is also toxic to humans). In normally growing cells the microtubules are thought to be continuously degraded and reformed. Consistent with this, microtubules can polymerize by self-assembly in cell-free systems. This is a two-stage process: in the first stage, a molecule of the protein **α-tubulin** combines with a

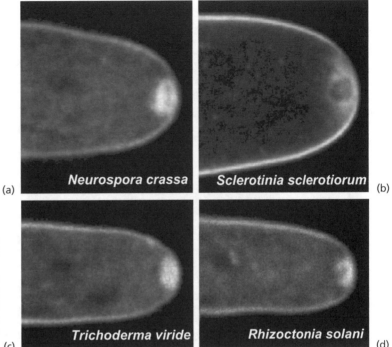

Fig. 3.16 (a–d) Dye labeling (FM4-64) of the cytoplasmic membrane and Spitzenkörper of different fungi, including Ascomycota (*Neurospora crassa*), Basidiomycota (*Sclerotinia* and *Rhizoctonia*), and mitosporic fungi (*Trichoderma viride*). *Phycomyces* (Zygomycota) also showed similar labeling (not shown in this image).

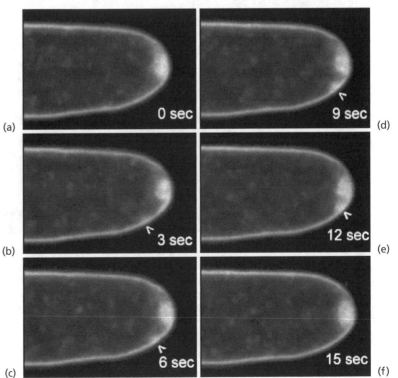

Fig. 3.17 (a–f) Confocal images of an FM4-64-stained satellite Spitzenkörper (arrowhead) of *Botrytis cinerea* showing a time course (seconds) of different stages in its formation, migration, and fusion with the main Spitzenkörper. (Courtesy of N.D. Read; from Fisher-Parton *et al.* 2000.)

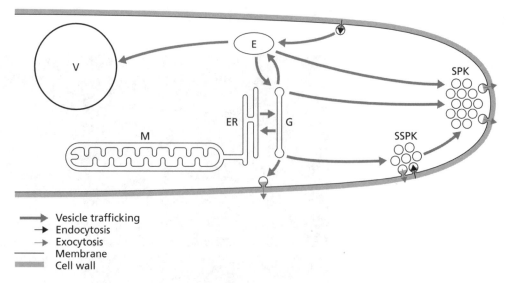

Vesicle trafficking
Endocytosis
Exocytosis
Membrane
Cell wall

Fig. 3.18 Hypothetical model of the organization of the vesicle-trafficking network in a growing hypha, based on the pattern of FM4-64 staining. E = endosome; ER = endoplasmic reticulum; G = Golgi cisterna; M = mitochondrion; SPK = Spitzenkörper; SSPK = satellite Spitzenkörper; V = vacuole. (From Fisher-Parton *et al.* 2000.)

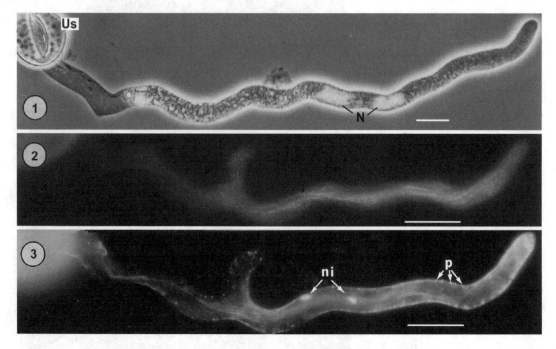

Fig. 3.19 Young (3 hour) germlings from a uredospore (Us) of the rust fungus *Uromyces phaseoli*, treated with different compounds to reveal the distribution of hyphal components. Scale bars = 10 μm. (Image 1) Treated with DAPI, a compound that binds to A/T-rich regions of DNA, and observed by a combination of phase-contrast and fluorescence microscopy. The two nuclei (N) in the germ-tube are seen by DAPI fluorescence. (Image 2) The same germling treated with fluorochrome-labeled, anti-tubulin antibodies and observed by fluorescence microscopy. The many microtubules that run longitudinally in the hypha are clearly seen. (Image 3) The same germling observed by fluorescence microscopy but treated with rhodamine-conjugated phalloin (a deadly toxin from toadstools of the "death cap" fungus *Amanita phalloides* which exerts its effects by binding to actin). A conspicuous actin cap is seen in the hyphal tip. Actin is also seen as peripheral plaques (p) and nuclear inclusions (ni) in zones similar to those in which microtubules are seen. (Courtesy of H.C. Hoch; from Hoch & Staples 1985.)

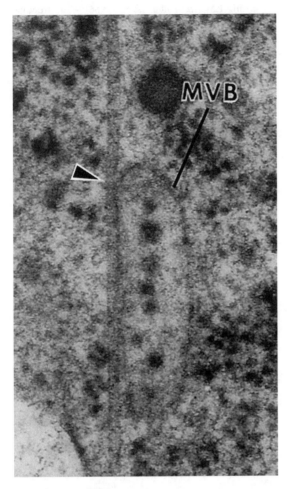

Fig. 3.20 A fungal microtubule (arrowhead) closely associated with a membrane-bound cellular organelle termed a multivesicular body (MVB). Such close association indicates a potential role for moving the organelle, through the activities of motor proteins. (Courtesy of R. Roberson; from Roberson & Fuller 1988.)

molecule of a sister protein β-tubulin to form a dimer; then the dimers polymerize to form tubulin chains. The microtubules formed by this process interact with two mechanochemical enzymes (motor proteins), **kinesin** and **dynein**, and probably help to transport organelles within the fungal hypha.

Actin microfilaments are much narrower than microtubules, being about 5–8 nm diameter. In the slime mould *Physarum polycephalum* the actin microfilaments are known to function in cytoplasmic contraction, when actin associates with its motor protein, **myosin**. This could also be true of filamentous fungi, where myosin-like proteins have recently

been detected. However, the most compelling evidence for a role of the actin cytoskeleton in organizing fungal growth has come from studies on the yeast, *Saccharomyces cerevisiae*. This fungus has microtubules but, in contrast to mycelial fungi, the microtubules of yeast **do not seem to be involved** in the transport of major organelles; instead, they seem to function mainly in orientation of the mitotic spindle. Yet, actin cables do play a major role in directing secretory vesicles to growth sites in yeast, and by analogy actin might play a similar role in distributing the apical vesicles to sites of wall growth in hyphal tips.

In later chapters we return to the roles of the cytoskeleton in fungal growth (Chapter 4), differentiation (Chapter 5), and the behavior of fungal zoospores (Chapter 10). But we end this chapter by noting that fungal tubulins differ from those of plants and animals. They are inhibited by the antibiotic griseofulvin and by the benzimidazole fungicides, whereas plant and animal tubulins are insensitive to these compounds. This is why griseofulvin can be used to treat the dermatophytic (ringworm) fungal infections of humans, and why the benzimidazoles can be used to treat fungal infections of plants (Chapter 17). Conversely, the fungal tubulins (with the exception of Oomycota) are unaffected by **colchicine** (the toxin from the autumn crocus) which inhibits nuclear division in plant and animal cells.

Cited references

Bartnicki-Garcia, S., Bracker, C.E., Reyes, E. & Ruiz-Herrera, J. (1978) Isolation of chitosomes from taxonomically diverse fungi and synthesis of chitin microfibrils *in vitro*. *Experimental Mycology* **2**, 173–192.

Casadevall, A. (1995) Antibody immunity and *Cryptococcus neoformans*. *Canadian Journal of Botany* **73**, S1180–S1186.

Fisher-Parton, S., Parton, R.M., Hickey, P., Dijksterhuis, J., Atkinson, H.A. & Read, N.D. (2000) Confocal microscopy of FM4-64 as a tool for analysing endocytosis and vesicle trafficking in living fungal hyphae. *Journal of Microscopy* **198**, 246–259.

Grove, S.N. & Bracker, C.E. (1970) Protoplasmic organization of hyphal tips among fungi: vesicles and Spitzenkörper. *Journal of Bacteriology* **104**, 989–1009.

Heath, I.B. (1978) *Nuclear Division in the Fungi*. Academic Press: New York.

Heath, I.B. (1994) The cytoskeleton in hyphal growth, organelle movements, and mitosis. In: *The Mycota*, vol. 1 (Wessels, J.G.H. & Meinhardt, F., eds), pp. 43–65. Springer-Verlag, Berlin.

Hoch, H.C. & Staples, R.C. (1985) The microtubule cytoskeleton in hyphae of *Uromyces phaseoli* germlings: its relationship to the region of nucleation and to the F-actin cytoskeleton. *Protoplasma* **124**, 112–122.

Howard, R.J. & Aist, J.R. (1979) Hyphal tip cell ultrastructure of the fungus *Fusarium*: improved preservation by freeze-substitution. *Journal of Ultrastructural Research* **66**, 224–234.

Hunsley, D. & Burnett, J.H. (1970) The ultrastructural architecture of the walls of some hyphal fungi. *Journal of General Microbiology* **62**, 203–218.

McCabe, P.M., Gallagher, M.P. & Deacon, J.W. (1999) Microscopic observation of perfect hyphal fusion in *Rhizoctonia solani*. *Mycological Research* **103**, 487–490.

Read, N.D. & Kalkman, E.R. (2003) Does endocytosis occur in fungal hyphae? *Fungal Genetics and Biology* **39**. 199–203.

Rees, B., Shepherd, V.A. & Ashford, A.E. (1994) Presence of a motile tubular vacuole system in different phyla of fungi. *Mycological Research* **98**, 985–992.

Roberson, R.W. & Fuller, M.S. (1988) Ultrastructural aspects of the hyphal tip of *Sclerotium rolfsii* preserved by freeze substitution. *Protoplasma* **146**, 143–149.

Roberson, R.W. & Fuller, M.S. (1990) Effects of the demethylase inhibitor, Cyproconazole, on hyphal tip cells of *Sclerotium rolfsii*. *Experimental Mycology* **14**, 124–135.

Seviour, R.J., Kristiansen, B. & Harvey, L. (1984) Morphology of *Aureobasidium pullulans* during polysaccharide elaboration. *Transactions of the British Mycological Society* **82**, 350–356.

Torralba, S. & Heath, I.B. (2002) Analysis of three separate probes suggests the absence of endocytosis in *Neurospora crassa* hyphae. *Fungal Genetics & Biology* **37**, 221–232.

Chapter 4

Fungal growth

The key to the fungal hypha lies in the tip (Noel Robertson)

This chapter is divided into the following major sections:

- apical growth of fungal hyphae
- spore germination and the orientation of hyphal tip growth
- the yeast cell cycle
- kinetics of fungal growth
- commercial production of fungal biomass: Quorn™ mycoprotein

In this chapter we focus on the mechanisms of fungal growth, with special reference to the organization of growth and wall synthesis at the hyphal tip, which is central to understanding the biology of fungi. We also discuss the ways in which hyphal branches arise and orientate themselves for maximum efficiency of nutrient capture. And, we consider the kinetics of fungal growth in batch culture and continuous culture systems, relating this to industrial-scale processes, including the commercial production of **Quorn™ mycoprotein** – a highly successful fermentation system for producing "single-cell protein." Several topics in this chapter are covered in depth by Gow & Gadd (1995) and Howard & Gow (2001).

Apical growth of fungal hyphae

Apical growth is the hallmark of fungi. Apart from the fungus-like Oomycota, which have adopted apical growth by a remarkable degree of convergent evolution (Latijnhouwers *et al.* 2003), no other organisms grow continuously as a tube that extends at the extreme tip by the localized synthesis of wall components, and, arguably, no other group shows such extreme plasticity. The hyphal apex can swell into a balloon-like structure such as a spore or yeast cell, or it can taper to such a degree that it can penetrate a layer of inert gold film or the wall of a host plant by exerting turgor pressure alone. In other circumstances, the fungal hypha can give rise to complex tissues and infection structures, discussed in Chapter 5.

Figure 4.1 illustrates part of this plasticity, when hyphae of *Neurospora crassa* are observed by placing a coverslip over the margin of a colony on an agar plate. The sequence of nine frames was taken over a 1-hour period, starting from the time when the coverslip was added. In the first frame (a) the hyphal tip was growing normally, and two lateral branches had arisen behind the growing tip. Soon afterwards (b and c) the hyphal tips began to swell (a response to disturbance caused by the coverslip) and then branched repeatedly from the tips before resuming a more normal pattern of apical growth.

By taking any convenient reference points, such as the branching points shown as v1 and v2 in Fig. 4.1, it is seen that the length of hypha already formed remains unchanged, and all new growth occurs from the original hyphal tip or from the branch tips. In fact, the incorporation of new wall material is mainly confined to the extreme tip. This is illustrated in Fig. 4.2, where growing hyphae were exposed to a short pulse of radiolabeled wall precursors such as ^{3}H-*N*-acetylglucosamine (from which chitin is synthesized) or ^{3}H-glucose (from which wall glucans are made) and then autoradiographed. The radiolabel is incorporated maximally at the hyphal tip, and the rate of label incorporation falls off sharply over the first few micrometres – the apical dome of the hypha.

Rapid rates of hyphal tip extension (such as 40 μm min^{-1} in *N. crassa*) can only be possible if the apex is supplied with vesicles and other cytoplasmic components from behind. For this reason, what we term apical growth is actually apical extension, because the true rate of growth, defined as increase in biomass per unit of time, is much slower. The length of hypha needed to support an extending apex can be estimated by making a diagonal cut across a colony margin with a scalpel, so that individual hyphae are severed at different distances from their tips. Hyphae that are cut too close to the tip die from physical damage. Hyphae cut further back continue to extend but more slowly than usual, and eventually a point is reached at which the cut is so far back that it has no effect on the apical extension rate. This distance is termed the **peripheral growth zone** of a fungal colony, defined as the length of hypha needed to maintain the **maximum extension rate** of the **leading hyphae** at the colony margin; it varies between fungi, from below 200 μm up to several millimeters for the fastest-extending fungi.

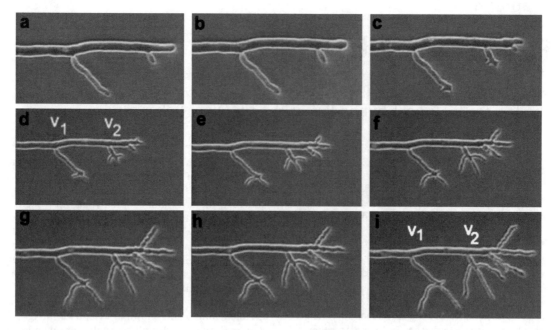

Fig. 4.1 (a–i) Sequence of frames from a videotape of *Neurospora crassa* growing over a 1-hour period beneath a coverslip on an agar plate. (Frames (a–c) are shown at higher magnification than the other frames.)

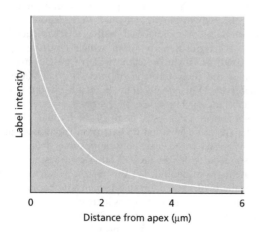

Fig. 4.2 Incorporation of radiolabeled wall precursors during a brief (5-minute) exposure.

Early experiments on the mechanism of apical growth in fungi

Robertson (1958, 1959) did many of the key early experiments on apical growth of fungi, using extremely simple methods coupled with truly remarkable insight. He grew colonies of *Fusarium oxysporum* or *N. crassa* on nutrient-rich agar, flooded them with water and then observed the behavior of the tips. As shown in Fig. 4.3, roughly half of the tips stopped growing immediately and then resumed growth within a minute, but from a narrower apex than before. The other tips stopped growing for several minutes, swelled into a diamond shape during this time, and eventually regrew by producing one or more narrow tips just behind the original apex. He then repeated the experiments, again flooding the colonies with water but replacing this within 40 seconds by a solution of the same osmotic potential as the original agar (an isotonic solution). This caused all the tips to stop for several minutes, but they swelled during this time and eventually regrew from narrow subapical branches (Fig. 4.3).

To interpret these findings, Robertson hypothesized that the normal pattern of apical growth involves two independent processes: (i) continuous extension of a plastic, deformable tip and (ii) rigidification of the wall behind the extending tip. He envisaged these two processes as occurring at the same rate, but with rigidification always slightly behind the tip, like two cars travelling along two lanes of a motorway at exactly the same speed but one is always slightly behind the other. Then, if extension growth is halted by an osmotic shock (adding water) the process of rigidification will continue and tend to "overtake" the

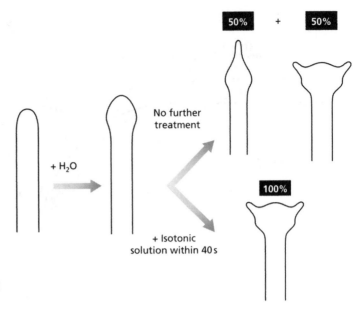

50% + 50%

No further
treatment

+ H₂O

100%

+ Isotonic
solution within 40 s

Fig. 4.3 Robertson's experiments on hyphal
tip growth. See text for details.

apex. If the tip readjusts to the new osmotic conditions in time it can grow on, but now from a thinner region of the apex where the wall has not yet rigidified. Roughly half of the tips in the original experiment seemed to be able to do this. However, if the tip cannot adjust in time then the apex will be sealed off by rigidification, and growth will only occur when new tips have been produced – in this case behind the original apex and by a process that takes several minutes. This would explain why all the tips stopped for several minutes when water was replaced by the isotonic solution, because the tips would need to make two separate osmotic adjustments and could not do so before the apex had rigidified.

We shall see later that this explanation was essentially correct. It is now supported by many lines of evidence from wall enzymology and ultrastructural studies. But at this stage we should note some further points. Sometimes the hyphal tips swell and burst in response to flooding with water, perhaps because the wall at the extreme apex is too fragile to adjust to rapid changes in osmotic potential or perhaps because the wall at the extreme apex is continuously being degraded by wall-lytic enzymes. Sometimes the tips grow on as usual after being flooded with water, but a branch develops later from the position where the apex had reached at the time of flooding. In any case, growing hyphal tips are very sensitive to many types of disturbance and they tend to respond in the same way – by a "stop–swell–branch" sequence as shown in Fig. 4.3. This response can be elicited by mild heat or cold shock, by exposure to an intense light beam, or

even when hyphal tips encounter physical barriers. Some morphological mutants of *Neurospora* and *Aspergillus* even show this pattern regularly during growth on agar plates. In Chapter 5 we shall see that the stop–swell–branch sequence occurs during the production of several differentiated structures, including the pre-penetration structures of fungal parasites of plants and animals.

Assembly of the wall at the hyphal apex

Wall synthesis at the hyphal apex is a complex process, the details of which are still not fully known, but from various lines of evidence we can construct a composite picture of wall growth and subsequent wall maturation (Figs 4.4, 4.5). Some of the main components of this system are discussed below.

Chitin synthase

Chitin synthase catalyses the synthesis of chitin chains, and is therefore one of the principal enzymes involved in fungal wall growth. Chitin is known to be formed *in situ* at the apex, rather than arriving in membrane-bound vesicles. When hyphal homogenates are tested for enzyme activity *in vitro*, chitin synthase is found in at least two forms: as an inactive **zymogen** in chitosomes (Chapter 3), and as an **integral membrane protein**. We saw in Chapter 3 that chitosomes resemble some of the microvesicles in the hyphal apex (see Fig. 3.13). However, the "shell"

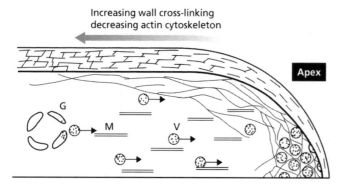

Increasing wall cross-linking
decreasing actin cytoskeleton

G

M

V

Apex

Fig. 4.4 Representation of the possible organization of wall growth at the hyphal apex. Only half of the hypha is shown. Vesicles (V) derived from the endoplasmic reticulum and Golgi body (G) are transported to the apex, probably by microtubule (M)-associated motor proteins. The vesicles can then be directed to the plasma membrane, perhaps by actin-associated motor proteins. The newly formed wall at the extreme hyphal tip is thin and has few cross-linkages, but becomes increasingly cross-linked further back. By contrast, the actin cytoskeleton is highly delevoped at the extreme tip (see Fig. 3.19) and might help to provide structural support, compensating for the lack of wall cross-linking at the tip. The concentration of actin progressively decreases behind the tip.

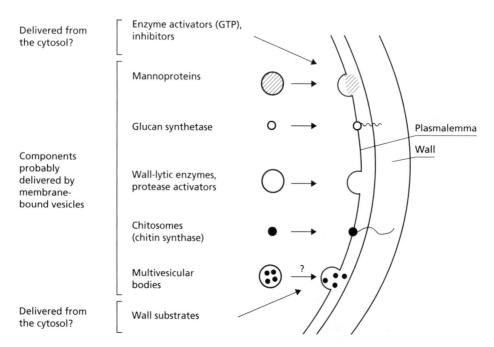

Delivered from the cytosol?

Enzyme activators (GTP), inhibitors

Mannoproteins

Glucan synthetase

Plasmalemma

Wall

Components probably delivered by membrane-bound vesicles

Wall-lytic enzymes, protease activators

Chitosomes (chitin synthase)

Multivesicular bodies

?

Delivered from the cytosol?

Wall substrates

Fig. 4.5 Diagram of some components of wall synthesis at the hyphal tip. Vesicles are thought to deliver the main wall-synthetic enzymes (chitin synthase and glucan synthase) to the tip, where they lodge in the plasma membrane as integral membrane proteins. Mannoproteins and other glycoproteins are transported in vesicles from the endoplasmic reticulum–Golgi secretory system (because the glycosylation of proteins occurs only in the Golgi). Multivesicular bodies, whose functions are still unclear, may be carried as vesicular cargoes along microtubules. Enzyme activators and inhibitors also are thought to be involved in the orchestration of tip growth, but the substrates for wall synthesis arrive from metabolic reactions in the cytosol.

around a chitosome is not a phospholipid membrane, so chitosomes might be packaged within phospholipid membranes for transport to the tip – perhaps in the multivesicular bodies that are sometimes seen in electron micrographs (see Fig. 3.20).

The zymogen form of chitin synthase, when inserted into the membrane, must be **activated by a protease** which probably arrives at the apex in other vesicles. Then the substrate is delivered from the cytosol to the inner face of the chitin synthase enzyme (anchored in the membrane), and chitin chains are synthesized and extruded from the membrane outer face (Fig. 4.5). The substrate for chitin synthesis is *N*-acetylglucosamine, but it is supplied as a sugar nucleotide, UDP-*N*-acetylglucosamine (where UDP = uridine diphosphate), with a high-energy bond required for chitin synthesis as explained in Chapter 6. Clearly, there must be mechanisms for regulating the activity of chitin synthase during wall growth. This could be achieved in a number of ways, including enzyme inhibitors, because the cytosol is known to contain a chitin synthase inhibitor, which might prevent any "spill-over."

Glucan synthase

Glucan synthase is the other major enzyme involved in wall growth. It catalyses the synthesis of β-1,3-glucan chains, which often comprise the bulk of the fungal wall. Like chitin synthase, glucan synthase is thought to arrive in vesicles and becomes inserted in the plasma membrane at the apex. The substrate for this enzyme is a sugar nucleotide (UDP-glucose), supplied from the cytosol. However, the activity of glucan synthase is regulated in a different way from chitin synthase. The enzyme is composed of two subunits, one of which (on the membrane outer face) contains the catalytic site, and the other is a guanosine triphosphate (GTP) binding protein. So the enzyme is thought to be activated when GTP arrives at the cytoplasmic face, then glucan chains are synthesized and extruded into the wall. The glucan chains seem to undergo further modification within the wall. In particular, short β-1,6-linked side chains develop and link the β-1,3-glucan chains. The number of these branched linkages increases progressively as the wall matures behind the apex.

Mannoproteins

Mannoproteins and other glycoproteins form a relatively small proportion of the total wall composition of fungal hyphae, but are more common in yeasts and in the yeast-like phase of dimorphic fungi. These glycosylated proteins are among the few wall components that are pre-formed in the endoplasmic reticulum–Golgi complex and are delivered to the apex in vesicles.

Cross-linking and maturation of the hyphal wall

Various types of cross-linkage occur between the major wall polymers after these have been inserted in the wall, and this seems to occur progressively back from the hyphal tip. For example, essentially pure glucans can be extracted from newly formed fungal walls by using hot alkali, but in the older wall regions an increasing proportion of the glucan is alkali-insoluble, apparently because it is complexed with chitin. In support of this view, the glucans can be extracted after treating walls with chitinase to degrade the chitin. The chitin and glucans are linked by covalent bonds. Little is known about the process except that amino acids may be involved, because the amino acid lysine is associated with up to half of the glucan–chitin linkages in walls of *Schizophyllum commune* (Basidiomycota). In addition to these intermolecular bonds, the chitin chains associate with one another by hydrogen bonding, to form microfibrils. The glucans also associate with one another. These additional bondings behind the growing apex could serve to convert the initially plastic wall into a progressively more cross-linked and rigidified structure.

Wall lytic enzymes

There are opposing views on whether wall lytic enzymes are necessary for apical growth. On the one hand, it has been suggested that the existing wall must be softened in order for new wall components to be inserted, in which case wall growth would involve a balance of wall lysis and wall synthesis. Consistent with this, **chitinase, cellulase** (in Oomycota), and **β-1,3-glucanase** activities can be found in hyphal wall fractions, although these enzymes might exist usually in a latent form. On the other hand, it has been argued that the substantial cytoskeleton of tubulins and actin could help to reinforce the hyphal tip, precluding the need for a rigid wall and therefore precluding the need for wall-degrading enzymes. However, there is no doubt that wall-lytic enzymes would be required for the production of new tips (new hyphal branches) that emerge from the previously rigid wall further back from the hyphal apex.

A steady-state model of wall growth

Wessels (1990) proposed a steady-state model of fungal tip growth that could make the role of tip-located wall-lytic enzymes unnecessary, and also could explain several other features of tip growth. According to this model, the newly formed wall at the extreme tip is suggested to be viscoelastic, so that the wall polymers flow outwards and backwards as new components

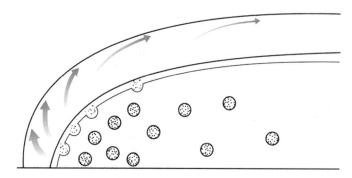

Fig. 4.6 Representation of the steady-state model of hyphal tip growth, in which the wall is envisaged as being viscoelastic. New wall polymers synthesized at the extreme tip are suggested to flow outwards and backwards as new components are continually added at the tip. The decreasing thickness of the arrows behind the tip signifies progressively reduced flow as the polymers become cross-linked. (Based on a diagram in Wessels 1990.)

are continually added at the extreme tip (Fig. 4.6). Then the wall rigidifies progressively by the formation of extra bonds behind the tip. This is reminiscent of Robertson's original idea of a plastic, deformable wall at the extreme tip, with subsequent rigidification occurring behind the tip. Thus, for example, if the supply of vesicles to the tip is slowed or halted (by an osmotic shock, etc.) the rate of hyphal extension would slow or stop, but wall cross-linking might be unaffected if bonding occurs spontaneously within the wall. We could therefore have at least some of the elements of a **unifying theory of fungal tip growth**, based on much biochemical and ultrastructural evidence. It is still necessary to explain how hyphae with an essentially fluid (viscoelastic) wall could resist turgor pressure, but the answer to this could lie in the structural support provided by the massive actin cytoskeleton at the hyphal tip, as we saw in Chapter 3.

Jackson & Heath (1990) investigated this for *Saprolegnia ferax* (Oomycota). They showed that treatment of hyphae with cytochalasin E (one of several cytochalasins, which disrupt cell dynamics by binding to actin microfilaments) caused disruption of the actin cap and led initially to an increase in the rate of tip extension, but then the tips swelled and burst. The weakest region of the tip, most susceptible to bursting, was not the extreme apex where the actin cap is densest but on the shoulders of the apex where the actin is less dense and where the wall presumably has not yet rigidified sufficiently to compensate for the weaker cytoskeleton. It will be recalled that the shoulder of the apex is where new tips originate when hyphae are flooded with water (Fig. 4.3).

Another important aspect of the steady-state model is that it could help to explain how fungi release enzymes into the environment for breakdown of complex polymers. As explained in Chapter 6, enzymes are relatively large molecules, commonly in the range of 30–50 kDa (kiloDaltons), and there is no evidence that fungal walls have continuous pores of this size through which enzymes could be released into the environment. The viscoelastic wall model could help

to resolve this problem if the enzymes are released from vesicles that fuse with the hyphal tip and these enzymes then flow outwards through the developing wall.

The driving force for apical growth

Having considered the dynamics of wall growth and wall rigidification, the remaining question concerns the driving force for apical extension. The cytoskeletal components have emerged as the strongest candidates for this, consistent with many studies on animal cells where protrusions such as pseudopodia are linked to the polymerization of actin.

Studies on *Saprolegnia* (Oomycota) have shown that the apex can extend even when hyphae have negligible turgor pressure, presumably because actin polymerization drives this process (reviewed by Money 1995). Actin is abundant in hyphal tips, and both tip extension and cytoplasmic streaming can be halted by treating fungi with cytochalasins ("cell-relaxers") which bind to actin. In *S. cerevisiae* there is strong evidence that F-actin is involved in the localization of bud formation and that it interacts with the motor protein myosin to transport vesicles to the bud site.

The question of whether microtubules are directly involved in fungal tip growth is more problematical. Hyphal extension can be halted by the benzimidazole fungicides, the related azole drugs, and griseofulvin (see Chapter 17), all of which interfere with microtubule function. Coinciding with this stoppage of growth, there is a progressive depletion of vesicles in the hyphal tip (Howard & Aist 1980). Thus, microtubules must in some way be involved in tip growth – perhaps by providing tramlines for vesicle cargoes. Calcium also seems to be intimately involved in tip growth (Jackson & Heath 1993) because the tips of several fungi, including *Neurospora*, and also *Saprolegnia* (Oomycota), require external calcium for continued tip growth. Moreover, the plasmalemma at the extreme tip is reported to have a high concentration of stretch-activated calcium channels, allowing the ingress of calcium when the membrane is stretched. The significance of this is that the intracellular levels of free calcium are always tightly regulated – cells maintain a low calcium level by sequestering excess calcium in intracellular stores such as the endoplasmic reticulum, mitochondria, and vacuoles. So any localized ingress of calcium through the plasma membrane would cause a perturbation, including an interaction with the cytoskeleton, because calcium is known to cause the contraction of F-actin. There is abundant evidence for a role of calcium-mediated signaling in many fungi and other organisms, but the details of calcium signaling and how it relates to tip growth remain unclear. Bartnicki-Garcia (2002) provides an excellent and thought-provoking review of this and other outstanding questions in hyphal tip growth.

Spore germination and the orientation of hyphal tip growth

Fungi respond to many types of environmental signal, including signals that trigger spore germination (i.e. the production of a hyphal tip where none existed before – see Chapter 10) and signals that change the orientation of hyphal tip growth. Below we consider several examples of these processes.

Studies on germinating spores

Some fungal spores, such as the uredospores of rust fungi (Basidiomycota), have a fixed point of germination termed the **germ pore**, where the wall is conspicuously thinner than elsewhere. Similarly, the zoospores (motile, flagellate cells) of Chytridiomycota, Oomycota, and plasmodiophorids have a fixed point of germination, and they settle and adhere to recept-

ive surfaces so that their future point of germ-tube outgrowth is located next to that surface (Chapter 10). However, many spores seem to be able to germinate from any point on the cell periphery. The germination process often follows a common pattern (Fig. 4.7). Initially, the spore swells by hydration, then it swells further by an active metabolic process and new wall materials are incorporated over most or all of the cell surface – the phase termed **nonpolar growth**. Finally a germ-tube (a young hypha) emerges from a localized point on the cell surface, and all subsequent wall growth is localized to this region. The first sign that an apex will emerge is the localized development of an apical vesicle cluster.

In the conidia of *Aspergillus niger* the transition from nonpolar to polarized growth is temperature-dependent (Fig. 4.7). At a normal temperature of about 30°C, the spore initially incorporates new wall material over the whole surface and then an apex is formed. However when the spores are incubated at 44°C they continue to swell for 24–48 hours, producing giant rounded cells up to 20–25 μm diameter (a 175-fold increase in cell volume) with walls up to 2 μm thick. At this stage the cells stop growing. But if these "giant cells" are shifted down to 30°C before they stop growing they will respond by producing a hyphal apex, and this behaves in an unusual way: instead of forming a normal hypha it produces a small spore-bearing head (Fig. 4.7). These observations suggest two things. First, that the transition from nonpolar to polar growth in *A. niger* is temperature-dependent – it is blocked at the restrictive temperature (44°C). Second, that the fungus can still "mature" at the restrictive temperature: it reaches a developmental stage at which it is committed to sporulate, and it does so as soon as the temperature is lowered.

The production of spores from germinating spores with a minimum of intervening growth is termed

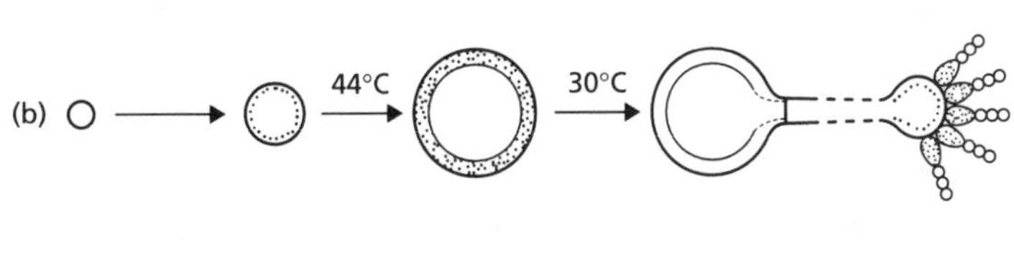

Fig. 4.7 Stages in germination of spores of *Aspergillus niger*. (a) In normal conditions (e.g. 30°C) the spore swells and incorporates new wall material over the whole of the cell surface (shown by stippling), then a germ-tube emerges and all new wall incorporation is localized to the hyphal tip. (b) At 44°C the spore continues to swell and incorporates wall material in a nonpolar manner, producing a giant cell with a thick wall. If the temperature is lowered to 30°C this cell produces an outgrowth, which immediately differentiates to produce a spore-bearing head. (Based on Anderson & Smith 1971.)

microcycle sporulation. It occurs naturally in some fungi, especially if they grow in water films in nutrient-limited conditions. For example, microcycle sporulation has been reported for some saprotrophs on leaf surfaces (e.g. *Cladosporium*, *Alternaria* spp., Chapter 11), some leaf-infecting pathogens (e.g. *Septoria nodorum*), several vascular wilt pathogens that colonize xylem vessels (e.g. *Fusarium oxysporum*, Chapter 14), and the rhizosphere fungus *Idriella bolleyi* which is a biological control agent of root pathogens. All these fungi will germinate to form normal hyphae in nutrient-rich conditions, so their microcycling behavior in nutrient-poor conditions might be a means of spreading to new and potentially more favorable environments.

Spore germination tropisms

A **tropism** is defined as a directional growth response of an organism to an external stimulus. The spores of some fungi show this very markedly, a classic example being the yeast-like fungus *Geotrichum candidum* which is a common cause of spoilage of dairy products. As shown in Fig. 4.8, the cylindrical spores of this fungus germinate typically from one or other pole, but the site of germ-tube emergence is influenced strongly by the presence of neighboring spores when the spores are seeded densely on agar and covered with a coverslip. In these conditions the germ-tubes always emerge from the end furthest away from a touching spore – a phenomenon termed **negative autotropism**. The causes of this behavior are still unclear. On the one hand, it has been suggested to involve the release of auto-inhibitors, which would accumulate maximally in the zone of contact of two spores but could diffuse away from the "free" ends, leading to germination there. On the other hand, oxygen depletion in the zone of spore contact could be a critical factor for *G. candidum*

because the spores always germinate towards an oxygen source (a small hole in a plastic coverslip placed over the spore layer) and this positive tropism to oxygen could overcome the negative autotropism of touching spore pairs.

The spores of *Idriella bolleyi* (a mitosporic fungus) also show negative autotropism, but they show an even more spectacular response when placed in contact with cereal root hairs (Fig. 4.9). The spores of *Idriella* always germinate **away from living root hairs but towards dead root hairs** and rapidly penetrate them. This behavior seems to be ecologically relevant because *I. bolleyi* is a weak parasite of cereal and grass roots. It exploits the root cortical cells as they start to senesce naturally behind the growing root tip, and in doing so it competes with aggressive root pathogens that otherwise would use the dead cells as a food base for infection. Thus, the spore germination tropisms of *I. bolleyi* help to explain its role as a biological control agent of cereal root pathogens, similar to the role of nonpathogenic strains of the take-all fungus, discussed in Chapter 12. The tropic signals for *I. bolleyi* spores seem to be quite specific, because *G. candidum* and some other fungi tested in the same conditions showed quite different responses; for example, *G. candidum* germinated towards both living and dead root hairs (Allan *et al.* 1992).

Fungal spores can also show orientation responses to electrical fields of sufficiently high strength (5–20 V cm^{-1}). For example, in one study the spores of *Neurospora crassa* and *Mucor mucedo* were found to germinate towards the anode, whereas spores of *Emericella nidulans* showed no significant orientation response. The somatic (older) hyphae of these and other fungi showed an array of orientation responses: *Neurospora* hyphae grew towards the anode and formed branches towards the anode; but hyphae of *E. nidulans* and *M. mucedo* grew and branched towards the

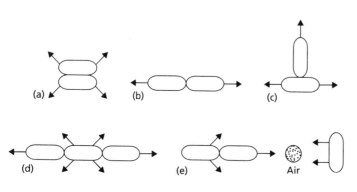

Fig. 4.8 Germination behavior of spores of *Geotrichum candidum*, when incubated in a thin water film beneath a coverslip. The spores always germinate from positions near their poles. Arrows indicate the positions of germ-tube outgrowth in different conditions. (a–d) Negative autotropism of spores touching in pairs or in groups – the spores always germinate from a position furthest from a touching spore. (e) The presence of oxygen (a small hole in the coverslip) negates the negative autotropism – the spores germinate from a point closest to the oxygen source. (Redrawn from Robinson 1973.)

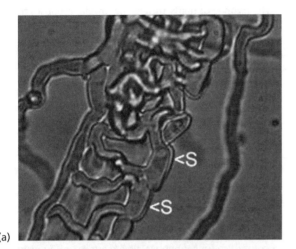

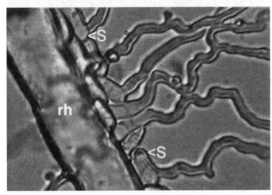

(a)

(b)

Fig. 4.9 Behavior of spores (S) of *Idriella bolleyi* on wheat root hairs in aseptic conditions. (a) Spores germinate towards a dead root hair, and then envelop and penetrate the root hair. (b) Spores germinate away from a living root hair (rh). (From Allan *et al.* 1992.)

example, towards a nutrient source or away from a potential inhibitor. Despite the fact that all fungi require nutrients, and therefore might be expected to orientate towards nutrient-rich zones, there is no evidence that the normal somatic hyphae of true fungi do this. Only the Oomycota show this nutrient-seeking behavior, and it can be strain-specific – some strains of *Saprolegnia* or *Achlya* spp. orientate towards mixtures of amino acids, whereas other strains show no response. Where it occurs, the response can be very striking. On a nutrient-poor medium the hyphae will turn through 180 degrees to an agar disk containing casein hydrolysate or other amino acid mixtures (Fig. 4.10) and the hyphae also branch from the side

(a)

(b)

Fig. 4.10 (a) Orientation of hyphal tips and orientated emergence of hyphal branches of *Achlya* and *Saprolegnia* spp. towards an agar disk (black circle) containing a mixture of amino acids. (b) Reorientation of hyphae from germinating spores of *Pythium aphanidermatum* towards a mixture of amino acids (Mitchell & Deacon 1986).

cathode. In a more recent study (Lever *et al.* 1994) the galvanotropic responses of somatic hyphae were found to be pH- and calcium-dependent. *Neurospora* hyphae even changed from being strongly cathodotropic at pH 4.0 to strongly anodotropic at pH 7. Given the range of different responses to electrical fields it is difficult to summarize this topic, except to say that fungal hyphae can be responsive to electrical/ionic fields. Gow (2004) recently reviewed this topic.

Hyphal tropisms

In all the examples above we have used the term tropism loosely to describe the position where a hyphal tip emerges. But, strictly, a tropism is a bending response that orientates a hypha in a particular direction – for

closest to the amino acid source. Manavathu & Thomas (1985) investigated this for a strain of *Achlya ambisexualis* and found that, of all the single amino acids tested, only the sulfur-containing amino acid methionine could elicit hyphal tropism. However, gradients of many other single amino acids would elicit tropism if the medium contained a uniform background of cysteine. As an explanation of this it was suggested that cysteine, when taken up by cells, can donate one of its sulphydryl (-SH) groups to other amino acids and thereby generate methionine. In bacteria the attraction of cells to several types of compound is mediated by chemoreceptor complexes, and this involves a role for methionyl derivatives which donate methyl groups to the interior domains of the receptor complexes (Armitage & Lackie 1990).

A similar system might be involved in chemotropism by the Oomycota, because Manavathu & Thomas found that several methyl-donor compounds could elicit a tropic response. In work with a strain of *Achlya bisexualis*, Schreurs *et al.* (1989) found that the hyphal tips orientate towards the tips of micropipettes containing either methionine or phenylalanine. Also, when micropipettes containing attractant amino acids were placed behind the hyphal tips the branches emerged from the hyphae and grew towards the attractants. Thus it seems that the initiation of branching and the tropism of hyphal tips are closely related responses to environmental signals, and in some Oomycota these responses might be mediated by plasma membrane receptors for specific amino acids.

The hyphae of many fungi show tropic responses to *non-nutrient factors* of potential ecological relevance. For example, germ-tubes arising from spores of the arbuscular mycorrhizal fungi (Glomeromycota; Chapter 2) can grow towards volatile metabolites (perhaps aldehydes) from roots; some wood-rotting fungi (e.g. *Chaetomium globosum*, Ascomycota) orientate towards volatile compounds from freshly cut wood blocks, and hyphae of the seedling pathogen *Athelia (Sclerotium) rolfsii* (Basidiomycota) orientate towards methanol and other short-chain alcohols from freshly decomposing organic matter (Chapter 14). Sexual pheromones also elicit orientation responses (discussed in Chapter 5).

The yeast cell cycle

In contrast to fungal hyphae, which grow continuously from a hyphal tip, yeasts typically grow by a repeated budding process to produce colonies of single cells. As shown for *Saccharomyces cerevisiae* in Fig. 4.11, in each "turn" of the yeast cell cycle a young bud emerges from a predictable point on the mother cell. The bud grows apically by channeling of wall components to the bud tip, but at a later stage the mode of growth switches and wall components are inserted uniformly over the cell, so that the bud becomes more swollen. Meanwhile, the cell has undergone mitosis, and one of the two daughter nuclei enters the bud. The final stage of cytokinesis (cell separation) occurs by the development of a primary septum, composed of

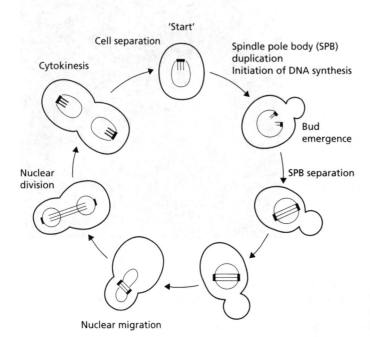

Fig. 4.11 Events in the cell cycle of *Saccharomyces cerevisiae*. (Based on a drawing by Hartwell 1974.)

chitin, between the mother and daughter cells. This plate of chitin is then overlaid by β-glucan and mannan (the major wall components) to form the secondary septum, and the cells finally separate, leaving the chitin plate on the mother cell.

The yeast cell cycle has been studied intensively as a model of the regulation of cell growth and division. Four stages in this cycle are recognized: **G1** (first gap), **S** (DNA synthesis), **G2** (second gap), and **M** (mitosis). In each turn of the cycle a bud emerges, grows to nearly full size, receives one of the daughter nuclei from nuclear division and then separates from the parent cell. At its fastest the cycle takes about 1.5 hour in *S. cerevisiae*, but the time can vary within wide limits, depending on the availability of nutrients. Almost all of this variation occurs in G1, because S, G2, and M together occupy a more or less constant time.

The most important checkpoint in this cycle is termed **start**. It occurs during G1 in *S. cerevisiae* and it is the stage where the cell integrates all the information from intracellular and environmental signals to determine whether the cell cycle will continue, or enter stationary phase, or the cell will undergo sexual reproduction (Chapter 5). Many **cell division cycle (CDC) genes** have been identified in the budding yeast *S. cerevisiae*, and homologous genes that regulate development have also been found in the distantly related **fission yeast**, *Schizosaccharomyces pombe*.

Analysis of the genes and gene products of these two organisms has helped us to understand how cells establish their polarity of growth (Fig. 4.12). In budding yeast (sequence shown in Fig. 4.12a) the first bud develops at one of the poles of the cell. When this bud has developed and separated from the mother cell, the next bud arises at a point adjacent to the bud scar. It grows initially by polar growth (stage 4) but then by wall growth over most of the bud surface (stage 5). At the time of bud emergence, a "tag" or **"landmark"** is laid down at the site where a new bud will form, and a ring of septin proteins is deposited at this point. At a later stage of development (stage 6) the cellular machinery, including actin microfilaments, will direct vesicles and wall precursors to this site, resulting in the localized development of a septum to separate the daughter cell. The fission yeast, *Schizosaccharomyces pombe* (Fig. 4.12b), grows in a different way from *Saccharomyces*, because it produces cylindrical cells that extend at both ends, with tags at the poles of the

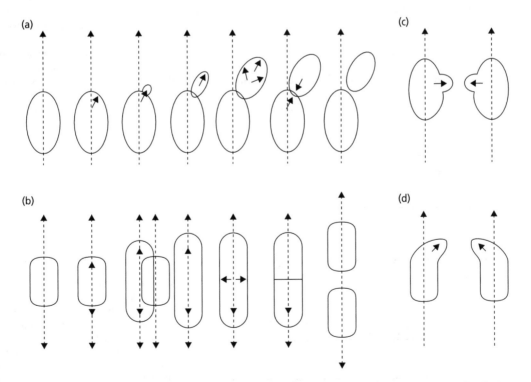

Fig. 4.12 Comparison of cell division in the budding yeast *Saccharomyces cerevisiae* (a,c) and the fission yeast *Schizosaccharomyces pombe* (b,d), showing how an axis of polarity is established at different points during cell development. (Reprinted from Mata, J. & Nurse, P. (1998) Discovering the poles in yeast. *Trends in Cell Biology* **8**, 163–167, copyright 1998, with permission from Elsevier.)

cell. When the cell has reached a critical volume the nucleus divides, and tags then direct the production of a septum at the site where the cell will divide (stages 5–7). Both *Saccharomyces* and *Schizosaccharomyces* can undergo mating under the influence of pheromones produced by cells of opposite mating types, so this represents yet another developmental stage involving the establishment of polar sites (Fig. 4.12c,d).

The *CDC42* gene is a key component in the establishment of polarized growth of *Saccharomyces*. It encodes a GTP (guanosine triphosphate)-binding protein, which acts in concert with the protein products of other genes (*CDC24* and *BEM1*) to recruit septins and actin microfilaments to the sites where localized wall growth will occur. Less is known about the establishment of polarity in fission yeast, but homologs of some of the key budding yeast genes have been identified in *Schizosaccharomyces* – for example, the fission yeast genes *ral1/sdc1* (homologous to *CDC24*) and *ral3/scd2* (homologous to *BEM1*). Some human genes also show base sequence homology with *CDC* genes of *Saccharomyces*, indicating a degree of gene conservation through evolutionary time. Thus, the yeast cell cycle can be used to investigate cellular events of much wider significance, including the regulation (or deregulation) of normal cell division.

Do mycelial fungi have a cell cycle?

The essence of the yeast cell cycle described above is that, during a certain length of time (which varies depending on nutrient availability and other environ-mental factors) one cell produces two cells, while the biomass doubles, and the number of nuclei also doubles. This raises the question of whether an equivalent cycle occurs in mycelial fungi. The first clear evidence that there is such a cycle in mycelial fungi was obtained with an unusual fungus, *Basidiobolus ranarum* (previously assigned to the Zygomycota, but now transferred to Chytridiomycota based on its small subunit rRNA gene sequence). This fungus usually grows as single large cells in the hind gut of frogs or other amphibiams, where it causes little or no damage. But it will grow as hyphae with complete, unperforated septa on agar plates. *B. ranarum* also has extremely large nuclei (about 20 μm diameter) which are easily seen by light microscopy and which are arranged regularly, one per cell. As shown in Fig. 4.13, the hyphae extend by tip growth on agar, synthesizing protoplasm and drawing it forwards as the tip grows. When a critical volume of protoplasm has been synthesized, the large central nucleus divides and a septum is laid down at the point of nuclear division, creating two cells, each with one nucleus. The new apical cell then grows on, and it repeats the whole process when enough protoplasm has been synthesized. The penultimate cell, which has been isolated by the complete septum, produces a new branch apex and its protoplasm flows into this. In effect, therefore, two hyphal tips (each with a given protoplasmic volume) are formed from the original single tip cell, and there is a clear relationship between cytoplasmic volume, nuclear division, and branching, exactly like the cell cycle of yeasts. However, it is termed a **duplication cycle** rather than a cell cycle, and the cells remain attached to one another.

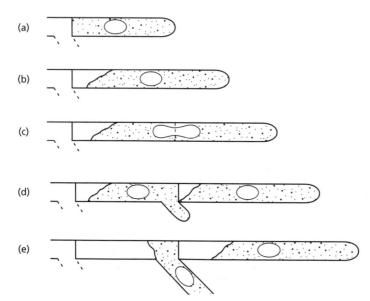

(a)

(b)

(c)

(d)

(e)

Fig. 4.13 The duplication cycle of *Basidiobolus ranarum*, a fungus that grows as hyphae with complete, unperforated septa on agar plates. (a,b) An apical cell extends, synthesizes new protoplasm, and continually draws the protoplasm forwards. (c) When the protoplasmic volume attains a critical size the nucleus divides and a septum is formed. (d,e) The new apical cell grows on and repeats the process; the subapical cell produces a branch, and the protoplasm and nucleus migrate into this, producing a second apical cell.

A similar duplication cycle has been shown to occur in many fungi, even those with normal, perforated septa (Trinci 1984). This was done by allowing the spores of various fungi to germinate on agar and the developing colonies were photographed at intervals (for an example of this, see Fig. 3.5). From the photographs, the number of hyphal tips and the total hyphal length were recorded at different times and used to calculate a **hyphal growth unit** (**G**) where:

$$G = \frac{\text{Total length of mycelium}}{\text{Number of hyphal tips}}$$

After initial fluctuations in the very early stages of growth, the value of **G** became constant and characteristic of each fungal species or strain. For example, a **G** value of 48 μm (equivalent to a hyphal volume of 217 μm^3) was calculated for hyphae of *Candida albicans*, and values of 32 μm (629 μm^3), 130 μm (4504 μm^3), and 402 μm (11,986 μm^3) were found for a wild-type strain and two "spreading" mutants of *Neurospora crassa*.

The constancy of these values for individual strains demonstrates that, as a colony grows, **the number of hyphal tips is directly related to the cytoplasmic volume**. For example, when a colony of *C. albicans* had produced an additional 48 μm length of hypha (or a hyphal volume of 217 μm^3) it had synthesized enough protoplasm to produce a new tip. We can therefore consider a fungal colony as being composed of a number of "units" (the **hyphal growth units**), each of which represents a hyphal tip plus an **average** length of hypha (or volume of cytoplasm) associated with it. They are not seen as separate units because they are joined together, but in some respects they are equivalent to the separate cells produced in the yeast cell cycle. In fact, the duplication cycle of a typical mycelial fungus, *Emericella nidulans*, has been shown to be closely associated with a nuclear division cycle. The apical compartment grows to about twice its original length, then the several nuclei in this compartment divide more or less synchronously and a septum is laid down near the middle of the apical compartment. After this, a series of septa are formed in the new subapical compartment to divide it into smaller compartments, each with just one or two nuclei, while the multinucleate tip grows on and will repeat the process in due course.

Earlier in this chapter we mentioned the **peripheral growth zone** of a fungal colony – the length of hypha needed to support the normal extension rate of tips at a colony margin. It can be estimated by cutting the hyphae at different distances behind the tips, and it can be as large as 5–7 mm. Clearly, this is quite different from the hyphal growth unit which ranges from about 30 to 400 μm. The difference is explained by the fact that the hyphal growth unit is measured in nutrient-rich conditions and is a true reflection of *growth* (increase in biomass, or numbers of tips) whereas the peripheral growth zone is a reflection of the rate of *extension* of a colony margin, and it applies to older colonies, where some of the hyphae become **leading hyphae**, which are much wider and have much faster extension rates than the rest.

The fungal mycelium as a nutrient-capturing system

The fungal mycelium is a highly efficient and adaptable device for capturing nutrients. Over a wide range of nutrient concentrations, a fungal colony will extend across an agar plate at the same rate – whether on water agar or a standard nutrient-rich medium. But on water agar the colony is very sparsely branched, whereas on nutrient-rich agar the branching pattern is dense. This high degree of adaptability is a key feature of fungi, especially in soil and bodies of water where nutrients are likely to occur in localized pockets.

Figure 4.14 is a classic demonstration of this behavior. A young larch seedling was inoculated with a

Fig. 4.14 The fungal nutrient-capturing system. (Courtesy of D.J. Read; from Read 1991.)

mycorrhizal fungus and grown in peat against the face of a transparent perspex box. The root system itself is quite limited: it consists of the region marked by double arrowheads (<<) where the roots are enveloped by a mycorrhizal sheath (Chapter 13). Most of the branching network that we see is a system of aggregated fungal hyphae, termed mycelial cords (Chapter 5) whch explore the soil for nutrients. When they find a localized pocket of organic nutrients (see the large arrowhead in Fig. 4.14) they produce a mass of hyphae to exploit the nutrient-rich zone.

Kinetics of fungal growth

Growth can be defined as an orderly, balanced increase in cell numbers or biomass with time. All components of an organism increase in a coordinated way during growth – the cell number, dry weight, protein content, nucleic acid content, and so on.

Figure 4.15 shows a typical growth curve of a yeast in shaken liquid culture, when the logarithm of cell number or dry weight is plotted against time. An initial **lag phase** is followed by a phase of **exponential** or **logarithmic growth**, then a **deceleration** phase, a **stationary** phase, and a phase of **autolysis** or cell death. During exponential growth one cell produces two in a given unit of time, two produce four, four produce eight, and so on. Provided that the culture is vigorously shaken and aerated, exponential growth will continue until an essential nutrient or oxygen becomes limiting, or until metabolic byproducts accumulate to inhibitory levels.

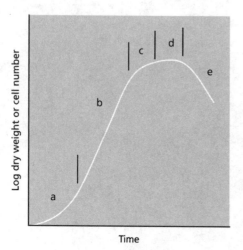

Fig. 4.15 Typical growth curve of a batch culture: a, lag phase; b, exponential or logarithmic growth phase; c, deceleration phase; d, stationary phase; e, phase of autolysis.

This type of curve is typical of a **batch culture**, i.e. a closed culture system such as a flask in which all the nutrients are present initially. The rate of growth during the exponential phase is termed the **specific growth rate** (μ) of the organism, and if all conditions are optimal then the **maximum specific growth rate**, μ_{max}, is obtained. This is a characteristic of a particular organism or strain.

The value of μ is calculated by measuring $\log_{10}$ of the number of cells (N_0) or the biomass at any one time (t_0) and $\log_{10}$ of cell number (N_t) or biomass at some later time (t), according to the equation:

$$\log_{10}N_t - \log_{10}N_0 = \frac{\mu}{2.303}(t - t_0)$$

where 2.303 is the base of natural logarithms. Rewritten, this equation becomes:

$$\mu = [(\log_{10}N_t - \log_{10}N_0)/t - t_0] \times 2.303$$

If, for example, $N_0 = 10^3$ cells ml^{-1} and $N_t = 10^5$ cells ml^{-1}, 4 hours later, then:

$$\mu = \frac{(5-3)\,2.303}{4} = \frac{2.303}{2} = 1.15\,\text{h}^{-1}$$

From this we can compute the **mean doubling time**, or **generation time** (g), of the organism, as the time needed for a doubling of the natural logarithm, according to the equation:

$$g = \frac{\log_e 2}{\mu} = \frac{0.693}{1.15}$$

So, in our example, g = 0.60 h. For *S. cerevisiae* at 30°C, near-maximum values of μ and g are 0.45 h^{-1} and 1.54 h respectively. For the yeast *Candida utilis* at 30°C, μ = 0.40 h^{-1} and g = 1.73 h.

Mycelial fungi also grow exponentially because they have a duplication cycle. Averaged for a colony as a whole, they grow as hypothetical "units," one producing two in a given time interval, two producing four, and so on. Representative values of μ_{max} and g for mycelial fungi are: 0.35 h^{-1} and 1.98 h for *Neurospora crassa* at 30°C; 0.28 h^{-1} and 2.48 h for *Fusarium graminearum* at 30°C, and 0.80 h^{-1} and 0.87 h for *Achlya bisexualis* (Oomycota) at 24°C.

These values compare quite favorably with those of yeasts. However, it is difficult to maintain exponential growth of mycelial fungi, because the hyphae do not disperse freely. Instead, they form spherical pellets in shaken liquid culture, and this leads to problems of nutrient and oxygen diffusion. This problem can be overcome to some degree by using compounds (**paramorphogens**) that alter the hyphal branching pattern, such as sodium alginate, carboxymethylcellulose, and other anionic polymers. They cause fungi to grow as more dispersed, loosely branched mycelia, perhaps by binding to hyphae and causing ionic repulsion. This can be desirable in some industrial processes but not in others. For example, dispersed filamentous growth favors the industrial production of fumaric acid by *Rhizopus arrhizus* and of pectic enzymes by *Aspergillus niger*, but pelleted growth is preferred for industrial production of itaconic acid and citric acid by *A. niger* (Morrin & Ward 1989).

Batch culture versus continuous culture systems

Batch culture systems are used commonly in industry because useful primary metabolites such as organic acids, and secondary metabolites such as antibiotics (Chapter 7), are produced in the deceleration and early stationary phases. Batch cultures are also used for brewing and wine-making, because the culture broth is the marketable product.

Continuous culture systems (Fig. 4.16) are an alternative to batch culture. In these systems, fresh culture medium is added at a continuous slow rate, and a corresponding volume of the old culture medium together with some of the fungal biomass is removed by an overflow device. Such culture systems are monitored automatically so that factors such as pH, temperature, and dissolved oxygen concentration are maintained at the desired levels. They are stirred vigorously to keep the organism in suspension and to facilitate diffusion of nutrients and metabolic byproducts.

There are various types of continuous culture system but the most common type is the **chemostat**. In this system the concentrations of nutrients are adjusted deliberately so that one essential nutrient is at a relatively low concentration while all other components

are present in excess. When the number of cells in the culture starts to increase, the rate of exponential growth becomes limited by the growth-limiting nutrient. At this stage the rate of growth of the culture can be controlled precisely by adjusting the rate at which fresh culture medium is supplied; this is termed the **dilution rate** of the culture. However, it is important to note that the fungus is always growing *exponentially* – only the *rate* of exponential growth is governed by the dilution rate. In theory, by adjusting the dilution rate the culture growth rate can be adjusted to any desired level, up to μ_{max} (any further increase in dilution rate would cause "wash-out" because the cells would be removed by overflow faster than they can grow). In practice, however, these cultures become unstable as they approach μ_{max} because then even a minor, temporary fluctuation in growth rate can cause wash-out.

Chemostats are useful for many experimental purposes, because the physiology of fungi can change at different growth rates or in response to different growth-limiting nutrients. These changes can be studied in detail in chemostats whereas they occur transiently in batch cultures. For example, when *Saccharomyces cerevisiae* is grown at low dilution rates (slow growth) in glucose-limited culture it uses a substantial proportion of the substrate for production of biomass. By contrast, *S. cerevisiae* produces ethanol at the expense of biomass at higher dilution rates. It switches from cell production to alcohol production in conditions favoring rapid metabolism, even though glucose is the growth-limiting nutrient in both cases. Chemostats also are useful for industrial processes, because cells or cell products (e.g. antibiotics) can be retrieved continuously from the overflow medium. In practice, however, most of the traditional industrial processes rely on batch cultures, either because the cost of converting to continuous culture systems does not justify the increased efficiency or because it is difficult to design and operate full production-scale chemostat systems. The batch cultures used industrially are often "fed batch" systems, in which nutrients or other substrates are added periodically to sustain the production of useful metabolites. As we will see in Chapter 7, by keeping cells perpetually growth limited in stationary phase and then "feeding" the culture with selected metabolic precursors, batch cultures can be used to produce substantial quantities of antibiotics or other commercial products.

Commercial production of fungal biomass (Quorn™ mycoprotein)

The most interesting recent application of continuous culture systems has been the development of an entirely

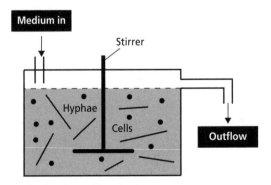

Fig. 4.16 Diagram of a continuous culture system to produce fungal biomass.

novel food product, termed Quorn mycoprotein. In fact, this is the sole survivor of the much-heralded "single-cell protein" revolution of the late 1900s, when scientists and international aid agencies attempted to develop protein-rich foods from microbial biomass, to meet the impending protein shortage in the developing world.

The development of Quorn is a major technological success, which took over 20 years to reach fruition. Now Quorn products are available in many supermarkets and are widely used as alternatives to meat products because of their nutritional profile (see Table 1.2). Quorn mycoprotein is produced commercially from chemostat cultures of the fungus *Fusarium venenatum*, which is grown at 30°C in a medium composed of glucose (the carbon source), ammonium (the nitrogen source), and other mineral salts. The fungal mycelium is retrieved continuously from the culture outflow,

then aligned to retain the fibrous texture, and vacuum-dried on a filter bed before being constituted into meat-like chunks. In this case a continuous culture system was deemed necessary in order to ensure a high degree of reproducibility of the product, and also for economic reasons because the yield of mycelium over a period of time was about five times higher than if a series of batch cultures were used. Glucose is used as the growth-limiting nutrient in the production system, and the dilution rate is set to give a doubling time (μ) of 0.17–0.20 h^{-1}, which is below the μ_{max} of 0.28 h^{-1}. The rate of substrate conversion to protein is extremely high, about 136 g protein being produced from every 1000 g sugar supplied. For comparison with this, the equivalent production of protein by chickens, pigs, and cattle would be about 49, 41, and 14 g respectively.

Trinci (1992) has described the many stages in the development of this fermentation technology.

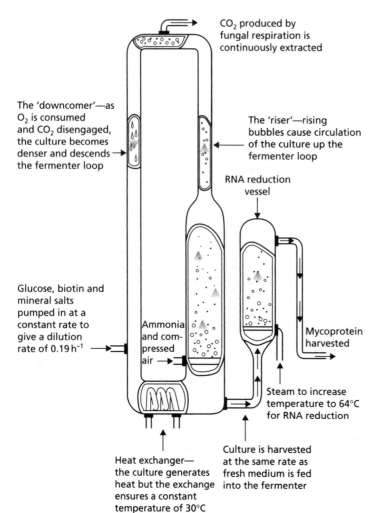

CO$_2$ produced by fungal respiration is continuously extracted

The 'downcomer'—as O$_2$ is consumed and CO$_2$ disengaged, the culture becomes denser and descends the fermenter loop

The 'riser'—rising bubbles cause circulation of the culture up the fermenter loop

RNA reduction vessel

Glucose, biotin and mineral salts pumped in at a constant rate to give a dilution rate of 0.19 h^{-1}

Ammonia and compressed air

Mycoprotein harvested

Steam to increase temperature to 64°C for RNA reduction

Heat exchanger—the culture generates heat but the exchange ensures a constant temperature of 30°C

Culture is harvested at the same rate as fresh medium is fed into the fermenter

Fig. 4.17 Diagram of the air-lift fermenter used by Marlow Foods for the production of mycoprotein in continuous flow culture. (From Trinci 1992.)

Initially, many potential fungi were screened to find a suitable organism for commercial use. Then a suitable large-scale fermenter system had to be designed because fungal mycelia have viscous properties in solutions, so the cultures are difficult to mix to achieve adequate oxygenation – the fungus uses 0.78 g oxygen for every 1 g biomass produced. The system used commercially is a 40 m^3 **air-lift fermenter** (Fig. 4.17) about the height of **Nelson's column**! Compressed air is used to aerate and circulate the culture. This avoids the heating that would be caused by a mechanical stirrer, saving on cooling costs. There is a potential problem with the high nucleic acid content of any type of microbial biomass when used as a human food source, because our bodies metabolize RNA to uric acid which can cause gout-like symptoms. So, the nucleic acid content of the harvested mycelium needed to be reduced while retaining as much of the protein as possible. This was achieved by exploiting the higher heat-tolerance of RNAases than of proteases. The culture outflow containing the biomass is collected in a vessel and its temperature is raised rapidly to 65°C and maintained at this level for 20–30 minutes. The growth of the fungus is stopped at this temperature and the proteases are destroyed so that relatively little protein is lost, but the ribosomes break down and the fungal RNAases degrade much of the RNA to nucleotides which are released into the spent culture filtrate. Then the mycelium can be harvested for drying.

A final problem has still not been overcome completely and it limits the efficiency of commercial production. During prolonged culture in fermenter vessels the fungus is subjected to selection pressure and it mutates to "colonial forms" with a high branching density but relatively low extension rate of the main, leading hyphae. These forms predominate over the wild-type after about 500–1000 hours of continuous culture. Their hyphal growth unit lengths range from 14 to 174 μm, compared with 232 μm for the wild-type, so they give a significantly less fibrous biomass, which is undesirable in the end product. The production runs have to be terminated prematurely to avoid this problem. Nevertheless, the development of Quorn is a significant technological achievement as well as a commercial success.

Cited references

Allan, R.H., Thorpe, C.J. & Deacon, J.W. (1992) Differential tropism to living and dead cereal root hairs by the biocontrol fungus *Idriella bolleyi*. *Physiological and Molecular Plant Pathology* **41**, 217–226.

Anderson, J.G. & Smith, J.E. (1971) The production of conidiophores and conidia by newly germinated conidia of *Aspergillus niger* (microcycle conidiation). *Journal of General Microbiology* **69**, 185–197.

Armitage, J.P. & Lackie, J.M., eds (1990) *Biology of the Chemotactic Response*. Society for General Microbiology Symposium 46. Cambridge University Press, Cambridge.

Bartnicki-Garcia, S. (2002) Hyphal tip growth: outstanding questions. In: *Molecular Biology of Fungal Development* (H.D. Osiewacz, ed.), pp. 29–55. Marcel Dekker, New York.

Gow, N.A.R. (2004) New angles in mycology: studies in directional growth and directional motility. *Mycological Research* **108**, 5–13.

Gow, N.A.R. & Gadd, G.M., eds (1995) *The Growing Fungus*. Chapman & Hall, London.

Hartwell, L.L. (1974) *Saccharomyces cerevisiae* cell cycle. *Bacteriological Reviews* **38**, 164–198.

Howard, R.J. & Aist, J.R. (1980) Cytoplasmic microtubules and fungal morphogenesis: ultrastructural effects of methyl benzimidazole-2-yl-carbamate determined by freeze-substitution of hyphal tip cells. *Journal of Cell Biology* **87**, 55–64.

Howard, R.J. & Gow, N.A.R., eds (2001) *Biology of the Fungal Cell. The Mycota*, VIII. Springer-Verlag, Berlin.

Jackson, S.L. & Heath, I.B. (1990) Evidence that actin reinforces the extensible hyphal apex of the oomycete *Saprolegnia ferax*. *Protoplasma* **157**, 144–153.

Jackson, S.L. & Heath, I.B. (1993) Roles of calcium ions in hyphal tip growth. *Microbiological Reviews* **57**, 367–382.

Latijnhouwers, M., de Wit, P.J.G.M. & Govers, F. (2003) Oomycetes and Fungi: similar weaponry to attack plants. *Trends in Microbiology* **11**, 462–469.

Lever, M.C., Robertson, B.E.M., Buchan, A.D.B., Miller, P.F.P., Gooday, G.W. & Gow, N.A.R. (1994) pH and Ca^{2+} dependent galvanotropism of filamentous fungi: implications and mechanisms. *Mycological Research* **98**, 301–306.

Manavathu, E.K. & Thomas, D. des S. (1985) Chemotropism of *Achlya ambisexualis* to methionine and methionyl compounds. *Journal of General Microbiology* **131**, 751–756.

Mata, J. & Nurse, P. (1998) Discovering the poles in yeast. *Trends in Cell Biology* **8**, 163–167.

Mitchell, R.T. & Deacon, J.W. (1986) Chemotropism of germ-tubes from zoospore cysts of *Pythium* spp. *Transactions of the British Mycological Society* **86**, 233–237.

Money, N.P. (1995) Turgor pressure and the mechanics of fungal penetration. *Canadian Journal of Botany* **73**, S96–102.

Morrin, M. & Ward, O.P. (1989) Studies on interaction of Carbopol-934 with hyphae of *Rhizopus arrhizus*. *Mycological Research* **92**, 265–272.

Read, D.J. (1991) Mycorrhizas in ecosystems – nature's response to the "Law of the Minimum". In: *Frontiers in Mycology* (Hawksworth, D.L., ed.), pp. 29–55. CAB International, Wallingford, Oxon, pp. 101–130.

Robertson, N.F. (1958) Observations of the effect of water on the hyphal apices of *Fusarium oxysporum*. *Annals of Botany* **22**, 159–173.

Robertson, N.F. (1959) Experimental control of hyphal branching forms in hyphomycetous fungi. *Journal of the Linnaean Society, London* **56**, 207–211.

Robinson, P.M. (1973) Oxygen – positive chemotropic factor for fungi? *New Phytologist* **72**, 1349–1356.

Schreurs, W.J.A., Harold, R.L. & Harold, F.M. (1989) Chemotropism and branching as alternative responses of *Achlya bisexualis* to amino acids. *Journal of General Microbiology* **135**, 2519–2528.

Trinci, A.P.J. (1984) Regulation of hyphal branching and hyphal orientation. In: *The Ecology and Physiology ot the Fungal Mycelium* (Jennings, D.H. & Rayner, A.D.M., eds), pp. 23–52. Cambridge University Press, Cambridge.

Trinci, A.P.J. (1992) Myco-protein: a twenty-year overnight success story. *Mycological Research* **96**, 1–13.

Wessels, J.G.H. (1990) Role of cell wall architecture in fungal tip growth generation. In: *Tip Growth in Plant and Fungal Cells* (Heath, I.B., ed.), pp. 1–29. Academic Press, New York.

Chapter 5

Differentiation and development

This chapter is divided into the following major sections:

- mould-yeast dimorphism
- infection structures of plant pathogens
- sclerotia
- nutrient-translocating organs: mycelial cords and rhizomorphs
- asexual reproduction
- sexual development

Differentiation can be defined as the regulated change of an organism from one state to another. These states can be physiological, morphological, or both. So the germination of fungal spores (Chapter 10) and the switch from primary to secondary metabolism (Chapter 7) are examples of differentiation. But here we focus on the developmental changes that lead to the production of a wide range of differentiated structures, such as the infection structures of fungal pathogens, the regulation of sexual and asexual development, the switch between hyphal and yeast forms of some human-pathogenic fungi, and other developmental processes. We consider both the underlying control mechanisms and the functions of the differentiated structures.

Mould-yeast dimorphism

Most fungi grow either as hyphae (the mycelial, mould, or M-phase) or as single-celled yeasts (the Y-phase). In general, yeasts and yeast phases are found in environments with high levels of soluble sugars that can diffuse towards the cells, or where the cells

can be dispersed in liquid films or circulating fluids to obtain nutrients. Yeasts have little or no ability to degrade polymers such as cellulose or proteins, etc., and they also have no penetrating power, unlike the mycelial fungi which commonly have these abilities. So, the yeast form and the mycelial form represent two different growth strategies, suited to particular environments and conditions.

However, some fungi can alternate between a mycelial form and a yeast-like form, in response to environmental factors. These **dimorphic fungi** (with two forms) include several pathogens of humans. For example, *Candida albicans* commonly grows as a yeast on the mucosal membranes of humans, but converts to hyphae for invasion of host tissues (see Fig. 1.4). This dimorphic switch can be induced experimentally by growing *C. albicans* in horse serum of low nutrient content (Chapter 16). Similarly, the fungi such as *Metarhizium* and *Beauveria* spp., which commonly parasitize insects, penetrate the insect cuticle by hyphae but then proliferate in a single-celled form in the circulating fluids of the host (Chapter 15). As a further example, the vascular wilt pathogens of plants (e.g. *Fusarium oxysporum, Ophiostoma novo-ulmi*) penetrate initially by hyphae but then spread as yeast-like forms in the xylem vessels (Chapter 14).

The switch between mycelial (M) and yeast-like (Y) growth of dimorphic fungi occurs in response to environmental factors, and can be reproduced experimentally as shown in Table 5.1. Several opportunistic pathogens of humans grow in the M phase as saprotrophs in plant and animal remains (their normal habitat, Chapter 16) and also grow as mycelia in laboratory culture at 20–25°C. But they convert to budding yeasts or swollen cells in the body fluids or when grown at 37°C in laboratory culture. This **thermally**

Table 5.1 Some environmental or genetic factors that cause transitions between mycelial growth or swollen, yeast-like growth.

Fungus	Conditions for mycelial growth	Conditions for swollen or yeast-like growth
Human pathogens		
Histoplasma capsulatum	20–25°C	37°C
Blastomyces dermatitidis	20–25°C	37°C
Paracoccidioides brasiliensis	20–25°C	37°C
Sporothrix schenckii	20–25°C	37°C
Coccidioides immitis	20–25°C	37°C
Candida albicans	Low nutrient levels	High nutrient levels
Saprotrophs		
Mucor rouxii and some other Zygomycota	Aeration	Anaerobiosis
Plant pathogens		
Ophiostoma ulmi	Calcium Some nitrogen sources	Low calcium
Phialophora asteris		Flooding with water
Ustilago maydis	Dikaryon	Monokaryon
Insect pathogens		
Metarhizium anisopliae	Solid media	Submerged liquid culture
Beauveria bassiana	Solid media	Submerged liquid culture

regulated dimorphism is a significant factor in human pathogenesis. By contrast, the dimorphic saprotrophic *Mucor* species (e.g. *M. rouxii*, *M. racemosus*) do not respond to temperature changes but respond to oxygen levels. They grow as budding yeasts in anaerobic conditions but as mycelia in the presence of even low concentrations of oxygen. *Ustilago maydis* and other plant-pathogenic smut fungi (Basidiomycota) are yeast-like in their monokaryotic phase but hyphal in the dikaryotic form (Chapter 2) so their transition is governed genetically.

This range of responses shows that there is no common environmental cue that governs the M–Y transition, and instead we need to consider the underlying control mechanisms (Gow 1994; Orlowski 1995). In this regard, the studies on cell polarity of budding yeasts and fission yeasts (see Fig. 4.12) are particularly important because several of the genes that regulate polarity have been identified.

Control of the dimorphic switch

The usual approach to identifying the underlying basis of dimorphism is to grow a fungus in conditions that are, as nearly as possible, identical except for one factor that changes the growth form. Then populations of M and Y forms can be compared for differences in biochemistry, physiology, or gene expression. Even so, it is difficult to establish an obligatory, causal relationship between these differences and a change of cell shape, because the altered environmental factor might cause coincidental changes in biochemistry and gene expression. In fact, most of the differences that have been found to date are quantitative rather than qualitative. Some examples are given below.

Differences in wall composition

The wall components of M and Y forms sometimes differ:

- in *Mucor rouxii* the Y form has more mannose than the M form;
- in *Paracoccidioides brasiliensis* the Y form has α-1,3 glucan, whereas the M form has β-1,3 glucan;
- in *Candida albicans* the M form has more chitin than the Y form;
- in *Histoplasma capsulatum* and *Blastomyces dermatitidis* the M form has less chitin than the Y form.

Perhaps these differences are not surprising, given that wall composition and wall bonding are intimately linked with cell shape (Chapter 4). However, the lack of consistency between the fungi suggests that wall composition alone cannot provide a common basis for understanding dimorphism.

Differences in cellular signaling and regulatory factors

Environmental signals often affect cellular behavior through a signal transduction pathway leading to altered metabolism or gene expression. The intracellular factors involved in this include calcium and calcium-binding proteins, pH, cyclic AMP, and protein phosphorylation mediated by protein kinases.

The complexing of calcium with the calcium-binding protein calmodulin was found to be essential for mycelial growth of *Ophiostoma ulmi* (Dutch elm disease); otherwise, the fungus grew in a yeast form. Consistent with this, the levels of calmodulin typically are low (0.02–$0.89\ \mu g\ g^{-1}$ protein, for 14 species) in fungi that always grow as yeasts, but tend to be higher (2.0–$6.5\ \mu g\ g^{-1}$ protein) in mycelial fungi (Muthukumar *et al.* 1987). An external supply of calcium is required for apical growth of *Neurospora crassa* and many other fungi, and calcium is needed in larger amounts for initiation of the M form than for budding in *C. albicans*. High intracellular levels of cyclic AMP are associated with yeast growth in *Mucor* species, *C. albicans*, *H. capsulatum*, *B. dermatitidis*, and *P. brasiliensis*, whereas low cAMP levels are associated with hyphal growth. The supply of cAMP externally can also cause a dimorphic switch. Changes in cytosolic pH have been associated with the M–Y transition of *C. albicans*, and with polar outgrowths in some other cell types. Thus, it seems clear that intracellular signaling compounds are associated with phase transitions, but they are the messengers and mediators not necessarily the direct cause of changes in cell morphology.

Differences in gene expression

Differences in gene expression in the M and Y phases can be detected by extracting messenger RNA and comparing the mRNA banding patterns by gel electrophoresis. Or, preferably, by using the messenger RNAs as templates to produce complementary DNA (cDNA), which is more stable. An example of this approach, though relating to the development of fungal fruitbodies, is discussed later in this chapter. In several cases it has been shown that a few polypeptides are constantly associated with only the M or the Y phase. But for dimorphic fungi there seems to be no case in which a gene or gene product is obligatorily involved in the generation of cell shape. Harold (1990), in a review of the control of cell shape in general, wrote: ". . . form does not appear to be hardwired into the genome in some explicit, recognizable fashion. It seems to arise epigenetically . . . from the chemical and physical processes of cellular physiology." In other words, there are no cell-shape genes, as such!

A potentially unifying theme – the vesicle supply center

Given that a wide range of environmental factors can influence cell shape, although in different ways in different organisms, Bartnicki-Garcia and his colleagues (see Bartnicki-Garcia *et al.* 1995) have adopted a different approach, using computer simulations as a basis for understanding the dimorphic switch and fungal morphogenesis in general. The central feature of the simulations is a postulated **vesicle supply center** (VSC) such as a Spitzenkörper, envisaged as releasing vesicles in all directions to "bombard" the cell membrane and synthesize the wall. As shown in Fig. 5.1, it is then possible to simulate almost any change in growth form by changing the spatial location and/or the rate of movement of the VSC. If the VSC remains fixed then the cell will expand uniformly as a sphere. If the VSC moves rapidly and continuously forward it will produce an elongated structure, like a hypha. If it moves more slowly it will produce an ellipsoidal cell; if it moves to one side of the hypha it will cause bending, and so on. Thus, the key to morphogenesis could be the rate of displacement of the VSC. Other variations in shape and size could occur if the rate of vesicle release from the VSC were altered relative to its rate of movement. We saw in Chapters 3 and 4 that the apical vesicle cluster is intimately associated with cytoskeletal components that are implicated in cellular movements, so it is feasible that the rate of displacement of a VSC could be altered by factors that affect the cytoskeleton. Video-enhanced microscopy of hyphal tips shows that the Spitzenkörper exhibits random oscillations in growing hyphal tips, associated with oscillations in the direction of growth, and the Spitzenkörper also can divide to leave a "daughter" VSC at a future branch point (Chapter 3).

We return to the subject of dimorphism in Chapter 16, when we discuss the human-pathogenic fungi.

Infection structures of plant pathogens

Fungal pathogens of plants can penetrate either through an intact host surface or through natural openings such as stomata, but in any case the invasion of a host is preceded by production of specialized infection structures of various types (Fig. 5.2). In this section we focus on plant-pathogenic fungi, but equivalent structures are produced by fungal

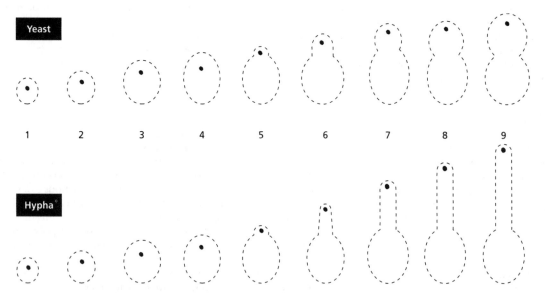

Fig. 5.1 Computer simulation of dimorphism in *Candida albicans*, based on the assumption that wall growth occurs by bombardment of the wall by apical vesicles generated from a vesicle-supply center (VSC, black dot). The yeast (top series) and hyphal shapes (lower series) were "grown" simultaneously at the same rate (10,000 vesicles per frame). Each frame indicates a unit of time. The VSC was moved at different rates to generate the shapes. Between frames 4 and 5 the speed of VSC movement was increased fourfold to produce a cell outgrowth. Then it was returned to its previous rate to produce the yeast bud, but continued at a fourfold rate to generate the hypha. (Based on Bartnicki-Garcia & Gierz 1993.)

pathogens of insects, discussed in Chapter 15. The simplest pre-penetration structures are terminal swellings called **appressoria** (singular: appressorium) if they occur on germ-tubes, or **hyphopodia** if they develop on short lateral branches of hyphae. More complex **infection cushions** can be formed if a fungus infects from a saprotrophic food base and needs to overcome substantial host resistance. In these cases the fungus penetrates from several points beneath the infection cushion, helping to overwhelm the host defenses.

All these pre-penetration structures serve to anchor the fungus to the host surface, usually by secretion of a mucilaginous matrix. Enzymes such as cutinase sometimes are secreted into this matrix. The penetration process is achieved by a narrow hypha, termed an **infection peg**, which develops beneath the pre-penetration structure. There has been much debate about the relative roles of enzymes and mechanical forces in the infection process. Enzymes probably are involved locally, to aid the passage of the penetration peg through the wall, but they do not cause generalized dissolution of host walls. Mechanical forces almost certainly are involved, and have been documented in detail for *Magnaporthe grisea*, the fungus that causes rice blast disease. The penetration pegs of this fungus can

penetrate inert materials such as Mylar and even Kevlar, the polymer used to manufacture bullet-proof vests. A turgor pressure of about 8 MPa (mega-Pascals) is generated in the appressorium of this fungus by the conversion of stored glycogen into osmotically active compounds before the penetration peg develops (Howard *et al.* 1991). This force is channeled into the narrow peg because the wall of the upper surface of the appressorium is heavily melanized and resists deformation. By contrast, the underside of the appressorium has a very thin wall, or perhaps none at all, but the adhesive released by the appressorium is extremely strong and forms an "O-ring" seal on the host surface so that the force generated by the infection peg is not dissipated.

Many appressoria and infection cushions develop a melanized wall, and this can enable the fungus to persist on the plant surface until the plant-host resistance declines. Classic examples of this are found in *Colletotrichum* spp. that cause leaf and fruit spots, including *C. musae* which causes the small brown flecks on the skins of ripe bananas (Figs 5.3, 5.4). These fungi are weak parasites that infect fruit tissues only after the fruit has ripened. But their splash-dispersed spores can land on the host surface at any time and, being thin-walled and hyaline (colorless), they cannot

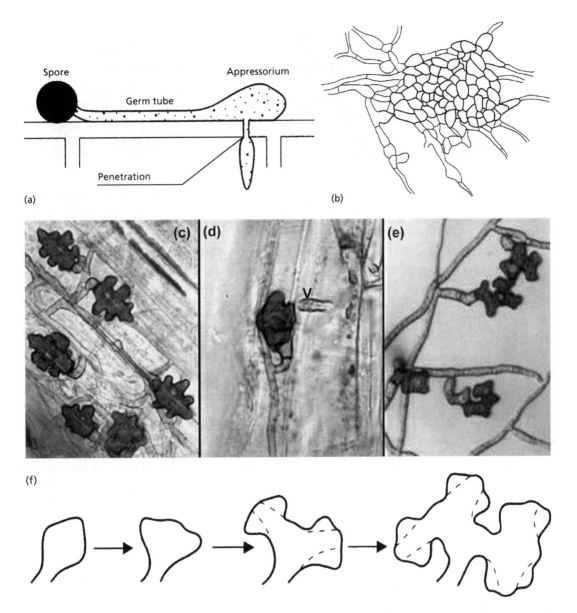

Fig. 5.2 Examples of infection structures of plant-pathogenic fungi. (a) Diagrammatic representation of an appressorium produced from a germinating spore. A narrow penetration peg develops from beneath the appressorium, to breach the host cell wall, and then expands to produce an infection hypha. (b) An infection cushion composed of a tissue-like mass of melanized hyphae, from which multiple penetration pegs invade the host plant. (c–e) Lobed hyphopodia of one of the take-all fungi (*Gaeumannomyces graminis* var. *graminis*) growing on a cereal stem base. (c) Shows the structures in surface view, whereas (d) is a hyphopodium in side view. The arrowhead in (d) is a papilla (a localized ingrowth of the wall of the plant cell that the fungus is attempting to penetrate) signifying that the plant cell is still alive and resisting invasion. (e) Lobed hyphopodia of *G. graminis* produced on the plastic base of a Petri dish. (f) The hypothesized mode of development of lobed hyphopodia by repeated stoppage, swelling and branching of a hyphal tip.

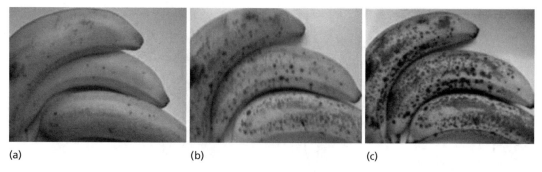

(a) (b) (c)

Fig. 5.3 (a–c) Bananas infected by *Colletotrichum musae*, photographed over a 10-day period after the fruit ripened. Small brown flecks on the fruit surface (a) result from the activation of single-celled, melanized appressoria that remained dormant for several months. The lesions progressively expand and coalesce, leading to softening and over-ripening of the fruit tissues.

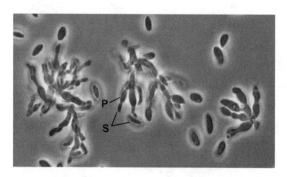

Fig. 5.4 Dense clusters of flask-shaped phialides (P) and spores (S) of *Colletotrichum musae* scraped from the surface of a brown lesion on a ripe banana.

survive desiccation or exposure to UV irradiation. So, the spores often germinate immediately and the germ-tube produces a melanized appressorium which persists on the host surface until the onset of host senescence.

Some of the fungi that depend on melanized infection structures can be controlled by antifungal antibiotics that specifically block the pathway of melanin biosynthesis. Both *Magnaporthe grisea* and *Rhizoctonia oryzae* (which causes sheath blight of rice) are examples of this, although the level of disease control in practice has been disappointing because these pathogens can easily mutate and become resistant to the antibiotics (Chapter 17).

The morphogenetic triggers for differentiation of infection structures

Several physical and chemical factors have been reported to influence the development of appressoria and other infection structures, but the main require-

ment is contact with a surface of sufficient hardness. *In vivo* this could be a leaf cuticle or an insect cuticle. *In vitro* it can be simulated by various artificial membranes. Thus, **contact-sensing** seems to be one of the key morphogenetic triggers.

Contact-sensing can be either **topographical** or **nontopographical**. In nontopographical contact-sensing the fungus merely responds to the presence of a hard surface, and this is true of the air-borne conidia of *Blumeria (Erysiphe) graminis* (powdery mildew of cereals). These spores have minute warts on their surface, and within a few minutes of landing on a leaf or of being placed on a glass surface, the warts in contact with the surface secrete an adhesive containing wall-degrading enzymes. This is a localized response because the warts on the rest of the spore surface do not secrete the adhesive. By contrast, topographical sensing is more specific, because the fungus responds to ridges or grooves of particular heights (or depths) or spacing on the host surface, and recognition of this surface topography is used to locate the preferred infection site. Examples of this are found in the germ-tubes produced from the uredospores of several cereal rust fungi (*Puccinia graminis*, *P. recondita*, etc.). As shown in Figs 5.5 and 5.6, the germ-tubes initially grow at random on cereal leaves, but when they encounter the first groove on the leaf surface (the junction of two leaf epidermal cells) they orientate perpendicular to this groove and then grow across the leaf surface. This is thought to maximize the chances of locating a stomatal pore, because the stomata are arranged in staggered rows along the leaf. These fungi show precisely the same behavior on inert grooved surfaces such as leaf replicas made of polystyrene, confirming that the response is to topographical signals and not to chemical stimuli.

The "nose-down" orientation of the germ-tube tips shown in Fig. 5.6 indicates that the Spitzenkörper is

Fig. 5.5 Scanning electron micrograph showing directional growth of hyphae arising from uredospores of *Puccinia graminis* on an inert replica of the lower surface of a wheat leaf. Note the perpendicular alignment of the hyphae to the contours of the leaf replica, and the short lateral branches that arise in the grooves. The hyphae arising from two spores have located the "stomata" on the leaf replica and produced appressoria directly over the stomatal pores. (Courtesy of N.D. Read; from Read *et al.* 1992.)

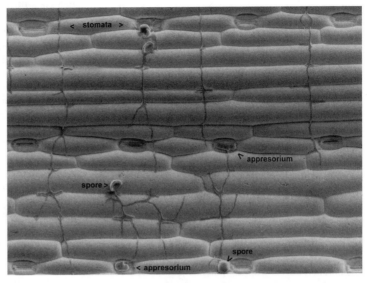

Fig. 5.6 Scanning electron micrograph of two hyphae of *Puccinia graminis* growing perpendicularly over the ridges and grooves of a polystyrene replica of a microfabricated silicon wafer. The germ-tube tips have a "nose-down" orientation which might facilitate topographical sensing. Arrowheads indicate the dried remains of mucilage that adhered the germ-tubes to the surface. Although it is not clearly shown in this image, the hyphae also form projections that grow into the grooves (arrows), equivalent to the short lateral branches shown in Fig. 4.6. (Courtesy of N.D. Read; from Read *et al.* 1992.)

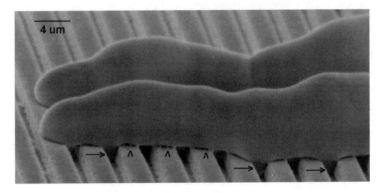

displaced towards the surface on which the fungus is growing, facilitating contact-sensing. Consistent with this, transmission electron micrographs show that parallel arrays of microtubules are abundant just beneath the plasma membrane of the lower surface of the germ-tube, where they could be involved in transducing the contact signals. All of this depends on close adhesion to the surface, mediated by the extracellular matrix, and in the rust *Uromyces appendiculatus* the digestion of this matrix by applying a protease, Pronase E, prevents the topographical signaling. By preparing protoplasts of this fungus and using the patch-clamp technique it has been shown that the plasma membrane at the germ-tube tip contains stretch-activated ion channels. So it is believed that stretching of the membrane when the germ-tube tip

encounters a ridge or groove leads to an ion flux (possibly Ca^{2+}) through these channels, that coordinates the orientation response.

The alignment of a germ-tube is the first stage in an intricate developmental sequence which ensures that infection from uredospores of rust fungi always occurs through the stomata of a leaf surface. This sequence is shown in Fig. 5.7, for the bean rust fungus *Uromyces appendiculatus*.

When a germ-tube locates a stomatal ridge, it stops growing after about 4 minutes, and the apex swells to form an appressorium. The original two nuclei in the germ-tube migrate into the appressorium, then divide, and a septum develops to isolate the appressorium from the germ-tube. About 120 minutes after contacting the stoma, an infection peg grows into

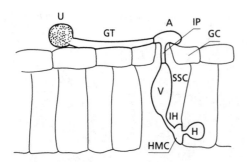

Fig. 5.7 Infection structures of the bean rust fungus, *Uromyces appendiculatus*, penetrating a stomatal opening from a germinating uredospore. A = appressorium; GC = stomatal guard cell; GT = germ-tube; H = haustorium; HMC = haustorial mother cell; IH = infection hypha; IP = infection peg; SSC = substomatal cavity; U = uredospore; V = substomatal vesicle. (Based on a drawing by H.C. Hoch & R.C. Staples; see Hoch *et al.* 1987.)

the substomatal cavity and produces a substomatal vesicle. Then an infection hypha develops from this to produce a haustorium mother cell on one of the leaf parenchyma cells. The infection process is completed when the fungus penetrates the host cell to form a **haustorium**, a specialized nutrient-absorbing structure (Chapter 14).

Studies of this sequence for *U. appendiculatus* have shown that all the events up to, and including, the development of the infection hypha are induced when germ-tubes locate the "stomata" on nail varnish replicas of leaf surfaces. These events can also be induced by

scratches on other artificial surfaces. But the formation of the haustorial mother cell usually depends on chemical recognition of a leaf cell wall. Thus, most of this developmental process is pre-programmed and it requires only an initial topographical signal. To investigate this further, Hoch and his co-workers exploited the techniques of microelectronics to make silicon wafers with precisely etched ridges and grooves of different heights and spacings. The wafers were then used as templates to produce transparent polystyrene replicas on which rust spores would germinate and the responses to surface topography could be studied. Initial studies with *U. appendiculatus* showed that appressoria were induced in response to single ridges or grooves of precise height (or depth), about 0.5 μm, but little or no differentiation occurred in response to ridges or grooves lower or higher than this (Hoch *et al.* 1987). The inductive height corresponds to the height of the lip on the guard cells of bean stomata, which probably is the inductive signal *in vivo*.

In a further study, a total of 27 rust species were tested on ridges of different heights (Allen *et al.* 1991), and this enabled the species to be categorized into four groups. **Group 1** included *U. appendiculatus* and seven other species, which produced appressoria in response to a single ridge or groove of quite precisely defined height; ridges or grooves higher or lower than this had little effect (Fig. 5.8a). **Group 2** included *Puccinia menthae* (mint rust) and three other species, which needed a minimum ridge height (0.4 μm for *P. menthae*) for production of appressoria but also responded to all ridge heights above this, up to at least 2.25 μm (Fig. 5.8b). **Group 3** was represented by a single species, *Phakopsora pachyrhizi* (soybean rust), which could form appressoria

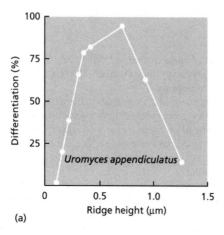

(a)

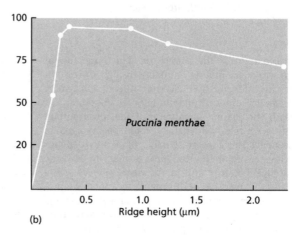

(b)

Fig. 5.8 Production of appressoria by two rust fungi in response to single ridges of different heights on polystyrene replicas of microfabricated silicon wafers. (a) *Uromyces appendiculatus* responded only to ridges of narrowly defined height. (b) *Puccinia menthae* responded to all ridges above a minimum height. (Based on Allen *et al.* 1991.)

even on flat surfaces. **Group 4** included many cereal rusts (*Puccinia graminis*, *P. recondita*, etc.) that did not respond to single ridges of any dimension. However, the cereal rusts have since been shown to differentiate in response to multiple, closely spaced ridges of optimal 2.0 μm height and 1.5 μm spacing. The finding that either ridges or grooves elicit the same response indicates a minimum requirement for two consecutive right angles as the topographical signal. It is suggested that the different requirements for ridge heights or spacings by different rust species could reflect adapta-

tion to the stomatal topography of the host, consistent with the high degree of host specificity of these biotrophic parasites (Chapter 14). Further details of contact sensing and the possible underlying mechanisms can be found in Read *et al.* (1992).

Sclerotia

Sclerotia (singular: sclerotium) are specialized hyphal bodies involved in dormant survival (Fig. 5.9).

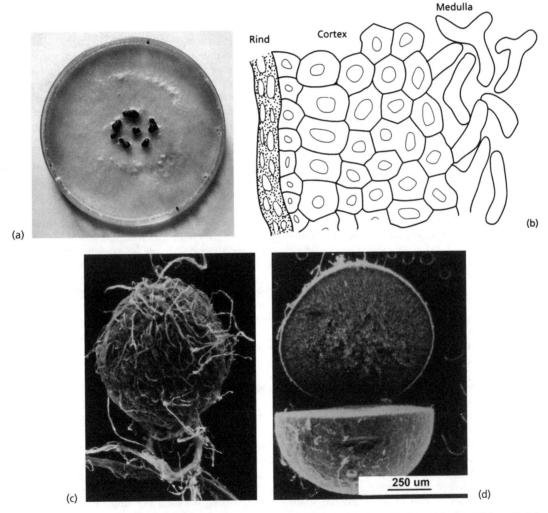

Fig. 5.9 Sclerotia: specialized multicellular dormant survival bodies. (a) Colony of *Sclerotinia sclerotiorum* that has produced a ring of sclerotia on an agar plate. (b) Diagram of part of a cross-section of a sclerotium of *Athelia rolfsii*, showing a crushed, melanized rind, a tissue-like cortex of thick-walled cells, and a central medulla of normal hyphae. (c) Scanning electron micrograph of a mature sclerotium of the mycorrhizal fungus *Paxillus involutus*. This sclerotium developed from nutrients that were translocated along a mycelial cord (an aggregated mass of hyphae) shown at the bottom of the image. (d) Scanning electron micrograph of a cut sclerotium of *Cenococcum geophilum* (a mycorrhizal fungus), showing the internal zonation of the tissues. ((c,d) Courtesy of F.M. Fox; see Fox 1986.)

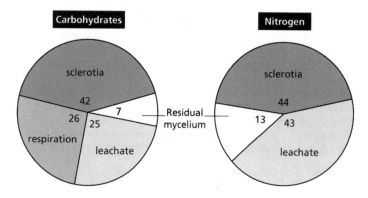

Fig. 5.10 Conservation of mycelial carbon (glucose equivalents) and nitrogen (glycine equivalents) into newly formed sclerotia of *Athelia rolfsii*, 4 days after mycelial mats were transferred to starvation conditions. The values shown are percentages of the original carbohydrate or nitrogen in the mycelial mats that became incorporated into sclerotia or that were lost by respiration or by leakage into the glass beads on which the mycelial mats were incubated. (From Christias & Lockwood 1973.)

They are formed by relatively few fungi – mainly Basidiomycota. At their most complex they can be up to 1 cm diameter, with clearly defined internal zonation. Sclerotia of this type are produced by omnivorous plant pathogens such as *Athelia (Sclerotium) rolfsii* and *Sclerotinia sclerotiorum*, by the ergot fungus *Claviceps purpurea*, and by some mycorrhizal fungi such as *Cenococcum geophilum* and *Paxillus involutus* (although the sclerotia of these mycorrhizal fungi are quite small). Some other fungi produce microsclerotia less than 100 μm diameter; these are merely clusters of melanized chlamydospore-like cells – for example, in the plant pathogen *Verticillium dahliae*. Between these extremes lie a range of types, such as the spherical or crust-like sclerotia of *Rhizoctonia solani* (sexual stage: *Thanatephorus cucumeris*) commonly seen as brown, scurfy patches on the surface of potato tubers.

All sclerotia develop initially by repeated, localized hyphal branching, followed by adhesion of the hyphae and anastomosis of the branches. As the sclerotial initials develop and mature, the outer hyphae can be crushed to form a sheath, while the interior of the sclerotium differentiates into a tissue-like **cortex** of thick-walled, melanized cells, and a central **medulla** consisting of hyphae with substantial nutrient storage reserves of glycogen, lipids, or trehalose (Chapter 7). Sclerotia can survive for considerable periods, sometimes years, in soil. They germinate in suitable conditions, either by producing hyphae (the **myceliogenic** sclerotia of *A. rolfsii*, *Cenococcum*, *R. solani*, etc.) or by producing a sexual fruiting body (the **carpogenic** sclerotia of *Claviceps* and *Sclerotinia sclerotiorum*).

Nutrient depletion is one of the most important triggers for sclerotial development. In these conditions, sclerotia develop rapidly from pre-existing mycelia or from sclerotial initials laid down at an earlier time, and a large proportion of the mycelial reserves of the fungal colony are remobilized and conserved in the developing sclerotia. Christias & Lockwood (1973) demonstrated this by growing four sclerotium-forming fungi in potato-dextrose broth, collecting the mycelial mats before they had produced sclerotial initials, then washing the mats and placing them either on the surface of normal, unsterile soil or on sterile glass beads through which water was percolated continuously to impose a nutrient-stress equivalent to that in soil (Chapter 10). In all cases the mycelia responded by initiating sclerotia within 24 hours, and the sclerotia had matured by 4 days. When these sclerotia were harvested and analyzed for nutrient content, they contained up to 58% of the original carbohydrate in the mycelia and up to 78% of the original nitrogen. An example is shown in Fig. 5.10, for *A. rolfsii* on leached glass beads. Such high levels of carbohydrate conservation could only be explained if some of the wall polymers of the fungal hyphae are broken down and the products are remobilized into the developing sclerotia. As we shall see later, wall polymers can be degraded by controlled lysis and used as nutrient reserves to support differentiation of many types of structure, including the "fruiting bodies" of mushrooms and toadstools.

Nutrient-translocating organs

All fungi translocate nutrients in their hyphae, but some fungi produce conspicuous differentiated organs for bulk transport of nutrients across nutrient-free environments. Depending on their structure and mode of development, these translocating organs are termed **mycelial cords** or **rhizomorphs**. They are quite common among wood-rotting fungi, and also among the ectomycorrhizal fungi of tree roots, where carbohydrates are transported from the roots to the mycelium in soil, and mineral nutrients and water are translocated back towards the roots. Mycelial cords are also found at the bases of the larger mushrooms and toadstools, serving to channel nutrients for fruitbody development (Fig. 5.11).

Fig. 5.11 Fruitbodies and mycelial cords of a puffball (*Lycoperdon* sp.) growing on decayed wood.

Mycelial cords

Mycelial cords have been studied most intensively in *Serpula lacrymans* (Basidiomycota) which causes **dry-rot** of timbers in buildings (Chapter 7). Once this fungus is established in the timbers, it can spread several meters beneath plaster or brickwork to initiate new sites of decay. It spreads across non-nutritive surfaces as fans of hyphae, which draw nutrients forwards from an established site of decay. The hyphae differentiate into mycelial cords behind the colony margin.

The early stage of differentiation of mycelial cords occurs when branches emerge from the main hyphae and, instead of radiating, they branch immediately to form a T-shape and these branches grow backwards and forwards close to the parent hypha. The branches produce further branches that repeat this process, so the cord becomes progressively thicker, with many parallel hyphae. Consolidation occurs by intertwining and anastomosis of the branch hyphae and by secretion of an extracellular matrix which cements them together. Some of the main hyphae then develop into wide, thick-walled **vessel hyphae** with no living cytoplasm, while some of the narrower hyphae develop into **fiber hyphae** with thick walls and almost no lumen. Interspersed with these types of hyphae are normal, living hyphae rich in cytoplasmic contents. The cords of other fungi, such as the mycorrhizal species *Leccinum scabrum* (Fig. 5.12), do not have fibre hyphae but otherwise show a similar pattern of development. In mature hyphal cords there is evidence of a large degree of degeneration of hyphal contents and of the deposition of large amounts of cementing material between the hyphae (Fig. 5.13).

The factors that control the development of mycelial cords are poorly understood, but studies on *S. lacrymans*

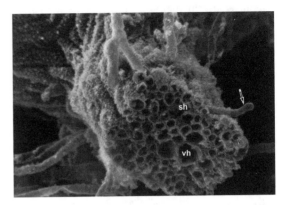

Fig. 5.12 Scanning electron micrograph of a mycelial cord of the mycorrhizal fungus *Leccinum scabrum*, broken to show the internal distribution of hyphae. The wide, central vessel hyphae (vh) are surrounded by narrower sheathing hyphae (sh). The surface of the mycelial cord is covered with extracellular matrix materials, and hyphae (e.g. arrow) radiate into the soil to explore for nutrients. (Courtesy of F.M. Fox; from Fox 1987.)

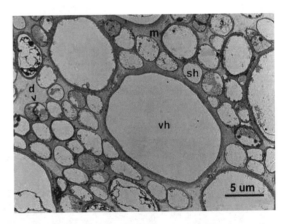

Fig. 5.13 Transmission electron micrograph of a section of a mycelial cord of *Leccinum scabrum*, showing wide, thick-walled, empty vessel hyphae (vh), thin-walled sheathing hyphae (sh), and abundant intercellular matrix material (m). Some hyphae (labeled d and also near the top right) have been sectioned through dolipore septa. (Courtesy of F.M. Fox; from Fox 1987.)

suggest that the availability of nitrogen is a key factor. Cords were found to develop on media containing inorganic nitrogen (e.g. nitrate) but not on media containing amino acids. Also, cords growing from a mineral nutrient medium onto an organic nitrogen medium gave rise to normal, diverging hyphal branches. So it was suggested that cords develop when the parent hyphae leak organic nitrogen in nitrogen-poor conditions, causing branch hyphae to grow close to the parent hyphae in the nitrogen-rich zone. Regulatory control by nitrogen seems logical for wood-decay fungi, because wood has a very low nitrogen content and these fungi could have evolved special mechanisms for conserving and remobilizing their organic nitrogen (Chapter 11). This could apply also to the cords of ectomycorrhizal fungi, because these fungi have a significant role in degrading organic nitrogen in otherwise nitrogen-limiting soils (Chapter 13). In terms of function, mycelial cords have been shown to translocate carbohydrates, organic nitrogen, and water over considerable distances between sources and sinks of these materials. The vessel hyphae seem to act like xylem vessels of plants, transporting water by osmotically driven mass flow (Chapter 7). The combination of their thick walls, the extensive extrahyphal matrix and reinforcement by fiber hyphae could enable vessel hyphae to withstand considerable hydrostatic pressure.

Rhizomorphs

Rhizomorphs serve similar functions to mycelial cords but have a more clearly defined organization. A notable example is the rhizomorph of *Armillaria mellea*, a major root-rot pathogen of broad-leaved trees. It spreads from tree to tree by growing as rhizomorphs through the soil, and it also spreads extensively up the trunks of dead trees by forming thick, black rhizomorphs beneath the bark. These rhizomorphs resemble boot laces, hence the common name for this fungus – the boot-lace fungus (Fig. 5.14).

As shown in Fig. 5.15, the rhizomorph has a specially organized apex or growing point similar to a root tip, with a tightly packed sheath of hyphae over the apex, like a root cap. Behind the apex is a fringe of short hyphal branches. The main part of the rhizomorph has a fairly uniform thickness and is differentiated into zones: an outer cortex of thick-walled melanized cells in an extracellular matrix, a medulla of thinner-walled, parallel hyphae, and a central channel where the medulla has broken down, serving a role in gaseous diffusion. Rhizomorphs branch by producing new multicellular apices, either behind the tip or by bifurcation of the tip.

Rhizomorphs extend much more rapidly than the undifferentiated hyphae of *A. mellea*, and they can grow for large distances through soil. However, they need to be attached to a food base because their growth depends on translocated nutrients, so one of the traditional ways of preventing spread from tree to tree is by trenching of the soil to sever the rhizomorphs. Almost nothing is known about the developmental triggers of rhizomorphs, except that ethanol and other small alcohols can induce them; similarly, almost nothing is known about their mode of development because they originate deep within an established colony in laboratory culture. However, the behavior of rhizomorphs is of considerable interest, as shown by the work of Smith & Griffin (1971) on *Armillariella*

(a)

(b)

Fig. 5.14 *Armillaria mellea*, "boot-lace fungus", growing on a decaying tree trunk. (a) Typical fruiting bodies of *Armillaria*. (b) Thick, black rhizomorphs that grow beneath the bark of dead trees.

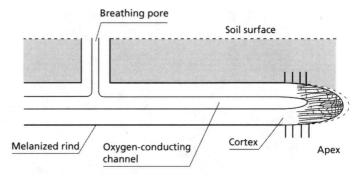

Fig. 5.15 Diagram of a rhizomorph of *Armillariella*.

elegans (related to *A. mellea*). In this fungus, the rhizomorph apex will only grow if it remains hyaline, and this means that the partial pressure of oxygen at the surface of the apex must be 0.03 or less (compared with about 0.21 in air). Above this level, the apex rapidly becomes melanized, stopping its growth. Yet growth of the apex is strongly oxygen-dependent, and the fungus seems to resolve this dilemma by a combination of factors. A high respiration rate is maintained at the apex, supported partly by diffusion of oxygen along the central channel, while the surface of the apex is covered by a water film which limits the rate of oxygen diffusion: at 20°C, oxygen diffuses about 10,000 times more slowly through water than through air. The dependence on a water film ensures that rhizomorphs grow naturally at a specific depth in soil, depending on the soil type and the climate. If a tip grows too close to the soil surface then the width of the water film is reduced and oxygen diffuses to the tip more rapidly, causing melanization. These tips near to the soil surface then break down to produce "breathing pores" connected to the central channel. Conversely, if the apex grows too deeply into moist soil then the water film increases and the rate of growth becomes oxygen-limited. Thus, the peculiar organization of a rhizomorph helps to regulate growth to specific zones in the soil, and these zones are where tree roots occur, maximizing the opportunities for infection.

Asexual reproduction

The enormous diversity of asexual spores of fungi is discussed in Chapter 10. Here we focus on the developmental processes leading to spore formation, and in this respect there are two fundamentally different patterns. **Sporangiospores** are formed by cleavage of the protoplasm within a multinucleate sporangium (Chytridiomycota, Oomycota, and Zygomycota). **Conidia** develop directly from hyphae or special hyphal cells (Ascomycota, mitosporic fungi, and some Basidiomycota) but never within a sporangium.

Sporangiospores

Sporangiospores are usually formed within a thin-walled sporangium, like the well-known sporangia of *Mucor* and other Zygomycota (Fig. 5.16). Within a sporangium the nuclei undergo repeated mitotic divisions, then the cytoplasm is cleaved around the individual nuclei by the alignment and fusion of membranes. This is followed by production of a wall around each of the spores in the Zygomycota, or by the development of flagella in the wall-less spores of Chytridiomycota and Oomycota. Finally, the spores are released by controlled lysis of all or part of the sporangium wall.

The process of cytoplasmic cleavage to produce sporangiospores seems to occur in several ways. In *Gilbertella persicaria* (Zygomycota; Fig. 5.16) and *Phytophthora cinnamomi* (Oomycota) a large number of cleavage vesicles are produced and these migrate around the nuclei so that they are aligned; then they fuse with one another so that their membranes become the plasma membranes of the spores. However, in *Saprolegnia* and *Achlya* (Oomycota) a large central vacuole develops in the sporangium and forms radiating arms between the nuclei, then the arms fuse with the plasma membrane of the sporangium to delimit the spores. The cytoskeleton is intimately involved in these processes, as Heath & Harold (1992) showed by the use of an actin-specific fluorescent stain (phalloidin conjugated to rhodamine). The cleavage planes in these sporangia of *Saprolegnia* and *Achlya* were associated with sheet-like arrays of actin which only appeared at the beginning of cleavage and disappeared when cleavage was complete.

In a study of cleavage in the zoosporic fungus *Phytophthora cinnamomi* (Hyde *et al.* 1991) the flagellar axonemes (microtubular shafts) were found to develop in a different class of vesicle – the flagellar vacuoles near each nucleus, rather than the normal cleavage vesicles. Then the flagellar vacuoles fused with the cleavage vesicles so that the flagella were located on the outside of

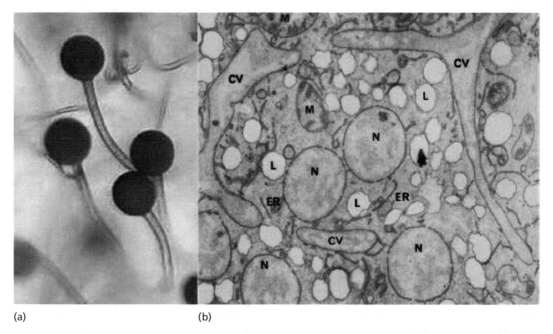

(a)   (b)

Fig. 5.16 (a) Sporangia of *Mucor* containing darkly pigmented spores. (b) Mid-cleavage stage in the sporangium of *Gilbertella persicaria* (Zygomycota). Cleavage vesicles (CV) fuse and extend between the nuclei (N) to separate the sporangium contents into uninucleate spores. ER = endoplasmic reticulum; L = lipid; M = mitochondrion. (Courtesy of C.E. Bracker; from Bracker 1968.)

the spores and were surrounded by **flagellar** membranes. This difference in origin of the flagellar membrane and the plasma membrane that surrounds the rest of the zoospore body could be significant for zoospore function, because there is evidence that different putative chemoreceptors occur on the flagella compared with on the body of a zoospore (Chapter 10).

The cleavage and release of zoospores of the Oomycota is strongly influenced by environmental factors; typically, the sporangia must be washed to remove nutrients and then flooded with water to induce the cleavage and release of zoospores. Most members of the Oomycota have remained essentially aquatic, producing zoospores from sporangia that remain attached to the hyphae; but some of the plant-pathogenic *Phytophthora* species and the related downy mildew fungi have detachable, wind-dispersed sporangia. In the case of *Phytophthora infestans* (potato blight; Fig. 5.17) and *P. erythroseptica* (pink rot of potatoes) these wind-dispersed sporangia will undergo cleavage and release their zoospores when incubated in water at a temperature of 12° or lower, but at higher temperatures (around 20°) the detached sporangia exhibit "direct" germination by producing a hypha.

This temperature-regulated development seems to be functionally significant. In Britain and other cool regions, **zoospores** are thought to be the main infect-

ive agents of *P. infestans* because the sporangia are produced, dispersed and land on potato leaves early in the growing season when the cool, wet conditions would favor zoospore production. But "direct" germination of sporangia (by hyphal outgrowth) might be more important for infection of the tubers later in the growing season when the temperatures are warmer. Some of the downy mildew pathogens such as *Pseudoperonospora humuli* on hops and *Plasmopara viticola* on grapevine typically produce zoospores for infection through the host stomata. However, some other downy mildew pathogens (*Bremia lactucae* on lettuce, *Peronospora parasitica* on cruciferous hosts) have sporangia that usually or always germinate by hyphae, which then invade through the host epidermal walls.

Several Oomycota are sensitive to the antibacterial agent streptomycin. At high concentrations this compound is toxic, but at sublethal concentrations it interferes with zoospore cleavage, causing the sporangium contents to be released as a multinucleate mass with several flagella and incapable of coordinated swimming. For this reason, streptomycin was used at one time to control *Pseudoperonospora humuli* (hop downy mildew), although now it has been replaced by conventional fungicides (Chapter 17). The mode of action of streptomycin on sporangia is unclear, but it does not necessarily involve the inhibition of

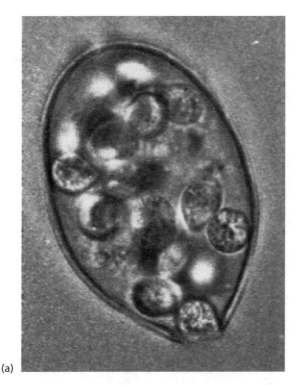
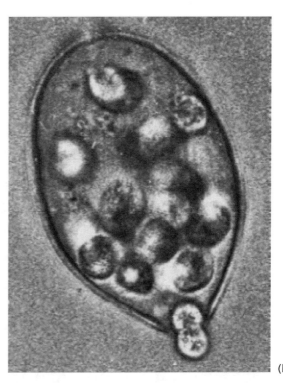

(a)

(b)

Fig. 5.17 (a,b) Detached sporangia of *Phytophthora infestans*, incubated at 12°C. The sporangial contents cleave to produce zoospores, then the apical papillum of the sporangium breaks down and the zoospores are released by squeezing through the narrow opening.

protein synthesis on 70S ribosomes, as occurs in bacteria. Griffin & Coley-Smith (1975) investigated this by adding radiolabeled streptomycin to *P. humuli*. Up to 95% of the label remained on or near the cell surface and could not be removed with water, but it was readily displaced by calcium ions. Consistent with this, streptomycin acts like a divalent cation in solution, so it might interfere with calcium-mediated processes by competing for calcium-binding sites on or near the cell surface.

Conidia

Conidia are formed in a variety of ways (see Fig. 2.12; Fig. 5.18) but always "externally" on a hypha or a conidiophore rather than by cytoplasmic cleavage within a sporangium. For example, conidia can be produced by the swelling of a hyphal tip followed by septation (*Thermomyces lanuginosus*; Fig. 5.18), by a sequential budding process (the "monilinia" stage of *Sclerotinia fructingena*; Fig. 5.18), by hyphal fragmentation, or by successive extrusion from flask-shaped cells (phialides) leading to chains of spores (*Penicillium expansum*, Fig.

5.18). In a few fungi the conidia develop on or in more complex structures such as a coremium (an aggregated mass of conidiophores; Fig. 5.18) or within a flask-shaped pycnidium, or on a pad of tissue (an acervulus – for example, *Colletotrichum musae*; Fig. 5.4). These developmental (ontogenic) patterns have been studied intensively, at least partly in an attempt to find natural relationships and thus a natural approach to the classification of the mitosporic fungi. Details can be found in Cole & Samson (1979).

Although the patterns of conidium development are diverse, a basic distinction can be made between **blastic** conidia which are formed by a budding or swelling process and then become separated from the parent cell, and **thallic** conidia which are formed essentially by a fragmentation process.

Neurospora crassa can be used as an example of blastic conidial development. This fungus sporulates by producing aerial hyphae (conidiophores) that grow for some distance away from the substrate then swell at their tips. The tip of the swelling produces a broad, bud-like outgrowth which swells and repeats this process, so that a branched chain of **proconidia** is formed, resembling the *Monilinia* stage of *Sclerotinia fructigena*

(a)   (b)

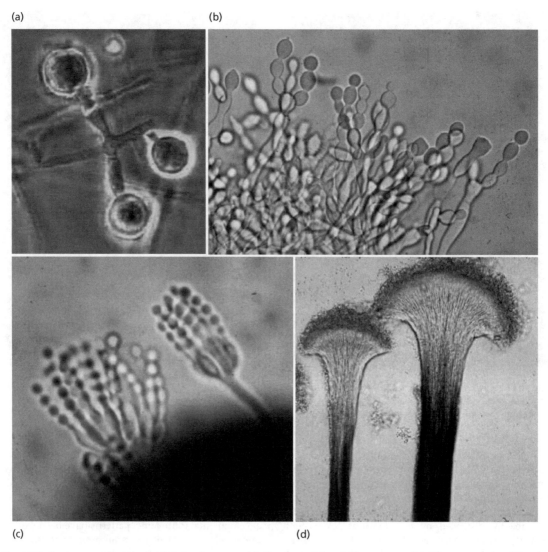

(c)   (d)

Fig. 5.18 Some examples of conidial development. (a) Single conidia of *Thermomyces lanuginosus*, produced at the ends of short hyphal branches. (b) Branched chains of conidia of *Monilinia* (the asexual stage of *Sclerotinia fructigena*) produced by a repeating budding-like process before the spores separate by the development of septa. (c) *Penicillium expansum* conidiophores with phialides that produce chains of spores. (d) Coremia of *Ophiostoma ulmi*; the conidiophores are aggregated to form stalk-like structures that produce masses of spores at their tips (but most of the spores have been dislodged during preparation of this specimen).

in Fig. 5.18. The proconidia become conidia when septa develop to separate them, starting at the base of the chain. At this stage the spores of *Neurospora* develop their characteristic pink color. Each conidium of *Neurospora* contains several nuclei because it has developed by the budding of a multinucleate hyphal (conidiophore) tip. However, in some other types of blastic development the conidia are characteristically uninucleate. For example, in *Penicillium*, *Aspergillus*, and

Trichoderma the conidiophore produces flask-shaped phialides (Fig. 2.33) each of which is uninucleate. The phialide extrudes a spore from its tip, and during this process the nucleus divides so that one daughter nucleus enters the developing spore while the other nucleus remains in the phialide to repeat this process. Essentially the same pattern occurs in *Fusarium* species (Fig. 9.5) which produce a succession of spores from the phialides. But in this case the single nucleus that enters the spore often undergoes repeated divisions, while the spore elongates into a banana shape; the resulting macroconidia of *Fusarium* have several septa and one nucleus in each spore compartment. These different sporulation strategies have important consequences in fungal genetics, which are discussed in Chapter 9.

Dipodascus geotrichum (Fig. 2.33) is a good example of thallic conidial development. In this case a hyphal branch grows to some length, then stops and develops multiple septa which separate it into short compartments. The septal pores are then plugged and the middle zone of each septum is enzymatically degraded to separate the spores.

Regulation and control of conidiation

Asexual sporulation occurs during normal colony growth, but in zones behind the extending colony margin or in the aerial environment rather than on the substrate. The factors that control conidial development are difficult to study in these conditions because development is not synchronized over the whole colony. Ng *et al.* (1973) overcame this problem by growing *Aspergillus niger* in a chemostat so that the hyphae grew synchronously, then the culture conditions could be adjusted to trigger the developmental stages. Conidiophores were found to be produced only in nitrogen-limited growth conditions but when the medium was carbon-rich. Apparently, this is the trigger that switches this fungus from vegetative growth to sporulation. However, the conidiophores did not develop further unless the medium was changed to contain nitrogen and a TCA cycle intermediate such as citrate. Then the tips of the conidiophores swelled into vesicles (large swollen heads – see Fig. 2.33) which produced phialides. The production of conidia from the phialides required yet another change – to a

medium that contained both nitrogen and glucose. On agar plates this whole developmental sequence occurs in a zone of about 1–2 mm diameter, located a few millimeters behind the colony margin. Presumably, there is a succession of physiological changes in the hyphae that co-ordinates the developmental sequence, but only by growing the fungus in chemostat culture was it possible to show that each stage is differentially regulated.

An even more comprehensive study of sporulation has been made with *Emericella nidulans*, where the sequential activation of many sporulation-related genes has been demonstrated (Adams 1995). The details are complex, but essentially the genes are suggested to fall into three categories – those involved in the switch from somatic growth to sporulation, those that regulate the developmental stages of sporulation, and those that govern secondary aspects such as spore color.

Although many fungi such as *Aspergillus* and *Penicillium* can sporulate in darkness, some require a light trigger (Schwerdtfeger & Linden 2003). The most common response is to near-ultraviolet irradiation (NUV; 330–380 nm wavelength) which can induce sporulation after a short (1 hour) exposure if the colony is then kept in darkness. However, a subsequent exposure to blue light can reverse the process because the photoreceptor exists in alternating forms, one responsive to NUV and one responsive to blue light. *Botrytis cinerea* (Fig. 5.20), which causes gray mould disease of strawberries and other soft fruits, behaves in this way. It never becomes wholly committed to sporulation, as Suzuki *et al.* (1977) demonstrated by subjecting the colonies to 1-hour exposures of NUV and blue light in different sequences. As shown in Fig. 5.19, at almost any stage during sporulation an exposure to blue light caused the fungus to form hyphal outgrowths instead of continuing the developmental pathway.

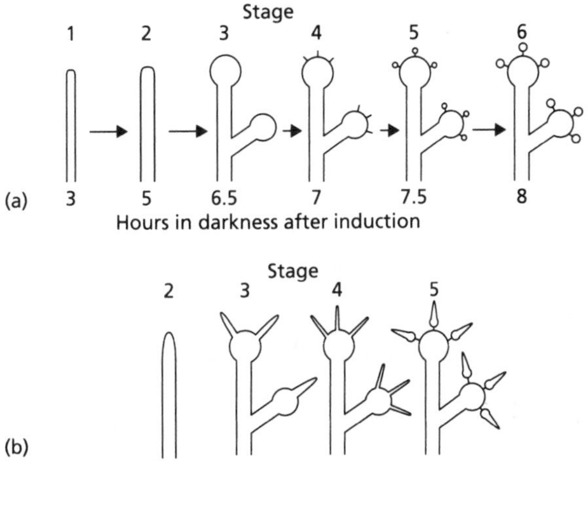

Fig. 5.19 Diagrammatic representation of stages in the development of conidia of *Botrytis cinerea*. (a) After exposure to near-ultraviolet (NUV) irradiation the aerial hyphae are transformed into branched conidiophores which swell at the tips to form small globose vesicles. Conidia are formed on minute projections from these vesicles, so the mature clusters resemble bunches of grapes – see Fig. 5.20. (b) A short exposure to blue light at different times after triggering of development by NUV causes development to be arrested. The fungus then switches back to a hyphal growth form – for example, the swelling of the immature conidia stops and the fungus produces narrow hyphal outgrowths from the immature conidia. (Based on Suzuki *et al.* 1977.)

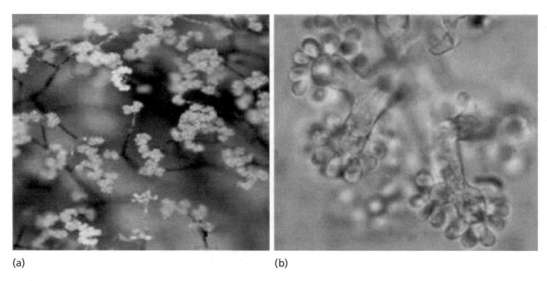

(a)   (b)

Fig. 5.20 *Botrytis cinerea* (sexual stage: *Botryotinia fuckeliana*), which commonly causes "grey mould" of soft fruits. (a) Low-power view of the characteristic branching conidiophores which bear clusters of grey conidia at their tips. (b) Close-up view showing the bulbous tips of the conidiophore branches, bearing immature conidia.

Roles of hydrophobins in fungal differentiation

Hydrophobins are a class of small secreted proteins that were discovered relatively recently and that seem to be unique to fungi. These proteins were discovered by chance, when Wessels and his colleagues were investigating the transcriptome (messenger RNA profiles) of monokaryotic and dikaryotic strains of a small bracket-producing fungus, *Schizophyllum commune*. Among the most abundantly expressed genes were a group that encoded hydrophobic proteins. Because of their strongly hydrophobic nature, these proteins cannot be detected by conventional protein purification. Subsequent studies have revealed similar proteins in a wide range of fungi and have shown that they have important roles in fungal differentiation.

All the hydrophobins studied to date have similar structure and properties. They consist of about 100 amino acids, including eight cysteine residues that occur in a specific pattern so that the protein can fold to produce a highly hydrophobic domain. The hydrophobins are soluble in water, but at an interface with air they self-assemble into a film with a hydrophobic face and a hydrophilic face. As shown in Fig. 5.21, this is thought to happen when a hypha forms an aerial branch so that the aerial portion is coated by a hydrophobic film. The potential significance for differentiation is that hyphae with hydrophobic surfaces might interact with one another in specific ways, as discussed later. For conidial fungi the most obvious significance is that hydrophobins could affect the surface properties and thus the functions of the spores (Chapter 10). The conidia of *Emericella nidulans*,

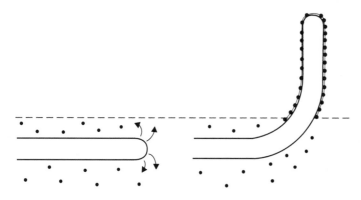

Fig. 5.21 Release of hydrophobin proteins from hyphae of *Schizophyllum commune* submerged in a culture medium (left), and spontaneous polymerization of hydrophobins into water-repellent films (right) when the hyphae emerge into an aerial environment. (Based on a diagram in Wessels 1996.)

A. fumigatus, and *N. crassa* are seen to have a pattern of hydrophobic rodlets on their surface, but the rodlets are absent when the hydrophobin genes are inactivated by targeted gene disruption. Identical rodlets are formed when solutions of pure hydrophobins are allowed to dry. Talbot (2001) recently reviewed the structure and roles of hydrophobins. We return briefly to this subject later in this chapter.

Sexual development

Sexual reproduction in all organisms involves three fundamental events: the fusion of two haploid cells (**plasmogamy**) so that their nuclei are in a common cytoplasm; nuclear fusion (**karyogamy**) to form a diploid; and **meiosis** to produce recombinant haploid nuclei. Depending on the fungus, these events can occur in close succession or separated in time; also, they occur at different stages of the life cycle according to whether the fungus is normally haploid or diploid. These points were outlined in Chapter 2 (see Figs 2.1, 2.8, 2.14, 2.15, 2.16).

Two further points must be made. First, some fungi are **homothallic** (self-fertile) but many are **heterothallic** (outcrossing). in which case sexual reproduction is governed by **mating-type** (compatibility) genes. Often there are two mating types, governed by a single gene locus (**bipolar compatibility**); in these cases the alternative genes are not an allelic pair but usually are quite different from one another so they are called **idiomorphs**. Some Basidiomycota have two mating-type loci (**tetrapolar compatibility**) with multiple idiomorphs at each locus. In these cases a successful mating occurs between two fungal strains that differ from one another at each gene locus. In most fungi the mating-type genes are regulatory genes, producing protein products that bind to DNA and control the expression of several other genes.

The second point is that the sexual spores of many fungi function as dormant spores (e.g. zygospores, oospores, ascospores, and some basidiospores). So sexual reproduction serves an important role in survival, and the sexual spores are typically produced at the onset of unfavorable conditions for growth. Fungi that do not undergo regular sexual reproduction sometimes produce alternative survival structures such as chlamydospores, sclerotia, or melanized hyphae. Other fungi adopt different strategies. For example, *Pythium oligandrum* produces sexual spores by parthenogenesis, and *Saccharomyces cerevisiae* and a few other Ascomycota undergo regular **mating-type switching** to ensure that there will always be a mixture of the two mating types (**a** or **α**) in a population. In this case, every time that an **a** cell buds, the parent cell switches to **α** while the daughter cell remains **a**. In fact, wild populations of *S. cerevisiae* are always diploid because the **a** and **α** cells fuse and undergo karyogamy, then the diploid cell buds to form further diploids which can respond rapidly to unfavorable conditions by undergoing meiosis and producing ascospores. For this reason, all the laboratory strains used by yeast geneticists have been mutated so that they do not undergo mating-type switching, and this enables the sexual crosses to be controlled experimentally.

Against this background, the rest of this chapter will focus on two topics: the roles of the mating-type genes, especially in hormonal regulation of mating (reviewed by Gooday & Adams 1993), and the development of fruiting bodies of Basidiomycota because these are the most advanced differentiated structures in the fungal kingdom.

Mating and hormonal control

Chytridiomycota

Most of the information on the control of sexual reproduction in Chytridiomycota has come from studies on *Allomyces* spp. (see Fig. 2.1 for the life cycle). These fungi are homothallic (self-fertile), but they produce motile male and female gametes (sex cells) from different gametangia. The female gametes are larger, hyaline and they release a pheromone, **sirenin** (Fig. 5.22), to attract the male gametes. The male gametes are small, and orange colored due to the presence of a carotenoid pigment. Sirenin is a powerful attractant, active at concentrations as low as 10^{-10} M, but optimally at 10^{-6} M. Compounds that attract cells at these concentrations can be assumed to alter the swimming pattern by binding to a surface-located receptor (Chapter 10). Although there is no information on this possible receptor, the male gametes are known rapidly to inactivate sirenin, which could aid their movement up a concentration gradient. It is interesting that both sirenin (from the female gametes) and the carotenoid pigment of the male gametes are produced from the same precursor isoprene units $[CH_2=C(CH_3)-CH=CH_2]$ (see Fig. 7.14). So this is an example where cells of the same genetic make-up show different biochemical properties because they have been produced in different gametangia, separated by a complete cross wall.

Oomycota

The Oomycota can be homothallic (e.g. most *Pythium* species) or heterothallic with two mating types. But in all cases a single colony produces both the "male" and "female" sex organs (antheridia and oogonia – see Fig. 2.34), so the mating-type genes govern compatibility,

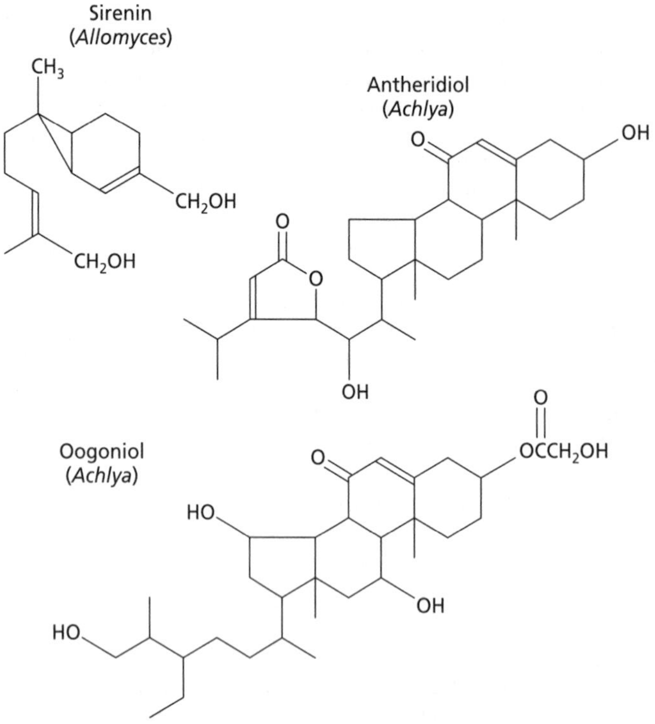

Sirenin
(*Allomyces*)

Antheridiol
(*Achlya*)

Oogoniol
(*Achlya*)

Fig. 5.22 Three pheromones that regulate sexual reproduction in zoosporic organisms. Sirenin is released by female gametes of *Allomyces* spp. Antheridiol and oogoniols regulate sexual attraction and mating in *Achlya* spp. (Oomycota) and possibly in other Oomycota.

not the development of the sex organs themselves. In a few cases, notably in *Achlya*, a single colony can show "relative sexuality" – it will behave as a male (fertilizing the oogonium from its antheridium) or "female" depending on the strain with which it is paired.

The hormonal control of mating has been studied intensively in *Achlya*. The hormones are steroids (Fig. 5.22) derived from the isoprenoid pathway, like sirenin discussed above. They were discovered by growing "strong female" and "strong male" strains in separate dishes of water, then passing the water over colonies of the opposite strain. The hyphae of the female produce the hormone **antheridiol**, which causes the male strain to increase its rate of cellulase enzyme production and to form many hyphal branches (the antheridial branches – see Fig. 2.34). Once it has been triggered by antheridiol, the male strain produces other steroid hormones, termed **oogoniols**, which trigger the development of oogonia (sex organs) in the female strain. In normal conditions the antheridial hyphae then grow towards the oogonia, clasp onto them, and produce fertilization tubes to transfer the "male" nuclei into the oogonium (see Fig. 2.34). Similar hormonal systems might occur in other Oomycota such as *Pythium* and *Phytophthora* spp., but they have not been characterized. A notable feature of *Pythium* and *Phytophthora* is that they cannot synthesize sterols

from nonsterol precursors, and often do not need sterols for somatic growth, but they always require trace amounts of sterols for both sexual reproduction and asexual reproduction. They would be able to obtain them from a plant host or other natural sources.

Zygomycota

The Zygomycota can be homothallic, or heterothallic with two mating types (termed "plus" and "minus"). One of the roles of the mating-type genes is to regulate the production of hormone precursors (**prohormones**) from β-carotene, which is another product of the isoprenoid pathway already mentioned. This hormonal system is shared by many members of the subgroup Mucorales because, for example, a plus strain of *Mucor* can elicit a sexual response from a minus strain of *Rhizopus*, even though the development of a hybrid is blocked at a later stage due to incompatibility. As shown in Fig. 5.23, the prohormones of the plus and minus strains are similar molecules but they differ slightly, owing to the different enzymes of the two strains. The prohormones are volatile compounds which diffuse towards the opposite mating type; then they are absorbed and converted to active hormones, the **trisporic acids**, by the complementary enzymes of the opposite strain. Initially, only low levels of

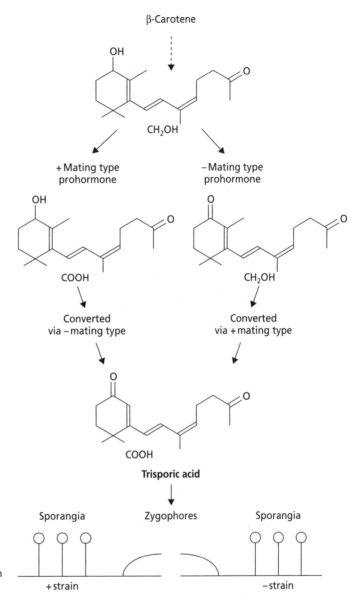

Fig. 5.23 The trisporic acid hormonal system of Zygomycota. See text for details.

the prohormones are produced by each strain, but trisporic acids cause gene derepression so that increasing levels of prohormones are produced. Trisporic acids produced by this mutual escalation cause the production of **zygophores** (aerial sexual branches) which grow towards one another and lead to sexual fusion (see Fig. 2.8).

Ascomycota

The Ascomycota usually have two mating types, termed **A** and **a**, but **a** and **α** in *Saccharomyces*. These mating-type genes are regulatory elements, controlling the activities of other genes, but only the system in yeasts has been well characterized. One of the chromosomes (III) of the yeast *Saccharomyces cerevisiae* has a *MAT* gene locus, which is flanked by two other loci termed *MATa* and *MATα*. Haploid cells will normally behave as **α** mating type, but can change to **a** mating type when the information at the *MATa* locus is copied and transferred into the *MAT* locus (the expression of the *MATα* locus is then shut down). This is a natural transposition event, which causes the switching of mating type every time a cell produces a bud.

The *MAT* locus is a regulatory locus, governing the expression of several genes. These include:

- The genes responsible for producing hormones termed **a-factor** and **α-factor**, consisting of 12 and 13 amino acids respectively:

  - **a-Factor**: NH$_2$-Trp-His-Trp-Leu-Gln-Leu-Lys-Pro-Gly-Gln-Pro-Met-Tyr-COOH;
  - **α-Factor**: NH$_2$-Tyr-Ile-Ile-Lys-Gly-Val-Phe-Trp-Asp-Pro-Ala-Cys(S-farnesyl)-COOCH$_3$.

- The genes that code for hormone receptors on the cell surface.
- The genes that code for cell surface agglutinins.

The **α** strains produce **α**-factor constitutively and it diffuses to **a** cells where it is recognized by a specific receptor. This receptor binding causes the growth of **a** cells to be arrested at "start" during G1 of the cell cycle – the only stage at which they are competent to mate (Chapter 4). The **a** strain then produces **a**-factor which diffuses to the **α** strain where it is, again, recognized by a specific receptor. Receptor binding in both cases leads to other changes in the cells: they produce short outgrowths which function as conjugation tubes, and the surfaces of these are covered by a strain-specific glycoprotein, so that when an **a** cell and an **α** cell make contact they adhere tightly by their complementary agglutinins. The conjugation tubes then fuse, and the two nuclei fuse to form a diploid cell. At this stage the cell can go on to produce a diploid budding colony or it can undergo meiosis, depending on whether the environmental conditions are suitable for growth. An essentially similar hormone and receptor system to this is found in other yeasts such as *Pichia* and *Hansenula* spp, and in the fission yeast *Schizosaccharomyces pombe*.

The sex pheromones of several mycelial Ascomycota and Basidiomycota have now been characterized. They are **peptides**, such as the α-factor of *S. cerevisiae*, or more often **lipopeptides** such as the a-factor of *S. cerevisiae*. There is considerable variation in the amino acid substitutions of these pheromones, but their basic structures are similar. For example, the jelly fungus *Tremella mesenterica* (Basidiomycota; see Fig. 2.7) produces a pheromone termed **tremerogen**. This is a highly lipophilic molecule with the structure shown below. It closely resembles the **a**-factor of *S. cerevisiae*:

Tremerogen A-10: NH$_2$-Glu-His-Asp-Pro-Ser-Ala-Pro-Gly-Asn-Gly-Tyr-Cys(S-farnesyl)-COOCH$_3$

The tetrapolar compatibility system in Basidiomycota

Most Basidiomycota are heterothallic, with mating-type genes at either one locus or two loci, termed A and B. There are multiple idiomorphs at each locus,

Table 5.2 Roles of mating-type genes in basidiomycota.

Pairing of strains with	Events observed
Different A, different B idiomorphs (**dikaryon**)	1 Septal dissolution 2 Nuclear migration 3 Clamp branches arise and fuse with hypha
Common A, different B idiomorphs	1 Septa dissolve 2 Nuclei migrate
Common B, different A idiomorphs	1 Septa remain intact 2 No nuclear migration 3 Clamp branches arise but do not fuse
Common A, common B idiomorphs	1 No septal dissolution 2 No nuclear migration 3 No clamp connections

and a successful mating will occur between any two strains that differ from one another at each locus. This greatly increases the chances of finding a mate, compared with fungi that have only two idiomorphs.

The basic mating behavior of Basidiomycota was outlined in Chapter 2 (see Fig. 2.18). Strains derived from haploid basidiospores are monokaryons and they can fuse with other compatible monokaryons to form a dikaryon. All subsequent growth involves the synchronous division of the two nuclei in each hyphal compartment and their regular distribution as nuclear pairs throughout the mycelium. In several members of the group this regular arrangement is aided by the production of clamp connections (see Fig. 2.20). Eventually, the dikaryotic colony will produce a fruitbody, and nuclear fusion and meiosis occur in the basidia.

By experimentally pairing strains with the same A or the same B ideomorph, it has been possible to deduce the regulatory roles of the A and B loci (Casselton *et al.* 1995). As summarized in Table 5.2, pairings of fully compatible strains (with different A and different B) lead to reciprocal exchange of nuclei, because the dolipore septa which normally prevent nuclear migration are digested, and clamp branches are formed at each septum to fuse with the compartment behind. In pairings of strains with the same A and same B (common A, common B) there is no septal breakdown, no nuclear migration, no dikaryotization, and no clamp connections. Pairings of strains with different A but common B loci show nuclear pairing, synchronous division of the nuclei and formation of clamp branches; but the dolipore septa do not break down, and the clamp branches do not fuse with the parent hypha. Pairings of strains with different B but common A lead to septal dissolution but none of the other events. Thus, the **A locus** (common B pairings) controls the pairing and synchronous division of nuclei and also the formation of clamp branches, whereas the **B locus** controls septal dissolution and the fusion of clamp branches. Septal dissolution coincides with a marked increase in the activity of **β-glucanase** in the hyphae, indicating that the B locus controls the derepression of glucanase genes.

Development of fruitbodies

The toadstools, brackets and other fruitbodies of Basidiomycota are the largest and most complex differentiated structures in the fungal kingdom. Their development is correspondingly complex and still only poorly understood. Here we consider one example where a start has been made to dissect this process at the biochemical and molecular level, and we end with a discussion of commercial mushroom production because of its economic importance.

Schizophyllum commune

Schizophyllum commune is ideally suited for laboratory studies because it grows readily in agar culture and produces its small (about 1–2 cm) fan-shaped fruitbodies in response to light. Actually, this trigger leads only to the development of fruitbody primordia – compact clusters of hyphae which are overarched by other hyphae. Further development from the primordia occurs when carbon nutrients are depleted from the medium, and is then fuelled by carbon reserves within the mycelium. Early in this process the mycelial storage compounds such as glycogen are converted to sugars, which are translocated to the developing primordia. Then, as the sugar levels in the hyphae decline, the hyphal walls begin to break down and the breakdown products are translocated to the primordia. The wall glucans seem to provide the major source of sugars, because fruitbody development is associated with a marked rise in glucanase activity in the mycelia. We have already seen that synthesis of this enzyme is derepressed by the B mating-type locus, but it is still subject to catabolite repression by sugars; so its generalized activity in the hyphae, as opposed to its localized activity in degrading septa, depends on depletion of the mycelial sugar reserves. The breakdown of hyphal walls to recycle nutrients for differentiation is, in fact, quite common in fungi. An example was seen earlier in the production of sclerotia (Fig. 5.10). The breakdown of wall glucans also fuels the developing ascocarps of *Emericella nidulans*.

Wessels and his colleagues (see Wessels 1992) identified several differentiation-associated genes in *S. commune*. In order to do this, they crossed and re-peatedly back-crossed strains to generate monokaryons that were essentially isogenic except for the mating-type locus. Then the monokaryons, and the dikaryons synthesized from them, were compared for their production of polypeptides and messenger RNAs when grown in different conditions. Any differentiation-specific mRNAs in the dikaryon were confirmed by making complementary DNA (cDNA) and testing this for lack of hybridization to the mRNA from the monokaryons. All these comparisons were made in two sets of conditions: (i) for 2-day-old colonies, when the monokaryons and dikaryons were growing as mycelia with similar colony morphology, and (ii) for 4-day-old colonies grown in light, when the monokaryon had produced copious aerial hyphae but the dikaryon had produced numerous small fruitbodies.

The following principal findings emerged from this work.

- For the 2-day colonies, 20 or so proteins were found only in the monokaryon and 20 or so only in the dikaryon. Yet there was no detectable difference in the bands of mRNA, suggesting that the same mRNAs are transcribed but their products undergo different post-translational modification in monokaryons and dikaryons.
- For the 4-day colonies, eight proteins were found only in the monokaryon, and 37 only in the dikaryon. Some of these 37 occurred only in the fruitbodies; others were found in both the fruitbodies and the mycelium of the dikaryon. These proteins included some that were secreted into the growth medium.
- For the 4-day colonies, about 30 unique mRNAs were found only in the dikaryon, whereas no unique mRNA was found in the monokaryon. cDNA was used as a probe to assess the levels of the "fruiting-associated mRNAs" during development. They were scarce in young vegetative colonies of both strains, and they remained scarce in the monokaryon, but they increased in the dikaryon when this began to fruit.
- Some of the secreted proteins were the cysteine-rich **hydrophobins**, mentioned earlier; in fact, the hydrophobins were first discovered in this work on *S. commune*. The gene (*SC3*) for one of these hydrophobins was expressed by both the monokaryon and dikaryon during the emergence of aerial hyphae; this hydrophobin is now known to cover the aerial hyphae and the hydrophobic hyphae on the fruitbody surface, but the gene is not expressed by hyphae that make up the main fruitbody tissue. Three other hydrophobin genes (*SC1*, *SC4*, and *SC6*) were highly expressed only in the dikaryon and especially in the developing fruitbody tissues (Wessels 1996). Hydrophobin genes have been shown to contain a putative signal peptide sequence at the N-terminus, a feature associated with secretion from the hyphal tips. It is suggested that the specific properties of different hydrophobins might affect

the surface interactions of hyphae, contributing to features such as the hydrophobicity of cells lining the internal air spaces, to prevent water-soaking of the fruitbody tissues.

Commercial mushrooms: the exploitation of differentiation

Mushroom production is a substantial industry. The major cultivated "white button" mushroom *Agaricus bisporus* (or *A. brunnescens*) had an annual global retail value of more than $US 10 billion in 1999. But this species accounts for only about 40% of the total production of cultivated mushrooms. Other important species include the oyster mushroom *Pleurotus ostreatus* (about 20% of total production), the Shiitake mushroom *Lentinula edodes* (about 10%), and *Volvariella volvacea* (5% or more).

The commercial production process for *A. bisporus* is described by Flegg (1985). A mixture of composted straw and animal dung is pasteurized and placed in wooden trays, then inoculated with a commercially supplied "spawn" consisting of sterilized cereal grains permeated with hyphae of *A. bisporus*. The spawn is allowed to "run" for 10–14 days so that the fungus thoroughly colonizes the compost. Then a thin casing layer of pasteurized, moist peat and chalk is added to the compost surface. Over the next 18–21 days the fungus colonizes this casing layer by producing mycelial cords, and fruitbodies are produced on these cords. The cropping of fruitbodies is done over a 30- to 35-day period, because the fungus produces "flushes" of fruitbodies at 7- to 10-day intervals.

In terms of differentiation there are several interesting features of this system. The casing layer is essential for a high fruitbody yield, and part of its role involves the activities of pseudomonads which are stimulated to grow in the casing layer by volatile metabolites, including ethanol, released by *A. bisporus*. In experimental conditions the role of the casing layer can be replaced by using activated charcoal, suggesting that the fungus produces autoinhibitors of fruiting which are removed by pseudomonads in the normal mushroom-production process. The casing layer also provides a non-nutritive environment in which the fungus produces mycelial cords. As we saw earlier, these translocating organs develop in nutrient-poor conditions and they would be necessary for channeling large amounts of nutrients to the developing fruitbodies. The regular periodicity of fruiting is also of interest. In commercial conditions the crop must be harvested regularly at the "button" stage to achieve this, and any delay in harvesting until the fruitbodies have opened will cause a corresponding delay in the next flush. Yet, most of the fruitbody primordia are already present at the time of the first flush, and mushrooms at the button

stage have already received all or nearly all of their nutrients from the mycelium. So, the effect of delayed picking must be to delay the release of other primordia for further development. The mechanism of this is not fully known, but the cellulase activity of the mycelium in the compost increases markedly as each flush of fruitbodies develops. It seems that the expression of cellulase genes is closely linked to fruiting, presumably when the mycelial sugar reserves are depleted (Chapter 6). Removal of the existing fruitbodies might act as a signal for a further round of mycelial activity, providing extra nutrients for the next batch of fruitbodies.

A final point of interest concerns the mechanism of fruitbody expansion from the button stage to the fully expanded mushroom. This must involve the differential expansion of tissues, and studies on this have been done mainly with a much simpler experimental system – the expansion of stipes (stalks) of *Coprinus* spp. In field conditions these "ink-cap" toadstools elongate very rapidly from the button stage to the mature stage. In laboratory conditions the initially short stipes can be severed at both ends and incubated in humid conditions, when they will elongate to more than seven times their original length within 24 hours. Most of this increase occurs by cell expansion rather than cell division, and it correlates with an increase in chitinase activity, presumably for loosening of the existing hyphal walls, and an increase in chitin synthase activity for new wall synthesis. But, unlike the apical growth of normal hyphae, new wall material is inserted along the length of the existing hyphae in the stipes. So, the wall extension is mainly by **intercalary growth**. This form of growth is also found in other rapidly extending structures, such as the sporangiophores of *Phycomyces* (Zygomycota) on herbivore dung (Chapter 11). The rapid expansion of mushrooms from the button stage depends on the intake of water from the mycelium. This is probably driven by the conversion of storage reserves into osmotically active compounds (Chapter 7). Consistent with this, mushroom fruitbodies contain high levels of mannitol which accounts for 25% or more of the dry weight, compared with only 1.5–4.5% in the mycelia. Horgen & Castle (2002) discuss several molecular approaches for improving mushroom production. Moore *et al.* (1985) and Moore (1998) discuss many aspects of the developmental biology of Basidiomycota, and of fungal morphogenesis in general.

Cited references

Adams, T.H. (1995) Asexual sporulation in higher fungi. In: *The Growing Fungus* (Gow, N.A.R. & Gadd, G.M., eds), pp. 367–382. Chapman & Hall, London.

Allen, E.A., Hazen, B.A., Hoch, H.C., *et al.* (1991) Appressorium formation in response to topographical signals in 27 rust species. *Phytopathology* **81**, 323–331.

Bartnicki-Garcia, S., Bartnicki, D.D. & Gierz, G. (1995) Determinants of fungal cell wall morphology: the vesicle supply center. *Canadian Journal of Botany* **73**, S372–378.

Bartnicki-Garcia, S. & Gierz, G. (1993) Mathematical analysis of the cellular basis of dimorphism. In: *Dimorphic Fungi in Biology and Medicine* (Van den Bossche, H., Odds, F.C. & Kerridge, D., eds), pp. 133–144. Plenum Press, New York.

Bracker, C.E. (1968) The ultrastructure and development of sporangia in *Gilbertella persicaria*. *Mycologia* **60**, 1016–1067.

Casselton, L.A., Asante-Owusu, R.N., Banham, A.H., *et al.* (1995) Mating type control of sexual development in *Coprinus cinereus*. *Canadian Journal of Botany* **73**, S266–S272.

Christias, C. & Lockwood, J.L. (1973) Conservation of mycelial constituents in four sclerotium-forming fungi in nutrient-deprived conditions. *Phytopathology* **63**, 602–605.

Cole, G.T. & Samson, R.A. (1979) *Patterns of Development in Conidial Fungi*. Pitman, London.

Flegg, P.B. (1985) Biological and technological aspects of commercial mushroom growing. In: *Developmental Biology of Higher Fungi* (Moore, D., Casselton, L.A., Wood, D.A. & Frankland, J.C., eds), pp. 529–539. Cambridge University Press, Cambridge.

Fox, F.M. (1986) Ultrastructure and infectivity of sclerotia of the ectomycorrhizal fungus *Paxillus involutus* on birch (*Betula* spp.) *Transactions of the British Mycological Society* **87**, 627–631.

Fox, F.M. (1987) Ultrastructure of mycelial strands of *Leccinum scabrum*, ectomycorrhizal on birch (*Betula* spp.). *Transactions of the British Mycological Society* **89**, 551–560.

Gooday, G.W. & Adams, D.J. (1993) Sex hormones and fungi. *Advances in Microbial Physiology* **34**, 69–145.

Gow, N.A.R. (1994) Yeast-hyphal dimorphism. In: *The Growing Fungus* (Gow, N.A.R. & Gadd, G.M., eds), pp. 403–422. Chapman & Hall, London.

Griffin, M.J. & Coley-Smith, J.R. (1975) Uptake of streptomycin by sporangia of *Pseudoperonospora humuli* and the inhibition of uptake by divalent metal cations. *Transactions of the British Mycological Society* **65**, 265–278.

Harold, F.M. (1990) To shape a cell: an inquiry into the causes of morphogenesis of microorganisms. *Microbiological Reviews* **54**, 381–431.

Heath, I.B. & Harold, R.L. (1992) Actin has multiple roles in the formation and architecture of zoospores of the oomycetes, *Saprolegnia ferax* and *Achlya bisexualis*. *Journal of Cell Science* **102**, 611–627.

Hoch, H.C., Staples, R.C., Whitehead, B., Comeau, J. & Wolfe, E.D. (1987) Signaling for growth orientation and differentiation by surface topography in *Uromyces*. *Science* **235**, 1659–1662.

Horgen, P.A. & Castle, A. (2002) Application and potential of molecular approaches to mushrooms. In: *The Mycota X1. Agricultural Applications* (Kempken, F., ed.), pp. 3–17. Springer-Verlag, Berlin.

Howard, R.J., Ferrari, M.A., Roach, D.H. & Money, N.P. (1991) Penetration of hard surfaces by a fungus employing enormous turgor pressures. *Proceedings of the National Academy of Science, USA* **88**, 11281–11284.

Hyde, G.J., Lancelle, S., Hepler, P.K. & Hardham, A.R. (1991) Freeze substitution reveals a new model for sporangial cleavage in *Phytophthora*, a result with implications for cytokinesis in other eukaryotes. *Journal of Cell Science* **100**, 735–746.

Moore, D. (1998) *Fungal Morphogenesis*. Cambridge University Press, Cambridge.

Moore, D., Casselton, L.A., Wood, D.A. & Frankland, J.C. (1985) *Developmental Biology of Higher Fungi*. Cambridge University Press, Cambridge.

Muthukumar, G., Nickerson, A.W. & Nickerson, K.W. (1987) Calmodulin levels in yeasts and mycelial fungi. *FEMS Microbiology Letters* **41**, 253–255.

Ng, A.M.L., Smith, J.E. & McIntosh, A.F. (1973) Conidiation of *Aspergillus niger* in continuous culture. *Archives of Microbiology* **88**, 119–126.

Orlowski, M. (1995) Gene expression in *Mucor* dimorphism. *Canadian Journal of Botany* **73**, S326–S334.

Read, N.D., Kellock, L.J., Knight, H. & Trewavas, A.J. (1992) Contact sensing during infection by fungal pathogens. In: *Perspectives in Plant Cell Recognition* (Callow, J.A. & Green, J.R., eds), pp. 137–172. Cambridge University Press, Cambridge.

Schwerdtfeger, C. & Linden, H. (2003) VIVID is a protein and serves as a fungal blue light photoreceptor for photoadaptation. *EMBO Journal* **22**, 4846–4855.

Smith, A.M. & Griffin, D.M. (1971) Oxygen and the ecology of *Armillariella elegans* Heim. *Australian Journal of Biological Sciences* **24**, 231–262.

Suzuki, Y., Kumagai, T. & Oka, Y. (1977) Locus of blue and near ultraviolet reversible photoreaction in the stages of conidial development in *Botrytis cinerea*. *Journal of General Microbiology* **98**, 199–204.

Talbot, N.J. (2001) Fungal hydrophobins. In: *The Mycota VIII. Biology of the Fungal Cell* (Howard, R.J. & Gow, N.A.R., eds), pp. 145–159. Springer-Verlag, Berlin.

Wessels, J.G.H. (1992) Gene expression during fruiting in *Schizophyllum commune*. *Mycological Research* **96**, 609–620.

Wessels, J.G.H. (1996) Fungal hydrophobins: proteins that function at an interface. *Trends in Plant Science* **1**, 9–15.

Chapter 6

Fungal nutrition

This chapter is divided into the following major sections:

- the basic nutrient requirements of fungi
- carbon and energy sources of fungi
- fungal adaptations for nutrient capture
- the breakdown of cellulose: a case study of extra-cellular enzymes
- mineral nutrient requirements: nitrogen, phosphorus, and iron
- efficiency of substrate utilization
- fungi that cannot be cultured

Fungi have quite simple nutritional requirements. They need a source of organic nutrients to supply their energy and to supply carbon skeletons for cellular synthesis. But, given a simple energy source such as glucose, many fungi can synthesize all their other cellular components from inorganic sources – ammonium or nitrate ions, phosphate ions and trace levels of other minerals such as calcium, potassium, magnesium, and iron. Fungi that normally grow in a host environment or in other nutrient-rich substrates might require additional components, but still the nutrient requirements of most fungi are quite simple.

Having said this, fungi need to capture nutrients from their surroundings. The cell wall prevents fungi from engulfing food particles, so fungi absorb simple, soluble nutrients through the wall and plasma membrane. In many cases this is achieved by releasing enzymes to degrade complex polymers and then absorbing the nutrients released by these **depolymerase** enzymes. Fungi produce a huge range of these enzymes, to degrade different types of polymer. In fact, there is hardly any naturally occurring organic compound that cannot be utilized as a nutrient source by one fungus or another.

In this chapter we consider the many adaptations that fungi have evolved for nutrient capture.

The nutrient requirements of fungi

For routine laboratory culture, most fungi are grown on media with natural components, such as **potato-dextrose agar** (extract of boiled potatoes, with 2% w/v glucose), **malt extract agar** (usually containing 2% malt extract), or **cornmeal (or oatmeal) agar** – a culture medium often used to promote sexual development. Fungal growth media are usually slightly acidic (pH 5–6) and rich in carbohydrates. This contrasts with bacterial culture media such as "nutrient agar" (beef extract and peptone) which are usually neutral or alkaline (pH 7–8) and rich in organic nitrogen. This difference reflects the higher gross protein content of bacteria compared with fungi. Other types of agar are used to culture specific fungi or to promote specific developmental stages.

However, in order to determine the **minimum nutrient requirements**, a fungus must be grown in chemically defined liquid culture, because agar contains impurities. Table 6.1 shows a typical minimal medium, containing only mineral salts and glucose. Several common fungi will grow on this medium. The main exceptions are noted below.

- Some fungi need to be supplied with one or more vitamins, the most common requirements being for small amounts of thiamine, biotin, or both.

Table 6.1 A chemically defined liquid culture medium for fungi.

Chemical	Quantity
NaNO$_3$ or NH$_4$NO$_3$ or L-asparagine at equivalent nitrogen content	2 g
KH$_2$PO$_4$ (alone or in a buffered mixture with K$_2$HPO$_4$)	1 g
MgSO$_4$	0.5 g
KCl	0.5 g
CaCl$_2$	0.5 g
FeSO$_4$, ZnSO$_4$, CuSO$_4$	0.005–0.01 g each
Sucrose or glucose	20 g
Distilled water	1 liter

Note: Common supplements required by fungi include the vitamins **biotin** (10 µg) or **thiamine** (100 µg).

- Some fungi cannot utilize nitrate or ammonium as their sole nitrogen source; instead they require a source of organic nitrogen. In these cases the requirement usually can be met by supplying a single amino acid such as asparagine. Several Basidiomycota fall in this category.
- Some fungi have more specific individual requirements, which will be found in their natural habitats. For example, some Oomycota (e.g. *Phytophthora infestans*) need to be supplied with sterols for growth; some other Oomycota (e.g. *Leptomitis lacteus*) need sulfur-containing amino acids such as cysteine, because they cannot use inorganic sulfur. Some of the fungi from animal dung need to be supplied with the iron-containing haem group, a component of cytochromes in the respiratory pathway.

It should be emphasized that all these comments refer to the **minimum** requirements for growth, not to the optimal requirements. Also, they relate to somatic (vegetative) growth, and additional nutrients may be required to complete the life cycle. For example, several *Pythium* and *Phytophthora* spp. do not require sterols for hyphal growth but they need them for both sexual and asexual reproduction. Many fungi have extra nutrient requirements for growth in suboptimal environmental conditions. For example, *Saccharomyces cerevisiae* has relatively simple requirements for growth in aerobic culture but needs to be supplied with a wide range of vitamins and other growth factors in anaerobic conditions, because some of the basic metabolic pathways that normally supply these requirements do not operate in anaerobic conditions (Chapter 7).

The carbon and energy sources of fungi

An enormous range of organic compounds can be utilized by one fungus or another. This is illustrated diagrammatically in Fig. 6.1, where nutrients are arranged in **approximate order of structural complexity** from left to right, and in **approximate order of their degree of utilization** (vertical axis).

At one extreme, the simplest organic compound, **methane** (CH$_4$), can be used by a few yeasts. They convert methane to methanol by the enzyme methane monooxygenase. Then this is converted to formaldehyde by the enzyme methanol dehydrogenase. Then formaldehyde combines with ribulose-5-phosphate, and two further reaction steps generate fructose-6-phosphate.

$$CH_4 \rightarrow H_3C\!-\!OH \rightarrow \underset{\underset{O}{\|}}{H\!-\!C\!-\!H} \rightarrow \rightarrow \rightarrow \text{fructose-6-phosphate}$$

methane methanol formaldehyde

Several more yeasts (e.g. *Candida* spp.) and a few mycelial fungi can grow on the longer-chain hydrocarbons (C$_9$ or larger) in petroleum products. The main limitation with these compounds, as with methane, is that they are not miscible with water and so growth is restricted to the water–hydrocarbon interface. Two mycelial fungi are notorious for their growth on the long-chain hydrocarbons in aviation kerosene: *Amorphotheca resinae* (Ascomycota) and *Paecilomyces varioti* (a mitosporic fungus). *A. resinae* has the dubious distinction of having caused at least one major aircrash, by blocking filters and corroding the walls of fuel storage tanks, but this problem is easily avoided (Chapter 17).

Moving further along the spectrum in Fig. 6.1, a larger number of fungi can utilize the common alcohols such as methanol and ethanol. In fact, ethanol is an excellent carbon source for *Candida utilis*, *Emericella nidulans*, and *Armillaria mellea*, and it can even be their preferred carbon substrate. Glycerol and fatty acids will support the growth of several fungi, and can be the preferred substrates for a few fungi such as the common "sewage fungus," *Leptomitus lacteus* (Oomycota). Amino acids also can be utilized, but they contain excess nitrogen in relation to their carbon content, so ammonium is released during their metabolism and this can often lower the pH to growth-inhibitory levels unless the culture medium is strongly buffered.

The vast majority of fungi can utilize glucose and other monosaccharides or disaccharides. Some also can use sugar derivatives such as aminosugars (e.g. glucosamine), sugar acids (e.g. galacturonic acid), or sugar alcohols (e.g. mannitol). However, the utilization

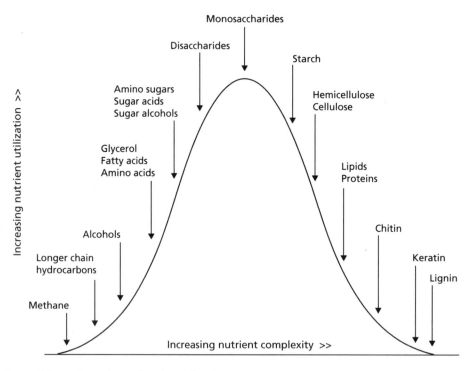

Fig. 6.1 Some of the major carbon substrates of fungi.

of any sugar or sugar derivative depends on the presence of appropriate membrane transport proteins, and these have different substrate-specificities. Fungi usually have a constitutive transport protein for glucose. It transports glucose in preference to other sugars in a mixture, but it has a relatively low binding specificity and so, in the absence of glucose, it will transport some other sugars. The uptake of these sugars leads to the induced synthesis of their specific carrier proteins. This can be demonstrated easily when *Saccharomyces cerevisiae* is grown in a shaken liquid culture medium containing a disaccharide such as lactose as the sole carbon source (Fig. 6.2). The fungus degrades lactose to glucose and galactose by means of a wall-bound enzyme, **β-galactosidase**, then the fungus grows rapidly by taking up the glucose preferentially, while galactose remains in the culture medium. When the glucose is depleted the growth rate slows for about 30 minutes while a galactose carrier protein is synthesized, and then the growth rate increases again, using galactose as the substrate. This type of biphasic growth curve (Fig. 6.2) is termed a **diauxic growth curve**. Sucrose is used in a similar way: it is cleaved to glucose and fructose by the wall-bound enzyme **invertase**, then the glucose is taken up before fructose. However, a few fungi cannot use sucrose as a substrate (e.g. *Rhizopus nigricans*,

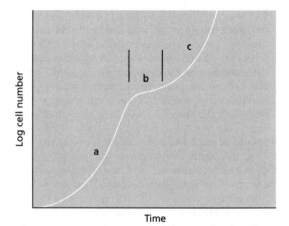

Fig. 6.2 A diauxic growth curve of a fungus grown on the disaccharide lactose: a, initial phase of growth on glucose; b, temporary slowing of growth while a galactose transporter is inserted in the membrane; c, resumed growth on galactose.

Zygomycota; *Sordaria fimicola*, Ascomycota) because they do not produce invertase.

The utilization of polymers requires the release of specific enzymes. For example, many fungi can utilize starch (by producing **amylase**), some can utilize lipids (by producing **lipases**) or proteins (by producing **proteases**), and some can grow on cellulose, hemicelluloses, pectic compounds (Chapter 14), or chitin. Often, these enzymes occur as complexes, consisting of two or more enzymes that attack the polymer at different sites, acting synergistically. We will see this in the case of cellulose breakdown, later in this chapter. At the extreme end of the spectrum in Fig. 6.1, a few highly specialized fungi degrade the most structurally complex, cross-linked polymers. For example, lignin is degraded by enzymes of a few wood-rotting Basidiomycota termed "white rot fungi" (Chapter 11). Similarly, "hard keratin" in the horns and hoofs of animals is degraded by a few **keratinophilic** fungi, including species related to the dermatophytic pathogens of humans (Chapter 16).

To summarize this section, fungi exploit a wide range of organic nutrient sources, but in all cases they depend on the uptake of simple, soluble nutrients which diffuse through the wall and enter the cells via specific transport proteins. These only allow the passage of small molecules such as monosaccharides, amino acids, and small peptides of two or three amino acids. Even disaccharides such as sucrose and cellobiose (a breakdown product of cellulose) may need to be degraded to monosaccharides before they are taken up by most fungi. Any larger molecules have to be broken down by extracellular enzymes (**depolymerases**) which are secreted by the fungus.

Fungal adaptations for nutrient capture

An electrical dimension to hyphal growth

The growing tips of fungi generate an electrical field around them. It can be mapped by placing a tiny vibrating electrode in the liquid film around a hypha, resulting in a trace like that shown in Fig. 6.3. Typically, the exterior of the hypha is more electronegative at the apex than further back, showing that current (which is positive by convention) enters at the tip and exits in the subapical regions. In most fungi that have been examined in sufficient detail – for example *Achlya* (Oomycota) and *Neurospora* (Ascomycota) – the current seems to be carried by protons (H^+) because it corresponds with a gradient of pH along the hyphal surface and the current still occurs when other candidate ions are reduced or eliminated from the growth medium. However, in marine fungi the current might be carried by K^+. For several years it was speculated that the electrical field might be the driving force for apical growth, perhaps by affecting the activities of membrane proteins or the cytoskeleton. However, this now seems unlikely because the direction of current flow can be reversed without affecting tip growth (Cho *et al.* 1991). Instead, the electrical field seems to be intimately involved in nutrient uptake.

The uptake of organic nutrients is an energy-dependent process. Ions (e.g. H^+) are pumped to the

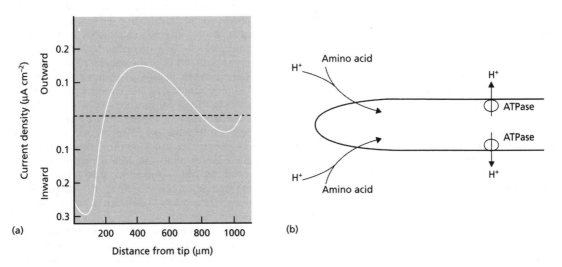

Fig. 6.3 (a) Typical current profile around an individual hypha of a growing fungus (*Achlya* sp.). (b) Interpretation of the current, involving proton export through ATPase-driven proton pumps behind the hyphal tip, and uptake of protons at the hyphal tip through symport proteins that simultaneously transport amino acids. ((a) From Gow 1984. (b) Based on Kropf *et al.* 1984.)

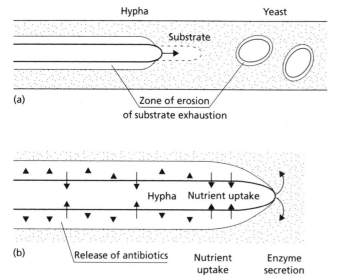

Fig. 6.4 Fungal strategies for growth on insoluble polymers. (a) Hyphae extend continuously at the apex, drawing protoplasm forward to evacuate the zones of enzyme erosion of the substrate. Yeasts do not utilize insoluble (nondiffusible) polymers because they would become trapped in their own substrate erosion zones. (b) Suggested defense of a substrate by a hypha of a polymer-degrading fungus. Enzymes are secreted at the apex to degrade the polymer, and the soluble nutrients released by these enzymes are absorbed subapically. Antibiotics or other inhibitors (shown as arrowheads) may be released subapically into the substrate erosion zone to prevent competing organisms from using the enzyme digestion products.

outside of the cell by membrane-located ion pumps, using energy derived from the dissociation of ATP. The ions then re-enter the cell through specific types of membrane proteins termed **symporters**, which simultaneously transport an organic molecule into the cell. This seems to apply generally in fungi, because their plasma membranes have been shown to have H^+-coupled transport systems for sugars, amino acids, nitrate, ammonium, phosphate, and sulfate, as well as other ion channels such as stretch-activated channels for calcium, and nonselective channels for both cations and anions (Garrill 1995).

But, fungi are thought to differ from most other organisms in one significant respect: in most cell types the ion pumps and the symport proteins probably occur in close proximity to one another, whereas in tip-growing fungi the ion pumps are thought to be most active behind the apex, whereas the symport proteins are active close to the tip. This spatial separation would explain the external electrical field (Fig. 6.3) and it makes sense intuitively. The nutrient-uptake symporters would be of most value at the hyphal tip, which extends continuously into fresh zones of nutrients, while immediately behind the tip there is a dense zone of mitochondria that could supply the ATP needed for the proton pumps (see Figs 3.2, 3.4, 3.15).

Enzyme secretion

The need to obtain nutrients has, literally, shaped the way that fungi grow. The rate of diffusion of soluble organic nutrients to a stationary cell would always tend to be growth-limiting, and the need to release degradative enzymes imposes even greater constraints. Enzymes are large molecules, about 20,000–60,000 Daltons in the case of fungal cellulases, so they do not diffuse far from the hyphal surface. As a result, fungi create localized zones of substrate erosion (nutrient depletion) when growing in substrates such as cellulose, and the hyphae must extend continuously into fresh zones (Fig. 6.4). This explains why yeast cells never produce depolymerase enzymes, because they would have no way of escaping from the erosion zones they had created. Instead, yeasts occur in environments rich in simple soluble nutrients – on leaf, fruit and root surfaces, or on mucosal membranes in the case of *Candida albicans*. They depend either on a continuous supply of nutrients from the underlying substratum, or on water currents to bring fresh supplies of nutrients.

The large size of enzymes creates a potential problem for their release through the hyphal wall, because this would require the presence of **continuous pores** of sufficiently large size. Estimates of wall porosity can be obtained by studying the permeability of walls to molecules of known molecular mass such as commercially available polyethylene glycols and dextrans. These studies suggest a cut-off of wall porosity at between 700 and 5000 Da, which is much lower than the size of most enzymes. These findings apply mainly to yeasts, and might be different for mycelial fungi. However, recent studies indicate that enzymes are released mainly and perhaps exclusively in the regions of new wall growth. This was first shown when a cellulase gene was cloned into yeast (*S. cerevisiae*) and the enzyme was found to be released from the growth sites of the bud. Also, a glucoamylase has been shown to be secreted exclusively at the tips of *Aspergillus niger*. So it seems that enzymes destined for release through the wall are transported to the hyphal tip, where they are released by exocytosis and might flow outwards with the newly formed wall components, according to the steady-state model of wall growth described in Chapter 4 (Wessels 1990). Some of these enzymes would reach the cell surface and be released into the external environment, whereas others might become locked in the wall and serve as wall-bound enzymes – for example, degrading disaccharides and other small molecules (oligomers) to monosaccharides, amino acids, etc.

Defense of territory

The release of enzymes to degrade a polymer represents an investment of resources, so we can expect it to be coupled with defense of the territory, preventing other organisms from sharing the breakdown products. There are a few known examples of noncellulolytic fungi that grow in close association with cellulose-degraders (Chapter 12), utilizing some of the enzyme breakdown products. But these fungi might grow in a mutualistic association that benefits the cellulolytic fungus.

In general, three factors might help polymer-degrading fungi to defend their territory.

1 The synthesis of depolymerases is tightly regulated by feedback mechanisms, discussed later, so the rate of enzyme production is matched to the rate at which the breakdown products can be utilized.
2 The final stages of polymer breakdown are achieved by wall-bound enzymes, so that the most readily utilizable monomers are not available to other organisms; we will see this later in the case of cellulose.
3 A polymer-degrading fungus might produce antibiotics or other suppressive metabolites. This is difficult to demonstrate at the scale of individual hyphae, but Burton & Coley-Smith (1993) reported that antibacterial compounds were released by hyphae of *Rhizoctonia* species – members of the Basidiomycota that are known to degrade cellulose.

It may be significant that antibiotics are produced mainly by polymer-degrading fungi (Chapter 12) and their production *in vitro* is associated with nutrient-limiting growth conditions (Chapter 7). These are the conditions that polymer-degrading fungi would experience most of the time, because depolymerases are produced only when more readily available nutrients are in short supply. Thus, antibiosis might have evolved not as an aggressive strategy but for **the defense of territory.**

The breakdown of cellulose: a case study of extracellular enzymes

Cellulose is the most abundant natural polymer on earth, representing about 40% of all the plant biomass that is produced and recycled on an annual basis. Fungi play the principal role in degrading cellulose and also in utilizing the cellulose breakdown products. So, we will consider the structure of cellulose and the role of cellulase enzymes in some detail as a model of how enzymes function in natural environments.

The chemistry of cellulose

In terms of its chemical structure, cellulose is a relatively simple polymer. It consists mainly of long, unbranched, chains of between 2000 and 14,000 glucose residues, linked by β-1,4 bonds. The β configuration means that each glucose unit is rotated 180 degrees to the next, and the whole chain is made up of these repeating pairs, shown between the brackets in Fig. 6.5. Such a molecule should be easily degraded by the extracellular enzymes of fungi, but the physical structure of cellulose presents problems of enzyme access to the substrate. The individual cellulose chains stack closely together in a near-crystalline manner to form **micelles**, which are reinforced by hydrogen-bonding, and the micelles themselves align to form microfibrils of about 10 nm diameter, visible in electron micrographs. These rigid, insoluble microfibrils

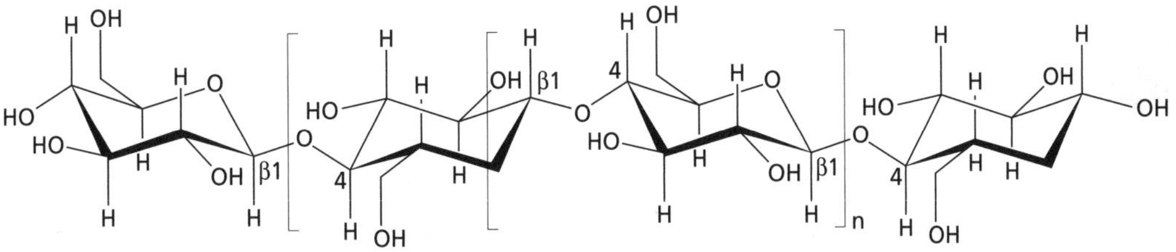

Fig. 6.5 Structure of a single cellulose chain, composed of thousands of repeated disaccharide (cellobiose) units; one disaccharide unit is shown within the brackets. The β1-4 bonds are also shown.

present a quite formidable obstacle to enzymatic degradation, but several types of fungus have evolved means of degrading them.

The cellulase enzymes

At least three types of enzyme are involved in the complete breakdown of a cellulose chain to its constituent glucose units, and some of these enzymes exist in multiple forms of different molecular mass. Collectively they are termed "**cellulase**" or, more precisely, the **cellulase enzyme complex**. The three main enzymes in this complex are:

1 An endoglucanase (**endo-β-1,4-glucanase**), which acts at random points within a cellulose chain, breaking the molecule into successively smaller fragments. This enzyme is found in multiple forms, ranging from about 11,000 to 65,000 Da.

2 An exo-acting enzyme, termed **cellobiohydrolase**, which acts only on the ends of cellulose chains, cleaving off successive disaccharide units (cellobiose). This enzyme is more uniform than the endoglucanase, ranging from about 50,000 to 60,000 Da.

3 **β-glucosidase** (or "cellobiase"), which cleaves the disaccharide cellobiose to glucose, for uptake by the fungus. This is a wall-bound enzyme, like many enzymes involved in the final stages of degradation of polymers.

The combined actions of these three types of enzyme are shown diagrammatically in Fig. 6.6. All three enzymes act synergistically and are tightly regulated, to ensure that a cellulose-degrading fungus does not release sugars at a faster rate than it can use them. The endoglucanases, by attacking the cellulose chains at random, progressively create more ends on which cellobiohydrolase can act. But the resulting cellobiose

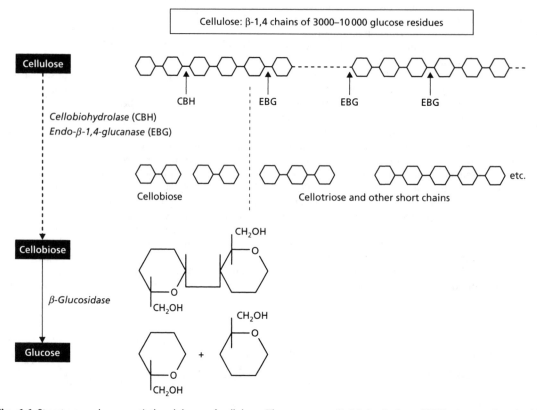

Fig. 6.6 Structure and enzymatic breakdown of cellulose. The enzyme **cellobiohydrolase** (CBH) cleaves disaccharide residues (cellobiose) from the nonreducing ends of the cellulose chains. The enzyme **endo-β-1,4-glucanase** (EBG) cleaves the chains at random to generate shorter chains, thereby providing further ends for the action of CBH. The enzyme **β-glucosidase** cleaves cellobiose to two glucose residues for uptake by the fungus.

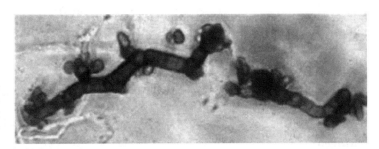

Fig. 6.7 *Sydowia polyspora* (*Aureobasidium pullulans*), one of the "sooty moulds" of leaf surfaces. It has short hyphal compartments and forms bud-like conidia at the septa.

can bind to the active site of cellobiohydrolase, and **competitively inhibit** the enzyme action. Thus, if cellobiose accumulates then the rate of cellulose breakdown will automatically be slowed, and this regulation links the rate of breakdown to the rate at which the fungus requires glucose for growth and metabolism. In addition to this, the regulation of cellulose breakdown is achieved by the common feedback system termed **catabolite repression**, whereby genes encoding the enzymes are repressed when more readily utilizable substrates like glucose are available.

The presence of cellulose is known to induce the synthesis of cellulase enzymes. It is difficult to imagine that cellulose itself could be the inducer because it is insoluble. Instead, cellulose-degrading fungi are thought to produce very low (constitutive) levels of cellulase enzymes so that the breakdown products such as cellobiose can act as signals for enzyme induction when cellulose is present in the environment. Cellobiose has been found to act as a weak inducer of cellulases in some fungi, and a derivative of cellobiose, termed sophorose, is a much stronger inducer for *Trichoderma reesei*, the fungus used commonly as a model organism for studies of cellulase action.

In summary, by a combination of **gene repression** (by high levels of glucose or cellobiose), **competitive inhibition** of enzyme action (by the partial breakdown products of cellulose, such as cellobiose), and **gene induction** (in the absence of glucose but presence of low levels of cellobiose or its derivatives), the rate of cellulose breakdown is matched closely to the rate at which a fungus can use the sugars released from the substrate.

Commercial and ecological aspects of cellulases

The ability to degrade "crystalline" cellulose (cotton fabrics, etc.) is found in relatively few fungi, but several fungi can degrade the "soluble substituted celluloses"

such as **carboxymethyl cellulose** in which about 30% of the glucose residues bear a substituent methyl group. These soluble celluloses form gels in water and are produced commercially for many purposes, such as wallpaper pastes and paint thickeners. The fungi that degrade these products secrete an endoglucanase but not cellobiohydrolase, and they include several common leaf-surface saprotrophs such as *Cladosporium* spp. and *Sydowia polyspora* (Fig. 6.7).

This raises the question of why some fungi produce only part of the cellulase enzyme complex if they cannot degrade natural forms of cellulose. A likely explanation was discovered more than 40 years ago by Taylor & Marsh (1963) but has been largely ignored. These workers found that several *Pythium* spp. (which are not generally considered to be cellulolytic fungi) could degrade cotton fibres taken from *unopened* cotton bolls on cotton plants, but could not degrade the cotton fibres once the cotton bolls had opened and the fibres had air-dried. Simple rewetting of the dried fibres did not render them degradable. But they could be degraded if they were "swollen" by treatment with KOH. The implication is that the cellulose in natural, moist plant cell walls may be susceptible to attack, providing a substrate that several fungi can use. But the cellulose chains bond together tightly when the tissues have dried, and then only a restricted group of fungi can degrade them.

Among the typical cellulose-degraders is *Trichoderma reesei*, originally isolated from rotting cotton fabric. By routine, repeated selection in laboratory conditions, some strains of this fungus have been obtained that can release as much as 30 g dry weight of cellulase enzymes per liter of culture broth (Penttila *et al.* 1991). The gene for a cellobiohydrolase of this fungus has been cloned into *Saccharomyces cerevisiae* under the control of a constitutive promoter. The yeast then secretes large amounts of a functional cellulase, even in the absence of the substrate. Such enzymes, whether produced by conventional or genetically engineered strains, have significant commercial roles, for example in abrading denim products.

Mineral nutrient requirements: nitrogen, phosphorus, and iron

Fungi need many mineral nutrients in at least trace amounts (Table 6.1) but nitrogen, phosphorus, and iron merit special mention because of their significance for fungal activities and interactions in nature.

Nitrogen

Of all the mineral nutrients, nitrogen is required in the largest amounts and can often be the limiting factor for fungal growth in natural habitats (Chapter 11). Fungi do not fix atmospheric N_2 but they can use many combined forms of nitrogen. We will discuss this in relation to the enzyme-mediated sequence below, which represents the normal pathway for assimilation of nitrogen by fungi and many other organisms.

$$NO_3^- \xrightarrow[\text{reductase}]{\text{nitrate}} NO_2^- \xrightarrow[\text{reductase}]{\text{nitrite}} NH_4^+$$
nitrate · · · · · · · · · · · nitrite · · · · · · · · · · · ammonium

$$\xrightarrow[\text{dehydrogenase}]{\text{glutamate}} \text{glutamate} \xrightarrow[\text{synthase}]{\text{glutamine}} \text{glutamine}$$

All fungi can use amino acids as a nitrogen source. Often they need to be supplied with only one amino acid such as glutamic acid or glutamine, and from this they can produce all the other essential amino acids by **transamination** reactions. The standard form of these reactions is shown below, using as an example, the production of alanine from glutamic acid, or vice versa:

$$HOOC.CH_2.CH(NH_2).COOH + CH_3.CO.COOH \leftrightarrow$$
glutamic acid · · · · + · pyruvic acid · $\leftrightarrow$

$$HOOC.CH_2.CO.COOH + CH_3.CH(NH_2).COOH$$
α-ketoglutaric acid · · + · · · · · · alanine

Most fungi can use ammonia or ammonium (NH_4) as a nitrogen source. After uptake, ammonia/ammonium is combined with organic acids, usually to produce either glutamic acid (from α-ketoglutaric acid) or aspartic acid; then the other amino acids can be formed by transamination reactions, as noted above. The relatively few fungi that cannot use ammonium and thus depend on organic nitrogen sources include some water moulds (*Saprolegnia* and *Achlya* spp.), several Basidiomycota, and the mycoparasitic *Pythium* spp. such as *P. oligandrum* (Chapter 11). However, ammonia/ammonium is often not an ideal nitrogen source in culture media, even for many of the fungi that can use

it. The reason is that NH_4^+ is taken up in exchange for H^+, and this can rapidly lower the pH of a culture medium to 4.0 or less, inhibiting the growth of many fungi.

Many fungi can use nitrate as their sole nitrogen source, converting it to ammonium by the enzymes **nitrate reductase** and **nitrite reductase**. The fungi that cannot use nitrate include *Saccharomyces cerevisiae* and many Basidiomycota, whereas some other fungi such as *Neurospora crassa* are induced to express the genes for nitrate reductase and nitrite reductase only when a preferred nitrogen source is unavailable.

By referring to the nitrogen assimilation pathway (above), it is clear that any fungus that can use NO_3 (or another nitrogen source towards the left of this pathway) must also be able to use the other forms of nitrogen towards the right. But, the regulatory controls on nitrogen uptake ensure that nitrogen sources are not necessarily used in the ways we might expect. If a fungus is supplied with a mixture of nitrogen sources, then ammonium is taken up in preference to either nitrate or amino acids. The reason is that ammonium, or glutamine which is one of the first amino acids formed from it, prevents the synthesis of membrane-uptake proteins for other nitrogen sources, and also prevents the synthesis of enzymes involved in nitrate utilization.

Phosphorus

All organisms need significant amounts of phosphorus, in the form of phosphates, for production of sugar phosphates, nucleic acids, ATP, membrane phospholipids, etc. But phosphorus is often poorly available in natural environments, because even soluble phosphate fertilizers are soon rendered insoluble when they complex with organic matter or with calcium and magnesium ions in soil. Plant roots, in particular, have difficulty in extracting phosphorus from soil, because they deplete the small pool of soluble phosphate in their immediate vicinity and then have to depend on the slow solubilization and diffusion of phosphate from further away. By contrast, fungi are highly adept at obtaining phosphorus, and they achieve this in several ways (Jennings 1989):

- They respond to critically low levels of available phosphorus by increasing the activity of their phosphorus-uptake systems;
- they release phosphatase enzymes that can cleave phosphate from organic sources;
- they solubilize inorganic phosphates by releasing organic acids to lower the external pH;
- their hyphae, with a high surface area/volume ratio, extend continuously into fresh zones of soil.

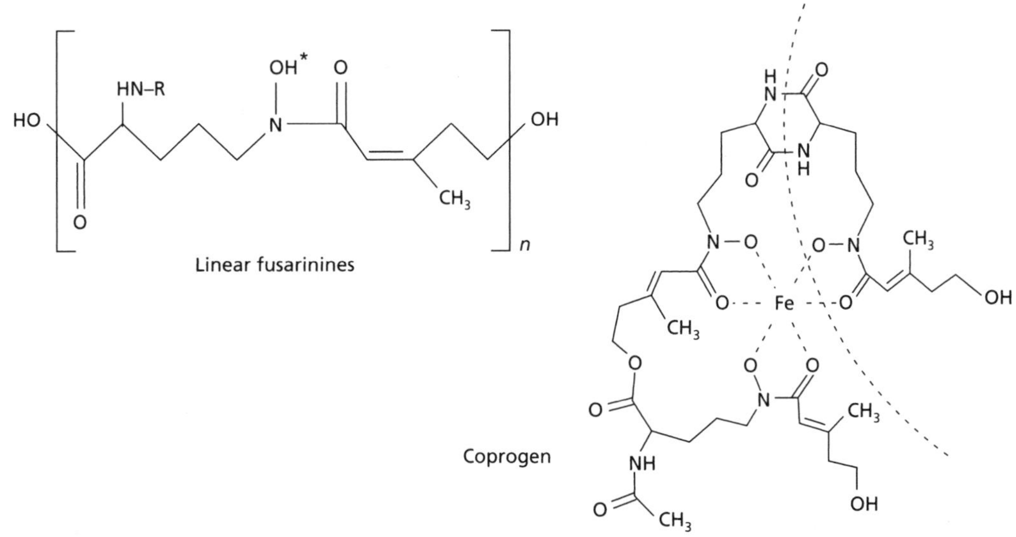

Fig. 6.8 The hydroxamate siderophores of fungi, used to capture ferric iron. The simplest of these are the linear fusarinines, widely found in species of *Fusarium*, *Penicillium*, and *Gliocladium*. They sequester iron at the position marked by the asterisk. More complex siderophores such as coprogen have several fusarinine units that bind iron in a ring-like structure (one of these units is shown by the broken line).

In addition to these points, fungi accumulate and store phosphates in excess of their immediate requirements, typically as polyphosphates in the vacuoles. It is not surprising that plants have mycorrhizal associations with fungi, because this is probably the most efficient way for the plant to obtain its phosphorus supply. The mycorrhizal fungal hyphae have a very high surface area for uptake of phosphorus and other mineral nutrients. The cost to the plant in terms of supplying sugars to the fungus would be much less than the cost of continuously producing new roots (Chapter 13).

Iron

Iron is needed in relatively small amounts but is essential as a donor and acceptor of electrons in cellular processes, including the cytochrome system in aerobic respiration (Chapter 7). Iron normally occurs in the ferric (Fe^{3+}) form, insolubilized as ferric oxides or hydroxides at a pH above 5.5, and it is taken up by a different process compared with other mineral nutrients. Iron must be "captured" from the environment by the release of iron-chelating compounds termed **siderophores**. These compounds chelate a ferric ion (Fe^{3+}) then they are reabsorbed through a specific membrane protein and Fe^{3+} is reduced to Fe^{2+} within the cell, causing its release because the siderophore has a lower affinity for Fe^{2+} than for Fe^{3+}. Finally, the siderophore is exported again to capture a further ferric ion. Siderophores and their specific membrane proteins are produced only in response to iron-limiting conditions.

All the fungal siderophores that have been characterized to date are of the **hydroxamate** type (Fig. 6.8). Their structures, functions, and applications are reviewed by Renshaw *et al.* (2002). Despite their high affinity for Fe^{3+}, these fungal siderophores have much lower affinity than do the siderophores (e.g. **pseudobactin** and **pyoverdine**) produced by fluorescent pseudomonads which are common on plant roots. This raises the possibility that fluorescent pseudomonads could be used for the control of plant-pathogenic fungi in the root zone of crops. For example, pseudobactin-producing pseudomonads can suppress germination of the chlamydospores of *Fusarium oxysporum* on low-iron media, whereas mutant pseudomonads, deficient in siderophore production, are ineffective. However, we shall see in Chapter 12 that competition for iron is only one of several ways in which *Pseudomonas* spp. can control plant-pathogenic fungi.

Efficiency of substrate utilization

Industrial microbiologists are specially concerned with the efficiency of substrate conversion into microbial cells or cell products. Microbial ecologists have been less concerned with this, although it has important implications for environmental processes. The two most commonly used criteria of the efficiency of substrate conversion are the **economic coefficient** and the **Rubner coefficient**:

Economic coefficient (expressed as a percentage) =

$$\frac{\text{Dry weight of biomass produced}}{\text{Dry weight of substrate consumed}}$$

Rubner coefficient (expressed as a proportion) =

$$\frac{\text{Heat of combustion of biomass produced}}{\text{Heat of combustion of substrate consumed}}$$

Typical values for the economic coefficient range from 20 to 35 if a fungus is grown in a dilute medium, but the values fall markedly if the medium is made progressively richer – fungi seem to be "wasteful" if supplied with excess substrate. Values for the Rubner coefficient typically are higher than for the economic coefficient. For example *Aspergillus niger* had a Rubner coefficient of 0.55–0.61 (equivalent to 55–61%) over a range of cultural conditions, compared with an economic coefficient of 35–46. The difference is explained mainly by the synthesis of lipid storage reserves during growth, because lipids have higher calorific values than the carbohydrates used as substrates in culture media. However, the important point revealed by all these values is that a considerable proportion of the substrate supplied to a fungus is consumed in energy production rather than being converted into biomass.

Substrate conversion efficiencies are difficult to obtain in natural systems, but Adams & Ayers (1985) did this in laboratory conditions by collecting all the spores produced by a mycoparasite (*Sporidesmium sclerotivorum*) when it was grown on sclerotia of its main fungal host, *Sclerotinia minor*. This experimental system mimics the conditions in nature, because sclerotia are produced as dormant survival structures on infected host plants and then overwinter in the soil, where they can be attacked by the mycoparasite. The reported substrate conversion efficiency was exceptionally high: an economic coefficient of 51–60 and a Rubner coefficient of 0.65–0.75. This mycoparasite has a remarkable way of parasitizing the host sclerotia: it penetrates some of the sclerotial cells initially but then grows predominantly **between** the sclerotial cells, scavenging small amounts of soluble nutrients that leak from them, and thereby creating nutrient stress. The host cells respond by converting energy storage reserves (principally glycogen) to sugars, which leak from the cells to support further growth of the mycoparasite. This essentially noninvasive mode of parasitism is also employed by **endophytic fungi** in plants (Chapter 14). They grow slowly and sparsely, between or within the plant cell walls, exploiting nutrients that leak from the host cells.

In terms of substrate efficiency we should also note that fungi can grow by **oligotrophy**, using extremely low levels of nutrients (*oligo* = few) on silica gel or glass. They seem to grow by scavenging trace amounts of volatile organic compounds from the atmosphere (Wainwright 1993).

Fungi that cannot be cultured

To close this chapter we should record that several fungi still cannot be grown in laboratory culture. They have been termed **obligate parasites** but now more commonly are termed **biotrophic parasites** (Chapter 14). Many of them are extremely important in environmental and economic terms, including the ubiquitous arbuscular mycorrhizal fungi (Glomeromycota), rust fungi (Basidiomycota), powdery mildew fungi (Ascomycota), and downy mildews (Oomycota). All of these produce nutrient-absorbing **haustoria** or equivalent structures in host cells (Chapter 14).

It remains to be seen if some of these fungi will ever be grown in **axenic** culture (i.e. separate from their hosts). However, significant progress has been made in culturing the rust fungi, starting with *Puccinia graminis* (black stem rust of wheat) and several other rust species (Maclean 1982). *P. graminis* was found to grow slowly, and only after a prolonged lag phase. Its linear extension rate on agar ranged from 30 to 300 µm day^{-1}, compared with rates from 1 to 50 mm day^{-1} for fungi that are commonly grown in laboratory conditions. *P. graminis* tends to leak vital nutrients, such as cysteine, into the growth medium, so a high spore density is needed in the inoculum to minimize the diffusion of nutrients away from individual sporelings. Also, *P. graminis* produces self-inhibitors which can result in "staling" of the cultures. This is true of several fungi if they are grown on sugar-rich media, and it causes a progressive reduction and eventual halting of growth. It can be overcome by using relatively weak media. Some other rust fungi need relatively high concentrations of CO_2 – for example, *Melampsora lini* which causes flax rust. If attention is paid to all these points then several rust fungi can be maintained in laboratory culture. However, in the process of being cultured (or of becoming culturable) some of them change irreversibly to a "saprotrophic" form that cannot reinfect plants.

General texts

Booth, C. (1971) *Methods in Microbiology*, vol. 4. Academic Press, London.

Garraway, M.O. & Evans, R.C. (1984) *Fungal Nutrition and Physiology*. Wiley, New York.

Griffin, D. (1994) *Fungal Physiology*, 2nd edn. Wiley-Liss, New York.

Cited references

Adams, P.B. & Ayers, W.A. (1985) Energy efficiency of the mycoparasite *Sporidesmium sclerotivorum* in vitro and in soil. *Soil Biology and Biochemistry* **17**, 155–158.

Burton, R.J. & Coley-Smith, J.R. (1993) Production and leakage of antibiotics by *Rhizoctonia cerealis, R. oryzae-sativae* and *R. tuliparum. Mycological Research* **97**, 86–90.

Cho, C., Harold, F.M. & Scheurs, W.J.A. (1991) Electrical and ionic dimensions of apical growth in *Achlya* hyphae. *Experimental Mycology* **15**, 34–43.

Garrill, A. (1995) Transport. In: *The Growing Fungus* (Gow, N.A.R. & Gadd, G.M., eds), pp. 163–181. Chapman & Hall, London.

Gow, N.A.R. (1984) Transhyphal electric currents in fungi. *Journal of General Microbiology* **130**, 3313–3318.

Jennings, D.H. (1989) Some perspectives on nitrogen and phosphorus metabloism in fungi. In: *Nitrogen, Phosphorus and Sulphur Utilization by Fungi* (eds Boddy, L., Marchant, R. & Read, D.J.), pp. 1–31. Cambridge University Press, Cambridge.

Kropf, D.L., Caldwell, J.H., Gow, N.A.R. & Harold, F.M. (1984) Transhyphal ion currents in the water mould *Achlya*. Amino acid proton symport as a mechanism of current entry. *Journal of Cell Biology* **99**, 486–496.

Maclean, D.J. (1982) Axenic culture and metabolism of rust fungi. In: *The Rust Fungi* (Scott, K.J. & Chakravorty, A.K., eds), pp. 37–120. Academic Press, London.

Penttila, M., Teeri, T.T., Nevalainen H. & Knowles, J.K.C. (1991) The molecular biology of *Trichoderma reesei* and its application in biotechnology. In: *Applied Molecular Genetics of Fungi* (Peberdy, J.F., Caten, C.E., Ogden, J.E. & Bennett, J.W., eds), pp. 85–102. Cambridge University Press, Cambridge.

Renshaw, J.C., Robson, G.D., Trinci, A.P.J., *et al.* (2002) Fungal siderophores: structures, functions and applications. *Mycological Research* **106**, 1123–1142.

Taylor, E.E. & Marsh, P.B. (1963) Cellulose decomposition by *Pythium. Canadian Journal of Microbiology* **9**, 353–358.

Wainwright, M. (1993) Oligotrophic growth of fungi – stress or natural state? In: *Stress Tolerance of Fungi* (Jennings, D.H., ed.), pp. 127–144. Academic Press, London.

Wessels, J.G.H. (1990) Role of cell wall architecture in fungal tip growth. In: *Tip Growth in Plant and Fungal Cells* (Heath, I.B., ed.), pp. 1–29. Academic Press, New York.

Chapter 7

Fungal metabolism and fungal products

This chapter is divided into the following major sections:

- how fungi obtain energy in different conditions
- coordination of metabolism: how the pathways are balanced
- mobilizable and storage compounds of fungi
- synthesis of chitin and lysine
- the pathways and products of secondary metabolism

In this chapter we discuss the basic metabolic pathways of fungi, as a basis for understanding how fungi grow on different types of substrate and in different environmental conditions. We also cover some of the distinctive and unusual aspects of fungal metabolism, including the production of a wide range of secondary metabolites of commercial and environmental significance, such as the **penicillin** antibiotics, and important mycotoxins like the highly carcinogenic **aflatoxins** and the toxic **ergot alkaloids**. Some of the material in this chapter will be familiar, basic biochemistry, but it is presented in the specific context of fungal biology.

How do fungi obtain energy in different conditions?

Figure 7.1 shows an overview of central metabolism and how this provides the precursors for synthesis of many other compounds. The spine of the diagram represents the central energy-yielding pathway, in which sugars are broken down via the **Embden–Meyerhof (EM) pathway (glycolysis)** and the **tricarboxylic acid (TCA) cycle**. Many of the intermediates can be drawn off to produce other essential metabolites or for synthesis of a wide range of specialized secondary metabolites.

Although fungi can obtain energy by oxidizing a wide range of compounds, it is convenient to begin by considering a fungus growing on a simple sugar such as glucose. As shown in Fig. 7.2, in the series of enzymic steps of the Embden–Meyerhof pathway, glucose is converted to glucose-6-phosphate, then to fructose-6-phosphate, and finally to fructose-1,6-biphosphate. These phosphorylation reactions occur at the expense of energy provided by two molecules of the energy-rich compound, adenosine triphosphate (ATP). Then fructose-1,6-biphosphate is split into two 3-carbon compounds, glyceraldehyde-3-phosphate and dihydroxyacetone phosphate. These two compounds are interconvertible, and in a further series of enzymic steps they are converted to **pyruvic acid**.

Pyruvic acid is one of the key intermediates of central metabolism, because it represents a branch-point: the reaction steps that follow will depend on whether the fungus is growing in the presence or absence of oxygen.

In the presence of oxygen, pyruvic acid (a 3-carbon compound) is transported to the **mitochondria**, where it is converted to the 2-carbon compound **acetyl-coenzyme A** (acetyl-CoA) by associating with a molecule of coenzyme-A and releasing a molecule of CO_2. Then acetyl-CoA combines with oxaloacetate (a 4-carbon compound) to produce citric acid (6-carbon). Finally, in the reactions of the TCA cycle, citric acid is converted back to oxaloacetate, with the loss of two molecules of CO_2 (Fig. 7.2).

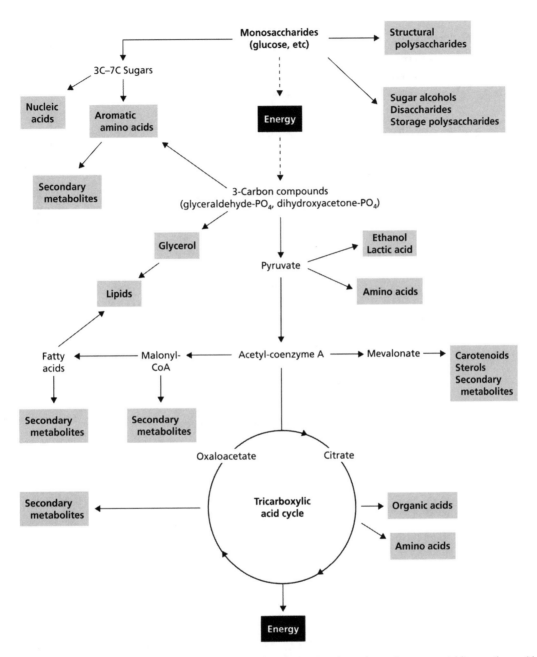

Fig. 7.1 Overview of the central metabolic pathways of fungi, showing how the main energy-yielding pathway (the Embden–Meyerhof pathway and the tricarboxylic acid cycle) provide the precursors for biosynthesis of various metabolic products (shaded boxes). **Note that only some of the intermediates of the central metabolic pathway are shown** – see Fig. 7.2 for more details. Also, note that secondary metabolites (including penicillins and mycotoxins) are produced from various precursors, but primarily from **acetyl coenzyme A.**

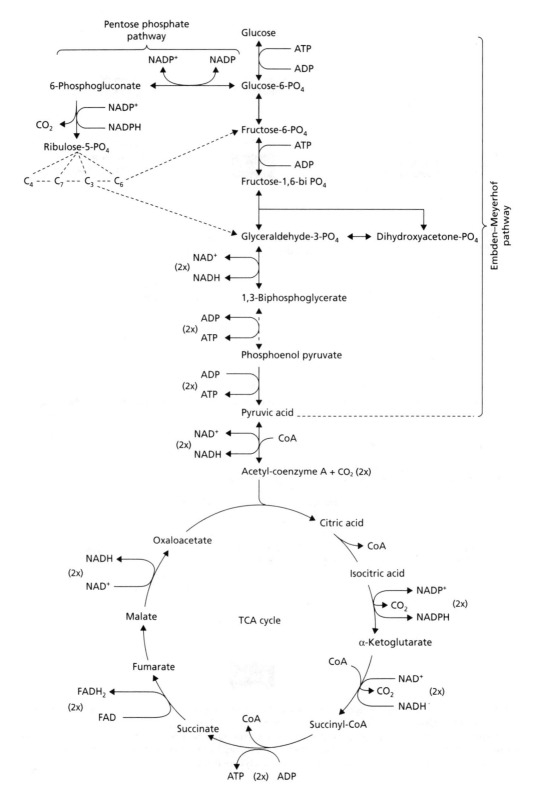

Fig. 7.2 Outline of the Embden–Meyerhof pathway and tricarboxylic acid cycle, which provide the major means of generating energy from sugars. Also shown (top left) is the pentose phosphate pathway which can provide some energy, but its major role is for biosynthesis, including synthesis of the 5-carbon sugars of nucleic acids.

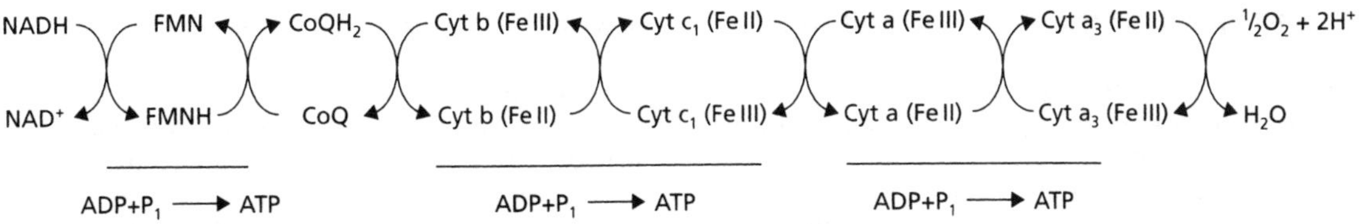

Fig. 7.3 Outline of the respiratory electron transport chain. CoQ = coenzyme Q; Cyt = cytochrome.

The net result of this whole sequence is that one molecule of glucose is completely oxidized to six molecules of CO_2 (and six molecules of water), according to the empirical equation for **aerobic respiration**:

$$C_6H_{12}O_6 + 6O_2 \rightarrow 6CO_2 + 6H_2O$$

remembering that every step after the cleavage of fructose-1,6-biphosphate into two 3-carbon compounds occurs **twice**, as shown in Fig. 7.2.

The oxidation of any substance must always be coupled with a corresponding reduction of another substance, and this role is served by three nucleotides: NAD^+ (nicotinamide adenine dinucleotide), $NADP^+$, and FAD (flavin adenine dinucleotide). As shown in Fig. 7.2, these compounds accept electrons and are correspondingly reduced to NADH, NADPH and $FADH_2$. These nucleotides then need to be reoxidized for the whole process to continue, and this is achieved by passing their electrons along an **electron transport chain**, where oxygen is the **terminal electron acceptor**.

The electron transport chain is shown in simplified, diagrammatic form in Fig. 7.3. It consists of a series of electron carriers which are located in the mitochondrial membrane, and in several cases span this membrane. They are aligned in a specific sequence. Initially, NADH [or NAD(P)H] is reoxidized to NAD^+ [or $NAD(P)^+$] by transferring its electrons to the carrier molecule, flavin mononucleotide (FMN), which becomes reduced to FMNH. (Note that, for our purposes, we can regard FMN/FMNH as equivalent to FAD/$FADH_2$; the essential point is that a flavin molecule acts as a carrier in the electron-transport chain.) The flavin nucleotide, in turn, transfers the electrons to coenzyme Q, and so on down the chain, until the final step when oxygen accepts the electrons and is, itself, reduced to water. The carriers in the electron transport chain generate a **proton motive force**, because protons (H^+ ions) are extruded to the outside of the mitochondrial membrane while OH^- ions accumulate inside. This polarization of the membrane is used to drive the synthesis of ATP (from ADP + inorganic phosphate) when protons re-enter through a membrane-located ATPase (also termed ATP synthase).

As shown in Fig. 7.3, the electron transport chain is, essentially, an electrochemical gradient, and at three stages along this gradient sufficient energy is released to synthesize a molecule of ATP from the oxidation of NADH/NAD(P)H. But only two molecules of ATP are produced from the oxidation of flavin nucleotides (e.g. FMN).

The energy yield from aerobic respiration

We can calculate the **theoretical energy yield from glucose during aerobic respiration** by calculating the number of ATP molecules that could be synthesized (Fig. 7.2):

2 ATP from the EM pathway down to pyruvic acid (4 ATP produced but 2 ATP used initially to phosphorylate glucose)

2 ATP from the TCA cycle (one in each turn of the cycle, but 2 molecules of pyruvate must be processed through this cycle)

30 ATP from the reoxidation of 10 pyridine nucleotides (NADH / NADPH)

4 ATP from the reoxidation of 2 flavin nucleotides

Total 38 ATP

However, the **actual ATP yield** will be much less than this, for at least two reasons:

1 Intermediates are continuously drawn from the pathways for biosynthetic reactions, so this represents a loss of potential ATP.
2 The major role of NADP/NADPH (as opposed to NAD^+/NADH) is to provide reducing power for biosynthetic reactions, rather than to generate ATP.

At this stage we should mention the special role of the **pentose-phosphate pathway** (Fig. 7.2). It can be used as an alternative to the EM pathway for generating energy from sugars (giving 1 ATP instead of the 2 ATP from the EM route). But its major role is in biosynthesis – it generates some important intermediates, such as ribose-5-phosphate for the synthesis of nucleic acids and erythrose-4-phosphate for the synthesis of aromatic amino acids.

What happens when oxygen is limiting?

In the absence of a terminal electron acceptor (usually oxygen), the electron transport chain cannot operate, so the pool of reduced nucleotides (NADH, etc.) accumulates and metabolism would rapidly cease. But fungi, and many other organisms, can still obtain some energy in the absence of oxygen by one of the following reactions:

1 $CH_3.CO.COOH \xrightarrow{\textit{pyruvic dehydrogenase}}$
   (**pyruvic acid**)

$$acetaldehyde + CO_2$$

then:

$CH_3.CHO \quad + NADH \xrightarrow{\textit{alcohol dehydrogenase}}$
(**acetaldehyde**)

$$CH_3CH_2OH + NAD^+$$
$$(\textbf{ethanol})$$

2 $CH_3.CO.COOH + NADH \xrightarrow{\textit{lactic dehydrogenase}}$
   (**pyruvic acid**)

$$CH_3.CHOH.COOH + NAD^+$$
$$(\textbf{lactic acid})$$

In both cases the end-product (ethanol or lactic acid) is more reduced than pyruvic acid, so the reactions are coupled with the reoxidation of NADH to NAD$^+$. This allows the Embden–Myerhof pathway to continue, and the cells release either lactic acid or ethanol into the surrounding medium. Most yeasts and mycelial fungi produce ethanol – this is the basis of the alcoholic drinks industry. But several Chytridiomycota produce lactic acid (e.g. *Allomyces, Blastocladiella*), as do humans when our tissue oxygen level is depleted.

Energy-yielding reactions of this type – where an **internal inorganic compound** is the terminal electron acceptor – are defined by the biochemical term **fermentation**. In the first equation above the terminal electron acceptor is acetaldehyde, because this accepts electrons from NADH and is, itself, reduced to ethanol. Similarly, in the second equation, pyruvic acid accepts electrons and is itself reduced to lactic acid.

In terms of energy yield, the conversion of pyruvic acid to either ethanol or lactic acid is very inefficient – only 2 moles of ATP are produced from every mole of sugar metabolized (compared with the potential 38 ATP from aerobic respiration). Fungi therefore need an abundant supply of sugars for growth in anaerobic conditions. We noted in Chapter 6 that fungi also need to be supplied with a wide range of other nutrients in anaerobic conditions, because the TCA cycle and several other reactions do not operate to provide the precursors for biosynthesis.

Alternative terminal electron acceptors

At least some fungi (e.g. *Neurospora crassa, Emericella nidulans*) have an alternative means of coping with anaerobic conditions: they use nitrate in place of oxygen as the terminal electron acceptor in the electron transport chain, so that the full TCA cycle operates. This provides a theoretical yield of 26 ATP from each molecule of glucose metabolized. The yield is lower than when oxygen is the terminal electron acceptor, because the energy differential along the electron-transport chain from NADH/NADPH to nitrate is sufficient to generate only 2 ATP (and only one ATP from flavin nucleotides to nitrate).

All energy-yielding pathways which (i) involve an electron-transport chain and (ii) use **external inorganic** substances as the terminal electron acceptor are described by the term **respiration**. However, a distinction is made between **aerobic respiration** (when oxygen is the terminal electron acceptor) and **anaerobic respiration** (when nitrate is the terminal electron acceptor).

Summary: the central role of the energy-yielding pathways

From Fig. 7.2 it can be seen that many types of substrate can be used as potential energy sources, provided that they can be fed into one of the energy-yielding pathways. For example, pentose sugars such as xylose (a major component of hemicelluloses) can be fed into the pentose-phosphate pathway and metabolized to give energy. An amino acid like glutamic acid can be deaminated to an organic acid (α-ketoglutaric acid in this case) which will feed into the TCA cycle and yield a theoretical 9 ATP during its conversion to oxaloacetate. Similarly, if acetate is supplied as a substrate it can be linked to coenzyme A, and acetyl-CoA can then combine with oxaloacetate (giving citrate) and be processed through the TCA cycle. Fatty acids can be degraded to acetyl-CoA by a process termed β-oxidation (Fig. 7.4) and can then be metabolized in the same way. The only limitation in all these cases is that oxygen (or perhaps nitrate) is required as a terminal electron acceptor. The only substrates that can supply energy through **fermentation** are sugars and sugar derivatives, through the Embden–Meyerhof pathway or the pentose-phosphate pathway.

Coordination of metabolism: balancing the pathways

We noted earlier that several intermediates of the basic energy-yielding pathways serve as precursor metabolites for biosynthesis. For example:

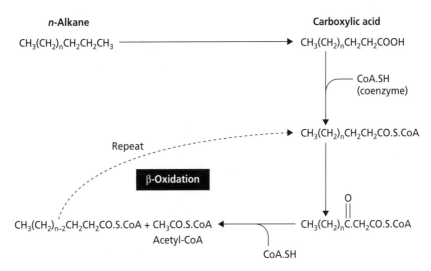

Fig. 7.4 Outline of the reactions in β-oxidation – a process that occurs in the mitochondria of fungi. Long-chain fatty acids are activated by combining with coenzyme A and then enter a repeating cycle in which a molecule of acetyl-coenzyme A is removed in each turn of the cycle. Since most fatty acids have an even number of carbon atoms, this results in the complete conversion of fatty acids to acetyl-CoA. Long-chain hydrocarbons (*n*-alkanes; top left) such as those in aviation kerosene (Chapter 6) can also be processed through this pathway, but first they need to be oxidized by **oxygenase** enzymes, which catalyze the direct incorporation of molecular oxygen into the molecule.

- Dihydroxyacetone phosphate (in the Embden–Meyerhof pathway) can be converted to glycerol for lipid synthesis.
- Acetyl-CoA is used for synthesis of fatty acids and sterols.
- α-ketoglutaric acid is used to produce amino acids of the important glutamate "family" (glutamate, proline, arginine).
- Oxaloacetate is used for the equally important aspartate family (aspartate, lysine, methionine, threonine, isoleucine).

Therefore, the question arises, **how can the pathways for energy production continue when intermediates are removed?**

The main problem would arise when intermediates of the TCA cycle are removed, thereby breaking the cycle. This problem is overcome by special **anaplerotic** reactions (literally "filling-up" reactions) which replenish the missing intermediates. One such reaction sequence is the glyoxylate cycle, described later in a different context. Another type of anaplerotic reaction involves the coupling of CO_2 to pyruvic acid, to give oxaloacetate, as follows:

$$\text{pyruvate} + \text{ATP} + \text{HCO}_3^- \xrightarrow{\textit{pyruvate carboxylase}}$$

$$\text{oxaloacetate} + \text{ADP} + \text{P}_i$$

The vitamin **biotin** serves a crucial role as the cofactor of pyruvate carboxylase, acting as a donor or acceptor of CO_2. Biotin also is the cofactor in other carboxylation reactions. This explains why several fungi need to be supplied with biotin in the growth medium if a fungus cannot synthesize it (see Table 6.1).

How are sugars generated from nonsugar substrates?

Sugars are always needed for the synthesis of fungal walls, nucleic acids, and storage compounds, so how are they produced when a fungus is growing on non-sugar substrates? This is done by a process termed **gluconeogenesis** (the generation of sugars anew) and is shown in Fig. 7.5. Many of the steps are simply a reversal of the Embden–Myerhof pathway, but the step between phosphoenolpyruvate and pyruvate in this pathway is **irreversible** and so must be bypassed.

Consider the case of a fungus growing on acetate (2-carbon) as the sole carbon source. After uptake, acetate is converted to acetyl-CoA and is used to generate oxaloacetate by the **glyoxylate cycle** – a short-circuited form of the TCA cycle. The first reaction step is the cleavage of isocitrate (a 6-carbon intermediate of the TCA cycle) to yield succinate (4-carbon) and glyoxylate (2-carbon). Then glyoxylate is condensed with

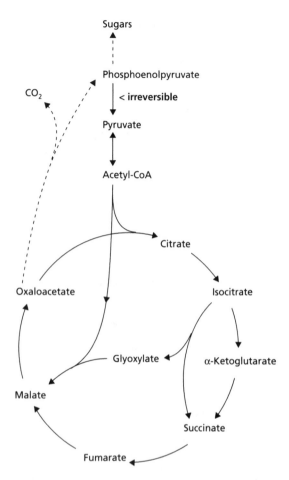

Fig. 7.5 Role of the glyoxylate cycle in generating sugars for biosynthesis when fungi are grown on nonsugar substrates such as acetate or organic acids.

substrates like fatty acids, organic acids, amino acids, ethanol, etc.

Secretion of organic acids as commercial products

Fungi are important commercial sources of organic acids (Chapter 1). For example, if *Aspergillus niger* is grown at high glucose levels (15–20%) and low pH (about 2.0) it will convert most of the sugar to **citric acid** and release this into the culture medium. Large amounts of oxaloacetate must be generated for this, by the carboxylation of pyruvic acid discussed earlier. Similarly *Rhizopus nigricans* (Zygomycota) produces large amounts of **fumaric acid**, another TCA cycle intermediate. In both cases the growth of the fungus is severely (and purposefully) restricted by low pH or some other factor, but the basic metabolic pathways continue to operate. The fungus behaves like a car with its engine ticking over: the fuel (substrate) is not used for growth, so a convenient metabolic intermediate is released as a kind of exhaust product. We shall see later that secondary metabolites such as antibiotics are produced in a similar way. These types of process are termed **energy slippage**; the fact that they occur at all indicates that they are necessary, perhaps because fungi need to keep their normal metabolic processes operating during periods when growth is temporarily halted.

Mobilizable and energy storage compounds of fungi

Fungi have a characteristic range of mobilizable and energy storage compounds, quite different from those of plants, but strikingly similar to those of insects. The main energy-storage compounds of fungi include lipids, **glycogen** (an α-linked polymer of glucose), and **trehalose** (a nonreducing disaccharide, Fig. 7.6).

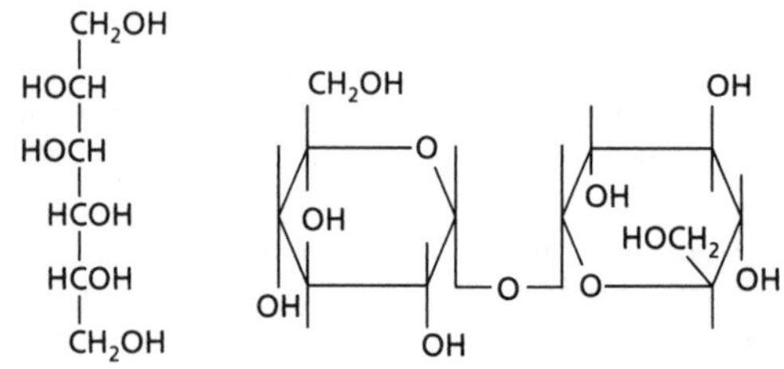

acetyl-CoA to produce malate and thence oxaloacetate. From this point, oxaloacetate is decarboxylated to yield phospoenolpyruvate (PEP), shown below, and sugars are formed from PEP by reversal of the rest of the Embden–Myerhof pathway:

$$\text{oxaloacetate} + \text{ATP} \xrightarrow{\begin{array}{c}\textit{phosphoenolpyruvate}\\ \textit{carboxykinase}\end{array}}$$

$$\text{phosphoenolpyruvate} + CO_2 + \text{ADP} + P_i$$

The important role played by the glyoxylate cycle is evidenced by the fact that its enzyme levels increase more than 20-fold when acetate is supplied as the sole carbon source, and mutant strains of fungi that cannot grow on acetate often are disrupted in one of the steps of this pathway. The same processes as described above could be used to generate sugars from other

Mannitol, derived from fructose phosphate

Trehalose, a non-reducing disaccharide composed of two glucose residues

Fig. 7.6 The structure of mannitol and trehalose, two characteristic "fungal sugars."

The main mobilizable carbohydrates are trehalose and straight-chain sugar alcohols (**polyols**) such as **mannitol** (Fig. 7.6) and **arabitol**, although **ribitol** is common in the Zygomycota. The Oomycota have none of these typical "fungal carbohydrates." Instead their storage compounds are lipids and soluble mycolaminarins (β-linked polymers of glucose), and they translocate glucose or similar sugars.

All these fungal carbohydrates can be derived from the more familiar sugars. For example, trehalose is a disaccharide of glucose, and mannitol is derived from fructose by a one-step reaction involving the enzyme **polyol dehydrogenase**. But a point of special interest is that these compounds are interconvertible, and seem to have different roles in different parts of a mycelial network. Brownlee & Jennings (1981) investigated this for the dry-rot fungus *Serpula lacrymans* (Basidiomycota). They devised a simple experimental system (Fig. 7.7) in which small blocks of wood were colonized by *S. lacrymans* and were then placed on a sheet of Perspex, so that the fungus grew across the Perspex, supported by nutrients translocated from the wood blocks. The fungus grew initially as a broad mycelial colony, but then formed mycelial cords (Chapter 5) behind the colony margin. Different regions of the colony could then be sampled to determine the different proportions of soluble carbohydrates that they contained.

The mycelial cords contained high levels of trehalose, but low levels of other sugars or sugar alcohols. The mid zone of the colonies (Fig. 7.7) contained high levels of trehalose and arabitol (a sugar alcohol), and these high levels were maintained in the submarginal and marginal zones of the colonies. In further experiments, ^{14}C-labeled glucose was added to the wood blocks, and the label was followed through different zones of the mycelial colonies. This again confirmed that arabitol and especially trehalose were the main forms in which sugars and sugar alcohols are translocated within the mycelial system, accumulating in the mid zone and submarginal zone. The conversion of trehalose (a disaccharide) to arabitol near the margin of the colony is likely to increase the osmotic potential, helping to draw water towards the colony margin so that the fungus can grow across nutrient-free zones.

Serpula lacrymans is well known to colonize the structural timbers in buildings. It needs water to colonize the timbers initially, but then its rapid rate of cellulose breakdown and the subsequent breakdown of glucose (to CO_2 and H_2O) generates "metabolic" water, which is sometimes exuded as droplets on the hyphae. In this way, the fungus can spread several meters across brickwork or plaster, drawing the water and nutrients forwards to colonize other timbers (Fig. 7.8).

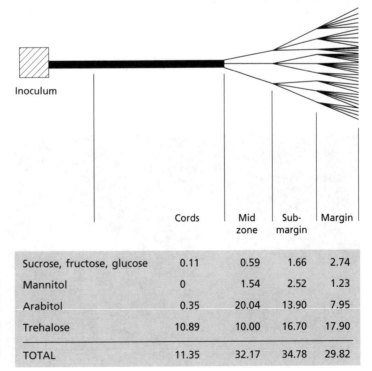

Fig. 7.7 Diagram of the soluble carbohydrate contents in different regions of a colony of *Serpula lacrymans* growing across a sheet of Perspex from a precolonized wood block. Data are per cent of dry weight of the sugars in different zones. (Based on Brownlee & Jennings 1981.)

	Cords	Mid zone	Sub-margin	Margin
Sucrose, fructose, glucose	0.11	0.59	1.66	2.74
Mannitol	0	1.54	2.52	1.23
Arabitol	0.35	20.04	13.90	7.95
Trehalose	10.89	10.00	16.70	17.90
TOTAL	11.35	32.17	34.78	29.82

Fig. 7.8 Part of a wooden door post heavily rotted by *Serpula lacrymans*. The surface of the timber is covered with dense yellow/white mycelial fronds of the fungus. The wood shows the characteristic block-like cracking (arrowheads) typical of dry rot.

Similar mycelial cords and fans of hyphae are seen in many ectomycorrhizal fungi of forest trees (Fig. 7.9), where they are responsible both for the spread of mycorrhizal development and for capturing mineral nutrients and translocating them back to the root system.

The "fungal carbohydrates" are also implicated in many plant-parasitic and symbiotic associations. The initial studies on this were made by Harley with the ectomycorrhizal fungi of beech trees. These fungi produce a substantial sheath of tissue around the root tips (Fig. 7.10), and this sheath can be dissected away for separate chemical analysis. When the leaves of tree seedlings were exposed to $^{14}CO_2$, the label was found mainly as sucrose in the leaf, stem, and root tissues,

Fig. 7.9 Mycelial cords and fans of hyphae produced by a mycorrhizal fungus, *Lactarius pubescens*, on an agar plate.

Fig. 7.10 Mycorrhizas of Basidiomycota on a tree root. The narrow roots are enclosed in a dense fungal sheath.

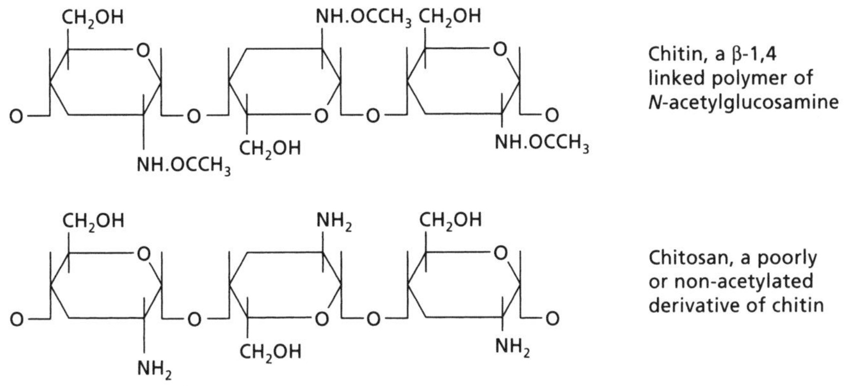

Fig. 7.11 Structure of chitin and its deacetylated derivative, chitosan.

but was then found in the form of mannitol and trehalose in the fungal sheaths. Similar findings have been made for plants infected by rust or powdery mildew fungi (the biotrophic plant pathogens discussed in Chapter 14). When leaves are exposed to $^{14}CO_2$, the label is found initially in the typical plant sugars (sucrose, glucose) but spores produced by the biotrophic fungi are found to contain the label in the form of fungal carbohydrates. Most plants do not metabolize these compounds (except in seeds and fruits), so the conversion of plant sugars to the typical fungal carbohydrates represents a "**metabolic valve**" – a one-way flow of nutrients from the host to the fungus. This may help to contribute to the success of biotrophic plant pathogens.

The actual pathways of nutrient translocation in fungi are unclear, but nutrients can move both forwards and backwards in hyphae, from regions of relative abundance to relative shortage. The tubular vacuolar system of fungi, described in Chapter 3, may be significant in this respect because it can transport fluorescent dyes against the general flow of cytoplasm. The typical fungal carbohydrates may have several other important roles in fungal physiology. For example, mannitol is a common constituent of fungal vacuoles, where it has a major role in regulating cellular pH (Chapter 8).

Chitin synthesis

Chitin is a characteristic component of fungal walls. The synthesis of this polysaccharide follows the pattern of synthesis of many polysaccharides and can be represented by the following general equation:

[Donor-sugar unit] + Acceptor →

Donor + [Acceptor-sugar unit]

For example, in the synthesis of chitin (Fig. 7.11), fructose-6-phosphate (from the Embden–Myerhof pathway) is initially converted to **N-acetylglucosamine** (GlcNAc) by successive additions of an amino group (from the amino acid glutamine) and an acetyl group (from acetyl-CoA). Then GlcNAc reacts with uridine triphosphate (UTP) to form UDP-GlcNAc plus inorganic phosphate. The high energy (activated) sugar unit is then added to the elongating chitin chain, by the enzyme **chitin synthase**, discussed in Chapter 4.

$$UTP + GlcNAc \rightarrow UDP\text{-}GlcNAc + P_i$$

then:

$$UDP\text{-}GlcNAc + n[GlcNAc] \rightarrow UDP + n+1[GlcNAc]$$

A poorly or nonacetylated form of chitin, termed **chitosan**, is found in Zygomycota such as *Mucor* spp. It is synthesized in the same way as chitin but is then deacetylated by the enzyme **chitin deacetylase**.

Lysine biosynthesis

Lysine is an essential amino acid that must be supplied as a dietary supplement for humans and many farm animals, because they are unable to synthesize it. Lysine is produced commercially by large-scale fermentation using the bacterium *Brevibacterium flavum*. The interesting feature of this amino acid is that it is synthesized by two specific pathways that are completely different from one another. These pathways are termed **DAP** and **AAA**, after their characteristic intermediates, α-diaminopimelic acid and α-aminoadipic acid (Fig. 7.12). The AAA pathway is found only in the chitin-containing fungi and some euglenids. All other organisms that synthesize lysine – the plants, bacteria and Oomycota – use the DAP pathway. Such a major

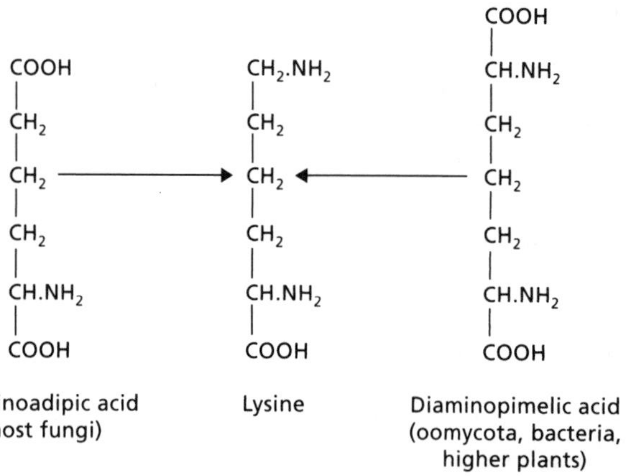

```
  COOH                CH₂.NH₂              COOH
   |                    |                   |
  CH₂                  CH₂                 CH.NH₂
   |                    |                   |
  CH₂ ──────────────▶  CH₂  ◀──────────── CH₂
   |                    |                   |
  CH₂                  CH₂                 CH₂
   |                    |                   |
  CH.NH₂               CH.NH₂              CH₂
   |                    |                   |
  COOH                 COOH                CH.NH₂
                                            |
                                           COOH

α-Aminoadipic acid    Lysine      Diaminopimelic acid
  (most fungi)                    (oomycota, bacteria,
                                    higher plants)
```

Fig. 7.12 Structure of the amino acid lysine, and the characteristic intermediates of the two distinct lysine biosynthesis pathways.

difference between fungi and other organisms indicates a clear, and ancient, evolutionary divergence.

Secondary metabolism

The term secondary metabolism refers to a wide range of metabolic reactions whose products are not directly or obviously involved in normal growth. In this respect, secondary metabolism differs from intermediary metabolism (the normal metabolic pathways discussed earlier in this chapter). Thousands of secondary metabolites have been described from fungi (Turner 1971; Turner & Aldridge 1983), and the only features that they have in common are:

• they tend to be produced at the end of the exponential growth phase in batch culture or when growth is substrate-limited in continuous culture (Chapter 4);
• they are produced from common metabolic intermediates but by special enzymatic pathways encoded by specific genes;
• they are not essential for growth or normal metabolism;
• their production tends to be genus-, species- or even strain-specific.

Interest in this diverse range of compounds stems mainly from their commercial or environmental significance. For example, the **penicillins** (from *Penicillium chrysogenum*), the structurally similar **cephalosporins** (from *Cephalosporium* or *Acremonium* species), and **griseofulvin** (from *P. griseofulvum*) are antibiotics produced commercially from fungi. The darkly pigmented melanins in some fungal walls are also secondary metabolites, as are the **carotenoid pigments** in the conidia of fungi such as *Neurospora crassa*. These compounds help to protect cells from damage caused

by reactive oxygen species, such as hydrogen peroxide and superoxide (Chapter 8). Some secondary metabolites are **plant hormones**, such as the **gibberellins** used commercially in horticulture. The initial discovery of gibberellins was made during studies of a plant disease – the bakanae disease of rice, caused by the fungus *Gibberella fujikuroi*. In this disease the shoots elongate markedly owing to the production of gibberellins by the parasitic fungus. Some secondary metabolites are involved in differentiation, examples being the fungal sex hormones discussed in Chapter 5. Others are marketed as pharmaceuticals, including **ciclosporin A** which suppresses organ rejection in transplant surgery. Further examples include the mycotoxins, such as the **aflatoxins** produced by *Aspergillus flavus* and *A. parasiticus* (Fig. 7.19) and the **ergot alkaloids** produced by *Claviceps purpurea* (Fig. 7.18) and related species.

This list of examples could go on, but many secondary metabolites have no obvious role in the life of the producer organism, and mutated strains that do not produce these compounds often grow as well as the wild-type strains in culture. This raises the question of why secondary metabolites are produced in such diversity and abundance, especially since they are encoded by cassettes of genes that could be expected to be lost if they confer no selective advantage.

One plausible suggestion is that **the process of secondary metabolism is necessary**, regardless of the end products. According to this view, secondary metabolism acts as an overspill or escape valve, to remove intermediates from the basic metabolic pathways when growth is temporarily restricted. This is similar to the explanation we noted earlier for the overproduction of organic acids: an organism needs to maintain the basic metabolic pathways during periods when growth is restricted, but the common metabolic intermediates cannot be allowed to accumulate because they would disrupt normal metabolism. So, it is argued that these intermediates are shunted into secondary metabolic pathways, whose products are either exported from the cell or accumulate as (predominantly) inactive compounds. Indeed, at one stage secondary metabolism was termed "shunt metabolism." It could then be argued that the genes encoding the secondary metabolic pathways are free to mutate (certainly more so than those encoding basic metabolism) and selection pressure would favor mutations that lead to products that benefit a fungus.

For example, antibiotics could be useful in defense of territory, mycotoxins as animal antifeedants, melanin for protection against UV damage, sex hormones for attracting partners, and flavor or odor components of toadstools for attracting insects for spore dispersal. If this view of secondary metabolism is correct, then it seems that many fungi have still to find useful roles for their secondary metabolites, or **we** have yet to find them!

In any case, secondary metabolism is under tight regulatory control. As noted above, secondary metabolites typically are produced towards the end of the exponential growth phase in batch cultures, when growth is limited by a critical shortage of a particular nutrient but when other nutrients are still available. In continuous-culture systems, secondary metabolites can be produced throughout the exponential growth phase. The critical factor is that the genes encoding secondary metabolism are repressed by high levels of particular nutrients, but in chemostats the culture medium can be designed so that these repressor substrates are the growth-limiting nutrients, always present at low concentration because they are utilized as soon as they enter the chemostat (Chapter 4).

The pathways and precursors of secondary metabolism

If we return to Fig. 7.1 at the start of this chapter, we see that a few key intermediates of the basic metabolic pathways provide the starting points for the pathways of secondary metabolism. Some of the more important examples are shown in Table 7.1.

The single most important secondary metabolic pathway is the **polyketide pathway**, which seems to

Table 7.1 Some secondary metabolites derived from different pathways and precursors.

Precursor	Pathway	Metabolites; representative organisms
Sugars	–	Few, e.g. muscarine (*Amanita muscaria*), kojic acid (*Aspergillus* spp.)
Aromatic amino acids	Shikimic acid	Some lichen acids
Aliphatic amino acids	Various, including peptide synthesis	Penicillins (*P. chrysogenum*, *P. notatum*) Fusaric acid (*Fusarium* spp.) Ergot alkaloids (*Claviceps*, *Neotyphodium*) Lysergic acid (*Claviceps purpurea*) Sporidesmin (*Pithomyces chartarum*) Beauvericin (*Beauveria bassiana*) Destruxins (*Metarhizium anisopliae*)
Organic acids	TCA cycle	Rubratoxin (*Penicillium rubrum*) Itaconic acid (*Aspergillus* spp.)
Fatty acids	Lipid metabolism	Polyacetylenes (Basidiomycota fruitbodies and hyphae)
Acetyl-CoA	Polyketide	Patulin (*Penicillium patulum*) Usnic acid (many lichens) Ochratoxins (*Aspergillus ochraceus*) Griseofulvin (*Penicillium griseofulvum*) Aflatoxins (*A. parasiticus*, *A. flavus*)
Acetyl Co-A	Isoprenoid	Trichothecenes (*Fusarium* spp.) Fusicoccin (*Fusicoccum amygdali*) Several sex hormones: sirenin, trisporic acids, oogoniol, antheridiol Cephalosporins (*Cephalosporium* and related fungi) Viridin (*Trichoderma virens*)

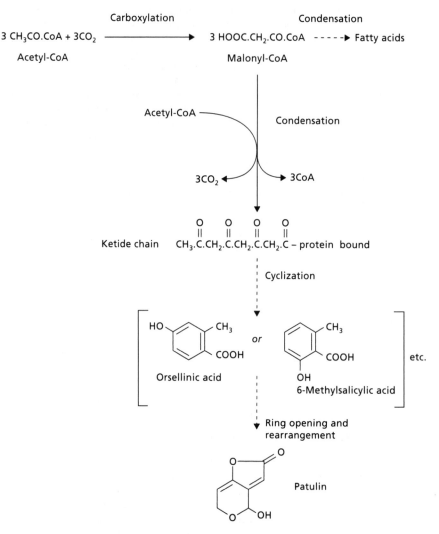

Fig. 7.13 Outline of the polyketide pathway. Various intermediates (shown in brackets) can be formed during cyclization, leading to different end-products. Patulin is derived from 6-methylsalicylic acid. Longer ketide chains than the one shown here can produce multiple ring systems, leading to products such as the aflatoxins (see Fig. 7.19).

have no other role except in secondary metabolism. It is shown in outline in Fig. 7.13. In this case the precursor is **acetyl-CoA**, which is carboxylated to form malonyl-CoA (a normal event in the synthesis of fatty acids), then three or more molecules of malonyl-CoA condense with acetyl-CoA to form a chain. This chain undergoes cyclization, then the ring systems are modified to give a wide range of products. These include the antibiotic **griseofulvin** (from *Penicillium griseofulvum*) used to treat dermatophyte infections of humans (Chapter 17), the potent **aflatoxins** (from *Aspergillus flavus* and *A. parasiticus*) discussed later, the

ochratoxins from various *Penicillium* and *Aspergillus* spp., and the antibiotic or mycotoxin **patulin** from *Penicillium patulum*.

Another important secondary metabolic pathway of fungi is the **isoprenoid pathway** for the synthesis of sterols. Again, **acetyl-CoA** is the precursor, but three molecules of this condense to form mevalonic acid (a 6-carbon compound) which is converted to a 5-carbon **isoprene** unit (Fig. 7.14). Then the isoprene units condense head-to-tail to form chains of various lengths, and the chains undergo cyclization and further modifications (Fig. 7.15). The products of

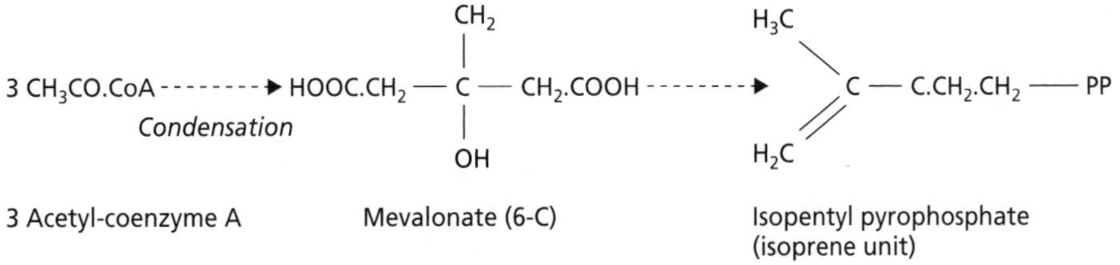

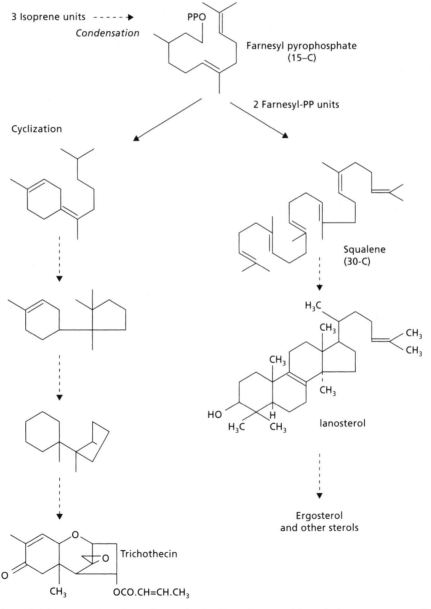

Fig. 7.14 Pathway from acetyl-coA to the production of isoprene units.

3 Acetyl-coenzyme A   Mevalonate (6-C)

Isopentyl pyrophosphate
(isoprene unit)

3 $CH_3CO.CoA$ ------→ $HOOC.CH_2$ — C — $CH_2.COOH$ ------→
             Condensation

CH_2

OH

H_3C

H_2C

C — $C.CH_2.CH_2$ —— PP

3 Isoprene units ----→
     Condensation

PPO

Farnesyl pyrophosphate
(15–C)

Cyclization

2 Farnesyl-PP units

Squalene
(30-C)

H_3C

CH_3

CH_3

CH_3

CH_3

CH_3

HO

H_3C H CH_3

lanosterol

Ergosterol
and other sterols

O

O

O

Trichothecin

CH_3 $OCO.CH=CH.CH_3$

Fig. 7.15 Outline of the isoprenoid pathway for synthesis of sterols, several fungal pheromones, several mycotoxins, and the cephalosporin antibiotics.

this pathway include the mycotoxins of *Fusarium* spp. that grow on moist grain, such as **T-2 toxin** and the **trichothecenes**.

The **shikimic acid pathway**, used normally for the production of aromatic amino acids, provides the precursors for the hallucinogenic secondary metabolites, **lysergic acid** and **psilocybin** in toadstools of *Psilocybe*, and the toxin **muscarine** in toadstools of *Amanita muscaria*. Other pathways lead from aliphatic amino acids to the **penicillins** (see below), **amatoxins** and **phallotoxins** (in toadstools of the "death cap," *Amanita phalloides*), from fatty acids to the volatile **polyacetylenes** produced by mycelia and fruitbodies of several Basidiomycota, and from intermediates of the TCA cycle to the **rubratoxins** of *Penicillium rubrum*.

Against this background, we now consider a few examples of secondary metabolites of special interest.

Penicillins

Penicillin was discovered by Alexander Fleming in 1929 as a metabolite of *Penicillium chrysogenum* (originally misidentified as *P. notatum*) that inhibited the growth of *Staphylococcus*. However, Fleming could not purify the compound in stable form, and this was only achieved in the early 1940s by two British scientists, Florey and Chain, working in the USA during the Second World War. For this work, they shared with Fleming the Nobel Prize for Medicine.

Penicillin is most active against Gram-positive bacteria. Its mode of action is to prevent the cross-linking of peptides during the synthesis of peptidoglycan in bacterial cell walls, so that the walls are weakened and the cells are susceptible to osmotic lysis. Remarkably, penicillin is still one of the front-line antibiotics after more than 60 years of usage. Also of interest is

the fact that the early fermentation systems that made penicillin production possible were initially developed for citric acid production, described in Chapter 1. However, the modern-day penicillins, like the related cephalosporins (peptide antibiotics), are semisynthetic products (Schmidt 2002).

Fig. 7.16 shows the basic structure of penicillin. It is a ring system derived from two amino acids, L-cysteine and D-valine, but is synthesized from a tripeptide precursor (α-aminoadipic acid–cysteine–valine) by the replacement of α-aminoadipic acid with an acyl group (shown as "R" in Fig. 7.16). This step is catalyzed by the enzyme **acyl transferase**. In the early years of commercial production, it was discovered that modification of the culture medium would produce penicillins with different acyl groups, conferring different properties on the penicillin molecule. So a range of penicillins were produced commercially by carefully controlling the supply of acyl precursors in the culture vessels. However, it was then found that several bacteria produce the enzyme **penicillin acylase**, which can be used to remove the acyl side chain and leave the basic molecule, **6-aminopenicillanic acid** (6-APA). Chemists could then attach any desired side chain to this molecule, with a high degree of precision. So the modern production method for **semisynthetic penicillins** involves three stages:

1 Culture of the fungus to produce maximum amounts of any type of penicillin, usually penicillin G which was the type first discovered. This is done by **fed-batch** culture (Chapter 4) in which glucose is added in stages to prevent suppression of the secondary metabolic genes, while the precursor amino acids are supplied in excess. Also, the pH and aeration are carefully controlled, because the penicillin molecule dissociates above pH 7.5 – a problem that

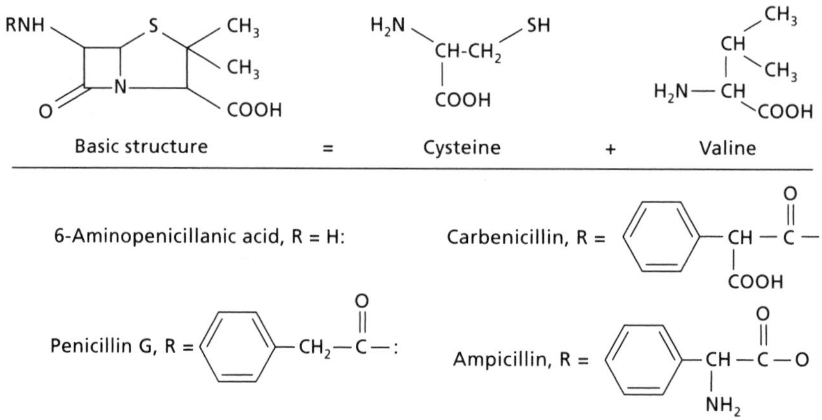

Fig. 7.16 Structure of penicillins.

Table 7.2 Some representative mycotoxins in foodstuffs.

Toxin	Representative fungi	Foodstuff	Effects
Aflatoxins	*Aspergillus flavus, A. parasiticus*	Peanuts, oilseeds	Nephrotoxic, hepatocarcinomas
Ergot alkaloids	*Claviceps purpurea*	Cereals, grasses	Neurotoxic
Fuminosins	*Fusarium moniliforme*	Maize	Human esophageal cancer in Africa?
Ochratoxin A	Some *Aspergillus* and *Penicillium* spp.	Grain crops	Nephrotoxic, kidney carcinoma
Patulin	*Penicillium expansum, Aspergillus clavatus*	Apples	Contact edema, hemorrhage
Sporidesmin	*Pithomyces chartarum*	Grass	Facial eczema of sheep, cattle
Sterigmatocystin	*Aspergillus* spp.	Grain, oilseeds	Hepatocarcinogen
Trichothecenes	*Fusarium* spp., *Stachybotrys chartarum*	Cereals	Abortive, blistering, estrogenic
Zearalenone	*Fusarium*	Cereals	Vulvovaginitis

prevented Fleming from producing penicillin consistently, but which Florey and Chain solved by rigorously controlling the fermentation conditions.

2 Recovery of penicillin from the culture filtrate and treatment with penicillin acylase to produce 6-APA.

3 Chemical addition of specific acyl groups (R in Fig. 7.16) to produce a range of "semisynthetic" penicillins with different properties. For example, **oxacillin** and closely related compounds are resistant to some bacterial β-lactamases – the enzymes that cleave the ring of penicillin G and thus inactivate the antibiotic. **Ampicillin** has significant activity against some Gram-negative bacteria, whereas the natural penicillins act mainly against Gram-positives. **Penicillin V** has enhanced resistance to degradation by stomach acid, so it is one of the best penicillins for oral administration, whereas penicillin G is susceptible to acid.

Despite these advances, all penicillins are susceptible to breakdown by the plasmid-encoded β-lactamases of some enteric bacteria, and penicillins cause allergic reactions in some patients. These problems have been approached by developing a structurally related group of antibiotics, the **cephalosporins**, originally discovered as products of a fungus called "Cephalosporium acremonium", but now produced commercially from strains of bacteria (*Streptomyces* spp).

Mycotoxins

Mycotoxins are a diverse range of compounds from different precursors and pathways. They typically cause toxicity when humans or higher animals ingest them over a relatively long period of time, from low concentrations in improperly stored food or animal feedstuffs. But we cannot exclude the possibility that even brief exposures to these toxins are hazardous.

The problem in establishing this lies in the fact that many years may elapse before the effects of exposure become evident. A few representative examples of mycotoxins are shown in Table 7.2. Edwards *et al.* (2002) review the molecular methods for detecting these compounds.

Many of the mycotoxin problems result from the improper storage of food and feed products, and therefore can be avoided (Chapter 8). But some *Fusarium* species grow on the grains of standing cereal crops in wet field conditions and can produce mycotoxins before the grain is harvested. The *Fusarium* toxins therefore pose particular problems. Several examples are given in Table 7.2. Some toadstools also contain deadly toxins, the classic examples being the "death cap" *Amanita phalloides*, and the "destroying angel" *Amanita virosa*. The toxicity of these is conferred by the **phallotoxins** and **amatoxins**, but these are acutely toxic compounds and are not normally classed as mycotoxins.

The ergot alkaloids and related toxins

The ergot fungus, *Claviceps purpurea* (Fig. 7.17) produces sclerotia that develop in place of the grain in infected cereals and grasses, then the sclerotia (termed **ergots**) fall to the ground and overwinter near the soil surface. In the following summer they produce minute fruiting bodies, resembling drumsticks, which contain many perithecia that release ascospores. The timing of release of these ascospores coincides with the time when the grass flowers open, so the fungus infects the developing ovary and then develops into a new sclerotium, to repeat the life cycle. The harvesting and milling of ergot-contaminated cereals has caused numerous deaths over the centuries. In one form of the disease, termed convulsive ergotism, the nervous system is affected and causes violent convulsions.

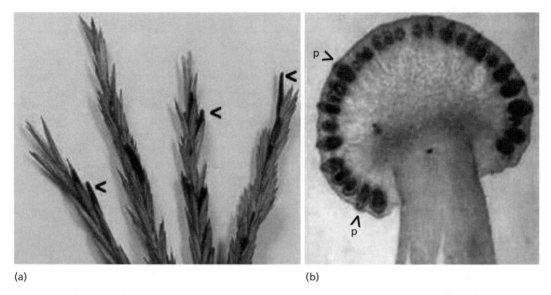

(a)   (b)

Fig. 7.17 (a) Ergots (sclerotia) of *Claviceps purpurea*, produced in place of the grains of a ryegrass plant (*Lolium* sp.). (b) Longitudinal section through the small, stalked fruiting body of *Claviceps*, which releases ascospores from perithecia (p) embedded in the surface of the fruiting body.

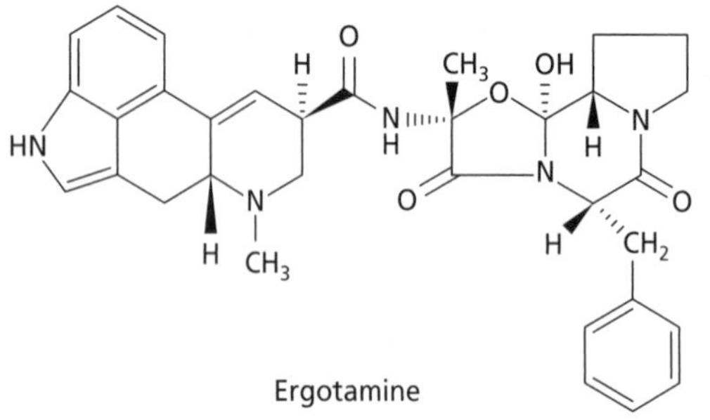

Ergotamine 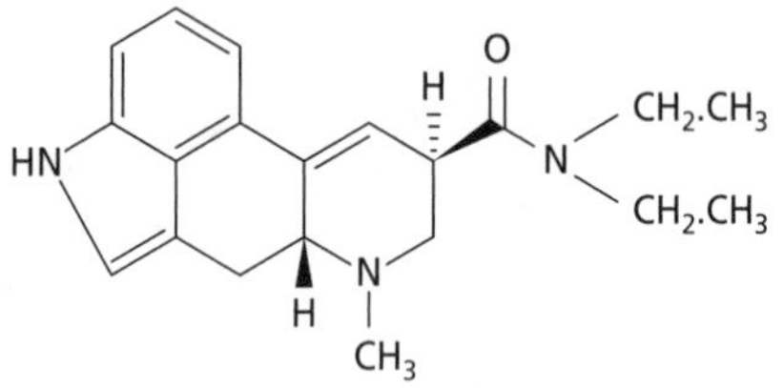   Lysergic acid dimethylamide (LSD)

Fig. 7.18 Structure of the ergot alkaloid ergotamine, and its derivative lysergic acid dimethylamide (LSD).

In another form, gangrenous ergotism, the blood capillaries contract, causing oxygen starvation and serious damage in the tissues. The ergot alkaloids have some important medical uses, to relieve some types of migraine and to control hemorrhaging after childbirth (Keller & Tudzynski 2002). They can also be abused – the indole nucleus of one of the ergot alkaloids, **ergotamine**, is lysergic acid, and this can be altered chemically to produce the hallucinogenic drug, **LSD** (lysergic acid diethylamide; Fig. 7.18). We return to this subject in Chapter 13.

In contrast to the ergot alkaloids, the health hazard posed by the many fungi that grow inconspicuously on stored food and feedstuffs was only recognized in relatively recent times. In the 1950s there were outbreaks

of canine hepatitis (associated with certain types of dog food) and of liver cancer in farmed rainbow trout, associated with cottonseed meal in the fish food. In 1959, 100,000 young turkeys died in Britain after being fed on fungus-contaminated cottonseed meal. The cause was traced to a new type of toxin, termed **aflatoxins**, some of which are among the most potent known carcinogens.

Aflatoxins

Aflatoxins are produced mainly by *Aspergillus flavus* (hence the name **afla**toxin) and *A. parasiticus*, both of which are common in tropical and subtropical conditions. They grow on the root systems and crop debris of groundnuts (peanuts), providing inoculum for colonization of the subterranean groundnut fruits. *A. flavus* can be isolated from the shells of almost any peanuts bought from a grocery store – it is seen as gray or black patches inside the shells. However, most strains do not produce aflatoxin, and even the strains that do produce these compounds require specific conditions for this (Chapter 8). Oil-rich crops such as peanut fruits and cottonseed are especially favorable for aflatoxin production, consistent with the finding that aflatoxin production in laboratory culture is stimulated by lipids. After breakdown of the lipids by lipases, the fatty acids are metabolized to acetyl-CoA by β-oxidation (Fig. 7.4), then the aflatoxins can be synthesized from acetyl-CoA by the polyketide pathway (Fig. 7.13). The review by Hicks *et al.* (2002) provides details of the genetics and biosynthesis of these compounds.

Aflatoxins can be detected in extracts of contaminated foods by thin-layer chromatography, because they show natural green or blue fluorescence under UV irradiation. This colour difference is related to the ring structure of the molecule (Fig. 7.19) and has led to the distinction between aflatoxins of the G (green-fluorescing) and B (blue) types. This is also linked to toxicity, because aflatoxin B_1 is more toxic than aflatoxin G_1.

The toxicity of all these compounds depends on the presence of a double bond on the end furan ring (marked * in Fig. 7.19). Aflatoxins B_1 and G_1 have this double bond and they are carcinogens, whereas alfatoxins B_2 and G_2 lack the double bond and are only weakly toxic. A similar difference is found in the milk of cows that are fed on aflatoxin-contaminated feed. Roughly 5% of the aflatoxin intake by a cow can be found as aflatoxin M in the milk, and it occurs as either aflatoxin M_1 or M_2 depending on whether the cows have ingested aflatoxin B_1 or B_2.

When aflatoxins are absorbed from the gut they are passed to the liver, where the aflatoxins with a double bond (at position *) are metabolized to a highly reactive but unstable form with an epoxide bridge at the position of the double bond on the end furan ring (Fig. 7.19, bottom left). This epoxide group enables the toxin to bind to DNA and depurinate it, causing damage to the genome which can lead to hepatomas. Alternatively (Fig. 7.19, bottom right), the epoxide can give rise to dihydrodihydroxyaflatoxin, which is acutely toxic. All developed countries now impose strict limits on the levels of aflatoxins and other common mycotoxins in foods intended for human consumption.

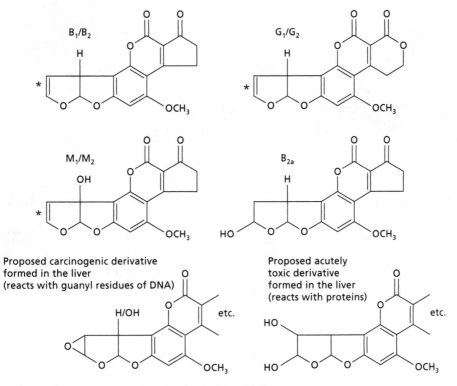

Fig. 7.19 Structures of some common aflatoxins. See text for details.

Sporidesmin

The mitosporic fungus *Pithomyces chartarum* is a saprotroph that grows on the accumulated dead leaf sheaths at the bases of pasture grasses. It is common in parts of New Zealand, Australia, and South Africa, where it causes a condition called **facial eczema** in sheep and cattle. The conspicuous symptoms are seen as blistering sores on parts of the body such as the face and udders that are exposed to sunlight. However this is a secondary symptom, and the primary cause is a mycotoxin, **sporidesmin**, which is present only in the spores of *Pithomyces*, not in the mycelium. The toxin is ingested when the animals graze on pastures where the fungus is sporulating, leading to necrosis of the liver, and scarring and partial blockage of the bile duct, so that the partial breakdown products of chlorophyll accumu-

late in the blood. They are photoactive compounds so they cause photosensitization of the skin where it is not protected by a covering of hair (Fig. 7.20).

P. chartarum requires quite specific conditions for sporulation – a combination of relatively high temperature and high humidity over a period of days. So, in countries where facial eczema is common there is a forecasting system so that farmers can bring the animals into enclosures when the risk of exposure to spores is high. Fungicides can also been used to control this problem.

Endnotes: some further toxins

More than 300 mycotoxins have been identified to date and, although many of them might not be a serious

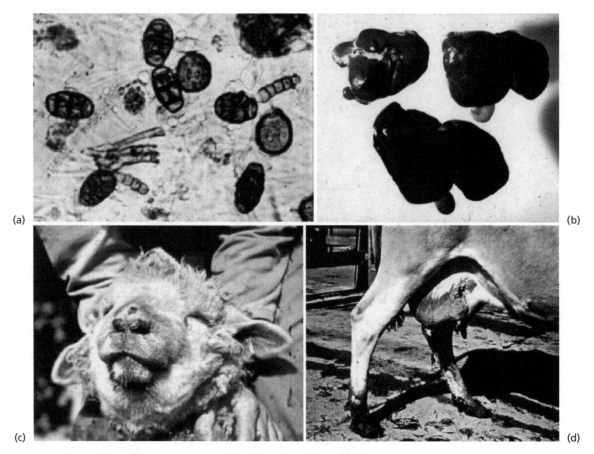

(a) (b) (c) (d)

Fig. 7.20 (a) Spores of *Pithomyces chartarum* collected in a spore sampling device. The darkly pigmented spores are multicellular and shaped like hand-grenades. They contain the toxin sporidesmin. (b) Comparison of a healthy liver (lower) and two damaged livers of animals suffering from facial eczema. (c) Blistering lesions on the nose and snout of a sheep suffering from facial eczema. (d) Photosensitization of the udder of a cow affected by the *Pithomyces* toxin. (Courtesy of E. McKenzie.)

cause for concern, they could present a potential hazard through long-term, low-dosage exposure. We cannot consider all of these compounds, but to end this chapter we will briefly consider three further examples of toxins and potentially toxic conditions.

Patulin

Patulin (Fig. 7.13) was originally discovered as a wide-spectrum antibacterial antibiotic but its development was abandoned because of its mammalian toxicity. It is a small, simple metabolite produced by many species of *Penicillium* and *Aspergillus*, including the common apple-rot fungus *Penicillium expansum* which causes a soft, watery rot when spores enter the apple skin through wounds (Chapter 14). Patulin has been detected in commercially produced apple juice and other fruit juices, especially if these are made from partly decaying apples. It can cause edema and hemorrhaging when ingested, and also seems to be carcinogenic in experimental animals. **It is unwise to eat any part of a rotted apple.**

Roquefort cheese

Roquefort cheese and other blue-veined cheeses are produced from goats' milk and inoculated with the fungus *Penicillium roqueforti*. Roquefort cheese contains low levels of the mycotoxin **roquefortine**, but these levels are not considered to be hazardous. **You make your choice!**

Sick building syndrome

This is a condition associated with dampness and condensation, which encourages the growth of mould fungi. The symptoms include headaches, fatigue, and general malaise, but the precise causes are not easy to determine. *Stachybotrys chartarum* is one of the darkly pigmented fungi that commonly occur in these environmental conditions. It is strongly cellulolytic and grows on substrates like the paper backing of plasterboard. This fungus was implicated in the death of thousands of horses in the Soviet Union, when the animals were fed on *Stachybotrys*-contaminated hay in the 1930s. *Stachybotrys* is one of several fungi (including *Fusarium* spp.) that produce the trichothecene toxins. There has been speculation that *Stachybotrys* is implicated in sick building syndrome, but careful research has consistently failed to provide any evidence that *Stachybotrys* or its toxins are involved in sick building syndrome (Kuhn & Ghannoum 2003).

Cited references

Brownlee, C. & Jennings, D.H. (1981) The content of carbohydrates and their translocation in mycelium of *Serpula lacrymans*. *Transactions of the British Mycological Society* 77, 615–619.

Edwards, S.G., O'Callaghan, J. & Dobson, A.D.W. (2002) PCR-based detection and quantification of mycotoxigenic fungi. *Mycological Research* **106**, 1005–1025.

Hicks, J.K., Shimizu, K. & Keller, N.P. (2002) Genetics and biosynthesis of aflatoxins and sterigmatocystin. In: *The Mycota XI. Agricultural Applications* (Kempken, F., ed.), pp. 55–69. Springer-Verlag, Berlin.

Keller, U. & Tudzynski, P. (2002) Ergot alkaloids. In: *The Mycota X. Industrial Applications* (Osiewacz, H.D., ed.), pp. 158–181. Springer-Verlag, Berlin.

Kuhn, D.M. & Ghannoum, M.A. (2003) Indoor mold, toxigenic fungi, and *Stachybotrys chartarum*: infectious disease perspective. *Clinical Microbiology Reviews* **16**, 144–172.

Schmidt, F.R. (2002) Beta-lactam antibiotics: aspects of manufacture and therapy. In: *The Mycota X. Industrial Applications* (Osiewicz, H.D., ed.), pp. 69–91. Springer-Verlag, Berlin.

Turner, W.B. (1971) *Fungal Metabolites*. Academic Press, London.

Turner, W.B. & Aldridge, D.C. (1983) *Fungal Metabolites*, vol. II. Academic Press, London.

Chapter 8

Environmental conditions for growth, and tolerance of extremes

This chapter is divided into the following major sections:

- some introductory comments
- temperature and fungal growth
- pH and fungal growth
- oxygen and fungal growth
- water availability and fungal growth
- effects of light on fungal growth

In common with all microorganisms, fungi are profoundly affected by physical and physicochemical factors, such as temperature, aeration, pH, water potential, and light. These factors not only affect the growth rate of fungi but also can act as triggers in developmental pathways. In this chapter we consider the effects of environmental factors on fungal growth, including the extremes of adaptation to environmental conditions.

Some introductory comments

A few introductory points must be made to put this chapter into perspective.

1 We will be concerned primarily with the effects of environmental factors on fungal growth, but we must recognize that almost all organisms can grow over a wider range of conditions than will support their full life cycle. For this reason, pure culture studies in laboratory conditions can be misleading – they do not necessarily predict the ability of a fungus to maintain its population in nature.

2 Fungi often can tolerate one suboptimal factor if all others are near optimal, but a *combination* of suboptimal factors can prevent fungal growth. For example, several fungi can grow at low pH (less than 4.0), and several can grow in anaerobic conditions, but few if any fungi can grow when low pH is combined with anaerobiosis.

3 Competitive interactions can restrict the growth of a fungus to a much narrower range than we find in laboratory conditions. A classic early example of this is shown in Fig. 8.1, where wheat plants were

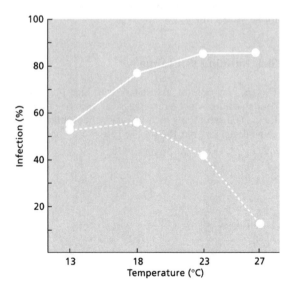

Fig. 8.1 Effect of temperature on infection of wheat seedlings by the take-all fungus of cereals, *Gaeumannomyces graminis*, in sterilized soil (top line) and unsterile soil (bottom line). (From Henry 1932.)

grown in **sterilized soil** and inoculated with the take-all fungus (*Gaeumannomyces graminis*), an aggressive root pathogen (see Fig. 9.11). The fungus caused progressively more disease as the soil temperature was raised from 13 to 23 or 27°C (its optimum for growth in pure culture). But in natural, **unsterile soil** the amount of disease declined as the temperature was raised above 18°. The main reason is that higher temperatures favor other microorganisms more than they favor the take-all fungus. Several antibiotic-producing fluorescent *Pseudomonas* spp. antagonize the take-all fungus, and they provide the basis for current attempts to develop an effective biological control strategy against the take-all disease (Chapter 12).

All these points caution against a simple approach to interpreting the effects of individual environmental factors on fungal activities.

Temperature and fungal growth

Microorganisms are often grouped into four broad categories in terms of their temperature ranges for growth: **psychrophiles** (cold-loving), **mesophiles** (which grow at moderate temperatures), **thermophiles** (heat-loving), and **hyperthermophiles**. A possible fifth category consists of **psychrotolerant** organisms, which can grow at low temperatures (at or below 5°C) but prefer more moderate temperatures.

However, these categories mean different things when applied to different types of microorganism. For example, most fungi are mesophiles and relatively few can grow at or above 37°C (human body temperature) or even above 30°C, whereas many bacteria can grow at this temperature. The upper limit for growth of any fungus (or any eukaryote) is about 62°C. By contrast, some bacteria thrive at 70–80°C, and some archaea can grow at over 100°C, the current record being for the archaeon *Pyrolobus fumarii*, which is found around natural thermal vents and has a temperature **optimum** of 106°C and a **maximum** of 113°C in culture.

The temperature ranges for growth of some representative fungi are shown in Fig. 8.2. By convention, **thermophilic fungi** are defined as having a minimum growth temperature of 20°C or above, a maximum growth temperature of 50°C or above, and an optimum in the range of about 40–50°C. Cooney & Emerson, in 1964, described all the thermophilic fungi that were known at that time. Haheshwari *et al.* (2000) provide updated information on the taxonomy and physiological features of the thermophilic fungi. The temperature ranges of three of these species – *Thermomyces lanuginosus*, *Chaetomium thermophile*, and *Thermomucor pusillus*. – are shown in Fig. 8.2. All three are very common in self-heating composts (Chapter 11) and can also be isolated from materials such as birds' nests and sun-heated tropical soils. The fungus with the highest recorded growth temperature (62°C) is *Thermomyces lanuginosus* (Fig. 8.3), which is very common in composts. However, far fewer than 100 thermophilic fungi have been described – an astonishingly low number, given the large number of habitats from which they can be isolated across the globe. This raises the question of whether different sampling approaches than those used to date might reveal new thermophilic species of biotechnological significance.

Aspergillus fumigatus is an extremely common fungus, found in a wide range of environments, but it is regarded as **thermotolerant** rather than thermophilic, because it can grow at temperatures as low as 12°C and its temperature optimum is less than 40°C. This fungus is remarkable for its very wide temperature range, 12–52°C, and its equally wide range of habitats. It grows commonly in composts, on mouldy grain, and on other decaying organic matter. It can also grow on the hydrocarbons in aviation kerosene. In addition to this, it is an opportunistic invader of the respiratory tract of birds and humans, where it can form persistent colonies termed **aspergillomas** in the lungs (Fig. 8.4). In recent years *A. fumigatus* has become a significant problem in surgical wards, especially in transplant units because it can colonize wounds and grow within the tissues of transplant patients. It is, however, an essentially saprotrophic species. We deal with the medical significance of this important fungus in Chapter 16, because it is unique in the sense that it can readily colonize humans as an opportunistic invader but it seems to lack any specific pathogenicity factors.

Most fungi are **mesophilic**, commonly growing within the range 10–40°C, though with different tolerances within this range. For routine purposes these fungi can usually be grown at room temperature (22–25°C). Two important examples shown in Fig. 8.2 are *Aspergillus flavus*, which produces the potent aflatoxins in stored grain products, such as peanuts and cotton-seed meal (Chapter 7), and *Penicillium chrysogenum*, used for the commercial production of penicillins (Chapter 7).

Psychrophilic fungi are defined as having optimum growth at no more than 16°C and maximum growth of about 20°C. In many cases they would be expected to grow down to 4°C or lower, whilst **psychrotrophic** fungi would be those that can grow at low temperatures but also above 20°C. There are many environments that could suit these organisms, including the polar and alpine regions, and the great oceans which have a stable temperature of about 5°C. Several yeasts have been found in polar and subpolar regions, including a basidiomycetous yeast isolated from clothing of the

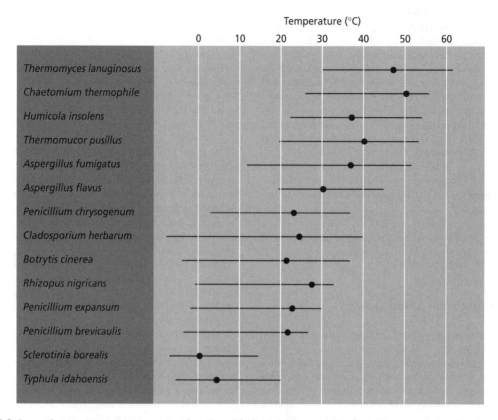

Fig. 8.2 Approximate temperature ranges and optima (black circle) for growth of some representative fungi.

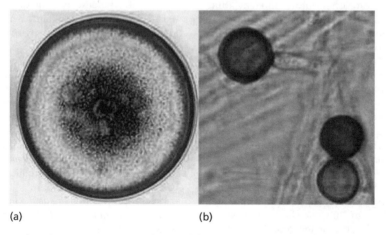

(a)   (b)

Fig. 8.3 *Thermomyces lanuginosus*, a mitosporic thermophilic fungus. (a) 10-day-old colony on malt-extract agar. (b) Large conidia which are darkly pigmented at maturity and are borne singly on short hyphal branches.

prehistoric iceman recently discovered in Siberia (Robinson 2001).

Several cold-tolerant fungi are found in more familiar environments. *Penicillium* species often grow on food residues in domestic refrigerators, at a temperature of around 4°C. Fungi with darkly pigmented hyphae, such as *Cladosporium herbarum*, are reported to grow on meat in cold storage at below 0°C, although this fungus normally occurs as a ubiquitous saprotroph on leaf surfaces. Another species, *Thamnidium elegans* (see

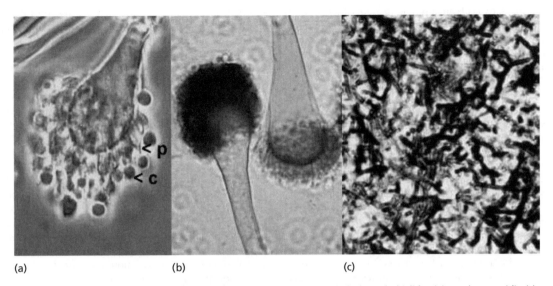

Fig. 8.4 *Aspergillus fumigatus*, a thermotolerant mitosporic fungus. (a,b) Flask-shaped phialides (p) produce conidia (c) on the inflated, club-shaped heads of conidiophores. (c) Clinical specimen, stained to show the hyphae of *A. fumigatus* in an aspergilloma of the lungs.

Fig. 2.9), normally grows on soil or dung, but also grows on cold-stored meat, which it degrades by producing proteases. Perhaps the best examples of psychrophilic fungi are the "snow moulds" such as *Sclerotinia borealis* and *Typhula idahoensis* which are found in the cold, northern states of the USA, where they cause serious damage to cereal crops or grass turf if there is prolonged snow cover. When the snow melts in spring, the cereal or grass leaves have rotted and are covered with sclerotia, which persist until the next snow cover, when they rot the plant tissues again. Up to 50% of the winter-sown cereal crops can be destroyed by these fungi each year. Psychrophilic *Pythium* spp. cause similar problems in Japan, whilst in Britain it is more common to see cereals or turf damaged by *Monographella nivalis*. This fungus is only weakly parasitic, but it invades and rots the plant tissues when their resistance is lowered by prolonged low temperature and low light. Late-season applications of nitrogenous fertilizer can predispose turf to attack because nitrogen promotes lush growth, rendering the plants susceptible to winter damage.

The physiological basis of temperature tolerance

There is no single feature that determines the different temperature ranges of fungi. Instead, there are a range of factors that contribute to temperature tolerance in different organisms. The one common theme is that the ability to grow in the more extreme environments involves adaptation of the whole organism, and the temperature limits will be set by the first cellular component or process that breaks down. Probably for this reason, the cellular complexity of all eukaryotes, including fungi, limits their upper temperature to about 60–65°C, whereas the simpler cellular organization of bacteria and archaea enables some to grow at 80°C or higher. The lower temperature limits for microbial growth are set by factors such as the reduced rates of chemical reactions at low temperatures, the increased viscosity of cellular water at subzero temperatures, and excessive concentrations of cellular ions leading to protein inactivation. As Robinson (2001) stated: "the lower growth temperature limit of psychrophiles is fixed, not by the cellular properties of cellular macromolecules, but instead by the physical properties of aqueous solvent systems inside and outside the cell." There are no substantiated reports of any microorganism growing below −12°C.

Studies on bacteria, yeasts, and filamentous fungi have revealed a general phenomenon – that changes in temperature lead to changes in the fatty acid composition of the membrane lipids. These changes help to ensure that membrane fluidity is optimal for the functioning of membrane transporters and enzymes. This phenomenon is termed **homeoviscous adaptation**. In some psychrophilic yeasts and filamentous fungi the fatty acids and membrane phospholipids are more unsaturated than in mesophiles, and the degree of unsaturation increases at lower temperatures.

Saturated fatty acids are less fluid than unsaturated fatty acids at any given temperature (compare the fluidity of margarine, which has a high content of unsaturated fatty acids, and butter which has a high content of saturated lipids). Consistent with this, an abundance of polyunsaturated fatty acids has been reported in the snow mould, *Monographella nivalis*. And, the thermophilic fungus *Thermomyces lanuginosus* has been found to have a twofold higher concentration of linoleic acid (unsaturated) when grown at 30°C than at 50°C.

Compounds such as the disaccharide trehalose (see Fig. 7.6) often occur in high concentrations in psychrophilic or psychrotrophic organisms, and fungi are reported to accumulate this sugar in response to low temperatures. Trehalose is thought to act as a general stress protectant in the cytosol, and is known to stabilize membranes during dehydration. Polyols (polyhydric alcohols) such as glycerol and mannitol (see Fig. 7.6) also tend to accumulate in response to stress conditions, and mannitol can act as a cryoprotectant.

The enzymes and ribosomal components of thermophiles are reported to be more heat-stable than those of mesophiles when extracted and tested in cell-free systems. This has been shown for thermophilic yeasts as well as for bacteria. The heat stability of enzymes is conferred by increased bonding between the amino acids near the enzyme active site, including bonds other than the heat-labile hydrogen bonds. Heat-stabilizing factors in the cytosol can also contribute to the thermostability of enzymes. In recent years, attention has focused on **heat shock proteins**, which can be synthesized at elevated levels in response to a brief (e.g. 1 hour) exposure to high temperatures, such as 45–55°C. They function as stress proteins, like the equivalent cold shock proteins or proteins produced in response to oxygen starvation. In fact, they are ubiquitous, being found

in organisms of all types, and they are also present in normal conditions. They act like chaperones, helping to ensure that the cell's proteins are correctly folded and that damaged proteins are destroyed. However, it is not clear that they have any specific relevance to the normal temperature ranges of fungi.

Finally, it may be asked whether thermophilic fungi have a particularly high rate of metabolism and a correspondingly high rate of substrate conversion into fungal biomass. In other words, do thermophilic fungi benefit specifically from being able to grow at high temperatures? The relevant growth parameter is the **specific growth rate**, "μ", defined in Chapter 4. Comparisons of thermophiles (e.g. *Thermomyces lanuginosus*) and mesophiles (e.g. *Aspergillus niger*) show no difference in the specific growth rate. So, it seems that thermophilic fungi occupy their high-temperature environments because they are specifically adapted to do so, but they are no more efficient in substrate utilization than are the mesophiles.

Hydrogen ion concentration and fungal growth

The responses of fungi to culture pH need to be assessed in strongly buffered media, because otherwise fungi can rapidly change the pH by selective uptake or exchange of ions. Mixtures of KH_2PO_4 and K_2HPO_4 are commonly used for this purpose. It is then found that many fungi will grow over the pH range 4.0–8.5, or sometimes 3.0–9.0, and they show relatively broad pH optima of about 5.0–7.0. However individual species vary within this "normal" range, as shown by the three representative examples in Fig. 8.5.

Several fungi are **acid-tolerant**, including some yeasts which grow in the stomachs of animals and some

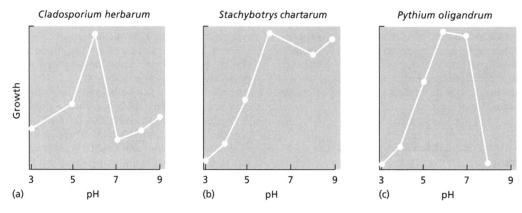

Fig. 8.5 (a–c) pH growth response curves of three representative fungi in laboratory culture (*Pythium oligandrum* is a member of the cellulose-walled Oomycota).

mycelial fungi (*Aspergillus*, *Penicillium*, and *Fusarium* spp.) which will grow at pH 2.0. But their pH optimum in culture is usually 5.5–6.0. Truly **acidophilic** fungi, able to grow down to pH 1 or 2, are found in a few environments such as coal refuse tips and acidic mine wastes; many of these species are yeasts. Among filamentous fungi, the most cited example of an acidophile is *Acontium velatum* which was isolated on laboratory media containing 2.5 N sulfuric acid. This fungus can initiate growth at pH 7.0 but it rapidly lowers the pH of culture media to about 3.0 which is probably close to its optimum.

Strongly alkaline environments, with a pH of about 10, are found in soda lakes and alkaline springs. The fungi that colonize these environments include specialized species of filamentous fungi, such as *Cladosporium*, *Fusarium*, and *Penicillium*, and some yeasts, but they are alkali-tolerant rather than alkalophilic. Some truly alkalophilic *Chrysosporium* species that grow up to pH 11 have been isolated from birds' nests. These fungi are specialized degraders of keratin – the protein found in skin, nails, feathers, and hair (Chapter 16).

The physiological basis of pH tolerance

In all cases that have been investigated, the fungi that grow at extremes of pH are found to have an internal, cytosolic pH of about 7. This intracellular pH can be assessed crudely in extracts of disrupted cells. But, the most accurate modern methods involve the insertion of pH-sensitive electrodes into hyphae, or loading hyphae with pH-sensitive fluorescent dyes that are permeabilized through the plasma membrane. These dyes show peaks of fluorescence at two wavelengths, and the relative size of the two peaks changes with pH, enabling changes of less than 0.1 pH unit to be measured accurately. The findings suggest that the fungal cytosol has strong buffering capacity. Even when the external pH is changed by several units, the cytosolic pH changes by, at most, 0.2–0.3 units. Fungal cells could achieve this homeostasis in several ways – for example, by pumping H^+ ions out through the cell membrane to counteract the inflow of H^+ in acidic environments, by exchange of materials between the cytosol and the vacuoles (which normally have acidic contents), and by the interconversion of sugars and polyols such as mannitol (Chapter 7) which involves the sequestering or release of H^+.

Because the cytosolic pH is so tightly regulated, any perturbation of cytosolic pH can act as an intracellular signal leading to differentiation or change of growth polarity, etc. There are several examples of this in plant and animal cells. One example relating to fungi (although, strictly, it relates to the Oomycota) stems from the fact that the cleavage of zoospores in the

sporangia of *Phytophthora* (Chapter 5) can be induced experimentally by a cold shock. By using a fluorescent pH-indicator dye, Suzaki *et al.* (1996) found that the cytosolic pH was raised transiently from 6.84 to 7.04 by this treatment, but no zoospore cleavage occurred if the sporangia had been microinjected with a buffer of pH 7.0, to prevent any change in cytosolic pH.

Ecological implications of pH

The effects of pH are much easier to investigate in laboratory conditions than in nature, because pH is **not a unitary factor**. In other words, a change of pH can affect many different factors and processes. For example, pH affects the net charge on membrane proteins, with potential consequences for nutrient uptake. It also affects the degree of dissociation of mineral salts, and the balance between dissolved carbon dioxide and bicarbonate ions. Soils of low pH can have potentially toxic levels of trace elements such as Al^{3+}, Mn^{2+}, Cu^{2+}, or Mo^{3+} ions. Conversely, soils of high pH can have poorly available levels of essential nutrients such as Fe^{3+}, Ca^{2+}, and Mg^{2+}. Nevertheless, in general the pH–growth response curves in laboratory culture seem to be relevant to natural situations. For example, *Pythium* spp. are generally intolerant of very low pH but occur in soils above pH 4–5, consistent with the data for *Pythium oligandrum* in Fig. 8.5. Similarly, *Stachybotrys chartarum* is found predominantly in near-neutral and basic soils, again consistent with the data in Fig. 8.5.

Fungi can alter the pH around them and thus to some degree create their own environment. The form in which nitrogen is made available can be a key factor in this respect. If nitrogen is supplied in the form of NH_4^+ ions, which almost all fungi can use in laboratory culture or in nature, then H^+ ions are released in exchange for NH_4^+ and the external pH can be lowered to a value of 4 or less, leading to growth inhibition of the more pH-sensitive fungi such as *Pythium* spp. Conversely, the uptake of NO_3^- can cause the external pH to rise by about 1 unit. Fungi also release organic acids (Chapter 7) which can lower the external pH. Some aggressive tissue-rotting pathogens of plants, such as *Athelia rolfsii* and *Sclerotinia sclerotiorum*, release large amounts of oxalic acid in culture or in plant tissues, lowering the pH to about 4.0. This seems to contribute significantly to pathogenicity, because these fungi also secrete pectic enzymes with acidic pH optima. Oxalic acid can combine with Ca^{2+} in the plant tissues, removing Ca^{2+} from the pectin in plant cell walls, so the walls are more easily degraded by the pectic enzymes (Chapter 14).

Relatively small pH gradients can help to orientate fungal growth, as Edwards & Bowling (1986) found by

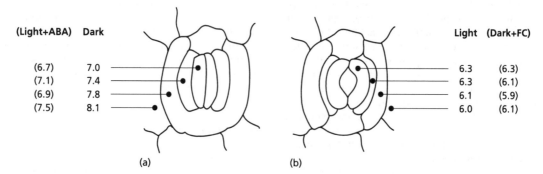

Fig. 8.6 pH of the leaf surface on individual cells around the stomata of *Commelina communis*, measured with a pH microelectrode. (a) When stomata were closed – either by incubation in darkness or (in brackets) by treatment with abscisic acid (ABA) in the light. (b) When stomata were open – either by incubation in the light or (in brackets) by treatment with fusicoccin (FC) in darkness. (Based on Edwards & Bowling 1986.)

mapping the pH around stomata on leaf surfaces with microelectrodes. A pH gradient of more than 1 unit was found around closed stomata, but little or no gradient was detected around open stomata (Fig. 8.6). This was true when the opening of stomata was controlled naturally by light/darkness and also when it was controlled experimentally by chemicals: the plant hormone abscisic acid causes stomata to close in the light, whereas the fungal metabolite fusicoccin (produced by the plant pathogen *Fusicoccum amygdali*) causes stomata to open in darkness.

As we saw in Chapter 5, several plant pathogens infect through stomata and they can be guided by topographical signals. Edwards & Bowling found that pH gradients might also be involved, because germ-tubes of the rust fungus, *Uromyces viciae-fabae*, frequently terminated over open stomata but not over closed stomata. To test the relevance of this, they made nail-varnish replicas of leaf surfaces with open stomata and placed these replicas (of surface pH 6.5) on agar of either pH 6.0 or pH 7.0 so that this was the pH of the (artificial) stomatal pore. When rust spores germinated on the leaf replicas, a significantly higher proportion of the germ-tubes were found to locate the artificial "stomatal pores" of pH 6 than of pH 7, suggesting that the germ-tubes grow down a pH gradient that acts as a cue for locating the stomatal pores.

Oxygen and fungal growth

Most fungi are strict aerobes, in the sense that they require oxygen in at least some stages of their life cycle. Even *Saccharomyces cerevisiae*, which can grow continuously by fermenting sugars in anaerobic conditions (Chapter 7), needs to be supplied with several preformed vitamins, sterols and fatty acids for growth in

the absence of oxygen. *Saccharomyces* also requires oxygen for sexual reproduction. Having established these points, we can group fungi into four categories in terms of their oxygen relationships.

1 Many fungi are **obligate aerobes**. Their growth is reduced if the partial pressure of oxygen is lowered much below that of air (0.21). For example, growth of the take-all fungus of cereals is reduced even at an oxygen partial pressure (Po_2) of 0.18. The thickness of water films around the hyphae can be significant in such cases, because oxygen diffuses very slowly through water, as we saw for the rhizomorphs of *Armillaria mellea* in Chapter 5. Aerobic fungi typically use oxygen as their terminal electron acceptor in **respiration**. This gives the highest energy yield from the oxidation of organic compounds.

2 Many yeasts and several mycelial fungi (e.g. *Fusarium oxysporum*, *Mucor hiemalis*, *Aspergillus fumigatus*) are **facultative aerobes**. They grow in aerobic conditions but also can grow in the absence of oxygen by fermenting sugars. The energy yield from **fermentation** is much lower than from aerobic respiration (Chapter 7), and the biomass production is often less than 10% of that in aerobic culture. However, a few mycelial fungi can use nitrate instead of oxygen as their terminal electron acceptor. This **anaerobic respiration** can give an energy yield at least 50% of that from aerobic respiration.

3 A few aquatic fungi are **obligately fermentative**, because they lack mitochondria or cytochromes (e.g. *Aqualinderella fermentans*, Oomycota) or they have rudimentary mitochondria and low cytochrome content (e.g. *Blastocladiella ramosa*, Chytridiomycota). They grow in the presence or absence of oxygen, but their energy always comes from fermentation. In this respect they resemble the lactic acid bacteria

which always ferment. Fungi of this type are found in nutrient-enriched waters, where fermentable substrates are abundant.

4 A few **obligately anaerobic** Chytridiomycota occur in the rumen and are discussed below.

The microbial consortium of the rumen and other gut environments

A specialized group of Chytridiomycota, protozoa, and bacteria grow in the rumen of animals such as cows, sheep, goats, and camels. These rumen microbes (Fig. 8.7) are **obligate anaerobes** which are killed by exposure to oxygen. They live as part of an intimate and complex microbial community, or **consortium**, in the animal's foregut, which has been modified to form a series of fermentation chambers. Here the intake of food is repeatedly regurgitated, rechewed, and reswallowed, so that it is broken up to expose a maximum surface area for digestion by the community of rumen organisms. After a residence time of 1–2 days, the rumen contents are passed along the digestive tract where many of the microbes are destroyed, providing amino acids for the animal host. The breakdown products from plant constituents such as cellulose and hemicelluloses are similarly absorbed. However, the end products of digestion in the rumen are not simple sugars but instead are a range of short-chain volatile fatty acids (VFAs), principally acetic acid, propionic acid, and butyric acid, together with CO_2, methane, ammonia, and occasionally lactic acid.

Bacteria dominate the rumen community in terms of both their numbers and types. Methanogenic (methane-producing) archaea also are prevalent and generate large amounts of methane as a metabolic end-product. In addition, there are many protozoa which ingest bacteria and other small particles. The final components are the rumen fungi, all of which are obligately anaerobic members of the Chytridiomycota and currently are classified in five genera: *Neocallimastix*, *Caecomyces*, *Piromyces*, *Orpinomyces*, and *Ruminomyces*. These fungi have a major role in degrading the plant structural carbohydrates, such as cellulose and hemicelluloses. Their zoospores show chemotaxis to plant sugars in culture, and rapidly accumulate on chewed herbage in the rumen. Then they encyst and germinate to produce rhizoids that penetrate the plant tissue and release enzymes to degrade the plant cell walls.

Mixed acid fermentation

The rumen chytrids are unusual among fungi because they have a **mixed acid fermentation**, similar to that of the lactic acid bacteria (Theodorou *et al.* 1996). The main products of this fermentation are formic acid ($HCOO^-$), acetic acid, lactic acid, ethanol, CO_2, and H_2, shown in the shaded boxes in Fig. 8.9. The initial fermentation of plant sugars such as glucose is carried out by the Embden–Myerhof pathway, leading to the production of ethanol and lactic acid (derived from pyruvate). This occurs in the cytosol of the chytrid cells. But some of the pyruvate is converted to malate (malic acid) which then enters a special type of organelle, termed the **hydrogenosome** because it produces molecular hydrogen. Hydrogenosomes are characteristic of rumen chytrids and anaerobic protozoa. They are functionally equivalent to the mitochondria of aerobic organisms, and are involved in the generation of ATP by electron transfer (Fig. 8.8). The main end-products released from the hydrogenosome are acetate, formate, CO_2, and H_2. The molecular hydrogen can then be used as an energy source for the

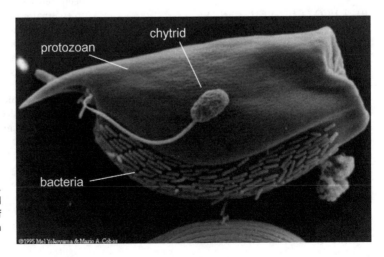

Fig. 8.7 Part of a rumen consortium, comprising protozoa, chytrids, and bacteria. (Reproduced by courtesy of M. Yokayama & M.A. Cobos; Michigan State University.)

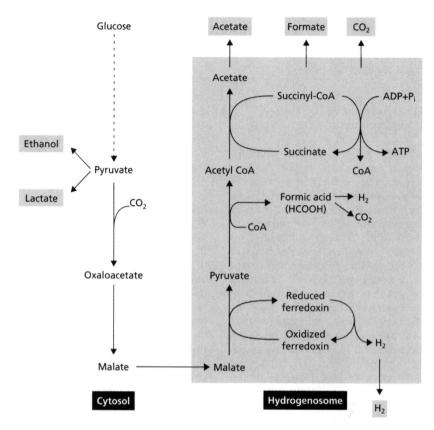

Fig. 8.8 Diagram of the mixed-acid fermentation of the rumen chytrid *Neocallimastix*. The end-products of this fermentation are shown in the small shaded boxes. Part of the fermentation occurs in the cytosol, part in the hydrogenosome. Some of the details are known (Orpin 1993; Marvin-Sikkema *et al.* 1994); others details are assumed, based on knowledge of the mixed-acid fermentation of some enteric bacteria. (After Trinci *et al.* 1994.)

rumen archaea, which generate methane, according to the equation:

$$4H_2 + HCO_3^- + H^+ \rightarrow CH_4 + 3H_2O$$

This methane is repeatedly belched from the animal's gut. Although we have accounted for all the main products of the mixed acid fermentation, it should be recognized that the balance of these products is variable and depends on the types of organism present and other conditions in the rumen.

Comparison of hydrogenosomes and mitochondria

It is now almost universally accepted that eukaryotic (nucleate) cells originated from prokaryotic cells by the process of **endosymbiosis** – when one cell engulfed another, and the engulfed cell progressively lost all or part of its genome and independent existence, to become a cellular organelle. There are likely to have been repeated instances of endosymbiosis, leading to the major organelles that are found in present-day eukaryotes, and resulting in the rapid expansion of unicellular eukaryotes which began about 2 billion years ago. The evidence for this is still retained in eukaryotic cells today. For example:

- Organelles such as mitochondria and chloroplasts are surrounded by double membranes, the outer membrane almost certainly being an enclosing membrane to "contain" the engulfed cell (as happens in more recent endosymbioses such as *Rhizobium* cells in the root nodules of legumes).
- Mitochondria and chloroplasts of present-day eukaryotes contain a residual genome. It is not sufficient for an independent existence, but codes for elements of the electron-transport chain in mitochondria and some of the photosynthetic functions of chloroplasts.

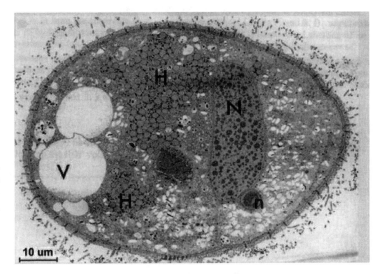

Fig. 8.9 Electron micrograph section of a cell of the anaerobic ciliate, *Nyctotherus ovalis*, showing: macronucleus (N), micronucleus (n), vacuoles (V), and hydrogenosomes (H). The small black dots within the hydrogenosomes are immunogold particles that were tagged to bind to DNA. (Reproduced from Akhmanova *et al.* 1998, with permission; © Macmillan Publishing.)

- Mitochondria and chloroplasts contain ribosomes, which closely resemble the ribosomes of prokaryotes in their sensitivity to antibacterial antibiotics and in the DNA sequence homology of the ribosomal RNA genes.

The origin of hydrogenosomes has remained an unsolved problem until recently. They are bounded by double membranes, have an inner membrane with cristae-like projections, and contain ribosome-like particles resembling those of methanogenic archaea. Some of these features indicate a similarity with mitochondria, but until recently no hydrogenosomal genes had ever been found. However, in 1998 the first evidence of hydrogenosomal genes was found in an anaerobic ciliate, *Nyctotherus ovalis*, which occurs in the hindgut of cockroaches. In electron micrograph sections of this organism the hydrogenosomes showed immunogold labeling when treated with a commercial antiserum against DNA (Fig. 8.9). This evidence was reinforced by using the polymerase chain reaction, with primers directed against conserved regions of the genes coding for **mitochondrial** small subunit (SSU) ribosomal RNA. A homologous SSU rRNA gene was obtained from the anaerobic ciliate, *N. ovalis*, indicating commonality between hydrogenosomes and mitochondria.

Physiology of oxygen tolerance

The existence of **strictly anaerobic organisms** such as rumen chytrids and ciliated protozoa indicates that oxygen can be toxic – it kills these organisms when they are exposed to even low levels of oxygen. The reason for this is well known: several highly reactive forms of oxygen such as O_2^- (**superoxide anion**), H_2O_2 (**hydrogen peroxide**), and OH· (**hydroxyl radical**) are produced inadvertently when oxygen reacts with some of the common cellular constituents such as flavoproteins and quinones. These reactive oxygen species would, ordinarily, damage cellular components such as macromolecules, in the same way as the peroxides of common disinfectants are used to kill microorganisms. So, all organisms that grow in the presence of oxygen need mechanisms for coping with the toxic effects of oxygen, and this is achieved in ways described below.

Superoxide is converted to hydrogen peroxide by the enzyme **superoxide dismutase**, according to the following equation:

$$O_2^- + O_2^- + 2H^+ \rightarrow 2H_2O_2 + O_2$$

Hydrogen peroxide is then converted to water and oxygen by the enzyme **catalase**, according to the following equation:

$$H_2O_2 + H_2O_2 \rightarrow 2H_2O + O_2$$

The combined effect of superoxide and catalase is, therefore, to convert the reactive oxygen species to water. All **obligate anaerobes** lack one or both of these enzymes. For example, *Neocallimastix* has superoxide dismutase but not catalase, so its inability to deal with peroxides probably accounts for its failure to tolerate the presence of oxygen.

The hydroxyl radical is the most toxic of all, but occurs only transiently – mainly as a consequence of ionizing radiation. The small amounts that are produced from H_2O_2 in normal conditions pose little threat if catalase is active.

Carbon dioxide

All fungi need carbon dioxide in at least small amounts for **carboxylation reactions** that generate fatty acids, oxaloacetate, etc. (Chapter 6). Fungi that grow in anaerobic conditions often have a high CO_2 requirement, whereas several aerobic fungi can be inhibited by high concentrations of CO_2.

In normal aerobic respiration, glucose is converted to CO_2 and water according to the familiar empirical equation:

$$C_6H_{12}O_6 + 6O_2 \rightarrow 6CO_2 + 6H_2O$$

However, oxygen and carbon dioxide behave differently from one another in solution, and this can have significant effects on fungal growth. CO_2 dissolves in water to form carbonic acid, which dissociates to bicarbonate ions in a pH-dependent manner. At pH 8, the equilibrium is approximately 3% CO_2 (equivalent to carbonic acid) with 97% HCO_3^- (bicarbonate ion). But at pH 5.5 the equilibrium is approximately 90% CO_2 and 10% HCO_3^-. Studies of fungi in laboratory culture at different pH levels suggest that fungi are more sensitive to the bicarbonate ion than to CO_2 as such. Even so, it can be questioned whether CO_2 (or bicarbonate) is a major growth inhibitor in nature. CO_2 is much more soluble than is oxygen in water, and when the different diffusion coefficients of oxygen and CO_2 are taken into account (the coefficient for CO_2 is actually lower) it can be calculated that CO_2 diffuses about 23 times more rapidly than O_2 in water. Thus, in normal aerobic respiration, when a fungus generates one mole of CO_2 for every mole of O_2 consumed, the oxygen will be depleted in a water film before the CO_2 level reaches a level of even 1%. In short, fungi that grow in undisturbed water, or even 1 millimeter below the surface of an agar plate, are likely to experience significant oxygen depletion. This is one of the major reasons why liquid culture media – and many industrial fermenter-based systems like those for the production of Quorn™ mycoprotein – need to be vigorously aerated (Chapter 4).

Water availability and fungal growth

General principles

All fungi need the physical presence of water for uptake of nutrients through the wall and cell membrane, and often for the release of extracellular enzymes.

Fungi also need intracellular water as a milieu for metabolic reactions. However, water can be present in an environment and still be unavailable because it is bound by external forces. So, in order to understand how the availability of water affects the growth of fungi, we need to establish some basic principles.

The sum of all the forces that act on water and tend to restrict its availability to cells is termed the **water potential**, denoted by φ and defined in terms of energy (negative MegaPascals), where one MPa is equivalent to 9.87 atmospheres, or 10 bar pressure. As familiar reference points, ultra-pure water has a potential of 0 MPa, normal sea water has a potential of about –2.8 MPa, and most plants reach "permanent wilting point" in soils of about –1.5 MPa. The units are negative because the environment exerts a pull on water.

The total water potential consists of the sum of several different potentials: **osmotic potential** (solute binding forces, denoted by φ_π), **matric potential** (physical binding forces, denoted by φ_m), **turgor potential** (φ_p), and **gravimetric potential** (φ_g). So, by summing all these forces:

$$\varphi \text{ (water potential)} = \varphi_\pi + \varphi_m + \varphi_p + \varphi_g$$

It follows that, in order for a fungus to **retain** its existing water, it must generate a potential **equal** to the external water potential φ, and in order to **gain** water from the environment a fungus must generate a (negative) potential **greater** than φ (Papendick & Mulla 1986).

How fungi respond to water potential

Most fungi are highly adept at obtaining water, even in environments that exert a significant water stress. However, the water moulds (Oomycota such as *Saprolegnia* and *Achlya* spp.) are exceptions to this. They have little ability to maintain their turgor against external forces, probably because they grow only or predominantly in freshwater habitats. It used to be thought that fungi needed to remain turgid in order to grow, but hyphae of the water moulds can continue to grow even when they have lost turgor. This is probably because the extension of hyphal tips is achieved by continuous extension of the cytoskeletal components, similar to the extension of pseudopodia in amoeboid organisms. Nevertheless, even the water moulds need to be turgid in order to **penetrate** solid surfaces (Harold *et al.* 1996; Money 2001). Since the penetrating ability of hyphal tips is one of the key features of mycelial fungi, a fungus that lacks turgor is essentially crippled.

Almost all fungi of soil and other terrestrial habitats can grow readily in media of –2 MPa. If the water stress is increased beyond this then the aseptate fungi (Zygomycota) and Oomycota are the first to stop growing (their lowest limit is about −4 MPa). But many septate fungi will grow at between −4 MPa and −14 MPa, and are not considered to be particularly stress-tolerant. In fact, the most stress-tolerant fungi will grow at near-maximum rates at −20 MPa and will make at least some growth at −50 MPa. These fungi with high stress tolerances include the yeast *Zygosaccharomyces rouxii*, used in the traditional production of soy sauce; its lower limit for growth is −69 MPa in sugar solutions.

From these comments it is clear that fungi as a whole can grow in environments where few other organisms can grow. This ability to tolerate water stress is one of the special features of fungi. But, there is an important qualification because the response of a fungus to water stress depends on how this stress is generated. Most fungi tolerate sugar-imposed osmotic stress better than salt-imposed stress – they are inhibited by salt toxicity long before they are inhibited by osmotic potential as such.

Ecological and commercial aspects of water-stress tolerance

Water-stress-tolerant fungi have major economic significance in the spoilage of stored food products, including cereal grains. No fungus can grow on stored grain that has been dried to 14% (w/v) moisture content, but even a slight rise to 15–16% water content will enable the stress-tolerant *Aspergillus* spp. (sexual stage, *Eurotium*) to grow. For example, *Aspergillus amstelodami* will initiate spoilage at −30 MPa in a slightly moist pocket of a grain store. This can set off

a chain reaction, when the spoilage fungi degrade starch to glucose, and thence to CO_2 and water. The metabolic heat generated in this process can cause the water to evaporate and condense elsewhere in the grain mass, so that moulding spreads progressively and eventually paves the way for the growth of less-stress-tolerant fungi.

Figure 8.10 shows how predictive models of post-harvest grain spoilage can be developed in laboratory studies, by combining different environmental factors – in this case, temperature and water potential. The lines on these graph are growth rate **isopleths** (the combinations of temperature and water potential at which different growth rates are seen). The broken white lines represent "minimal" growth rates of 0.1 mm per day on agar plates, the broken black lines represent 2.0 mm per day, and the solid black areas represent 4.0 mm or more per day. Areas that lie outside all of these zones represent potentially safe storage conditions. The data accord with the known biology of the grain-storage fungi. *Aspergillus amstelodami* is very tolerant of water stress and, together with another stress-tolerant fungus, *A. restrictus*, often initiates moulding in grain stores. *Aspergillus fumigatus* is less water-stress-tolerant but the isopleths are skewed, indicating its preference for higher temperatures (maximum 52°C). *Fusarium culmorum*, like many *Fusarium* species, is regarded as a "field fungus". It causes rotting of grains in field conditions if there is a wet harvest season and the grain cannot be dried quickly enough to a safe level.

Similar models can predict the conditions for myco-toxin production (Fig. 8.11). For example, *Aspergillus flavus* produces **aflatoxin** over most of the range of temperature/water potential that supports growth of this fungus. *Penicillium verrucosum* produces **ochratoxin A**

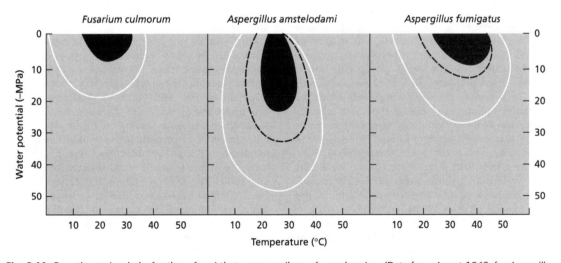

Fig. 8.10 Growth rate isopleths for three fungi that cause spoilage of cereal grains. (Data from Ayerst 1969, for *Aspergillus* spp., and from Magan & Lacey 1984, for *F. culmorum*.)

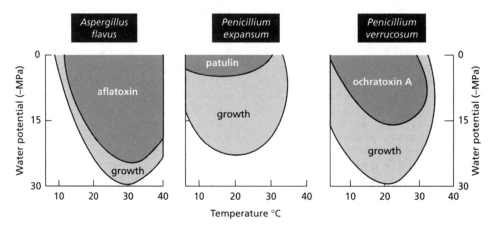

Fig. 8.11 Combinations of temperature and water potential that support mycotoxin production by *Aspergillus flavus*, *Penicillium expansum*, and *P. verrucosum*. (From Northolt & Bullerman 1982.)

(a nephrotoxin) over a narrower part of its growth range. *Penicillium expansum*, which causes soft rot of apples, produces patulin (which causes hemorrhage of the lungs and brains of experimental animals) over a much narrower range.

Physiological adaptations to water stress

Fungi typically respond to low (negative) external water potentials by generating an even lower internal osmotic potential, so that the cells remain turgid. Sometimes this is achieved by selective uptake and accumulation of ions from the environment, an example being the common accumulation of K^+ by marine fungi. However, high ionic levels are potentially damaging to cells, and even the marine fungi seem to take up K^+ primarily as a means of preventing the more toxic Na^+ ion from entering the cell. A more common method of balancing a high external osmotic environment is to accumulate sugars or sugar derivatives that do not interfere with the central metabolic pathways. These osmotically active compounds are termed **compatible solutes**. Glycerol is the most common compatible solute in the highly xerophilic (drought-loving) yeasts and filamentous fungi (Hocking 1993).

Comparisons between water-stress-tolerant fungi and stress-intolerant fungi have shown that both types produce compatible solutes in response to water stress, but differ in their ability to retain the solutes. For example, glycerol is the compatible solute of both *Saccharomyces cerevisiae* (stress-intolerant) and *Zygosaccharomyces rouxii* (stress-tolerant), and both fungi produce it to the same degree when subjected to water stress. But glycerol leaks from *S. cerevisiae* into the culture medium whereas *Z. rouxii* retains glycerol.

This has also been found in a comparison of the stress-tolerant fungus *Penicillium janczewskii* and the stress-intolerant species *P. digitatum*. Membrane fluidity thus seems to be implicated, and there is evidence of a higher content of saturated lipids in the membranes of water-stress-tolerant yeasts.

In recent years, increasing attention has focused on the compatible solutes in fungal spores, and particularly the spores of insect-pathogenic fungi. These fungi have the potential to be developed as commercial **biological control agents** of insects, in place of some of the toxic insecticides currently used (Chapter 15). One of the main limitations at present is that the spores need a sustained high humidity in order to germinate and penetrate an insect cuticle. There is evidence that spores with a high solute content can germinate faster and at somewhat lower humidities than do spores with low solute contents.

With this in mind, attempts are being made to increase the levels of compatible solutes in spores of insect-pathogenic fungi. These solutes can either be derived from nutrient-storage reserves or from nutrients taken up by the cells. Fig. 8.12 illustrates this for two insect-pathogenic fungi, *Beauveria bassiana* and *Metarhizium anisopliae*, grown on media adjusted to different levels of **osmotic stress** with either glucose or trehalose. The data are difficult to interpret, because many of these solutes are interconvertible, as we saw earlier for the mycelia of the dry-rot fungus (see Fig. 7.7). In the case of *Beauveria*, the compatible solute content in the fungal spores increased markedly as the solute concentration of the medium was increased, but mannitol was the main compatible solute in the spores when the fungus was grown on glucose, whereas trehalose accumulated in the spores when the fungus was grown on trehalose. A different

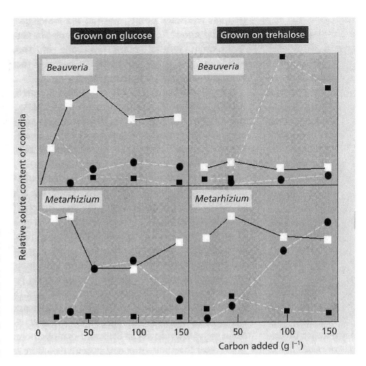

Fig. 8.12 Changes in the compatible solutes of the conidia of two insect-pathogenic fungi (*Beauveria bassiana* and *Metarhizium anisopliae*). Conidia were harvested from agar plates that contained increasing levels of glucose or trehalose. The compatible solutes found in the conidia were: mannitol (white squares), erythritol or arabitol (black circles), and trehalose (black squares). (Adapted from Hallsworth & Magan 1994.)

and more variable pattern was found in *Metarhizium*, but mannitol was again one of the main compatible solutes in the spores, supplemented by erythritol and arabitol as the solute concentration of the medium increased.

A different type of adaptation to water stress is found among the fungi that commonly grow as saprotrophs on the surfaces of living or senescing plant leaves – the environment termed the **phyllosphere** (Fig. 8.13). These fungi (*Cladosporium, Alternaria, Sydowia*, etc.) have darkly pigmented (melanized) hyphae and spores. They do not grow at low (negative) water potentials but they have a remarkable ability to withstand periodic wetting and drying, which few other fungi can tolerate. Park (1982) investigated this by growing these fungi on sheets of transparent cellulose film (Cellophane) placed on top of malt extract agar plates. Then he removed the pieces of film bearing the fungal colonies and suspended them over saturated solutions of $NaNO_2$ or KNO_3 in closed containers. These solutions generate equilibrium relative humidities of 66% and 45%, respectively, equivalent to about –70 MPa and –95 MPa. Even after 2 or 3 weeks in these severe drought conditions, the fungi started to regrow within an hour when the Cellophane films were returned to the agar plates, and they did so from the **original hyphal tips**. By contrast, a range of common soil fungi (e.g. *Fusarium, Trichoderma, Gliocladium*) or typical

food-spoilage fungi (*Penicillium* spp.) never regrew from their original hyphal tips, although many of them could regrow after 24 hours, from spores or surviving hyphal compartments behind the tips. Clearly, the phyllosphere fungi are naturally and specially adapted to the fluctuating moisture conditions in their normal habitat. They are, of course, the same fungi that grow as sooty moulds on kitchen and bathroom walls, where they experience the same wide fluctuations in moisture levels.

Light

Light in the near-ultraviolet (NUV) and visible parts of the spectrum (from about 380 to 720 nm) has relatively little effect on vegetative growth of fungi, although it can stimulate **pigmentation**. In particular, blue light induces the production of carotenoid pigments in hyphae and spores of several fungi, including *Neurospora crassa*. These carotenoids, which also occur in algae and bacteria, are known to quench reactive oxygen species, discussed earlier. The pigments serve to minimize photo-induced damage. Melanins similarly protect cells against reactive oxygen species and ultraviolet radiation.

Light has a much more profound effect on fungal differentiation, acting as a trigger for the production

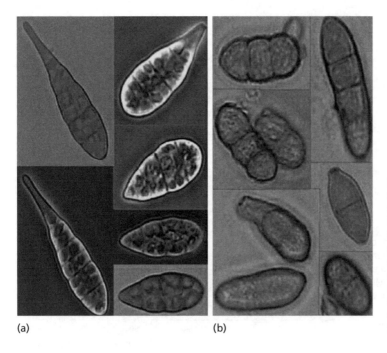

(a)   (b)

Fig. 8.13 Examples of darkly pigmented spores of fungi that commonly grow on senescing leaves and stems of plants, but also grow on bathroom and kitchen walls. These fungi are often termed **dematiaceous hyphomycetes** (with darkly pigmented hyphae and spores). (a) *Several spores of Alternaria* spp. (30–40 μm) viewed by phase-contrast or bright-field microscopy. (b) Spores of *Cladosporium* spp. and related fungi (about 10–15 μm).

of asexual sporing structures or sexual reproductive structures in several (but not all) fungi. For example, the toadstools and similar fruitbodies of many Basidiomycota are formed in response to light, but often with an additional requirement for a low level of CO_2. Often these photoresponses are elicited by NUV or blue light (about 450 nm), implicating a flavin-type photoreceptor. But there is considerable variation in the photoresponses of different fungi, almost certainly related to habitat requirements. For example, *Alternaria* spp. are induced to sporulate by UV irradiation (280–290 nm), and in *Botrytis cinerea* the triggering by NUV is reversed by subsequent exposure to blue light (Chapter 5). The sporulation of some other fungi is regulated by exposure to red/far red light, but this is less common than the blue light responses. Light also has other effects on fungal reproductive structures, notably in eliciting phototropism of the sporangiophores of some Zygomycota and of the ascus tips of some Ascomycota (Chapter 10).

Genetic dissection of blue light perception in *Neurospora crassa*

Neurospora crassa has always been an important model eukaryote, owing to its relatively small genome (about 40 megabases), its rapid growth and manipulability, its ease of genetic manipulation by random and stable integration of foreign DNA, and an abundance of well-characterized mutants. It is also a preferred model organism for investigating light perception, for two reasons. First, it perceives light only in the blue/UV range, and relatively few genes (which are well-characterized) seem to be involved in this response. Second, it shows a pronounced circadian rhythm – a molecular clock that has an innate period length close to 24 hours and that is compensated against temperature and nutrition, but can be reset by environmental (light) cues. In fact, there is a strong interaction between the circadian clock of *Neurospora* and the perception of light and subsequent signal transduction pathways. The recent publication of a high-quality draft genome sequence of *N. crassa* (Galagan *et al.* 2003) should provide further insights into light perception.

Turning specifically to the light response, a recent series of papers have characterized the first fungal blue light photoreceptor. It is a regulatory protein termed White Collar 1 (WC-1) linked to a chromophore – the yellow-pigmented flavin adenine dinucleotide (FAD). The WC-1 protein interacts with DNA to initiate gene transcription. Another protein (WC-2) also acts as a transcription factor and forms a complex with WC-1. This WC-1/WC-2 complex is localized in the nucleus and targets the light signal to the promoters of the blue-light-regulated genes. The wild-type genes for both of these proteins had been known for some time, and strains carrying mutant WC genes were known to be "blind" – they were unable to induce carotenoid synthesis in response to blue light. The remarkable feature of this blue light response in *N. crassa* is that it involves only two protein components and a very short signaling cascade.

Ordinarily, the blue-light-regulated genes that induce carotenoid synthesis in *N. crassa* are downregulated after about 2 hours – a phenomenon termed photoadaptation. But another mutant gene, termed *vivid*, causes a sustained expression of carotenoid genes in the light (hence its name, "vivid"). Analysis of the VIVID protein, which is located in the cytoplasm rather than the nucleus, showed that it also binds to a flavin-type chromophore and that it represents a second blue light photoreceptor (Linden 2002). The VIVID protein seems to be involved particularly in responses to different light intensities, and in modulating the circadian clock. Thus, there seems to be a dual light-perception system, with at least two photoreceptors that serve different roles. Initial light perception, involving WC-1/WC-2, is responsible for dark-to-light transitions. A second system involving VIVID enables *Neurospora* to detect changes in light intensity, and thereby regulate the production of carotenoids for protection against photodamage (Schwerdtfeger & Linden 2003).

Cited references

Akhmanova, A., Voncken, F., Van Alen, T., *et al.* (1998) A hydrogenosome with a genome. *Nature* **396**, 527–528.

Ayerst, G. (1969) The effects of moisture and temperature on growth and spore germination in some fungi. *Journal of Stored Products Research* **5**, 127–141.

Cooney, D.G. & Emerson, R. (1964) *Thermophilic Fungi.* Freeman, San Francisco.

Edwards, M.C. & Bowling, D.J.F. (1986) The growth of rust germ tubes towards stomata in relation to pH gradients. *Physiological and Molecular Plant Pathology* **29**, 185–196.

Galagan, J.E. and 76 other authors (2003) The genome sequence of the filamentous fungus *Neurospora crassa*. *Nature* **42**, 859–868.

Haheshwari, R., Bharadwaj, G. & Bhat, M.K. (2000) Thermophilic fungi: their physiology and enzymes. *Microbiology and Molecular Biology Reviews* **64**, 461–488.

Hallsworth, J.E. & Magan, N. (1994) Effect of carbohydrate type and concentration on polyhydric alcohol and trehalose content of conidia of three entomopathogenic fungi. *Microbiology* **140**, 2705–2713.

Harold, R.L., Money, N.P. & Harold, F.M. (1996) Growth and morphogenesis in *Saprolegnia ferax*: is turgor required? *Protoplasma* **191**, 105–114.

Henry, A.W. (1932) Influence of soil temperature and soil sterilization on the reaction of wheat seedlings to *Ophiobolus graminis* Sacc. *Canadian Journal of Research* **7**, 198–203.

Hocking, A.D. (1993) Responses of xerophilic fungi to changes in water activity. In: *Stress Tolerance of Fungi* (Jennings, D.H., ed.), pp. 233–256. Academic Press, London.

Linden, H. (2002) Blue light perception and signal transduction in *Neurospora crassa*. In: *Molecular Biology of Fungal Development* (Osiewacz, H.D., ed.), pp. 165–183. Marcel Decker, New York.

Magan, N. & Lacey, J. (1984) Effect of temperature and pH on water relations of field and storage fungi. *Transactions of the British Mycological Society* **82**, 71–81.

Marvin-Sikkema, F.D., Driessen, A.J.M., Gottschal, J.C. & Prins, R.A. (1994) Metabolic energy generation in hydrogenosomes of the anaerobic fungus *Neocallimastix*: evidence for a functional relationship with mitochondria. *Mycological Research* **98**, 205–212.

Money, N.P. (2001) Biomechanics of invasive hyphal growth. In: *The Mycota VIII. Biology of the Fungal Cell* (Howard, R.J. & Gow, N.A.R., eds), pp. 3–17. Springer-Verlag, Berlin.

Northolt, M.D. & Bullerman, L.B. (1982) Prevention of mould growth and toxin production through control of environmental conditions. *Journal of Food Production* **45**, 519–526.

Orpin, C.G. (1993) Anaerobic fungi. In: *Stress Tolerance of Fungi* (Jennings, D.H., ed.), pp. 257–273. Academic Press, London.

Papendick, R.I. & Mulla, D.J. (1986) Basic principles of cell and tissue water relations. In: *Water, Fungi and Plants* (Ayres, P.G. & Boddy, L., eds), pp. 1–25. Cambridge University Press, Cambridge.

Park, D. (1982) Phylloplane fungi: tolerance of hyphal tips to drying. *Transactions of the British Mycological Society* **79**, 174–178.

Robinson, C.H. (2001) Cold adaptation in Arctic and Antarctic fungi. *New Phytologist* **151**, 341–353.

Schwerdtfeger, C. & Linden, H. (2003) VIVID is a flavoprotein and serves as a fungal blue light photoreceptor for photoadaptation. *EMBO Journal*, **22**, 4846–4855.

Suzaki, E., Suzaki, T., Jackson, S.L. & Hardham, A.R. (1996) Changes in intracellular pH during zoosporogenesis in *Phytophthora cinnamomi*. *Protoplasma* **191**, 79–83.

Theodorou, M.K., Zhu, W-Y., Rickers, A., Nielsen, B.B., Gull, K. & Trinci, A.P.J. (1996) Biochemistry and ecology of anaerobic fungi. In: *The Mycota VI. Human and Animal Relationships* (Howard, D.H. & Miller, J.D. eds), pp. 265–295. Springer-Verlag, Berlin.

Trinci, A.P.J., Davies, D.R., Gull, K., *et al.* (1994) Anaerobic fungi in herbivorous animals. *Mycological Research* **98**, 129–152.

Chapter 9

Fungal genetics, molecular genetics, and genomics

This chapter is divided into the following major sections:

- overview: the place of fungi in genetical research
- *Neurospora* and classical (Mendelian) genetics
- structure and organization of the fungal genome
- genetic variation in fungi
- applied molecular genetics of fungi
- returning to the genome
- expressed sequence tags and microarray technology

In this chapter we cover the basic and applied genetics of fungi, including the features that continue to make fungi important model organisms for genetical research. The chapter includes recent molecular approaches in a range of fields such as the analysis of fungal pathogenicity determinants and the development of fungi as "factories" for foreign gene products. It also covers the roles of extrachromosomal genes in aging-related senescence and the effects of fungal viruses (hypoviruses) in suppressing pathogenic virulence.

Overview: the place of fungi in genetical research

For more than 60 years fungi have been major tools for classical genetical research because they have a combination of features unmatched by other eukaryotes:

- They are easy to grow in laboratory conditions and they complete the life cycle in a short time.

- Most fungi are haploid so they are easy to mutate and to select for mutants.
- They have a sexual stage for analysis of the segregation and recombination of genes, and all the products of meiosis can be retrieved in the haploid sexual spores.
- They produce asexual spores so that genetically uniform populations can be bulked up and maintained.

In addition to these points, fungi are eminently suitable for biochemical studies because of their simple nutrient requirements, and because "classical genetics" has provided excellent physical maps of the chromosomal genes. Studies on one fungus in particular – *Neurospora crassa* – led to the classical concept of "one gene, one enzyme", for which Beadle & Tatum received the Nobel Prize in 1945. However, it is more accurate to say that "one gene can encode one enzyme" – the situation is complicated because gene splicing occurs to remove noncoding introns in the pre-messenger RNA.

At the time of writing, the genomes of nearly 200 organisms have been sequenced – mainly bacteria and archaea, but also the genomes of "mouse and man." The "high-quality draft" genome sequences of ten fungi have been published, including *Saccharomyces cerevisiae*, *Neurospora crassa*, *Emericella nidulans*, *Schizosaccharomyces pombe*, the rice blast pathogen *Magnaporthe grisea*, and a wood-rotting fungus, *Phanerochaete chrysosporium*. The first four of these are Ascomycota with well-mapped chromosomes, providing a basis for combining classical and molecular genetics.

* See Online resources for websites that publish updated genome sequences.

Neurospora and classical (mendelian) genetics

Neurospora crassa is one of the most intensively studied fungi for genetical research. It is one of four (possibly five) *Neurospora* species that are **heterothallic**, requiring haploid strains of two different mating types, termed **A** and **a**, for sexual reproduction. An outline of this process was shown in Chapter 2 (see Fig. 2.15). Essentially, it begins when a female receptive hypha termed a trichogyne is fertilized by a "male" spore, the spermatium, of opposite mating type. The cells that will eventually become the asci are separated by septa. These **ascus mother cells** contain two haploid nuclei, one of each mating type. The nuclei fuse to form a diploid, and the ascus elongates. Meiosis within each ascus results in the production of four haploid nuclei, and each of these undergoes one round of mitosis to produce eight nuclei, which are packaged into eight ascospores, linearly arranged within each ascus (Fig. 9.1).

The pattern of gene segregation in an ascus can be followed by making crosses between strains that differ

in biochemical features or spore coat color. For example, Fig. 9.1a shows the pattern of meiosis in a fungus heterozygous at a locus that determines spore coat color. The allele **B** codes for dark spores and the allele **b** codes for pale spores. (Note that each chromosome consists of two chromatids, attached to a centromere, but only one arm of each chromosome is shown.) During the first meiotic division the chromosomes separate. The chromatids then separate in the second meiotic division, and this is followed by mitosis, leading to an ascus containing a linear arrangement of four black ascospores and four pale ascospores.

Figure 9.1b shows a different pattern of segregation of spore color, resulting from crossing over, in which two homologous chromatids break and rejoin, with reciprocal exchange of DNA. The subsequent pattern of spore coat color is different in the final ascus.

Normally, there would be several crossover events (chiasmata, singular chiasma) on any one arm of a chromosome, but this would best be detected by crossing strains that differ at three different loci – for example, loci **X**, **Y**, and **Z** – on one arm of the chromosome. Broadly, the chance of a crossover event occurring between any two gene loci on a chromosome depends on the physical distance between these loci. Similarly, the chance of crossing-over between a gene locus and a centromere depends on the distance between these. So it is possible to construct physical maps of the relative positions of different gene loci on any one arm of a chromosome (**chromosome mapping**) by making repeated crosses involving different gene loci. (This is not exactly true because the frequency of crossing-over tends to be lower near the centromere and higher near the ends of the chromosomes – the telomeres – but it does allow the order of genes to be determined.)

To provide a "real" example of the patterns of gene segregation and recombination in an ascus, Fig. 9.2 shows the results of a cross between two parental strains of *Sordaria brevicollis*, a fungus closely related to *Neurospora crassa*. One strain has the wild-type alleles for buff-colored (**b**) and yellow-colored (**y**) ascospores on one arm of the chromosome. The other strain has mutations at both of these gene loci, indicated as b_m and y_m, where the subscript *m* denotes a mutation (see label 1 in Fig. 9.2).

In this particular example, the outcome of such a cross would normally produce either buff spores (with the alleles **b** and y_m) or yellow spores (with the alleles b_m and **y**). But the **b** and **y** loci are sufficiently far apart that there is about 20% probability of a crossover occurring between these loci (see label 2) so that after meiosis and a subsequent round of mitosis, the eight mature ascospores will display four different colors (see label 3). Spores with the **b** and **y** alleles are white, those with **b** and y_m are buff-colored; those with b_m and y_m are black, and those with b_m and y are yellow-

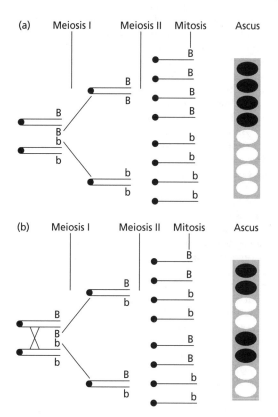

Fig. 9.1 Illustration of the segregation of spore color genes during the first division (a) or second division (b) of meiosis in Ascomycota. See text for details.

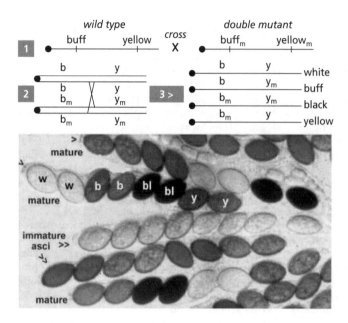

Fig. 9.2 Part of a crushed perithecium of *Sordaria*, showing several asci containing eight ascospores. In the normal, intact perithecium the ascospores would be released through small pores (arrowheads) at the ascus tips. (Courtesy of C. Charier & D.J. Bond.)

colored. Three of the asci in Fig. 9.2 are labeled "mature" – the spores are labeled w (white), b (buff), bl (black), and y (yellow). Each of these asci shows a different pattern of spore segregation, depending on which chromatids were involved in chiasma formation.

Structure and organization of the fungal genome

The genome of an organism includes all the genetic information, not just the genes encoded by the nucleus. In fungi the genome often includes four separate components: the **chromosomal genes**, the **mitochondrial genes**, **plasmids** (and mobile genetic elements), and **fungal virus genes**, which are truly resident genetic elements. Each of these can contribute significantly to the phenotype of fungi.

Chromosomes and chromosomal genes

The main features of fungal nuclei were described in Chapter 3. To recap briefly, most fungi are haploid, but the Oomycota are diploid, a few fungi can alternate between haploid and diploid somatic phases, and some yeasts (e.g. *Candida albicans*) are permanently diploid. Some fungi and fungus-like organisms have polyploid series (e.g. *Allomyces* spp. and *Phytophthora* spp., including *P. infestans*). This can be shown by staining the hyphae with a fluorochrome such as DAPI, which intercalates in the A-T-rich regions of DNA, and then measuring the fluorescence of individual nuclei

under a microscope. Fluorescence increases in a stepwise manner with each increase in the number of chromosome sets (e.g. Tooley & Therrien 1991).

The chromosomes of nearly all fungi are small and highly condensed. They are difficult to count by conventional microscopy of stained cells because the nuclear membrane persists during most of the mitotic cycle. However counts have now been obtained for several fungi by a combination of cytology, linkage analysis (which enables genes to be assigned to particular chromosomes), and pulse-field electrophoresis of extracted chromosomes (Oliver 1987; Sansome 1987). As shown in Table 9.1, the haploid chromosome count of most fungi and fungus-like organisms seems to lie between 6 and 16, but can be as low as 3 or as high as 40.

The nuclear genome size of fungi is small in comparison with many other eukaryotes. For example, the genome size of *Saccharomyces cerevisiae* is just over 12 Mb (12 megabase pairs, or 12 million base pairs), and that of *Schizophyllum commune* (Basidiomycota) is about 37 Mb. Some reported genome sizes of other fungi are given in Table 9.2. The values for *S. cerevisiae* and *Schizophyllum commune* are only about three and eight times larger than the genome of *Escherichia coli* (4 Mb) and much smaller than the genome of the fruit fly *Drosophila* (165 Mb) or humans (about 3000 Mb).

Part of the reason for the small genome size of fungi is that they have little multicopy (reiterated) DNA. It represents only about 2–3% of the genome in *Emericella nidulans* and about 7% in *S. commune*. The reiterated DNA codes mainly for cell components that are needed in large amounts – ribosomal RNA, transfer RNA and chromosomal proteins. However, an abnormally large amount of the genome of the downy mildew pathogen *Bremia lactucae* (Oomycota) is repetitive (65% of the total genome of about 50 Mb); the reason for this is unknown.

Fungi transcribe a substantial amount of the nuclear DNA into messenger RNA – an estimated 33% in *S. commune* and 50–60% in *S. cerevisiae*. Compared with other eukaryotes, therefore, they have relatively little noncoding (redundant) DNA. Fungi resemble other eukaryotes in that their protein-encoding genes contain noncoding DNA sequences termed **introns**. The introns are transcribed into mRNA but are excised before the mRNA is translated into proteins. However, the introns of fungi are very short (often about 50–200 base pairs) compared with those of higher eukaryotes (often 10,000 base pairs or more), and *S. cerevisiae* is unusual because it has very few introns.

Table 9.1 Reported chromosome counts in some representative fungi.

Fungi	Chromosome count
Oomycota	
Phytophthora spp. (many)	9–10
Achlya spp.	3, 6, 8
Saprolegnia spp.	8–12
Pythium	commonly 10 or 20
Chytridiomycota	
Allomyces arbuscula	16
A. javanicus	14 (variable in hybrids and polyploids)
Ascomycota	
Schizosaccharomyces pombe	3
Neurospora crassa	7
Saccharomyces cerevisiae	16
Emericella (*Aspergillus*) *nidulans*	8
Coccidioides posadasii	4
Trichophyton rubrum	4
Magnaporthe grisea	7
Basidiomycota	
Filobasidiella neoformans	11
Schizophyllum commune	11
Coprinus cinereus	13
Puccinia kraussianna	30–40

Mitochondrial genes: normal functions and involvement in aging

Mitochondria contain a small circular molecule of DNA. The size of the mitochondrial genome varies, from as little as 6.6 kb (kilobase pairs) in humans to more than 1 Mb in plants. Fungal mitochondrial genomes are often in the range of 19–121 kb; for example 70 kb in *S. cerevisiae*, and 50 kb in *Schizophyllum commune*. Any variations are due mainly to the amount of noncoding material, because all mitochondrial DNAs code for the same things: some components of the electron-transport chain (including cytochrome c and ATPase subunits), some structural RNAs of the mitochondrial ribosomes, and a range of mitochondrial transfer-RNAs. Both the nuclear and the mitochondrial genes are needed to produce complete, functional mitochondria.

The mitochondrial DNA of fungi has received special attention in relation to aging, because in several filamentous fungi (*Podospora*, *Neurospora*, *Aspergillus*) a single mutation in a single mitochondrion can lead to senescence of the whole colony, when the mutant

Table 9.2 Some reported (approximate) genome sizes of fungi and fungus-like organisms.

Fungi/fungus-like organisms	Genome size (Mb)
Aspergillus fumigatus (potential human pathogen)	30
A. niger (industrially important: citric acid, enzyme production)	30
Candida albicans (human commensal and potential pathogen)	16
Filobasidiella (*Cryptococcus*) *neoformans* (human pathogen)	21
Emericella (*Aspergillus*) *nidulans* (experimental model fungus)	28
Neurospora crassa (experimental model fungus)	40
Phanerochaete chrysosporium (wood-decay basidiomycota)	40
Phytophthora infestans (plant pathogen; Oomycota)	240
Phytophthora sojae (pathogen of soybean; Oomycota)	62
Pneumocystis jiroveci (pathogen of immunocompromised humans)	7.7
Saccharomyces cerevisiae (brewing and breadmaking yeast)	12
Schizosaccharomyces pombe (experimental model; fission yeast)	14

gene causes the gradual displacement of wild-type mitochondrial DNA (reviewed by Esser 1990; Bertrand 1995). The strains of *Podospora* that exhibit the senescence phenotype can be maintained indefinitely as repeatedly subcultured young colonies, but they stop growing, become senescent and die after they have been grown continuously for about 25 days. It has long been known that a non-nuclear "infective factor" is involved, because nonsenescent strains (which never undergo senescence) acquire the ability to senesce when their hyphae anastomose with senescence-prone strains. Moreover, mitochondria were implicated because the onset of senescence could be postponed indefinitely by growing strains in the presence of sublethal doses of inhibitors of mitochondrial DNA synthesis or mitochondrial protein synthesis, but senescence occurred when the inhibitors were removed. More recent studies showed that DNA from strains that were undergoing senescence could be transformed into protoplasts of healthy strains, and the protoplast progeny senesced immediately. The cause of this seems to be a 95 kb plasmid which normally exists as an integral part of the mitochondrial DNA of healthy strains or of juvenile (pre-senescent) cultures of senescent strains. But as the senescent strains age this DNA is excised from the mitochondrial genome, becomes a closed circular molecule, and self-replicates, causing senescence.

Precisely how it does this is still in doubt, but the plasmid shows DNA homology with an intron in one of the mitochondrion genes – the gene that codes for a subunit of cytochrome-c-oxidase, an enzyme essential for normal function of the respiratory electron transport chain. Senescent strains of *Podospora* lack cytochrome-c-oxidase activity, perhaps because the plasmid inserts in the mitochondrial DNA, leading to disruption of gene function or causing mitochondrial gene rearrangements. Dysfunction of cytochrome-c-oxidase or other components of electron transport would be lethal for *Podospora* because this fungus seems unable to grow anaerobically by fermenting sugars.

Plasmids and transposable elements

Plasmids usually are closed-circular molecules of DNA with the ability to replicate autonomously in a cell. However, they can also be linear DNA molecules if the ends are "capped" (like chromosomes) to prevent their degradation by nucleases. Plasmids or plasmid-like DNAs have been found in several fungi. The most notable example is the "two-micron" plasmid of *S. cerevisiae*, so-called because of its 2 μm length as seen in electron micrographs. This plasmid is a closed circular molecule of 6.3 kb, and it is unusual because it is found in the nucleus, where it can be present in up to 100 copies. It has no known function, but in the

past it was used to construct "vectors" for gene cloning in yeast.

Most other plasmids of fungi are found in the mitochondria. The best characterized are the linear DNA plasmids of *Neurospora crassa* and *N. intermedia*. They show a degree of base sequence homology to the mitochondrial genome, suggesting that they are defective, excised segments of the mitochondrial genes. However, some other mitochondrial plasmids of *Neurospora* are closed circular molecules with little or no homology to the mitochondrial genome. They have a variable "unit" length of about 3–5 kb (in different cases) and the units can join head-to-tail to form larger repeats. None of these fungal plasmids has any known function, so they are not like bacterial plasmids that code for antibiotic resistance, pathogenicity, or the ability to degrade pesticides, etc.

Transposons (transposable elements) are short regions of DNA that remain in the chromosome but encode enzymes for their own replication. They produce RNA copies of themselves, and they encode the enzyme, **reverse transcriptase**, which synthesizes new copies of DNA from this RNA template, similar to the action of retroviruses such as HIV. The new copies of DNA can then insert at various points in the same or other chromosomes, leading to alterations in gene expression. Transposons seem to be rare in filamentous fungi, but there are several types in *S. cerevisiae*. The best studied of these are the chromosomal **Ty elements**, present in about 30 copies in yeast cells. Oliver (1987) described the known and possible roles of Ty elements (Fig. 9.3). In addition to a role in altering gene expression, they could have significant effects on chromosomal rearrangements when the "delta sequences" on the ends of these elements combine with one another. These transposable elements seem to have no function, except for self-perpetuation.

The mating-type genes of *S. cerevisiae* are transposable casettes, causing mating-type switching as discussed in Chapter 5. However, mating-type switching cannot occur in *Neurospora crassa*, because individual haploid strains consist of only one mating type.

Viruses and viral genes

Fungal viruses were first discovered in the 1960s, associated with "La France" disease of the cultivated mushroom *Agaricus bisporus*. (The name was coined by British mushroom growers, reflecting the *entente cordiale* that has long characterized Anglo-French relations!) In this disease the fruitbodies are distorted and the fruitbody yield is poor. Electron micrographs of both the hyphae and the fruitbodies showed the presence of many isometric virus-like particles (VLPs), assumed to be the cause of the problem. VLPs were then

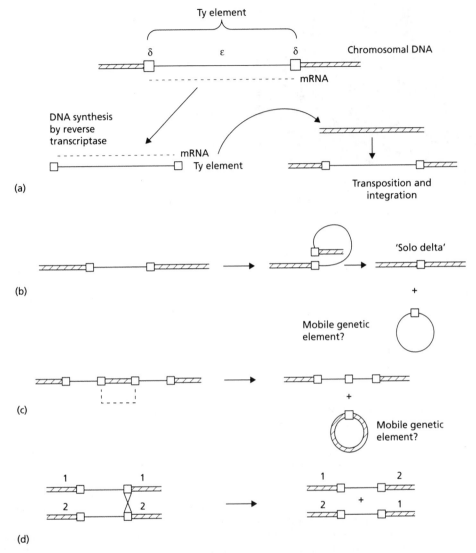

Fig. 9.3 Ty elements of yeast. (a) The Ty element consists of an epsilon region which encodes its own replication, flanked by delta sequences. Copies of the transposon are then inserted elsewhere in the same or in other chromosomes. (b) The delta regions of one Ty element can undergo homologous recombination, leading to a "solo delta" in the chromosome. (c) Homologous recombination between two Ty elements on a chromosome could lead to excision of part of the chromosome. (d) Homologous recombination between Ty elements on different chromosomes could create hybrid chromosomes. (Based on Oliver 1987.)

discovered in other fungi, and by the 1980s they were known in over 150 species, including representatives of all the major fungal groups (Buck 1986). With a few notable exceptions, however, the presence of VLPs is not associated with any obvious disorder, so most fungal viruses seem to be symptomless.

Studies on a range of fungi have shown that fungal viruses (or VLPs) have similar basic features (Fig. 9.4).

- They are isometric particles, 25–50 nm diameter, with a genome of double-stranded RNA (dsRNA), a capsid composed of one major polypeptide, and they code for a dsRNA-dependent RNA polymerase for replication of the viral genome.
- The genome size is variable. Even within a single fungus it ranges from about 3.5 to 10 kb. In some cases this variation is due to internal deletions of a

Fig. 9.4 Isometric virus-like particles extracted from hyphae of *Colletotrichum* sp. The particles aggregate in crystalline arrays *in vitro*. (Courtesy of Rawlinson *et al.* 1975.)

full-length molecule, but in other cases the genome is divided between different capsid particles.

- In most fungi the VLPs are found infrequently in hyphal tips, but they can occur as crystalline arrays in the cytoplasm of older hyphal regions, often closely associated with sheets of endoplasmic reticulum that enclose the aggregates.

- The natural means of transmission of VLPs is via the cytoplasm during hyphal anastomosis (hyphal fusions) or by passage into the asexual spores. VLPs can also enter the sexual spores of some Basidiomycota and in *Saccharomyces*, but this seems to be rare in the sexual spores of mycelial Ascomycota.

VLPs are **resident genetic elements** of fungi because they have no natural mechanism for crossing species barriers. For several years this created a problem in determining their functions, because the association between VLPs and phenotypic characters was only correlative. However, two major developments have changed this and opened the field to critical investigation. First was the discovery that virus-like dsRNA can be present in fungi even when VLPs are absent. In these cases it can be assumed that the virus has lost the ability – and the need – to produce a capsid. Second, protoplasting techniques and transformation systems have now been developed for several fungi, so that dsRNA can be extracted, purified and introduced into protoplasts, or complementary DNA (cDNA) can be derived from dsRNA *in vitro* and then transformed into protoplasts. These approaches, discussed later in this chapter, have shown that the viral dsRNA of *Saccharomyces* and several other yeasts can cause the cells to produce **killer toxins** which act on other strains of the same species.

In the chestnut blight fungus *Cryphonectria parasitica*, also discussed later, the dsRNA causes a marked reduction in pathogenic virulence, creating **hypovirulent** strains that can potentially be used to control the serious **chestnut blight disease**.

In recognition of the unique properties of viral dsRNA and its role in reducing pathogenic virulence, a new name has been approved for this group of viruses – the **hypovirus** group (Hypoviridae).

Genetic variation in fungi

Nonsexual variation: the significance of haploidy

Mutation is the basis of all variation, but mutations are expressed and recombined in different ways depending on the biology of an organism. One of the most significant features of fungi is that they have a haploid genome, whereas all other major groups of eukaryotes are diploid.

Haploid organisms typically expose all their genes to selection pressure. Any mutation will either cause a loss of fitness, or an increase in fitness (e.g. antibiotic or fungicide resistance). This can be beneficial in the short term but the disadvantage is that haploid organisms cannot accumulate mutations that are not of immediate selective value. Diploid organisms have exactly the opposite features. Mutations often are recessive to the wild type, so they are not immediately expressed; instead they accumulate and can be recombined in various ways during sexual crossing.

However, mycelial fungi typically have several haploid nuclei in a common cytoplasm (Chapter 3), and so recessive mutations can be shielded from selection pressure, being complemented by the wild-type nuclei. Mycelial fungi can also expose their genes to selection pressure periodically – whenever they produce uninucleate spores or when hyphal branches develop from only one "founder" nucleus. In other words, **mycelial fungi have many of the advantages of both haploidy and diploidy**. This is not true for the Oomycota, which are diploid. The situation is different again for yeasts because these grow as uninucleate cells. Several yeasts (e.g. *Candida* spp.) are permanently diploid, and even *Saccharomyces* grows as a diploid yeast in nature, owing to mating-type switching (Chapter 5).

Nonsexual variation: heterokaryosis

Heterokaryosis is defined as the presence of two or more genetically different nuclei in a common cytoplasm (*hetero* = different; *karyos* = kernel, or nucleus). Fungi that exhibit this are termed **heterokaryons**, in contrast to **homokaryons** which have only one nuclear type.

Table 9.3 Effects of composition of the growth medium on the ratio of nuclear types in a heterokaryon of *Penicillium cyclopium*. (Data from Jinks 1952.)

Composition of medium (%)		% of nuclei in the heterokaryon		Relative growth rates of homokaryons A and B
Minimal nutrients	Apple pulp	Type A	Type B	A : B
0	100	8.6	91.4	0.47 : 1
20	80	7.8	92.2	0.53 : 1
40	60	11.1	88.9	0.54 : 1
60	40	12.7	87.3	0.67 : 1
80	20	13.5	86.5	1 : 1
100	0	51.8	48.2	1.56 : 1

Heterokaryons can arise in two ways. First, when a mutation occurs in any of the nuclei of a hypha and the mutated nucleus proliferates along with the wild-type nuclei. This must happen very often, but a stable, functional heterokaryon will develop only if the genetically different nuclei proliferate in the apical cells so that all the newly formed hyphae contain both types. A second way in which heterokaryons arise is by tip-to-tip fusion (anastomosis) of the hyphae of two strains (see Fig. 3.6). Again, the nuclei would need to proliferate in the apical cells to form a stable heterokaryon.

Most experimental studies on heterokaryosis have involved the pairing of strains with defined mutations, such as amino acid auxotrophs. The heterokaryon will then behave as a prototroph, capable of growing on minimal medium. The most interesting feature in these cases is that the ratio of nuclear genotypes can vary within wide limits and is influenced by environmental conditions. So, at least in theory, a single heterokaryotic strain can change the frequency distribution of its nuclear types in response to selection pressure. This has been demonstrated experimentally in classic experiments by Jinks (1952), as shown in Table 9.3. A heterokaryon of the apple-rot fungus, *Penicillium cyclopium* was constructed by allowing two different homokaryons to fuse. Then the heterokaryon was grown on agar containing different proportions of apple pulp or minimal nutrient medium. The ratio of nuclear types (which we will call A and B) in the heterokaryon was assessed by testing random samples of uninucleate (homokaryotic) spores from the colony.

When the heterokaryon was grown on apple-pulp medium the proportion of B-type nuclei was very high, but as the amount of apple pulp was lowered so the proportion of A-type nuclei increased, and dramatically so when the heterokaryon was grown on minimal medium. Other experiments of this type have shown that the nuclear ratio in a heterokaryon can vary by up to 1000 : 1 in either direction. If this happens at all commonly in nature it would contribute significantly to **continuous variation** and selection of the best adapted nuclear ratio.

How do heterokaryons break down?

Heterokaryons can break down in two ways (Fig. 9.5) – either during the production of uninucleate spores, or when branches arise that contain only one nuclear type. Many common fungi produce uninucleate spores, often by repeated mitotic division of a "mother" nucleus in a phialide – *Aspergillus*, *Penicillium*, *Trichoderma*, *Gliocladium*, etc. (Fig. 9.5a(i)). Multinucleate spores can also be produced from phialides, if a single nucleus enters the developing spore and then divides to produce several nuclei. For example, this is seen in many *Fusarium* species (Fig. 9.5b). But some other fungi (e.g. *Neurospora*, Fig. 9.5c) produce conidia directly from multinucleate hyphal tips or buds, and these spores will be either homokaryotic or heterokaryotic, depending on whether the cells that produced them were homokaryotic or heterokaryotic.

Heterokaryons also break down if a branch arises that, by chance, contains only one nuclear genotype. This branch can produce further branches and eventually give rise to a homokaryotic sector of the colony (Fig. 9.5a(ii)). If the homokaryon is favored more than the heterokaryon in the prevailing environment then it will expand to occupy progressively more of the colony margin; if not favored it will be suppressed. Figure 9.5d shows an example of this, where a fungal colony was initially darkly pigmented (with daily zones of white aerial hyphae). Two light-colored sectors soon developed and because of their faster growth they progressively expanded. This type of **sectoring** is quite often seen in fungal colonies – either as

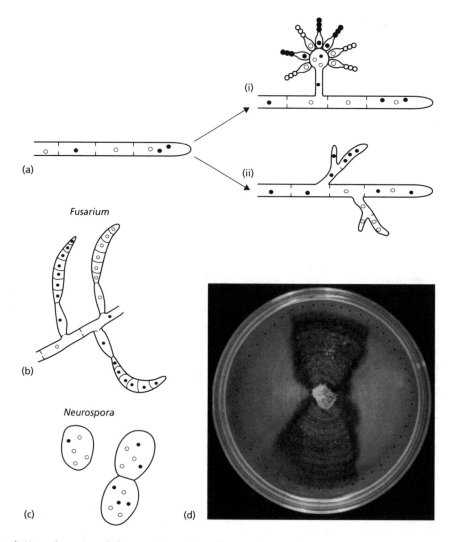

Fusarium

Neurospora

Fig. 9.5 (a–d) Heterokaryosis and the reversion to homokaryons. See text for details.

differences in pigmentation or sporulation. But sectors can also differ in pathogenicity or biochemical properties, and these often go undetected.

The significance of heterokaryosis

Heterokaryosis is a potentially powerful phenomenon, but caution is required because the extent of heterokaryosis in natural environments is largely unknown. Many laboratory studies have involved paired auxotrophic mutants, and these can be regarded as "forced heterokaryons" – there is a strong selection pressure to maintain the heterokaryotic condition. Also, there are significant barriers to the creation of

heterokaryons in nature, because many fungi have nuclear-encoded "heterokaryon incompatibility" (*het*) gene loci. For example, *Neurospora crassa* has at least 10 such loci, with two alleles at each locus, so there are potentially 2^{10} (1024) different **"vegetative" compatibility groups** (VCGs). Pairings of strains of different VCGs lead to hyphal fusion at the points of contact, followed by different degrees of cytoplasmic incompatibility, depending at least partly on the number of *het* loci that any two strains have in common. Typically, the cytoplasm of the fused cells dies (Fig. 9.6) resulting in a clear demarcation zone between opposing colonies (Fig. 9.7). Further details can be found in Glass & Kaneko (2003).

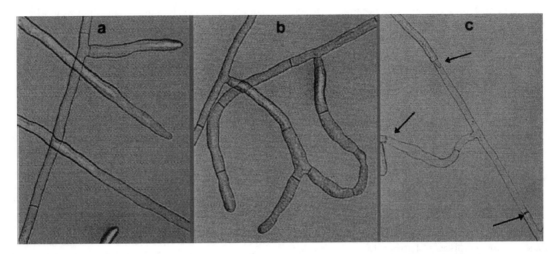

Fig. 9.6 Anastomosis reactions of *Rhizoctonia solani*. (a) No reaction, when strains of different anastomosis groups show no hyphal attraction. (b) Compatible reaction, when strains of the same anastomosis group orientate towards one another and fuse to form a continuous hyphal network. (c) Incompatible reaction, when strains of the same anastomosis group but with different vegetative compatibility genes undergo hyphal fusion, followed by cell death of the fused hyphal compartments. The fused hyphae between the three arrows are dead. (Courtesy of H.L. Robinson.)

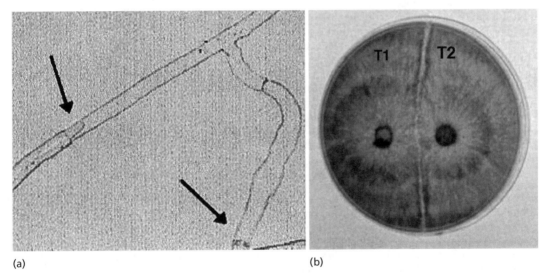

(a)   (b)

Fig. 9.7 (a) Part of Fig. 9.6c at higher magnification, showing regrowth of a hyphal tip into the dead hyphal compartment. (b) Vegetative incompatibility between two strains (T1 and T2) of *Rhizoctonia solani* (Basidiomycota) opposed on an agar plate. The clear demarcation zone between the colonies resulted from post-fusion death of the hyphae where the colony tips fused (Courtesy of P.M. McCabe.)

Nonsexual variation: parasexuality

Many common mitosporic fungi, such as *Aspergillus*, *Penicillium*, *Fusarium*, and *Trichoderma*, seem largely to have abandoned sexual reproduction, because their sexual stages are not seen in laboratory conditions and are found only infrequently, if at all, in nature. For example, the common fungi, *Aspergillus fumigatus* and *A. niger* have never been found to produce a sexual stage. This leads us to ask whether they have alternative mechanisms of genetic recombination, and the answer seems to be that they do have this potential through a mechanism termed **parasexuality** (or the parasexual cycle).

Parasexuality was discovered by Pontecorvo (Pontecorvo 1956), during studies on heterokaryosis in *Emericella* (*Aspergillus*) *nidulans*. He constructed a heterokaryon from two parental strains that had markers at two gene loci (we will call the strains Ab and aB) and was analyzing the homokaryotic spores produced by the heterokaryon. As expected, most of the spores had nuclei of the "parental" types, either Ab or aB, but a significant number were found to be recombinants (AB or ab) and their frequencies were too high to be explained by mutation. Evidently, the genes had recombined in the heterokaryon, although this cannot occur by heterokaryosis alone because the nuclei remain as distinct entities regardless of how they are mixed in the cytoplasm. Further investigation led Pontecorvo to propose a parasexual cycle, involving three stages:

1 **Diploidization.** Occasionally, two haploid nuclei fuse to form a diploid nucleus. The mechanism is largely unknown, and this seems to be a relatively rare event, but once a diploid nucleus has been formed it can be very stable and divide to form further diploid nuclei, along with the normal haploid nuclei. Thus the heterokaryon consists of a mixture of the two original haploid nuclear types as well as diploid fusion nuclei.

2 **Mitotic chiasma formation.** Chiasma formation is common in meiosis, where two homologous chromosomes break and rejoin, leading to chromosomes that are hybrids of the parental types. It can also occur during mitosis but at a much lower frequency because the chromosomes do not pair in a regular arrangement. Nevertheless, the result will be the same when it does occur – the recombination of genes.

3 **Haploidization.** Occasionally, nondisjunction of chromosomes occurs during division of a diploid nucleus, such that one of the daughter nuclei has $2n + 1$ chromosomes and the other has $2n - 1$ chromosomes. Such nuclei with incomplete multiples of the haploid number are termed **aneuploid** (as opposed to euploid nuclei, with n or complete multiples of n). They tend to be unstable and to lose further chromosomes during subsequent divisions. So the $2n + 1$ nucleus would revert to $2n$, whereas the $2n - 1$ nucleus would progressively revert to n. Consistent with this, in *E. nidulans* ($n = 8$) nuclei have been found with 17 ($2n + 1$), 16 ($2n$), 15 ($2n - 1$), 12, 11, 10, and 9 chromosomes.

It must be emphasized that each of these events is relatively rare, and they do not constitute a regular cycle like the sexual cycle. But the outcome would be similar. Once a diploid nucleus has formed by fusion of two haploid nuclei from different parents, the parental genes can potentially recombine. And, the chromosomes that are lost from an aneuploid nucleus during its reversion to a euploid could be a mixture of those in the parental strains.

Significance of parasexuality

Parasexuality has become a valuable tool for industrial mycologists to produce strains with desired combinations of properties. However its significance in nature is largely unknown and will depend on the frequency of heterokaryosis, determined by cytoplasmic incompatibility barriers. Assuming that heterokaryosis does occur, we can ask why several (obviously successful) fungi have abandoned an efficient sexual mechanism of genetic recombination in favor of a more random and seemingly less efficient process. The answer might be that the parasexual events can occur at any time during normal, somatic growth and with no preconditions like those for sexual reproduction. Although each stage of the parasexual process is relatively rare, there are many millions of nuclei in a single colony, so the chances of the parasexual cycle occurring within the colony as a whole may be quite high.

Sexual variation

Sex is the major mechanism for producing genetic recombinants, through crossing-over (chiasma formation) and independent assortment of homologous chromosomes during meiosis. A fungus such as *Emericella* (*Aspergillus*) *nidulans*, with eight chromosomes, could generate 2^8 (i.e. 256) different chromosome combinations by independent assortment alone. This would depend on an efficient outcrossing mechanism. As noted in Chapter 5, many fungi are heterothallic (outcrossing), requiring the fusion of cells of two different mating types. But some are homothallic (e.g. most *Pythium* spp., about 10% of Ascomycota, and a few Basidiomycota) and some exhibit **secondary homothallism** – the sexual spores are binucleate with one nucleus of each mating type (e.g. *Neurospora tetrasperma* and *Agaricus bisporus*). Another variation is seen in *Saccharomyces cerevisiae* and the distantly related fission yeast, *Schizosaccharomyces pombe*. Both of these undergo mating-type switching (Chapter 5). These variations on the normal mechanisms of outcrossing have probably evolved because the sexual spores of fungi function as dormant spores to survive adverse conditions. At least in the short term, **survival is more important than sex!**

Molecular approaches to population structure

Fungal species are dynamic entities. Their populations fragment by geographical isolation or the development of somatic incompatibility barriers, then the fragments diverge by genetic drift or in response to local selection pressure. This is particularly true for the clonal mitosporic fungi, because strains of different vegetative compatibility groups (VCGs) are isolated permanently from one another. Many sexual species also have VCGs but the mating-type genes override the somatic incompatibility genes so that strains of different VCGs can mate.

In general, fungi have too few morphological features for identification of population subunits, so biochemical and molecular tools must be used for this. One approach is to compare the electrophoretic banding patterns of proteins on gels, using either total protein extracts or different forms (isozymes) of particular enzymes such as pectic enzymes, visualized on the gels by color reactions with the enzyme substrate. These **zymograms** (e.g. Fig. 9.8) reflect random mutations in the DNA encoding the enzyme, although not at the enzyme active site which is highly conserved. About 30% of the amino acid changes resulting from mutations will affect the net charge on the protein and thus alter its electrophoretic mobility. Since these changes are random, they tend to accumulate over time and thus reflect the history of a population. As one practical example, MacNish *et al.* (1993) used a combination of pectic zymogram and VCG typing to identify different subgroups of the soil-borne fungus *Rhizoctonia solani* that causes bare patch (stunting) disease of wheat in Australia (Chapter 14). All fungal isolates from within each disease patch were of identical VCG + zymogram group, but different patches often represented different VCG + zymogram groups.

Figure 9.9 shows how this approach can be used to understand the population biology of the fungus.

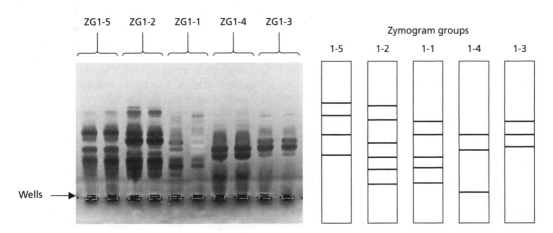

Fig. 9.8 Five distinctive zymogram groups (ZGs) of *Rhizoctonia* strains that cause bare patch disease of wheat in Australia (similar to crater disease, shown in Fig. 14.3). Protein extracts were run on acrylamide gels containing pectin then stained to develop the bands of pectic enzymes. (Courtesy of M. Sweetingham; from MacNish & Sweetingham 1993.)

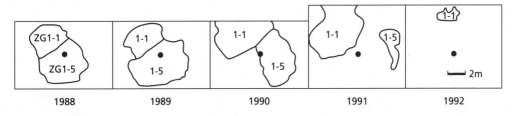

Fig. 9.9 Mapped positions of patches of stunted wheat plants caused by *Rhizoctonia* in a field trial site in Australia. Two adjacent patches were caused by different pectic zymogram groups (ZG1.1 and ZG1.5) which never invaded the territory of the other. The patches expanded and contracted in successive years. The central dot is a fixed reference point. (After MacNish *et al.* 1993.)

The patches can merge, but never overlap because the subgroups seem unable to invade one another"s territory. The patches (and therefore the fungal populations) are also dynamic – they can expand, contract or even disappear in different cropping seasons. When the disease patch disappears the fungus can no longer be found in the soil.

Fungal population structure can also be analyzed by the use of restriction enzymes (endonucleases) on DNA extracted from cells. Any one restriction enzyme (e.g. BamH1) will cut the DNA at specific "target" points. For example, BamH1 cuts DNA at sites where the nucleotide sequence GGATCC occurs (with CCTAGG on the complementary strand of DNA), giving fragments of different lengths that band on gels according to their size. The banding patterns are termed **restriction fragment length polymorphisms** (RFLPs). They are like fingerprints, reflecting accumulations of point mutations that change single nucleotides within the sequence recognized by the restriction enzyme, or chromosomal rearrangements that change the relative positions of these sequences. Different enzymes give different RFLP patterns because they cut the DNA at different sites. They also give different numbers of fragments, because some recognize four-base sequences which are more common than, say, six-base sequences. Also, these enzymes can be used on mitochondrial DNA, which is more highly conserved than the total DNA, so different levels of sensitivity can be selected to analyze both minor and major changes within a fungal population. Kohn (1995) described a good example of this approach for comparing the intercontinental and intracontinental clonal subgroups of the plant pathogen *Sclerotinia sclerotiorum* in both wild and agricultural plant communities.

The **polymerase chain reaction** (PCR) can be used to develop diagnostic probes for specific fungi. The simplest and most common approach involves the random amplification of DNA from a crude DNA extract by adding a random primer composed of, for example, 10 nucleotides. This anneals to a complementary nucleotide sequence on the extracted DNA. Then a DNA polymerase extends along the DNA, reading the sequence of bases along from the primer. This DNA is then amplified during 25–40 successive rounds of PCR. The technique is known as **RAPD**, pronounced "rapids" (random amplified polymorphic DNA). Alternatively, the DNA that codes for particular proteins can be targeted with primers based on knowledge of the partial amino acid sequence of the protein. In any case, selected fragments of the amplified DNA (in single-stranded form) can be suitably tagged and used as diagnostic probes that will bind to equivalent single-stranded DNA sequences in extracts of a sample.

Probes of this type are available commercially for detecting several individual plant pathogens (Fox 1994). For analysis of population structure, both RFLP and RAPD have been used to distinguish the different pathogenic strains of *Ophiostoma* spp. that cause Dutch elm disease, helping to trace the origin and progress of the recent Dutch elm disease epidemics (Pipe *et al.* 1995; Chapter 10). As a further example, PCR-based methods have been developed to detect, and quantify, the levels of several important mycotoxins in food products (Edwards *et al.* 2002). Molecular tools are being developed rapidly in hospital research laboratories, for the molecular typing of fungi that cause human diseases. One of the main drivers for this is to be able to distinguish between individual strains or subgroups within a fungal species, and thereby to aid epidemiological studies. Many examples of these techniques are described in Domer & Kobayashi (2004).

The genes encoding **ribosomal RNA** are widely used for identifying fungi and for constructing phylogenetic trees (see Fig. 1.1). Ribosomal RNA (rRNA) is needed in such large amounts to produce the cellular ribosomes that there are multiple copies of the rRNA genes, often arranged in tandem but separated from one another by untranscribed spacers (Fig. 9.10a). Each single rRNA gene (Fig. 9.10b) has coding information for the three types of rRNA found in eukaryotic ribosomes (18S, 5.8S and 28S), but also contains other valuable information, especially in the **internally transcribed spacers** (ITS) and **externally transcribed spacer** (ETS). The rRNA gene initially produces a pre-rRNA (Fig. 9.10c), which then undergoes processing, including the excision of the spacer regions, to produce the three "mature" rRNAs (Fig. 9.10d). The 18S rRNA has changed sufficiently over evolutionary time to be used as a kind of "molecular clock." But the ITS regions are even more variable, and can often be used to distinguish different species. This is done through PCR, using primers that bind to the 5' end of each DNA strand and progress towards the 3' end. The nucleotide sequences in the ITS regions of the bulked DNA can then be compared with the known sequences of different fungi.

In Chapter 11, we consider some further techniques in the **"biochemical and molecular toolbox"** for fungi.

Applied molecular genetics of fungi

In this section we consider some examples of molecular approaches for understanding fungal behavior or for direct, practical applications. The examples can only be illustrative but many of them represent groundbreaking work and cover some inherently interesting issues.

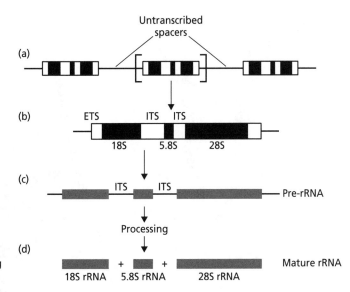

Fig. 9.10 (a–d) Organization and processing of the eukaryotic rRNA genes.

Production of heterologous proteins in *Saccharomyces cerevisiae*

The availability of efficient vectors has enabled *S. cerevisiae* to be used as a "factory" for the products of many foreign (heterologous) genes. One of the most significant products to have emerged from this is a vaccine against hepatitis B. It consists of one of the viral coat proteins, Hepatitis B surface Antigen (HBsAg), and was produced by transforming yeast cells with the viral gene encoding this protein. Hepatitis B vaccine was the first genetically engineered vaccine to be approved for use in humans (it contains no genes). Professor Sir Ken Murray was knighted for his work in developing this vaccine.

Many other proteins have been produced experimentally from yeast, including cellulases, amylases, interferon, epidermal growth factor, and β-endomorphin. However, there have also been problems in using *Saccharomyces* for heterologous protein production. In particular, yeast has a relatively poor ability to remove introns from foreign genes (its own introns are few and small), so it is most efficient when transformed with complementary DNA (cDNA) derived *in vitro* from the messenger RNA of a protein (the introns are spliced out during the processing of mRNA). Yeast also fails to recognize the promotor regions of the genes of other fungi, and it does not always faithfully glycosylate foreign proteins. This can be important because several bioactive proteins are glycoproteins that depend on the sugar chains for their activity.

These difficulties have served to demonstrate that *S. cerevisiae* is genetically quite different from the mycelial fungi. For example, *E. nidulans* can recognize the promoter sequences of the genes of other fungi and also can excise their introns. It may be possible to use *E. nidulans* or the fission yeast *Schizosaccharomyces* as an alternative to *Saccharomyces* for heterologous protein production. But, in general, it is now thought that the best approach is to use cell lines related to the natural producer organism – mammalian cell lines for mammalian gene products, and so on.

Identification of genes for plant pathogenicity and differentiation

We saw in Chapter 4 how differentiation-specific mRNAs were identified in *Schizophyllum commune* by comparing the mRNA profiles of cultures grown in conditions where fruitbodies were or were not produced. The specific mRNAs can then be used as templates to produce cDNA *in vitro*. This cDNA, produced from labeled nucleotides, becomes a probe for binding to complementary sequences of extracted chromosomal DNA. In this way a specific gene can be identified even if nothing is known about the basic genetics of a fungus. By using such techniques, Wessels and his co-workers were able to characterize a unique group of proteins, the hydrophobins, which have major roles in fungal biology and differentiation. These techniques of "reverse genetics" have also made it possible to perform **targeted gene disruption**. For the fungal gene of interest, a cDNA is produced *in vitro* and disrupted to make it nonfunctional. Then it is transformed into the fungus, to replace the original gene by **homologous recombination**. The following example shows the power of this technique.

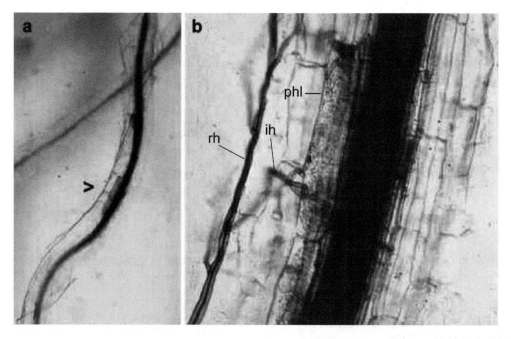

Fig. 9.11 A wheat root at different magnifications, infected by the take-all fungus, *Gaeumannomyces graminis*. The fungus grows on the root surface as dark "runner hyphae" (rh) then invades the root cortex by infection hyphae (ih), destroying the phloem (phl) and causing intense discoloration and blockage of the xylem.

Roles of pre-formed inhibitors in plant resistance to pathogens

The take-all fungus of cereals, *Gaeumannomyces graminis* (Fig. 9.11), grows on the surface of cereal roots by dark "runner hyphae" then penetrates the root cortex and enters the vascular system, causing disruption of the sugar-conducting phloem cells, and blockage of the water-conducting xylem vessels. If enough roots are killed in this way during the growing season then the plants die prematurely, with serious loss of grain yield.

G. graminis has two main pathogenic forms – variety *tritici* (GGT) which attacks wheat roots but not oat roots, and variety *avenae* (GGA) which attacks both wheat and oats. This difference was explained in the 1960s when oat roots were found to contain pre-formed inhibitors. The most potent of these inhibitors is **avenacin A** (Fig. 9.12), a saponin (soap-like compound) which combines with sterols in the fungal membrane, creating ion-permeable pores. In laboratory culture both GGT and GGA grow readily in aqueous extracts of wheat roots, but GGT is totally suppressed by aqueous extracts of oat roots whereas GGA is unaffected by them. The reason was suggested to be that GGA detoxifies avenacin, cleaving the terminal sugars from the molecule (Fig. 9.12), by producing a glycosidase enzyme termed **avenacinase**. This enzyme would thus be a key

pathogenicity determinant, allowing GGA to extend its host range to oats.

This was shown to be the case by targeted disruption of the avenacinase gene (Osbourne *et al.* 1994). GGA was transformed with a marked, disrupted cDNA of the gene, which inserted at the gene site by homologous recombination and caused the fungus to lose its pathogenicity to oats, while retaining its normal pathogenicity to wheat. But when the marked cDNA had inserted elsewhere in the genome (nonhomologous recombination) the fungus was still fully pathogenic to oats. Extending from this work, Osbourne *et al.* examined other host–pathogen systems where saponins have been implicated as plant resistance factors. In particular, tomatoes are known to contain the saponin **α-tomatine** (Fig. 9.12), and a pathogen of tomatoes, *Septoria lycopersici*, is known to detoxify this by cleaving a single sugar from the molecule. The enzyme responsible (**tomatinase**) was found to be very similar to avenacinase: it was recognized by an anti-avenacinase antibody, and cDNA of avenacinase hybridized with DNA components of *S. lycopersici*, presumably by recognizing the gene for tomatinase. This might be explained by the fact that all the saponin-detoxifying enzymes are β-glycosidases, with perhaps some common structure, reflected in the DNA sequences. However, avenacinase (from GGA) had very little ability to

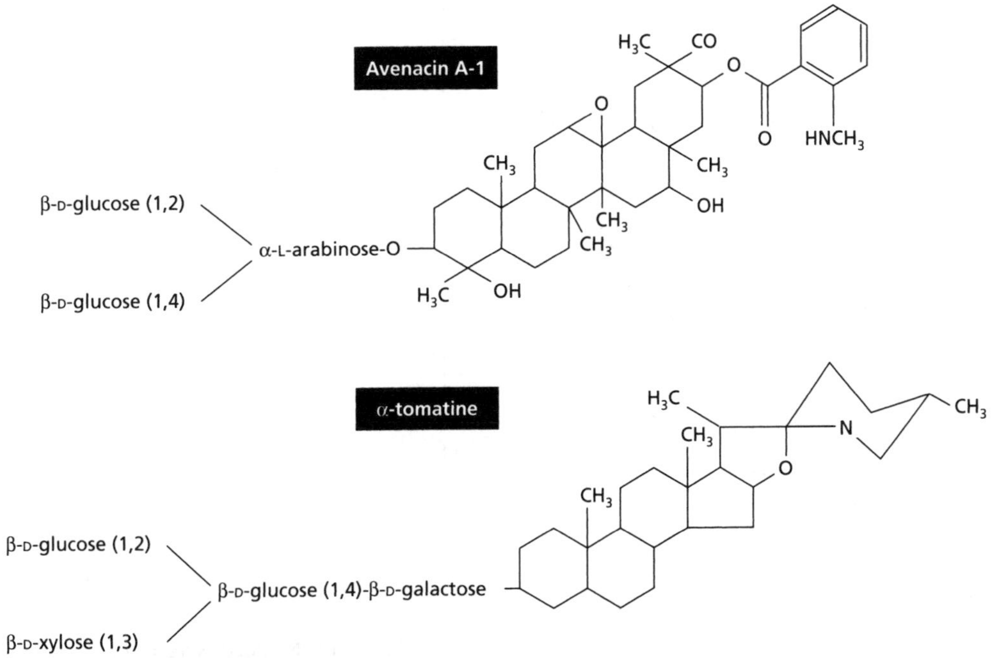

Avenacin A-1

β-D-glucose (1,2)

α-L-arabinose-O

β-D-glucose (1,4)

α-tomatine

β-D-glucose (1,2)

β-D-glucose (1,4)-β-D-galactose

β-D-xylose (1,3)

Fig. 9.12 Structures of two saponins that are pre-formed resistance compounds in plants: avenacin in roots of oats, and α-tomatine in tomato. Pathogens with the appropriate enzymes can detoxify these compounds by cleaving some of the terminal sugar residues.

detoxify tomatine, and tomatinase (from *S. lycopersici*) had negligible ability to detoxify avenacin. So it seems that these pathogenic fungi have evolved saponin-detoxifying enzymes with quite specific activity against the saponins of their hosts.

Yeast killer systems

With this topic we return to the role of viral dsRNA. Individual species of at least eight genera of yeasts (*Saccharomyces*, *Candida*, *Kluyveromyces*, etc.) have been found to contain **killer strains**. These strains secrete a protein that kills other strains of the same species but not of unrelated species. The toxin binds to a receptor on the wall of susceptible cells, then passes to the membrane where it causes leakage of H$^+$, and the cells die owing to loss of transmembrane potential and disruption of amino acid uptake, K$^+$ balance, etc. The toxins are stable only at low pH and are thought to give a significant advantage to killer strains over nonkiller strains of the same species in acidic environments.

In *Kluyveromyces lactis* the toxin is encoded by a linear DNA plasmid, but in all other cases it is encoded by dsRNA. Both the killer and nonkiller strains can contain virus-like particles (VLPs) so this feature alone does not correlate with killer activity. However, these particles are found to be of two types: the dsRNA in the "L type" encodes the virus coat protein, whereas the dsRNA in the "M type" codes for the toxin. The M dsRNA depends on the L dsRNA for the coat protein, but the L dsRNA can occur alone with no effect on the cells.

The molecular biology of the killer system in *S. cerevisiae* has been studied intensively and has shown why the toxin producers are not affected by their own toxin. They produce the toxin as a large precursor protein which undergoes changes during passage through the secretory system to produce a **protoxin**. This protoxin is finally cleaved at the cell membrane, to release the active toxin but leave part of the molecule in the membrane. This residual part seems to interact with a toxin receptor, making the cell resistant to active toxin in the external environment.

Similar dsRNA killer systems are found in the yeast phase of *Ustilago maydis* (Basidiomycota) which causes smut disease of maize. A killer system might also occur in the take-all fungus *Gaeumannomyces graminis*, because some dsRNA-containing strains can markedly inhibit the growth of other strains at low pH. However the mechanism in this case remains unknown.

Fig. 9.13 Recorded spread of chestnut blight caused by *Cryphonectria parasitica* after it was first recorded in New York in 1904.

Hypovirulence of plant pathogens: the control of chestnut blight

Chestnut blight, caused by the fungus *Cryphonectria parasitica* (Ascomycota), has destroyed many millions of trees in North America, following the first report of its occurrence in the New York Zoological Garden in 1904 (Fig. 9.13). Its progressive spread across the eastern half of the USA reduced the native American chestnut (*Castanea dentata*) almost to the status of an understorey shrub, whereas it was once a magnificent "high forest" tree. However, recent studies raise the prospect that chestnut blight might be controlled by genetic manipulation of hypovirus-associated dsRNA, discussed below.

Cryphonectria infects through wounds in the bark of chestnut trees, and then spreads in the cambium, progressively girdling the stem and killing the plant above the infection point (Fig. 9.14). This disease was also a significant problem in European sweet chestnut trees (*Castanea sativa*) in the 1940s, but in Italy in the early 1950s some heavily infested sweet chestnut plantations began to recover spontaneously. The lesions in these trees stopped spreading round the trunks, and strains of *C. parasitica* isolated from them showed abnormal features. They grew slowly and erratically on agar, were white rather than the normal

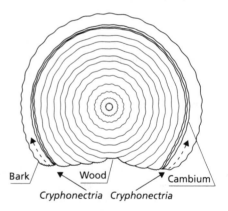

Fig. 9.14 Diagram of a cross-section of a trunk with a spreading canker caused by *Cryphonectria parasitica*. This fungus produces airborne spores that infect through wounds, then progressively girdles the trunk by growing in the cambium.

orange color, produced significantly fewer conidia, and showed only low virulence when wound-inoculated into trees. This low virulence (**hypovirulence**) could be transmitted to other strains during hyphal anastomosis on agar, so it was coined **transmissible hypovirulence**. None of the hypovirulent strains

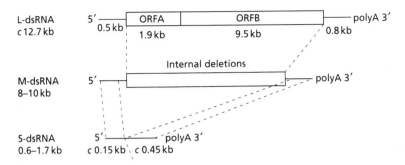

Fig. 9.15 Double-stranded RNA of hypovirulent strains of *C. parasitica*. Large (L) dsRNA is the full-length molecule comprising two conserved end regions and a central coding region of two open reading frames (ORFA and ORFB) which confer hypovirulence. Medium (M) and small (S) dsRNAs are also commonly found in hyphae. They are internally deleted copies of the L-dsRNA.

contained "conventional" virus-like particles, but all the hypovirulent strains contained dsRNA whereas virulent strains had no dsRNA. Electron micrographs show that this dsRNA is contained in rounded or club-shaped, membrane-bound vesicles in the cytoplasm (Newhouse *et al.* 1983) and that, unlike many VLPs, these can occur in significant amounts in the hyphal apices of *C. parasitica*.

The spontaneous disease decline in Italy led French workers to develop a highly successful biological control program. Hypovirulent strains of *C. parasitica* were cultured in the laboratory and inoculated at the expanding margins of cankers in the field. Within a short time the cankers stopped growing, and only hypovirulent strains could be recovered where once the virulent strain had been. This **transmissible change of phenotype** was always accompanied by the transmission of dsRNA.

Following the success of this program in Europe, attempts were made to introduce hypovirulent strains of *Cryphonectria* into the USA. But they gave only partial and localized disease control, because the pathogen population in the USA consists of numerous vegetative compatibility groups (VCGs) that limit the natural transfer of dsRNA, owing to cytoplasmic death when the strains anastomose. In one early study about 0.5 hectare of natural chestnut forest in the USA was found to contain at least 35 VCGs. More recently the VCGs have been shown to be in a continuous state of flux: samplings of identical trees over several years showed that some of the predominant VCGs declined, while new ones arose (Anagnostakis 1992). The European population of *C. parasitica* is much more uniform in terms of VCGs, and this probably accounts for the success of the biocontrol program in Europe.

Progress in understanding the role of dsRNA in *Cryphonectria* was significantly delayed by the lack of a suitable transformation system for this fungus.

However, when a system was eventually developed it led to rapid progress (Nuss 1992). The dsRNA of *C. parasitica* was found to be highly variable, with lengths falling into three broad size ranges – S (small), M (medium), and large (L, about 12.7 kb). As shown in Fig. 9.15, all these types have the same terminal regions – a poly-A tail at the 3′ end and a 28-nucleotide conserved sequence at the 5′ end. They differ mainly in the degree of internal deletion. The L-form seems to be the full-length molecule, and the smaller forms are defective (presumably functionless) derivatives which accumulate in the hyphae. This high degree of variability of RNA genomes is not unusual because there is no effective proofreading system for RNA, to ensure that it is copied faithfully.

With the development of a transformation system, Nuss and his colleagues were able to produce a cDNA from the full-length dsRNA and transform it into virulent strains. It caused the virulent strains to become hypovirulent and to exhibit all the features typical of hypovirulent strains – slow growth, pale coloration, and reduced sporulation (the hypovirulence-associated traits). Thus, dsRNA was shown unequivocally to be the cause of hypovirulence. The cDNA could also be used for molecular analysis of the dsRNA, which was shown to consist of two open reading frames (ORFs) – ORF A of 622 codons (nucleotide triplets) and ORF B of 3165 codons. When a cDNA copy of ORF A was introduced into *Cryphonectria*, it caused **a loss of pigmentation, a reduction of sporulation** and **a reduction of laccase activity**. But the transformed strain remained **virulent**. By contrast, ORF B codes for a polyprotein that was found to contain motifs characteristic of RNA-dependent RNA polymerases and RNA helicases. So, the primary function of ORF B could be the replication and maintenance of the dsRNA, causing the loss of pathogenic virulence, whereas ORF A encodes many of the hypovirulence-associated traits (reduced sporulation, etc.) that are potentially disadvantageous in a biocontrol strain, reducing its environmental fitness. It may be possible, therefore, to manipulate the cDNA *in vitro* so that it has only the most desirable traits for biocontrol.

It is notable that the cDNA, when transformed into *Cryphonectria* (causing the fungus to be hypovirulent), became stably integrated in the chromosomal genome so that it was replicated along with the other chromosomal genes. This cDNA was maintained throughout the life cycle, even entering the sexual spores, whereas dsRNA seldom enters the sexual spores of *Cryphonectria* or other Ascomycota. This would mean that biocontrol strains could be produced with permanent, stable hypovirulence, subject to proofreading like the rest of the chromosomal genes.

Recent developments

The question arising from the stable integration of "hypovirulence" cDNA is: can there still be **cytoplasmically transmitted hypovirulence**? The answer seems to be "yes" because the chromosomally integrated cDNA is transcribed into dsRNA rather than into single-stranded mRNA (as in the rest of the genome) and this dsRNA accumulates in the cytoplasm, where it can be transmitted between strains during hyphal anastomosis. This still leaves unanswered the question of how dsRNA suppresses virulence. Preliminary evidence suggests that it does so by downregulating some of the normal chromosomal genes, including the gene for production of laccase, an enzyme involved in lignin breakdown (Chapter 11) and which is likely to be significant for a fungus that colonizes wood.

When the hypovirulence cDNA (derived from dsRNA of *C. parasitica*) was transformed into other canker-forming *Cryphonectria* species, and into a less closely related *Endothia* species, it failed to convert these to hypovirulence. However, when the same cDNA was used as a template to produce RNA and this was transfected into the other fungi, it gave rise to full-length dsRNA in the cytoplasm and caused both a marked reduction of virulence and an alteration of the growth rate and pigmentation (Chen *et al.* 1994). The success achieved with RNA but not cDNA seems to be explained by the fact that *C. parasitica* produces RNA from the cDNA but then splices this RNA to delete a 73-base sequence before the RNA can act as a template for dsRNA production.

In summary, this work on the hypovirus system of *Cryphonectria* has raised many issues of fundamental interest in fungal genetics as well as holding the prospect of developing new approaches to plant disease control. However, recent studies in Switzerland (Hoegger *et al.* 2003) suggest the need for a cautious approach, because the only way of transmitting the hypovirus is by anastomosis, and both the nuclear DNA and the mitochondrial DNA of the donor (biocontrol) strain are introduced into the environment during this process. The mitochondrial DNA of the donor strain was found to be transmitted in nearly one-half of the treated cankers. The nuclear DNA also persisted in the treated cankers but it did not spread beyond them. The major issue that remains to be resolved is the long-term safety of introducing genetically modified strains of pathogenic fungi into the environment, especially if the nuclear DNA is transferred and increases the genetic diversity of the pathogen.

And back to the genome

The term **genomics** was coined in 1986 to describe the mapping, sequencing and analysis of genomes, the ultimate goal being to understand the structure, function, organization, relationships, and evolution of genes. While many sequenced genomes of bacteria and archaea have been published, there are relatively few published genome sequences of fungi, although several draft sequences are available. All the genome sequences generated by governmental or public funding bodies are made available on the internet (see Online resources for a list of sequenced genomes) so that researchers can compare different genomes and search for similarities and differences.

The basic techniques of genome sequencing are relatively simple, and many of the procedures are automated. Essentially, the DNA to be sequenced (termed the **template**) is denatured by heat or alkali to produce single-stranded DNA, and DNA polymerase is used to synthesize DNA based on the nucleotide sequence in the template. DNA polymerases require a region of double-stranded DNA to initiate synthesis. This is provided by adding a short single-stranded DNA molecule – a **primer**, with a DNA sequence complementary to the template DNA. The primer binds to the template to form a short region of double-stranded DNA, from which the rest of the template DNA is synthesized.

For sequencing projects, relatively long template DNA sequences are prepared by cutting the DNA randomly with restriction enzymes and inserting them into plasmids or other vectors such as cosmids (cloning vectors that resemble plasmids but are packaged into λ phage capsids which can carry inserts up to 40 kb). The many lengths of sequenced DNA from different regions of the genome are then searched for overlapping regions. In this way the DNA can be assembled into **contigs**, representing continuous coverage of the nucleotide sequence of whole chromosomes or regions of chromosomes. The genome sequence is then searched for characteristic features such as promoters, transcription initiation sites, and "stop" codons, to deduce the positions of protein-encoding genes. In practice, much of the generation of genome sequences relies heavily on computer programs.

Most genome sequencing projects involve the **whole genome shotgun** approach, in which short sequences are generated randomly rather than systematically and are then assembled into contigs. Typically, every part of the sequence is covered at least five times and often ten times (by overlapping of the sequenced regions). The result is a "high quality draft" of the whole genome, but often about 2% of the genome cannot be mapped accurately – where there are frequent repeat sequences and DNA regions of high G + C (Guanine + Cytosine) content.

What do we gain from whole genome sequences?

Many benefits accrue from comparing the genomes of different fungi, or from comparing fungal genomes with those of other organisms. To give just a few examples:

- The nucleotide sequences of genes can be used to predict the protein sequences, so automated searches such as BLAST (basic local alignment search tool) can identify homologous genes in different organisms (and any evolutionary changes in those genes over different periods of time).
- Nucleotide sequences that do not appear in current databases may represent genes with undiscovered functions – the basis of "gene mining" for potential new proteins of commercial interest.
- Many aspects of human, animal and plant disease are still unresolved, so the elucidation of genes controlling these processes could provide new directions for tackling these problems.

Above all, as more and more gene functions are discovered they add to the sum of knowledge, and since most of this information is freely available the discovery of a newly characterized gene in one organism can help to "fill the gaps" in the genomes of other organisms. The sequencing of the first eukaryotic genome (*Saccharomyces cerevisiae*) released in 1996, showed that about half of the open reading frames (ORFs) had no clear homologs in published databases. This pattern has been repeated time and again in the genomes sequenced since that time.

Genomics is "Big Science." The sequencing of *S. cerevisiae* involved 90 research laboratories, and the paper describing the genome sequence of *Neurospora crassa* had 77 authors. Given the scale of these projects, and the thousands of fungal genomes that could potentially be sequenced, it is important to prioritize and coordinate sequencing efforts. As one example, the fungal research community of the USA has, since the year 2000, undertaken broad consultation and published a series of "White Papers" on the **Fungal Genome Initiative**. Of 15 candidate fungi initially proposed, seven are currently being sequenced. In 2003 the second White Paper included a list of 44 additional fungi with emphasis on clusters of related species to promote comparative genome analysis: [http://www.broad.mit.edu/annotation/fungi/fgi]

In a field that is moving so rapidly, and where funding decisions have still to be made, the FGI website (address above) is the most reliable source of information. But Table 9.4 gives brief details of the original 15 submissions to illustrate the rationale behind such sequencing efforts. It included representatives of all the major fungal phyla (Chytridiomycota, Zygomycota, Ascomycota, and Basidiomycota) and organisms in three categories – those of medical significance, those of commercial significance, and those that would contribute to understanding of evolution and fungal diversity.

Significant findings from the *Neurospora crassa* genome sequence

The high quality draft sequence of the *N. crassa* genome was completed in 2003 (Galagan *et al.* 2003) and represents a milestone – the culmination of more than 60 years of research on one of the most genetically well characterized fungi. The sequence still needs further detailed work, to check potential discrepancies and to join the existing contigs, but already it has revealed new information, including the identification of genes potentially associated with light signalling and secondary metabolism.

The main features of the sequenced *N. crassa* genome are shown in Table 9.5 (from Galagan *et al.* 2003). Among the more notable points is the predicted presence of over 10,000 protein-encoding genes, most of which code for proteins of more than 100 amino acids. But 41% of the *Neurospora* proteins lack significant matches to any of the known proteins in public databases, and 57% of *Neurospora* proteins lack significant matches to genes in either *Saccharomyces cerevisiae* or *Schizosaccharomyces pombe*.

Another interesting feature of *Neurospora* is that it has the widest range of genome defense mechanisms known for any eukayotic organism. One of these is a process apparently unique to fungi, termed **repeat-induced point mutation (RIP)**.

RIP was discovered in *Neurospora* several years ago, as a process that effectively prevents genome evolution. The duplication of genes is widely recognized to be responsible for evolutionary development, because the

Table 9.4 Fungi initially proposed to form the basis of a coordinated genome sequencing effort in the USA.

Organisms by category	Significance (and chapter reference)[1–11]	Estimated genome (Mb)
Medicine		
Filobasidiella (Cryprococcus) neoformans serotype A	Basidiomycota. Encapsulated yeast; causes fatal meningitis in humans (Chapter 16)	24 Mb on 11 chromosomes
Coccidioides posadasii	Ascomycota. Soil fungus endemic to southwestern USA; causes fatal human infection; also a bioterrorism threat (Chapter 16)	29 Mb on 4 chromosomes
Pneumocystis carinii (human and mouse forms)	The leading opportunistic pathogen of AIDS patients; drug resistance is emerging (Chapter 16)	7.5, 6.5 Mb
Trichophyton rubrum	Ascomycota. The most common fungal infection in the world; adapted for growth on human skin (Chapter 16)	12 Mb on 4 chromosomes
Rhizopus oryzae	Zygomycota. Can cause infection of humans (zygomycosis)	36 Mb
Commerce		
Magnaporthe grisea	Causes rice blast disease. A model fungal plant pathogen (Chapter 5)	40 Mb on 7 chromosomes
Aspergillus flavus	Ascomycota/mitosporic fungus. Source of aflatoxin and one cause of human aspergillosis (Chapter 7)	40 Mb on 8 chromosomes
Emericella (Aspergillus) nidulans	Ascomycota. Key model system for genetics and cell biology. (Already part-sequenced)	31 Mb on 8 chromosomes
Aspergillus terreus	Mitosporic fungus. Major source of the cholesterol-lowering drug, lovastatin	30 Mb
Fusarium graminearum	Mitosporic fungus. Causes head blight on wheat and barley; produces mycotoxins. (Chapter 7)	40 Mb on 9 chromosomes
Evolution/fungal diversity		
Neurospora discreta	Ascomycota. Fungal model for population genetics and comparison with N. crassa (already sequenced)	40 Mb on 7 chromosomes
Coprinus cinereus	Basidiomycota. Model for fungal differentiation – produces toadstools (Chapter 5)	37.5 Mb on 13 chromosomes
Batrachochytrium dendrobatidis	Chytridiomycota. Recently described fungus that causes widespread population decline of amphibians (Chapter 2)	30 Mb on 20 (?) chromosomes
Ustilago maydis	Basidiomycota. Model for plant–pathogen interactions (Chapter 14)	20 Mb
Paxillus involutus	Basidiomycota. Symbiotic mycorrhizal fungus of many trees, easily manipulated in laboratory conditions (Chapter 14)	40 Mb

[1] *Filobasidiella neoformans* (asexual yeast phase: *Cryptococcus neoformans*). Serotype D is being sequenced because of its advanced genetic tools, but serotype A is more divergent and represents 90% of all clinical strains and 99% of strains from AIDS patients. Two virulence factors are of interest – the thick polysaccharide capsule that prevents phagocytosis, and melanin, which serves as an antioxidant.

[2] *Coccidioides posadasii*. One of two soil fungi endemic to desert regions of the Americas, infecting about 100,000 people each year in the USA, but only a small number of infections are fatal. Funding was already allocated to sequencing of the other species, *C. immitis*, so *C. posadasii* can provide comparative genomics.

[3] *Pneumocystis carinii*. An unusual fungus with cholesterol instead of ergosterol in the cell membrane. Consists of several strains adapted to specific hosts. Comparison of two species of the fungus (human and mouse) should allow comparative genomics relating to infection.

[4] *Trichophyton rubrum*. One of 42 known species of dermatophytes (of which 31 can infect humans). Understanding of this and similar dermatophytes is very limited.

[5] *Rhizopus oryzae*. A representative of Zygomycota and the fungus responsible for relatively rare cases of damaging human disease, especially in diabetics.

[6] *Magnaporthe grisea*. A serious plant pathogen of rice, estimated to cause yield losses sufficient to feed 60 million people per year. Also, a model for study of plant–fungal pathogenic interactions.

[7] *Aspergillus flavus*. An important fungus in its own right, and valuable for comparisons with other Aspergillus spp., including *A. nidulans*, which has a well-characterized genetic system, and *A. terreus*, which is one of the producers of "statins" that can lower blood cholesterol levels.

Table 9.5 *Neurospora crassa* genome features. (From Galagan *et al.* 2003.)

Feature	Measurement
Size (base pairs)	38,639,769
Chromosomes	7
Protein-coding genes	10,082
Transfer RNA genes	424
5S rRNA genes	74
Per cent coding	44
Per cent introns	6
Average gene size (base pairs)	1673 (481 amino acids)
Average intergenic distance (base pairs)	1953
Predicted protein-coding sequences:	
Identified by similarity to known sequences	13%
Conserved hypothetical proteins	46%
Predicted proteins (no similarity to known sequences)	41%

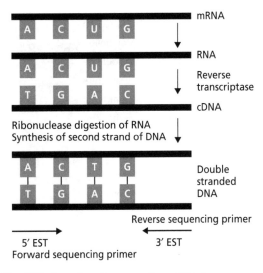

Fig. 9.16 Procedure for generating cDNA from messenger RNA, then producing expressed sequence tags (ESTs) from either the 5′ or 3′ end of cDNA.

gene copy is free to mutate and can eventually assume a new function while the original gene retains its function. But this process is blocked in *Neurospora* during the haploid dikaryotic phase of the sexual cycle (Chapter 2). The RIP process detects and mutates both copies of a duplicated gene by causing numerous mutations from G-C to A-T pairs, and often leads to DNA methylation which causes gene silencing in *Neurospora* (as DNA methylation also does in mammals). The effects of this can extend to adjacent genes beyond the duplicated sequences. Consistent with this role of RIP, *N. crassa* has an unusually low proportion of genes in mutigene families, in relation to its genome size, and it has almost no highly similar gene pairs. Several other lines of evidence support the view that genome evolution in *N. crassa* has been largely

arrested since the acquisition of RIP at some point in its evolutionary history.

It is suggested that RIP acts as a defense against "selfish DNA," thereby protecting the genome. Consistent with this is the fact that no intact mobile elements were detected in the genome sequence, and 46% of repetitive nucleotides can be identified as relics of mobile elements. Sequence comparisons with other *Neurospora* spp. would help to address the broader significance of these findings.

Even though *N. crassa* has an impressive array of genome defense mechanisms, it is not unique in having defense systems. For example, the many anastomosis groups and vegetative compatibility loci in *Rhizoctonia solani* and *Cryphonectria parasitica* probably serve similar functions in impeding the spread of potentially damaging mobile viruses or other genetic elements.

Table 9.4 (cont')

[8] *Fusarium graminearum*. Produces several mycotoxins and represents a major genus of food-spoilage organisms.

[9] *Neurospora discreta*. Would enable comparisons with *N. crassa* (a fungus that used to be a serious contaminant of bread, but nowadays is rare outside of the laboratory). An extensive collection of naturally occurring populations of *N. discreta* in North America would enable studies on population dynamics.

[10] *Batrachochytrium dendrobatitis*. A representative of Chytridiomycota – the earliest fungal lineage – and of aquatic fungi that degrade polymers. *B. dendrobatidis* is a recently described fungus thought to be the primary agent of the global amphibian decline. It invades the top layers of amphibian skin cells, causing thickening of the keratinized tissues and thereby limiting gas exchange.

[11] *Paxillus involutus*. A common mycorrhizal fungus of trees. Mycorrhizal fungi of various types form intimate mutualistic relationships with about 90% of plants worldwide.

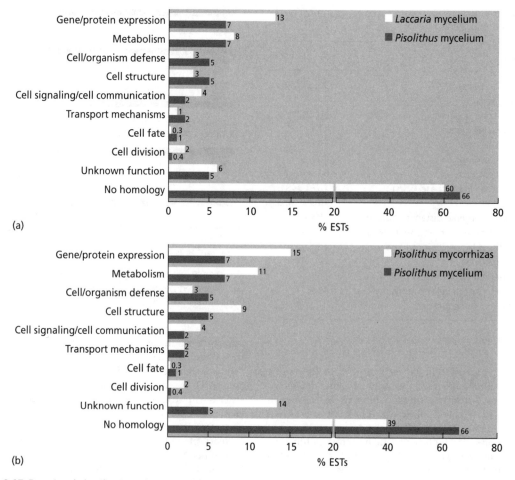

(a)

(b)

Fig. 9.17 Functional classification of expressed sequence tags (ESTs) from (a) mycelial cultures of two mycorrhizal fungi or from (b) symbiotic tissue of *Pisolithus* mycorrhizas compared with mycelial cultures of *Pisolithus*. (From Peter *et al.* 2003, with permission from the New Phytologist Trust.)

Expressed sequence tags and microarray technology

Expressed sequence tags (ESTs) provide an alternative to whole genome sequencing and are particularly useful for finding genes that are expressed in specific conditions (in other words, the genes that are "switched on" at different times or in different circumstances). This technology can also be used as a quick and inexpensive approach to finding new genes and for constructing genome maps.

ESTs are small lengths of DNA (about 200–500 nucleotides) produced by sequencing one or both ends of an expressed gene. The technique involves sampling the pool of messenger RNA (the product of gene expression) and using this mRNA as a template

to synthesize a complementary DNA (cDNA), using the enzyme **reverse transcriptase**. cDNA is much more stable than mRNA, and it has the additional advantage of containing only the coding regions, because introns have been removed by splicing during the natural processing of messenger RNA.

Once a single strand of cDNA has been formed from the RNA template, this RNA is digested by a ribonuclease. Then a complementary strand of DNA is synthesized, to produce double-stranded DNA, as shown in Fig. 9.16.

Large numbers of ESTs can be generated easily and can subsequently be assembled into contigs or compared with the many ESTs of various organisms now available on electronic databases, providing a means of identifying different types of gene. As an illustration of the power of such techniques, Martina *et al.* (2003)

used ESTs to compare the types of genes expressed by mycelial cultures of two mycorrhizal fungi (*Laccaria bicolor* and *Pisolithus microcarpus*) and also to compare the genes expressed in symbiotic tissues of *Pisolithus* mycorrhizas (Fig. 9.17). By comparison with EST databases, several of the ESTs could be assigned to functional groups such as genes involved in metabolism, cell defense, or cell structure. But the majority of ESTs showed no homology to known genes (as is also true for whole genome sequences).

A similar EST analysis was reported by Freimoser *et al.* (2003), comparing two subspecies of the common insect-pathogenic fungus *Metarhizium anisopliae* – one subspecies (*M. anisopliae anisopliae*) has a broad host range and the other (*M. anisopliae acridum*) is a specific grasshopper pathogen. Both strains were grown under conditions that maximize the secretion of insect cuticle-degrading enzymes, and a surprisingly high proportion of ESTs could be assigned to functional categories (Fig. 9.18). Both subspecies had ESTs for virtually all the pathogenicity-related genes cloned to date from *M. anisopliae*.

DNA microarray technology

Essentially, a DNA microarray is a small solid support such as a microscope slide or a nylon membrane onto which DNA is spotted (or printed by robotic techniques) to produce hundreds or thousands of tiny spots representing different types of DNA, arranged in a specific order. Messenger RNAs from a sample to be analyzed are then used to generate cDNAs, which are fluorescently labeled so that any binding of the cDNA to the immobilized spots on the microarray (through complementary base-pairing) can be scanned automatically by laser technology. There are several types of microarray, designed for different purposes, but the main ones usually measure one of the following:

- Changes in the level of gene expression, monitored automatically by mixing samples of a test DNA (tagged with red fluorescence) and control DNA (tagged with green fluorescence) so that the laser differentiates between the levels of expression of the test and control DNA.
- Changes in genomic gains and losses, again detected by different levels of red or green fluorescence depending on the number of copies of a particular gene.
- Mutations in DNA, often involving only one or a few nucleotides.

Most of these applications are used in medicine, including disease diagnosis, drug development and tracking disease development, but they are equally applicable to many aspects of basic biology and they are becoming ever more widely used because the same basic techniques can be done cheaply by manual spotting of DNA onto solid supports. They can reveal patterns of differential or coordinated expression of genes in almost any biological system.

Online resources

Genome News Network – a quick guide to sequenced genomes. http://w.w.w.genomesnetwork.org/sequenced_genomes/genome_guide_p1.shtml

Microarrays: National Center for Biotechnology Information. http://w.w.w.ncbi.nlm.nih.gov/About/primer/microarrays.html

The Whitehead Institute, Fungal Genome Initiative. http://www.broad.mit.edu/annotation/fungi/fgi

General texts

Bennett, J.W. & Arnold, J. (2001) *Genomics for fungi*. In: *The Mycota VIII. Biology of the Fungal Cell* (Howard, R.J. & Gow, N.A.R., eds), pp. 267–297. Springer-Verlag, Berlin Heidelberg.

Bennett, J.W. & Lasure, L.L. (1991) *More Gene Manipulations in Fungi*. Academic Press, San Diego.

Fincham, J.R.S., Day, P.R. & Radford, A. (1979) *Fungal Genetics*, 4th edn. Blackwell Scientific Publishers, Oxford.

Peberdy, J.F., Caten, C.E., Ogden, J.E. & Bennett, J.W. (1991) *Applied Molecular Genetics of Fungi*. Cambridge University Press, Cambridge.

Turner, G. (1991) Strategies for cloning genes from filamentous fungi. In: *Applied Molecular Genetics of Fungi* (Peberdy, J.F., Caten, C.E., Ogden, J.E. & Bennett, J.W., eds), pp. 29–43. Cambridge University Press, Cambridge.

Cited references

Anagnostakis, S.L. (1992) Diversity within populations of fungal pathogens on perennial parts of perennial plants. In: *The Fungal Community: its organization and role in the ecosystem* (Carroll, G.C. & Wicklow, D.T., eds), pp. 183–192. Marcel Dekker, New York.

Bertrand, H. (1995) Senescence is coupled to induction of an oxidative phosphorylation stress response by mitochondrial DNA mutations in *Neurospora*. *Canadian Journal of Botany* 73, S198–S204.

Buck, K.W. (1986) *Fungal Virology*. CRC Press, Boca Raton.

Chen, B., Choi, G.H. & Nuss, D.L. (1994) Attenuation of fungal virulence by synthetic hypovirus transcripts. *Science* 264, 1762–1764.

Domer, J.E. & Kobayashi, G.S., eds (2004) *The Mycota*, vol. XII. *Human Fungal Pathogens*. Springer-Verlag, Berlin.

Edwards, S.G., O'Callaghan, J. & Dobson, A.D.W. (2002) PCR-based detection and quantification of mycotoxigenic fungi. *Mycological Research* 106, 1005–1025.

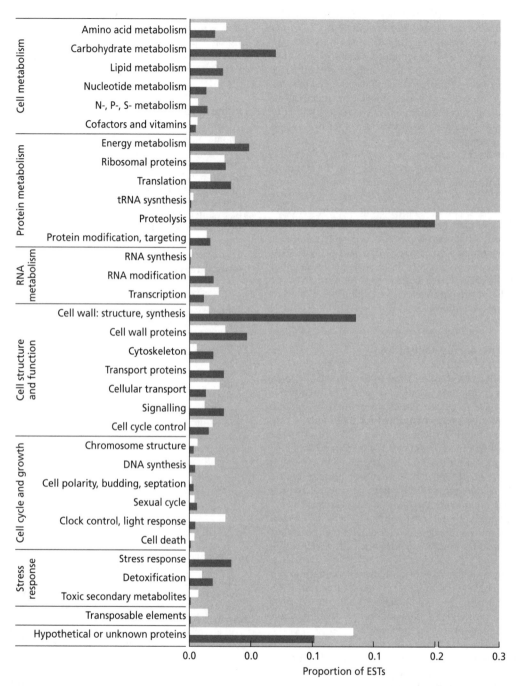

Fig. 9.18 Proportion of EST sequences of *Metarhizium anisopliae* (broad host range; white bars) and *M. anisopliae* (caterpillar-specific; black bars) with significant matches to genes of known functional categories. (From Freimoser *et al.* 2003.)

Esser, K. (1990) Molecular aspects of ageing: facts and perspectives. In: *Frontiers in Mycology* (Hawksworth, D.L., ed.), pp. 3–25. CAB International, Wallingford, Oxon.

Fox, R.T.V. (1994) *Principles of Diagnostic Techniques in Plant Pathology*. CAB International, Wallingford.

Freimoser, F.M., Screen, S., Bagga, S., Hu, G. & St Leger, R.J. (2003) Expressed sequence tag (EST) analysis of two subspecies of *Metarhizium anisopliae* reveals a plethora of secreted proteins with potential activity in insect hosts. *Microbiology* **149**, 239–247.

Galagan, J.E. and 76 other authors (2003) The genome sequence of the filamentous fungus *Neurospora crassa*. *Nature* **42**, 859–868.

Glass, N.L. & Kaneko, I. (2003) Fatal attraction: nonself recognition and heterokaryon incompatibility in filamentous fungi. *Eukaryotic Cell* **2**, 1–8.

Hoegger, P.J., Heiniger, U., Holdenrieder, O. & Rigling, D. (2003) Differential transfer and dissemination of hypovirus and nuclear and mitochondrial genomes of a hypovirus-infected *Cryphonectria parasitica* strain after introduction into a natural population. *Applied and Environmental Microbiology* **69**, 3767–3771.

Jinks, J.L. (1952) Heterokaryosis: a system of adaptation in wild fungi. *Proceedings of the Royal Society of London, Series B* **140**, 83–99.

Kohn, L.M. (1995) The clonal dynamic in wild and agricultural plant-pathogen populations. *Canadian Journal of Botany* **73**, S1231–S1240.

MacNish, G.C., McLernon, C.K. & Wood, D.A. (1993) The use of zymogram and anastomosis techniques to follow the expansion and demise of two coalescing bare patches caused by *Rhizoctonia solani* AG8. *Australian Journal of Agricultural Research* **44**, 1161–1173.

MacNish, G.C. & Sweetingham, M.W. (1993). Evidence of stability of pectic zymogram groups within *Rhizoctonia solani* AG-8. *Mycological Research* **97**, 1056–1058.

Martina, P. and 10 others (2003) Analysis of expressed sequence tags from the ectomycorrhizal basidiomycetes *Laccaria bicolor* and *Pisolithus microcarpus*. *New Phytologist* **159**, 117–129.

Newhouse, J.R., Hoch, H.C. & MacDonald, W.L. (1983) The ultrastructure of *Endothia parasitica*. Comparison of a virulent with a hypovirulent isolate. *Canadian Journal of Botany* **61**, 389–399.

Nuss, D.L. (1992) Biological control of chestnut blight: an example of virus-mediated attenuation of fungal pathogenesis. *Microbiological Reviews* **56**, 561–576.

Oliver, S.G. (1987) Chromosome organization and genome evolution in yeast. In: *Evolutionary Biology of the Fungi* (Rayner, A.D.M., Brasier, C.M. & Moore, D., eds), pp. 33–52. Cambridge University Press, Cambridge.

Osbourne, A., Bowyer, P., Bryan, G., Lunness, P., Clarke, B. & Daniels, M. (1994) Detoxification of plant saponins by fungi. In: *Advances in Molecular Genetics of Plant-Microbe Interactions* (Daniels, M.J., Downie, J.A. & Osbourn, A.E., eds), pp. 215–221. Kluwer Academic, Dordrecht.

Peter, M., Courty, P.-E., Kobler, A., *et al.* (2003) Analysis of expressed sequence tags from the ectomycorrhizal basidiomycetes *Laccaria bicolor* and *Pisolithus microcarpus*. *New Phytologist* **159**, 117–129.

Pipe, N.D., Buck, K.W. & Brasier, C.M. (1995) Molecular relationships between *Ophiostoma ulmi* and the NAN and EAN races of *O. novo-ulmi* determined by RAPD markers. *Mycological Research* **99**, 653–658.

Pontecorvo, G. (1956) The parasexual cycle in fungi. *Annual Review of Microbiology* **10**, 393–400.

Rawlinson, C.J., Carpenter, J.M. & Muthyalu, G. (1975) Double-stranded RNA virus in *Colletotrichum lindemuthianum*. *Transactions of the British Mycological Society* **65**, 305–308.

Sansome, E. (1987) Fungal chromosomes as observed with the light microscope. In: *Evolutionary Biology of the Fungi* (Rayner, A.D.M., Brasier, C.M. & Moore, D., eds). Cambridge University Press, Cambridge, pp. 97–113.

Tooley, P.W. & Therrien, C.D. (1991) Variation in ploidy in *Phytophthora infestans*. In: *Phytophthora* (Lucas, J.A., Shattock, R.C., Shaw, D.S. & Cooke, L.R., eds.). Cambridge University Press, Cambridge, pp. 204–217.

Fungal spores, spore dormancy, and spore dispersal

This chapter is divided into the following major sections:

- general features of fungal spores
- spore dormancy and germination
- spore dispersal
- dispersal and infection behavior of zoospores
- zoospores as vectors of plant viruses
- dispersal of airborne spores
- spore sampling devices and human health

Fungi are the supreme examples of spore-producing organisms. They produce millions of spores, with an astonishing variety of shapes, sizes, surface properties, and other features – all precisely matched to the specific requirements for dispersal and/or persistence in different environments. A small part of this diversity is illustrated in Fig. 10.1, for some of the more bizarrely shaped spores of the freshwater aquatic fungi that grow in fast-flowing streams. But even the common rounded spores of fungi have properties that determine whether they will be deposited on plant surfaces, or on soil, or in the human lungs, etc. In this chapter we discuss several examples of this fine-tuning, and we will see that the properties of a spore tell us much about the biology and ecology of a fungus.

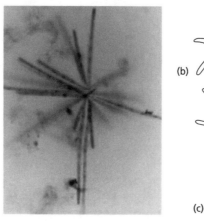
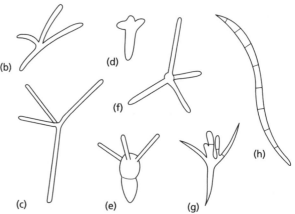

Fig. 10.1 Examples of tetraradiate, multiple-armed and sigmoid spores found in fast-flowing freshwater streams. Approximate spore lengths are shown in parentheses. (a) A single conidium of *Dendrospora* (150–200 μm); (b) conidium of *Alatospora* (30–40 μm); (c) conidium of *Tetrachaetum* (70–80 μm); (d) conidium of *Heliscus* (30 μm); (e) conidium of *Clavariopsis* (40 μm); (f) conidium of *Lemonniera* (60–70 μm); (g) conidium of *Tetracladium* (30–40 μm); (h) conidium of *Anguillospora* (150 μm).

General features of fungal spores

Because of their extreme diversity we can define fungal spores in only a general way, as *microscopic propagules that lack an embryo and are specialized for dispersal or dormant survival*. The spores produced by a sexual process (e.g. zygospores and ascospores) usually function in dormant survival whereas asexual spores usually serve for dispersal. However, many Basidiomycota do not produce asexual spores, or produce them only rarely, and instead the basidiospores are their main dispersal agents. Some fungi have an additional spore type, the **chlamydospore**. This is a thick-walled, melanized cell that develops from an existing hyphal compartment (or sometimes from a spore compartment) in conditions of nutrient stress.

The properties of these different spore types vary considerably but, in general, the spores of fungi differ from somatic cells in the following ways:

- The wall is often thicker, with additional layers or additional pigments such as melanins.
- The cytoplasm is dense and some of its components (e.g. endoplasmic reticulum) are poorly developed.
- Spores have a relatively low water content, low respiration rate, and low rates of protein and nucleic acid synthesis.
- Spores have a high content of energy-storage materials such as lipids, glycogen, or trehalose.

Spore dormancy and germination

Almost all spores are dormant, in the sense that their rate of metabolism is low. But they can be assigned to two broad categories in terms of their ability to germinate. Sexual spores often show **constitutive dormancy**. They do not germinate readily when placed in conditions that are suitable for normal, somatic growth (appropriate nutrients, temperature, moisture, pH, etc.). Instead, some of them require a period of aging (postmaturation) before they will germinate, and others require a specific activation trigger such as a heat shock or chemical treatment. By contrast, nonsexual spores show **exogenously imposed dormancy** – they remain dormant if the environment is unsuitable for growth, but they will germinate readily in response to the presence of nutrients such as glucose.

When triggered to germinate, all spores behave in a similar way. The cell becomes hydrated, there is a marked increase in respiratory activity, followed by a progressive increase in the rates of protein and nucleic acid synthesis. An outgrowth (the germ tube) is then formed, and it either develops into a hypha or, in the case of some sexual spores, it produces an asexual sporing stage. The germination process usually takes 3–8 hours, but zoospore cysts of the Oomycota can germinate much faster (20–60 minutes) and some sexual spores can take longer (12–15 hours).

Constitutive dormancy

Constitutive dormancy has been linked to several factors but is still poorly understood. The oospores of many *Pythium* and *Phytophthora* spp. seem to need a postmaturation phase before they can germinate. Initially the oospore wall is thick, about 2 μm diameter (Fig. 10.2), but it becomes progressively thinner (about 0.5 μm) by digestion of its inner layers. This process is hastened by keeping the spores in nutrient-poor conditions, at normal temperature and moisture levels. Then, after several weeks, the spores will germinate in response to nutrients or other environmental triggers. For example, *Pythium* oospores germinate in response to common nutrients (sugars and amino acids) or volatile metabolites (e.g. acetaldehyde) released from germinating seeds.

Ascospores can eventually become germination-competent by aging, but can be triggered to germinate at any time by specific treatments – in some cases a heat shock (e.g. 60°C for 20–30 minutes), cold shock (–3°C), or exposure to chemicals such as alcohols or furaldehyde. The ascospores of *Neurospora tetrasperma*

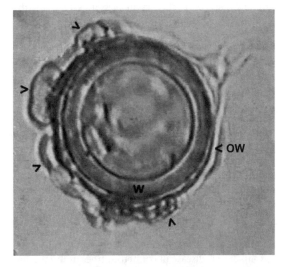

Fig. 10.2 A developing sexual spore (oospore) of *Pythium mycoparasiticum* (Oomycota). The spore has a very thick wall (w) and is contained in the outer wall (*ow*) of the oogonium (female reproductive cell). The arrows mark the positions of antheridia (male sex organs that fertilize the oogonium).

have been studied most thoroughly in this respect. Their dormancy cannot be explained in terms of a general permeability barrier, because they are permeable to radiolabeled oxygen, glucose, and water. Instead, their dormancy is linked to an inability to use their major storage reserve, trehalose, which is not metabolized during dormancy, but is metabolized immediately after activation. The enzyme **trehalase**, which cleaves trehalose to glucose, is found to be associated with the walls of the dormant spores, separated from its substrate. Activation somehow causes the enzyme to enter the cell, as one of the earliest detectable events in germination.

Constitutive dormancy of some other spores has been linked to endogenous inhibitors. For example, uredospores of the cereal rust fungus *Puccinia graminis* contain methyl-*cis*-ferulate, and those of bean rust, *Uromyces phaseoli*, contain methyl-*cis*-3,4-dimethoxycinnamate. Prolonged washing of spores can remove these inhibitors, and this perhaps occurs when the spores are bathed in a water film on a plant surface. At first sight it seems surprising that a spore adapted for dispersal should have an endogenous inhibitor. However, this might prevent the spores from germinating in a sporing pustule (they do not require exogenous nutrients for germination) and ensure that they germinate only after they have been dispersed.

Ecological aspects of constitutive dormancy

The behavior of constitutively dormant spores often has clear ecological relevance. For example, a characteristic assemblage of fungi occurs on the dung of herbivorous animals (Fig. 10.8). Their spores are ingested with the herbage, are activated during passage through the gut, and are then deposited in the dung where they germinate to initiate a new phase of growth. The spores of many of these **coprophilous** (dung-loving) fungi can be activated in the laboratory by treatment at 37°C in acidic conditions, simulating the gut environment. Examples include the ascospores of *Sordaria* and *Ascobolus*, the sporangiospores of several Zygomycota, and basidiospores of the toadstool-producing fungi, *Coprinus* and *Bolbitius*.

Heating to 60°C activates the spores of many thermophilic fungi of composts (Chapter 11). It also activates the ascospores of **pyrophilous** (fire-loving) fungi such as *Neurospora tetrasperma* which grows on burnt ground or charred plant remains. Most of these fungi are saprotrophs of little economic importance, but one of them, *Rhizina undulata* (Ascomycota), causes the "group dying" disease of coniferous trees in Britain and elsewhere. It infects trees replanted into clear-felled forests, and the foci of infection correspond to the sites where the trash from the felled trees was stacked and burned. The ascospores are heat-activated around or beneath the fires, then the fungus grows as a saprotroph

on the stumps and dead roots and produces mycelial cords that infect the newly planted trees. Once the cause of this disease had been recognized, the problem was easily solved by abandoning the practice of burning. However this is not possible in regions where lightning-induced fires are a periodic, natural occurrence. Some of the plants in these areas have become adapted to fire – their seeds remain dormant for years until they are heat-activated. Some of the mycorrhizal fungi are similarly adapted, an example being the ascospores of the mycorrhizal *Muciturbo* spp. in Australian eucalypt forests.

Basidiospores of the common cultivated mushroom, *Agaricus bisporus*, and of several mycorrhizal fungi (see below) show constitutive dormancy, but these spores will often germinate when placed next to growing colonies of the same fungus. For example, spores of *A. bisporus* germinate in the presence of isovaleric acid and isoamyl alcohol, which may be released from the "parent" hyphae. The significance of this behavior could be to increase the gene pool of the colony.

Dormancy and germination triggers in mycorrhizal successions

Many forest trees of temperate and boreal regions (e.g. pine, oak, beech, chestnut, etc.) form mycorrhizal associations with Basidiomycota or Ascomycota (Chapter 13). These mycorrhizal fungi produce a sheath around the root tips and extend into the soil as hyphae or mycelial cords (see Fig. 7.10). Several experimental studies have shown that a succession of mycorrhizal fungi occurs on young tree seedlings that are planted in forest nurseries or in previously tree-less sites. Some of these fungi establish mycorrhizas rapidly on tree seedlings from airborne spores. Classic examples of these "early (pioneer) colonizers" are *Laccaria* and *Hebeloma* spp. Other mycorrhizal fungi are slow to establish in new sites, but colonize after several years and eventually become dominant on the root systems. A classic example is the fly agaric (*Amanita muscaria*) on birch or pine trees. Figure 13.6 shows examples of some of these fungi.

These successional patterns have been studied in experimental field plots where the toadstools of mycorrhizal fungi appear above ground in autumn (Fig. 10.3) and where the range of different mycorrhizal fungi on the roots can be identified by features of the mycorrhizas themselves. When cores of soil were taken at different distances from the trees (Fig. 10.4) mycorrhizas of *Hebeloma* spp. predominated near the periphery of the tree root systems, mycorrhizas of *Lactarius* spp. predominated near the mid-zone of the root systems, and mycorrhizas of *Leccinum* (a polypore) and an unidentified mycorrhizal type predominated closest to the tree trunks. Thus, there was clear evidence of a

Fig. 10.3 Rings of toadstools of mycorrhizal fungi (mainly *Hebeloma* and *Lactarius* species) around young birch trees in late autumn in an experimental field plot.

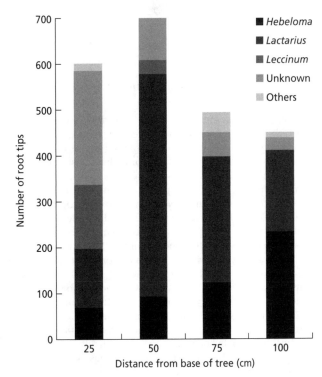

Fig. 10.4 The changing pattern of naturally occurring mycorrhizal fungi at different distances from the bases of birch trees in an experimental field plot. (Data from Deacon *et al.* 1983.)

succession, with "pioneer" mycorrhizal species colonizing the youngest parts of the root system and being replaced successively by other mycorrhizal species in the older root regions.

These distribution patterns, in which mycorrhizas of "pioneer" or "early" fungi are found at the periphery of the root system, but are replaced by later colonizers (characteristic of older trees) near the tree base, were explained by two types of experiment.

First, basidiospores were collected from fruitbodies of either "pioneer" or "later" mycorrhizal fungi and were added to pots of nonsterile soil. Then aseptically

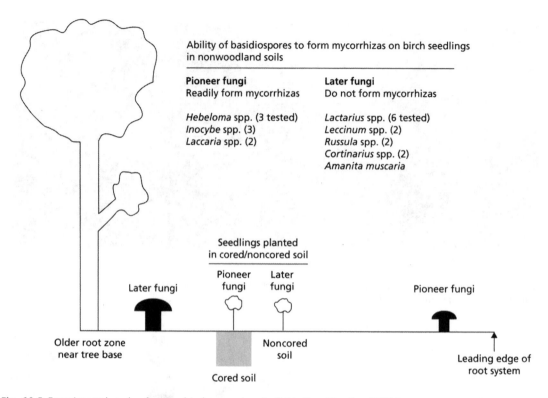

Fig. 10.5 Experimental study of mycorrhizal successions in field sites. (See Fox 1986.)

grown birch seedlings were planted in these soils. Only the pioneer fungi formed mycorrhizas in these conditions, and this was related to basidiospore germination, because only the pioneer fungi have basidiospores that germinate readily – the spores of the "later" fungi germinate extremely poorly (often less than 0.1% germination in any conditions that have been tested). Second, birch seedlings were raised aseptically (without mycorrhizas) then planted beneath older trees in a field site so that they would be infected by the fungi already established in the site. Some of these seedlings were planted directly into the undisturbed soil, but others were planted where the soil had first been removed with a corer (like the corers used for making holes in putting greens) and then replaced immediately (Fig. 10.5). All the seedlings in the cored positions became infected by pioneer fungi, presumably from spores in the soil. By contrast, all the seedlings planted in undisturbed positions were infected by the "later" fungi, presumably from hyphal networks which radiated through the soil.

So, the pattern of mycorrhizal establishment on birch in previously treeless sites can be summarized as follows. The pioneer fungi infect young seedlings in nurseries or in the field, from basidiospores that land on the soil surface and are washed into the root zone. They probably have annual cycles of infection from basidiospores as the root system grows and expands into new soil zones. The "later" fungi cannot establish initially because their spores germinate poorly. But they germinate eventually, especially in older parts of the root zone, and become dominant by spreading as mycelial networks to infect further root tips.

In natural woodlands and forestry plantations, seedlings are likely to be infected directly by the "later" fungi, but in new sites they will initially be infected by pioneer fungi. This has practical consequences, because only the pioneer fungi are suitable for mycorrhizal inoculation programs, commonly used in land-reclamation sites. The fungi most often used for this are the puffball, *Pisolithus tinctorius,* and the toadstool-forming fungus *Paxillus involutus.* Both are pioneer colonizers that tolerate relatively high levels of toxic minerals and the low water-retention properties of mine-spoil and other land-reclamation sites.

Exogenously imposed dormancy

In laboratory conditions, most asexual spores germinate readily at suitable temperature, moisture, pH and oxygen levels. Some germinate even in distilled water, although most require at least a sugar source, and a few have multiple nutrient requirements. However, in

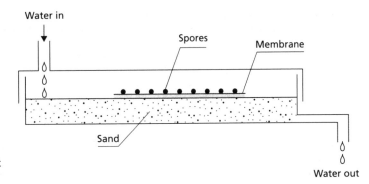

Fig. 10.6 A nutrient leaching system that mimics soil fungistasis.

nature all these spores can be held in a dormant state by the phenomenon termed **fungistasis** (or **mycostasis**). This is very common in soil (Lockwood 1977), and has also been reported on leaf surfaces.

Fungistasis is a microbially induced suppression of spore germination. For example, spores often fail to germinate in topsoil, where the level of microbial activity is high, but they germinate in sterilized soil or in subsoils of low microbial activity. The germination that occurs in sterilized soil can be prevented if the soil has been recolonized by microorganisms, and even single microorganisms – bacteria or fungi – will restore the suppression. This suggests that fungistasis is caused by nutrient competition or by general microbial metabolites (or both), but not by specific antibiotics or other inhibitors from particular microorganisms. A long history of research suggests that volatile germination inhibitors such as ethylene ($H_2C=CH_2$), allyl alcohol ($H_2C=CH.CH_2OH$), and ammonia can play a role in fungistasis. But the strongest evidence implicates **nutrient deprivation** as the key cause of fungistasis. Even spores that can germinate in distilled water are inhibited in soil because they leak nutrients into their immediate surroundings, and these nutrients are continuously metabolized by other soil organisms.

Lockwood and his colleagues (see Hsu & Lockwood 1973) devised a simple experimental system to test whether nutrient deprivation can mimic the fungistasis observed in soil (Fig. 10.6). In completely sterile conditions, spores were placed on membrane filters over a bed of washed sand or glass beads, then sterile water was percolated slowly through the sand or beads so that any nutrients released from the spores were continuously removed. Except for the special case of "activated" (heat-treated) ascospores of *Neurospora*, which germinated in all conditions, the spores did not germinate in the "nutrient-leaching" system, but they germinated if the flow of water was stopped for 24 hours or when a flow of glucose solution was used in place of water. By using very slow rates of water percolation it was possible to simulate the fungistatic effects of natural soils. The spores of different fungi have different

fungistatic sensitivities (related to spore size, spore nutrient reserves, and speed of germination), but with few exceptions there is remarkably good agreement between the sensitivity of spores to nutrient-leaching in the model system and their sensitivity to soil fungistasis (Table 10.1).

Ecological implications of fungistasis

Fungistasis causes spores to remain quiescent in soil or other natural environments until nutrients become available. Thus, saprotrophs can lie in wait for organic nutrients, and root-infecting pathogens or mycorrhizal fungi can wait for a root to pass nearby. In many cases the germination trigger is nonspecific. For example, the spores of many root pathogens can germinate in response to the root exudates of both host and non-host plants. This can be exploited for disease control, especially in traditional crop-rotation systems (organic farming) where the spores of parasitic fungi can be induced to germinate but then die because they cannot infect a plant. This phenomenon is termed **germination-lysis**.

In a few cases there is evidence of **host-specific** triggering of germination. Perhaps the best example is the triggering of sclerotia of *Sclerotium cepivorum* (Basidiomycota) which causes the economically important "white rot" disease of onions, garlic, and closely related *Allium* spp. (Coley-Smith 1987). These small sclerotia (about 1 mm diameter) are produced abundantly in infected onion bulbs and can survive for up to 20 years in soil until they are triggered to germinate by the host. The germination triggers are volatile sulfur-containing compounds (alkyl thiols and alkyl sulfides) such as diallyl disulfide (DADS), but the host plant releases the nonvolatile precursors of these compounds (alkyl sulfoxides and alkyl-cysteine sulfoxides) and these are converted to the volatile germination triggers by many common soil bacteria. Knowledge of this system has suggested some novel approaches for controlling *S. cepivorum*, but unfortunately with little success to date. One approach would be to breed

Table 10.1 Relationship between germination of fungal spores incubated on natural soil and on a nutrient-leaching system designed to mimic the continuous removal of spore nutrients by soil microorganisms. (Data from Hsu & Lockwood 1973.)

Fungus and spore type	Germination %		Leaching system	
	Distilled water	Natural soil	Water flowing	Flow stopped for 24 h
Conidia				
Verticillium albo-atrum	60	9	8	no data
Thielaviopsis basicola	89	4	5	89
Fusarium culmorum	94	20	9	91
Curvularia lunata	95	16	13	91
Cochliobolus sativus	97	21	19	91
Alternaria tenuis	95	54	71	no data
Activated ascospores				
Neurospora crassa	98	87	84	no data

cultivars of onions, garlic, etc. that do not produce the compounds that trigger sclerotial germination. But this is not a viable proposition because the germination triggers are important flavor and odor components of *Allium* spp. Another approach would be to trigger germination in the absence of the host crop. In this respect, artificial onion oil is used widely as a flavor component of processed foods (including cheese-and-onion flavored potato crisps) and it contains large amounts of DADS. When applied to soil, artificial onion oil triggers the germination of sclerotia, which then die by germination-lysis. Up to 95% of sclerotia can be destroyed in this way, but even the remaining few per cent can be sufficient to cause significant crop damage.

Spore dispersal

Fungi have many different methods of spore dispersal. Here we will focus on selected aspects, asking how the spores or spore-bearing structures of fungi are precisely tailored for their roles in dispersal. In doing so, we cover many topics of practical and environmental significance.

Ballistic dispersal methods of coprophilous fungi

Coprophilous (dung-loving) fungi grow on the dung of herbivores and help to recycle the vast amounts of plant material that are deposited annually by grazing animals. The spore dispersal mechanisms of these fungi are highly attuned to their specific lifestyles – their function is to ensure that the spores are propelled

from the dung onto the surrounding vegetation, where they will be ingested and pass through an animal gut to repeat the cycle. In several cases this is achieved by ballistic mechanisms of spore discharge.

In the case of *Pilobolus* (Fig. 10.7) each spore-bearing structure consists of a large black sporangium, mounted on a swollen vesicle which is part of the sporangiophore. At maturity the sporangiophore develops a high turgor pressure, the wall that encloses both the sporangium and the vesicle breaks down locally by enzymic means, and the vesicle suddenly ruptures, squirting its contents forwards and propelling the sporangium for 2 meters or more. Mucilage released from the base of the sporangium during this process serves to stick the sporangium to any plant surface on which it lands; then the spores are released from the sporangium and can be spread by water or other agencies. As a further adaptation for dispersal, the sporangiophore is phototropic, ensuring that the sporangium is shot free from any crevices in the dung. The light signal is perceived by a band of orange carotenoid pigment at the base of the vesicle, and the vesicle itself acts as a lens that focuses light on the pigment. A unilateral light signal is thereby translated into differential growth of the sporangiophore stalk, aligning the sporangium towards the light source.

Pilobolus, therefore exhibits three special adaptations that also are found, to different degrees, in several other coprophilous fungi (Fig. 10.8):

1 The spore-bearing structure is **phototropic**, an adaptation also seen in the tips of the asci of *Ascobolus* and *Sordaria*, which grow on dung.
2 There is an **explosive discharge** mechanism. This also is seen in the asci of *Ascobolus* and *Sordaria* because

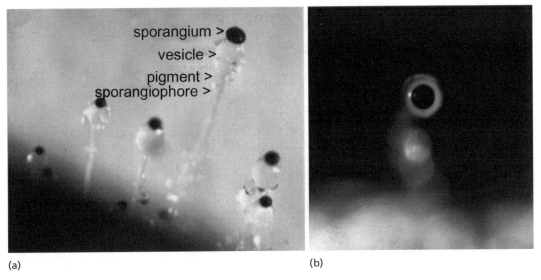

Fig. 10.7 *Pilobolus* (Zygomycota), a fungus with a ballistic method of spore discharge. (a) Several sporangia on a dung pellet. (b) A sporangium orientated towards a light source, showing how the subsporangial vesicle acts as a lens.

the asci act as guns, shooting spores up to 1–2 cm into the air. A different discharge mechanism is found in *Basidiobolus ranarum* (Chytridiomycota) which grows on the feces of lizards and frogs. In this case the sporangium is mounted on a subsporangial vesicle, as in *Pilobolus*, but the vesicle ruptures at its base, squirting the sap backwards and propelling the sporangium forwards, like a rocket. Yet another variation is found in *Sphaerobolus stellatus* (Basidiomycota) which produces basidiospores in a large ball-like structure within a cup-shaped fruitbody. At maturity, the inner layer of the cup separates from the outer layer and suddenly inverts, like a trampoline, springing the spore mass into the air.

3 There is a **large projectile**, based on the ballistic principle that large (heavy) objects travel further than small objects if released at the same initial velocity. *Sphaerobolus stellatus* has a spore mass about 1 mm diameter, allowing it to be thrown 2 meters vertically.

However not all coprophilous fungi employ ballistic mechanisms of spore discharge. In *Pilaira* (Zygomycota, Fig. 10.8), which is closely related to *Pilobolus* in the family Pilobolaceae, the sporangiophore merely extends several centimeters at maturity so that the sporangium collapses onto the surrounding vegetation.

Insect-dispersed fungi

Insects and other small arthropods can disperse several types of fungi, including spores that are produced in sticky, mucilaginous masses. This form of dispersal can be highly efficient because the fungus takes advantage of the searching behavior of the vector to reach a new site. There are many of these fungus–vector associations, ranging from cases where the association is almost incidental to cases of highly evolved mutualism. Here we consider one classic example – the dispersal of Dutch elm disease by a bark-beetle vector. We deal with other mutualistic associations in Chapter 13.

Dutch elm disease

Dutch elm disease (Figs 10.9, 10.10) is caused by two closely related fungi, *Ophiostoma ulmi* and *O. novo-ulmi* (Ascomycota). These fungi enter the plant through wounds made by bark beetles, then spread in the water-conducting xylem vessels by growing in a yeast-like budding phase. This causes reactions in the xylem vessels, leading to blockage and death of all or part of the xylem. In many respects, the symptoms and host reactions in Dutch elm disease resemble those caused by other **vascular wilt pathogens**. But bark beetles of the *Scolytus* and *Hylurgopinus* are specialized vectors of Dutch elm disease.

The disease cycle starts when young, contaminated beetles emerge from the bark of dead or dying elm trees in early spring, fly to neighboring healthy trees, and feed on the bark of the young twigs. During feeding, the beetles cause incidental damage to the xylem, thereby introducing the fungus into the tree. The fungus then spreads in the xylem, killing the whole tree or some of its major branches, and the bark of the newly killed trees is then used by the female beetles for egg

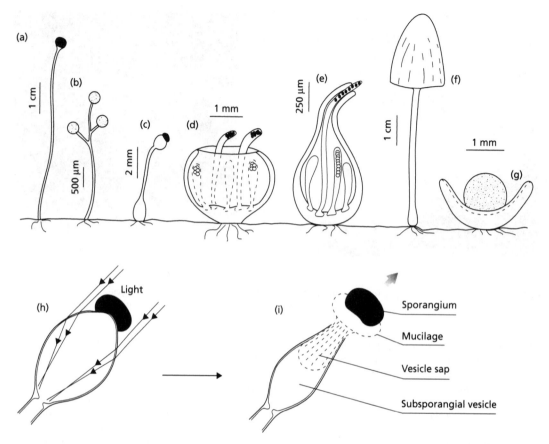

Fig. 10.8 Diagrammatic representation of some coprophilous fungi. (a) *Pilaira anomala* (Zygomycota): the sporangiophore elongates to several centimeters at maturity and the spores "flop" onto the surrounding vegetation. (b) *Mucor racemosus* (Zygomycota) with no special method of spore release. (c) *Pilobolus* (Zygomycota); see also (h) and (i). (d) *Ascobolus* sp. (Ascomycota); the tips of the mature asci project from the apothecium and are phototropic, shooting the ascospores towards a light source. (e) *Sordaria* sp. (Ascomycota); the neck of the perithecium is phototropic and the mature asci elongate up the neck to discharge the ascospores. (f) *Coprinus* sp. (Basidiomycota). (g) *Sphaerobolus* sp. (Basidiomycota); the large spore mass is shot from the cup-shaped fruitbody when the layers of this separate and the inner layer suddenly inverts. (h,i) *Pilobolus*, showing how the terminal vesicle of the sporangiophore acts as a lens to focus light and orientate the sporangiophore, and showing also the mechanism of sporangium discharge (see also Fig. 10.7).

laying. The female beetle tunnels into the inner bark and eats out a channel, depositing eggs along its length – the "brood gallery." The eggs hatch and the young larvae eat out a series of radiating channels before they pupate for overwintering. Meanwhile, the fungus that killed the tree grows from the xylem into the bark and sporulates in the beetle tunnels. In this way, the young adult beetles that emerge from the pupae in the following spring become contaminated with spores; they leave the bark and fly in search of new trees, repeating the disease cycle.

The fungus–vector relationship clearly benefits both partners, because the beetle carries the fungus to a new host, while the beetle is ensured of a fresh supply of breeding sites in the bark of newly killed trees – it will not lay eggs in older, dead bark. As shown in Fig. 10.10b, the fungus can produce a sexual stage in the bark – a cleistothecium with a long neck which extrudes ascospores in a mucilaginous matrix. This sexual stage can ensure the generation of recombinant strains, and again the beetle is involved in this because adult beetles feed on the bark late in the season, introducing

(a)   (b)

Fig. 10.9 Dutch elm disease caused by *Ophiostoma ulmi* and *O. novo-ulmi*. (a) Dying elm trees with thinning crowns. (b) Coremia of *Ophiostoma*, consisting of many aggregated conidiophores, bearing minute conidia in a large, sticky mass at the tip (much of this spore mass was lost during preparation). (c) Beetle bark galleries on the inside of the bark; the initial gallery caused by the adult female is marked, and radiating from this are galleries produced by the young larvae. (d) Section of a branch from a dying elm tree, showing a ring of blocked, discolored xylem vessels from a nonlethal attack in a previous growth season.

(c)   (d)

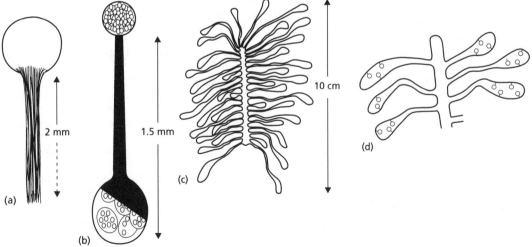

Fig. 10.10 Dutch elm disease. (a) Diagram of an intact coremium. (b) Sexual stage of *Ophiostoma* – a long-necked cleistothecium (closed fruitbody) containing oval asci each containing eight ascospores. The asci break down at maturity and the ascospores are extruded in mucilage at the tip. (c,d) Beetle brood gallery in the inner bark of a diseased tree. The adult female beetle deposits eggs along the gallery, then the emerging beetle larvae eat out a series of lateral galleries ending in chambers. The fungus sporulates in these chambers and contaminates the beetles that emerge in spring.

new strains of the pathogen, often different from the strain that killed the tree.

As testimony to the efficiency of this fungus–vector relationship, new strains of *C. novo-ulmi* swept across Britain and much of continental Europe in the late 1900s, decimating the native British elm population, after these strains were introduced on shipments of imported elm logs in the 1960s. A similar epidemic spread across the USA early in the last century. In both cases it seems that the epidemic arose from logs that had been imported and were not de-barked. If the bark had been removed from the logs (as is required by quarantine regulations) there would have been no problem because the beetle vectors would have been removed.

Molecular characterization of *Ophiostoma* has been used to trace the origins and histories of these epidemics (Mitchell & Brasier 1994; Brasier 1995). This revealed that *O. ulmi* has been present for many years in Britain and much of continental Europe, but as a heterogeneous population of nonaggressive strains comprising several vegetative compatibility (VC) groups. This population was in balance with the tree host, causing relatively little damage. The recent British epidemics have been caused by two aggressive subgroups of the fungus, one imported from North America (termed NAN) and one of Eurasian origin (EAN). These aggressive forms are sexually incompatible with the original nonaggressive population, so they have been described as a new species, *O. novo-ulmi*. At the advancing margins of the disease in Europe, the pathogen population is almost genetically pure and exists as a single VC "super group." We could expect this from the fungus–vector relationship, because the most aggressive strain will kill most of the trees at the advancing front, and the beetle population will proliferate in these trees, carrying the strain to new trees. However, behind the disease fronts the incidence of this super group declines to only some 20–30% of the fungal population, suggesting that the population is returning to a more stable form. One of the reasons may be that *Ophiostoma*, like *Cryphonectria* discussed in Chapter 9, can harbor virulence-suppressing dsRNA. A diversity of VC groups acts as a barrier to transmission of hypovirus genes – perhaps a natural defense against these extrachromosomal elements.

Dispersal of aquatic fungi: appendaged spores

Fungi that grow as saprotrophs in aquatic environments often have spores with unusual shapes and conspicuous appendages (Fig. 10.1). One of the more common types is the tetraradiate (four-armed) spore, often found in the fungi that grow on fallen tree leaves in well-aerated, fast-flowing streams (e.g. *Alatospora*,

Tetracladium, Tetrachaetum). Similar tetraradiate spores have been found in two marine fungi (Basidiomycota), while tetraradiate sporangia are produced by *Erynia conica* (Zygomycota), a fungus that parasitizes freshwater insects. There is even a yeast, *Vanrija aquatica*, which grows in mountain tarns, that produces tetraradiate cells instead of the normal ovoid yeast cells. In extreme cases, aquatic spores such as *Dendrospora* (Fig. 10.1) can have up to 20 radiating arms. And other aquatic spores are curved or sigmoid – for example, *Anguillospora* (Fig. 10.1).

In contrast to the freshwater fungi of Fig. 10.1, wood-rotting Ascomycota of estuarine and marine environments often have ascospores with flakes of wall material or mucilaginous appendages (Fig. 10.11). All these bizarrely shaped spores must be functionally significant, either in terms of their buoyancy or in terms of their entrapment or adherence onto substrates in aquatic environments.

Several lines of evidence suggest that the common tetraradiate spores of freshwater fungi may serve a range of different roles. For example, the yeast *Vanrija aquatica* produces appendages in response to nutrient-poor conditions in laboratory culture, suggesting that they might increase the surface area for nutrient absorption. The tetraradiate conidia of other fungi have been shown to sediment slowly in water, at about 0.1 mm s^{-1}, although differences in sedimentation rates are unlikely to be important in the turbulent, fast-flowing streams where these spores are commonly found. Perhaps more important is the role of spore shape in entrapment, because small air bubbles are often trapped between the arms of tetraradiate spores, causing the spores to accumulate in the "foam" of fast-flowing streams. In addition, these spores settle like a tripod on a natural or artificial surface, and then respond rapidly by releasing mucilage from the tips of the arms in contact with the surface, but not from the fourth (free) arm. This attachment is also followed rapidly by germination from the contact sites, so that the fungus establishes itself from three points, which is likely to increase the efficiency of colonizing a substrate in competition with other microorganisms. A quite different role was discovered quite recently: some of the appendaged fungal spores are produced by fungi that grow on the **living leaves of trees** that overhang fast-flowing streams. These conidia are easily dislodged from the conidiophores by raindrops. So, the tetraradiate or sigmoidal spore form might represent an adaptation to a multiplicity of needs in the habitats where these fungi grow.

The mucilaginous appendages of the marine Ascomycota function in attachment to surfaces. These fungi often colonize wood in estuarine environments (Moss 1986), where they have important roles as decomposers of woody materials.

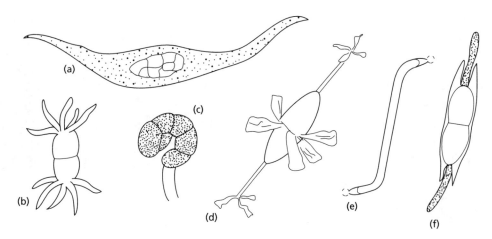

Fig. 10.11 Appendaged spores of estuarine and marine environments. (a) Ascospore of *Pleospora* with mucilaginous appendages (stippled), about 400 µm. (b) Ascospore of *Halosphaeria* with chitinous wall appendages (25 µm). (c) Conidium of *Zalerion* (25 µm). (d) Ascospore of *Corollospora* with membranous appendages (70 µm). (e) Ascospore of *Lulworthia* with terminal mucilaginous pouches (60 µm). (f) Ascospore of *Ceriosporiopsis* with mucilaginous appendages (stippled) (40 µm).

Dispersal and infection behavior of zoospores

Zoospores are motile, wall-less cells that swim by means of flagella. They are the characteristic dispersal spores of **Chytridiomycota**, **Oomycota**, and **plasmodiophorids**, although only the Chytridiomycota are true fungi.

Zoospores can swim for many hours using their endogenous energy reserves, and they show a remarkable degree of sensory perception, owing to the presence of receptors on the cell surface. These receptors enable zoospores to precisely locate the sites where they will encyst – whether on a host or an organic substrate. An example was shown in Chapter 2 (see Fig. 2.2) for the nematode parasite *Catenaria anguillulae*. Other important examples include the many Oomycota (*Pythium*, *Phytophthora*, and *Aphanomyces* spp.) that cause devastating diseases of crop plants (Chapter 14) or of salmonid fish, while a wide range of saprotrophic species play important roles as primary colonizers of organic substrates in natural waters. In this section we discuss the structure and function of zoosporic fungi. Many aspects of zoospore biology and infection have been reviewed by Deacon & Donaldson (1993) and Hardham (2001).

Structure and organization of zoospores

As shown in Fig. 10.12, the zoospores of Chytridiomycota are small, typically 5–6 µm diameter, and tadpole-shaped. Except for some rumen chytrids (which have several flagella) they have a single, smooth, posterior flagellum of the **whiplash** type. The most conspicuous feature of these zoospores is the nucleus, surmounted by a large nuclear cap, which is rich in RNA, protein, and ribosomes. The flagellar membrane is continuous with the cell membrane, and the core of the flagellum – the **axoneme** – consists of a ring of nine triplets of microtubules surrounding two central microtubules. This "9 + 2" arrangement is typical of most motile, flagellate cells. At the root of the flagellum is a **kinetosome** (a modified centriole derived from the nuclear division that preceded zoospore cleavage), and an array of tubular elements that probably provide anchorage and energy transfer to the flagellum. Surrounding the kinetosome is a large mitochondrion shaped like a doughnut ring, and a microbody–lipid complex which, presumably, supplies the energy for beating of the flagellum. When the zoospore comes to rest the flagellum is retracted into the cell by a reeling-in mechanism involving the rotation of the cell contents, then a cyst wall is formed. It is notable that, although the zoospores of Chytridiomycota are wall-less cells, they do not seem to have an osmoregulatory apparatus.

The plasmodiophorids also have small zoospores, about 5 µm diameter, but with two flagella – a short one directed forwards and a longer one directed backwards. By contrast, the zoospores of Oomycota (Fig. 10.13) are larger, typically 10–15 µm, and kidney-shaped, with two flagella inserted in a ventral groove. The longer flagellum is **whiplash type** and trails behind the swimming spore; the shorter flagellum projects forwards and is **tinsel-type**, with short glycoprotein hairs

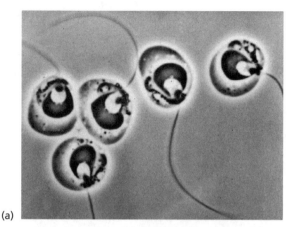

(a)

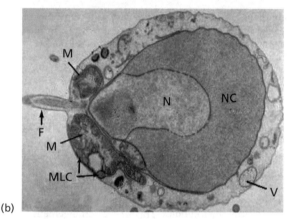

(b)

Fig. 10.12 (a) Zoospores of *Blastocladiella emersonii* (Chytridiomycota) with a smooth, posterior whiplash flagellum. (Courtesy of M. S. Fuller.) (b) Electron micrograph of a longitudinal section of the zoospore of *B. emersonii*. The zoospore plasma membrane is continuous with the flagellar membrane (F) but only part of the flagellum is seen in this section. M = mitochondrion; MLC = microbody–lipid globule complex; N = nucleus; NC = nuclear cap; V = vacuole. (Courtesy of M.S. Fuller; from Reichle & Fuller 1967.)

(a)

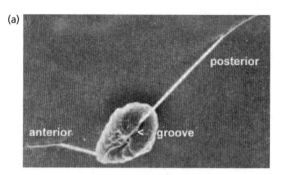

(b)

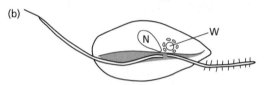

Fig. 10.13 (a) Scanning electron micrograph of a zoospore of *Phytophthora* (Oomycota) with a posterior whiplash flagellum and a shorter, anterior tinsel-type flagellum. (b) Diagrammatic representation of the zoospore, showing the insertion of the flagella in the ventral groove (shaded) and the location of the nucleus (N) and water-expulsion vacuole (W). ((a) Image courtesy of M.S. Fuller.)

(mastigonemes) projecting along its length. These act like a series of oars, pulling the zoospore forwards as the anterior flagellum generates a sine wave from its base to its tip. This is estimated to account for at least 90% of the forward swimming thrust. The posterior flagellum acts like a rudder. It periodically kicks at an angle of about 90°C, causing the cell to change the swimming direction.

The arrangement of organelles in zoospores of Oomycota is shown in Fig. 10.14. The top section passes through the region of flagellar insertion in the zoospore ventral groove and shows the nucleus (N)

extending to the base of the flagella. Beneath the plasma membrane are sheets of **peripheral cisternae (pc)**, **large peripheral vesicles, dorsal vesicles**, and **ventral vesicles**. The cell also contains mitochondria (M) and fingerprint vacuoles (FV) which contain glucans that probably serve as carbohydrate reserves. The asterisk marks a cavity probably resulting from extraction of lipid during preparation of the specimen. The lower section passes through the ventral groove where the osmoregulatory **water expulsion vacuole (WEV)** is located. The WEV consists of a central vacuole (CT) and surrounding vacuoles (SU). It contracts and expels water regularly every few minutes.

The significance of this complex ultrastructural organization of *Phytophthora* zoospores lies in the fact that the zoospore is a transitory phase of the life cycle, specialized for dispersal. When the zoospore locates a suitable site for infection it transforms rapidly into a walled cyst – a process that takes only a few minutes. Thus, the zoospore is a pre-programed cell, destined to undergo a rapid transition. The details of this are of much interest and can be followed by cytochemical methods.

The peripheral vesicles that lie just beneath the zoospore membrane are of at least three types, and can be distinguished by the binding of their proteinaceous contents to specific monoclonal antibodies or lectins. The **large peripheral vesicles** beneath most of the cell

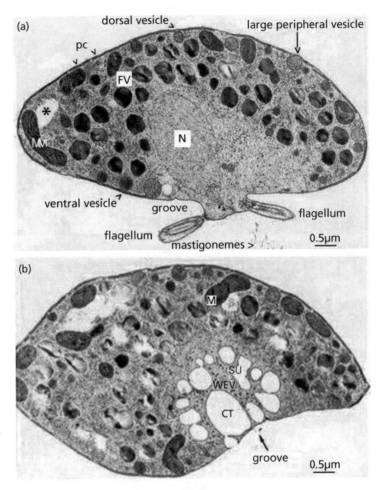

Fig. 10.14 (a,b) Two parallel cross-sections through a zoospore of *Phytophthora palmivora* (Oomycota). See text for details. (Courtesy of M.S. Fuller; from Cho & Fuller 1989.)

periphery contain a glycoprotein that is thought to serve as a protein store after encystment. These vesicles migrate towards the center of the cell when a zoospore encysts. The smaller **dorsal vesicles** contain a different glycoprotein which is released by exocytosis at an early stage of encystment, and it accumulates on the cell surface as a cyst coat. A third class of vesicles, the **ventral vesicles**, are found only around the zoospore ventral groove. They also contain a protein which is released during encystment, but it is deposited locally between the cyst and the surface on which encystment has occurred, and it acts as an adhesive.

A newly encysted zoospore has no wall, only an amorphous glycoprotein coat deposited by the dorsal vesicles. But a true wall is synthesized beneath the cyst coat in the first few minutes of encystment. This wall is derived from the peripheral cisternae (membrane lamellae) that lie immediately beneath the plasma membrane of the motile spore (Fig. 10.15). Once the

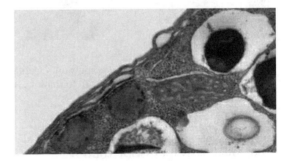

Fig. 10.15 Close-up view of a part of the zoospore surface of *Pythium aphanidermatum*, showing the laminate peripheral cisternae that will subsequently become vesicles for release of the cyst wall.

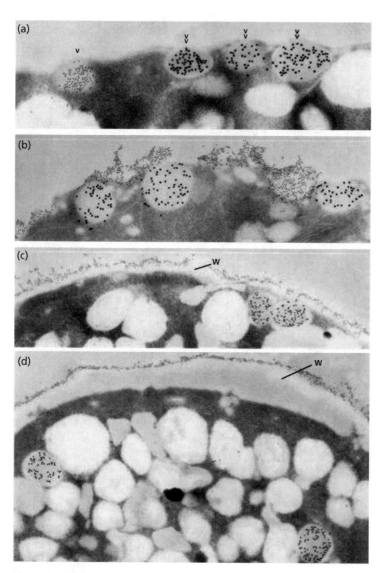

Fig. 10.16 Stages in encystment of *Phytophthora cinnamomi* zoospores (Oomycota) immunogold-labeled with monoclonal antibodies (mAbs) specific to the contents of different peripheral vesicles. (a) Outer region of a motile zoospore showing two classes of peripheral vesicles: the contents of large peripheral vesicles bind to a specific mAb that has been complexed to large (18 nm) gold particles (double arrowheads) whereas the contents of small dorsal vesicles bind to a different mAb complexed to 10 nm gold particles (single arrowhead). (b) Zoospore labeled at 1 minute after addition of pectin to induce encystment. Most of the small dorsal vesicles have released their contents onto the cell surface but the large peripheral vesicles still lie beneath the surface. (c) Spore labeled at 5 minutes after encystment. The spore has started to synthesize a cell wall (W), the surface of which is coated with material from the small dorsal vesicles. (d) Spore labeled after 10 minutes. The wall is now thicker, and the large peripheral vesicles have migrated towards the center of the cyst. They are thought to be a protein store for use during cyst germination and germ-tube growth. (Photographs courtesy of F. Gubler & A. Hardham; (a) from Hardham 1995; (b–d) from Gubler & Hardham 1988.)

cyst wall has developed, the water-expulsion vacuole disappears. The sequence of cyst wall formation is shown in Fig. 10.16.

Sensitivity of zoospores to lysis: a potential basis for disease control

Zoospores of all types, including Chytridiomycota (e.g. *Allomyces* spp.) and Oomycota (*Saprolegnia*, *Aphanomyces*, *Pythium*, and *Phytophthora*) lack a cell wall during their motile phase and the early stages of encystment. For this reason zoospores are highly susceptible to disruption by surface-active agents (surfactants) such as **rhamnolipids** which are pro-

duced by the bacterium *Pseudomonas aeruginosa*. The key feature of surface-active agents is that they have both a hydrophobic and a hydrophilic domain, so they can insert into the cell membrane of wall-less cells and disrupt the cells (Ron & Rosenberg 2001).

The roots of oat plants (*Avena sativa*) and the closely related wild grass, *Arrhenatherum elatius*, produce soap-like compounds (saponins) from a narrow zone just behind the root tips. In this case the saponin is termed avenacin (see Fig. 9.12), and it naturally fluoresces bright blue when viewed under ultraviolet illumination (Fig. 10.17). A similar saponin, β-aescin, is produced by the leaves of horse chestnut trees (*Aesculus hippocastanum*). When oat roots are placed in a suspension of fungal zoospores the spores rapidly

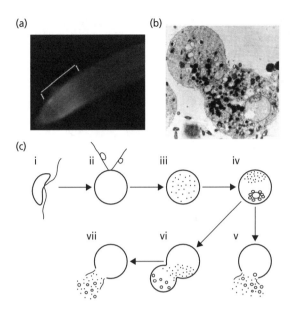

Fig. 10.17 (a) Oat root tips with natural blue auto-fluorescence caused by the presence of avenacin. (b) A *Pythium* zoospore undergoing disorganization and lysis in the presence of avenacin. (c) Responses of wall-less zoospores of Oomycota to saponins such as avenacin or β-aescin: (i) motile zoospore; (ii) immobilization and rounding-up; (iii) development of phase-dark granules; (iv) localization of granules and development of vacuoles; (v) lysis; (vi,vii) ballooning followed by lysis. Disruption of the zoospores usually occurs within 5–10 minutes. (From Deacon & Mitchell 1985.)

accumulate at the root tips in response to root tip nutrients and then lyse within a few minutes (Fig. 10.17).

The use of surfactants such as rhamnolipids, or even crude extracts of saponin-containing tissues such as oat roots, could provide disease control in hydroponic glasshouse-cropping systems where zoosporic fungi can cause serious diseases. This is now being investigated in several laboratories, to find environmentally safe alternatives to the use of fungicides.

Zoospore motility

In appropriate conditions zoospores of Oomycota can swim for 10 hours or more, at rates of at least 100 μm s^{-1} fuelled by endogenous nutrient reserves. So they could, in theory, swim as far as 3–4 meters for dispersal to new environments. However, the zoospores make frequent random turns (Fig. 10.18), and because of this the rate of dispersion by *Phytophthora* zoospores in still water has been found to be little more than the rate of diffusion of a small molecule such as HCl.

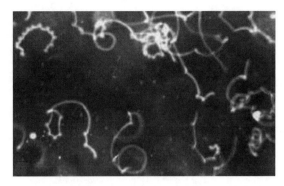

Fig. 10.18 Zoospore tracks of Oomycota, captured as negatives on photographic film during a 5-second exposure and showing frequent random turns in the absence of an attractant.

Clearly, the swimming activity of zoospores must serve other roles. One of these roles is that swimming zoospores can remain in suspension and be carried in moving water, whereas nonmotile spores tend to settle out. This has been demonstrated both in artificial soils and in field soils, where zoospores of Oomycota can escape entrapment in narrow, water-filled soil pores, so they can remain suspended and spread in surface run-off water, whereas zoospore cysts are easily trapped in soil. But the main role of zoospores is that their swimming is linked to sensory perception: they can swim towards attractants such as nutrients or oxygen (**positive chemotaxis**) or avoid unsuitable chemical environments (**negative chemotaxis**). Zoospores also can respond to pH gradients, to electrical or ionic fields (**electrotaxis**), and they can accumulate by autoaggregation. The extreme responsiveness of zoospores enables them to settle and encyst in environments that are most appropriate for subsequent development. For example, zoospores often accumulate in large numbers near root tips, at plant wound sites, or around individual stomata on a leaf surface. This "homing and docking" sequence of zoospores, discussed below, is as sophisticated and rapid as any that have been described in the biological world.

The "homing and docking" sequence of zoospores

The events in the "homing and docking" sequence of zoospores (Fig. 10.19) have been studied in detail for *Pythium* and *Phytophthora* species. The sequence begins when a zoospore detects a gradient of chemoattractant, which causes a partial suppression of random turns so that the spore tends to move up the attractant gradient. This phase of **zoospore taxis**, or zoospore kinesis, can

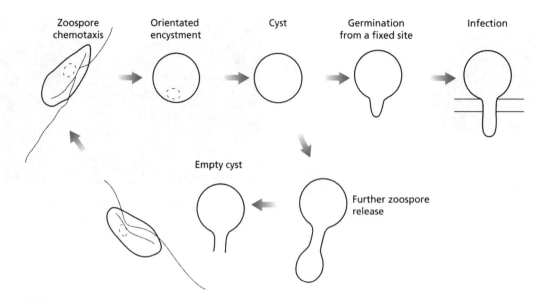

Fig. 10.19 The homing and docking sequence of zoospores.

be studied using capillaries filled with potential attractants. Often the zoospores of plant pathogens show chemotaxis to several individual sugars or amino acids, or to volatile compounds such as ethanol and aldehydes, which are likely to be released as fermentation products of roots in moist soil conditions. However, the strongest responses are usually seen with mixtures of compounds, such as seed and root exudates.

Most *Pythium* and *Phytophthora* species show taxis to the roots of both host and nonhost plants, but a few interesting examples of host-specific taxis have been reported. For example, the host-specific pathogen *Phytophthora sojae* shows chemotaxis *in vitro* to the flavonoids daidzein and genistein, which are known to be present in the soybean host. Similarly, *Aphanomyces cochlioides* shows strong chemotaxis to the flavonoid "cochliophilin A" from spinach plants. It should not be assumed that these compounds are the only factors involved in attraction to the host roots, but the

findings are notable because they parallel the behavior of *Rhizobium* spp., which show *in vitro* chemotaxis to the specific flavonoids of their hosts.

The next phase of the sequence seems to involve **recognition of a host surface component**, because the zoospores move over the host surface, with their flagella in contact with this. Some zoosporic fungi might have specific host surface requirements, but often they respond to pectin and other polyuronates (e.g. alginate) *in vitro*. For example, Fig. 10.20 shows the massing and encystment of *Pythium* zoospores on the root tip mucilage secreted by root cap cells of wheat. Similar massing and encystment is seen in the elongation zone of roots that are coated with calcium alginate gel – a treatment that masks any specific receptors on the root surface.

Recognition of a host surface component (perhaps coupled with a high concentration of root-derived nutrients) leads to **orientated encystment**, with the

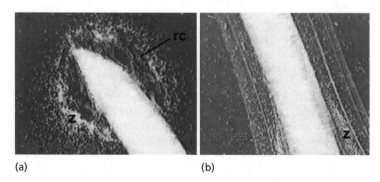

(a)   (b)

Fig. 10.20 Accumulation and encystment of *Pythium* zoospores on wheat roots. (a) Zoospores (z) accumulating in large numbers on the surface of a ball of root tip mucilage; some root cap cells are indicated (rc). (b) Zoospores (z) accumulating and encysting in the elongation zone of wheat roots. In this case the root had been coated with a double layer of calcium alginate gel to block any specific receptor compounds on the root surface.

ventral groove of the zoospore located next to the host. During this process the flagella are withdrawn but the zoospore remains adhered to the host by secretion of an adhesive. This stage of orientated encystment is important, because the zoospores of *Pythium* and *Phytophthora* spp. are known to have a fixed, predetermined point of germ-tube outgrowth. If the spore were to settle in a different orientation the germ-tube would not penetrate the host. There can also be a degree of host-specific encystment at this stage, because the *Pythium* spp. that parasitize grasses and cereals show significantly more encystment on grass roots than on the roots of dicotyledonous plants.

The precise orientation of encystment is also important because it ensures that the adhesive released from the zoospore's ventral vesicles will be deposited next to the host surface. Receptors on the flagella probably are involved in this process, because monoclonal antibodies that bind to **both** flagella of *Phytophthora* zoospores cause rapid encystment *in vitro*, whereas other monoclonal antibodies that bind to only the anterior or the posterior flagellum do not cause rapid encystment.

Cyst germination can be studied *in vitro* by agitating a zoospore suspension, causing the spores to encyst, and then adding individual amino acids or sugars to trigger germination. However, cyst germination *in vivo* (on the host plant) is suggested to be an autonomous process, triggered by events in the earliest stages of encystment. Consistent with this, major transmembrane fluxes of calcium occur in the early stages of encystment, and seem to coordinate cyst germination (Fig. 10.21).

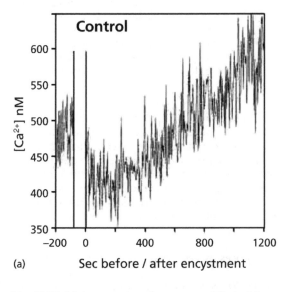

(a)

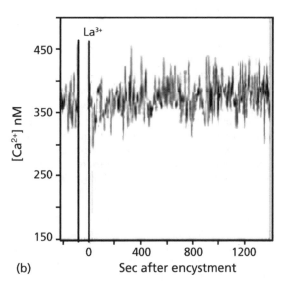

(b)

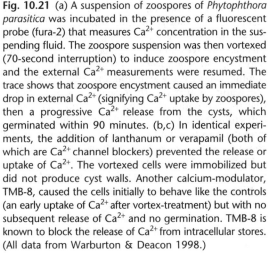

Fig. 10.21 (a) A suspension of zoospores of *Phytophthora parasitica* was incubated in the presence of a fluorescent probe (fura-2) that measures Ca^{2+} concentration in the suspending fluid. The zoospore suspension was then vortexed (70-second interruption) to induce zoospore encystment and the external Ca^{2+} measurements were resumed. The trace shows that zoospore encystment caused an immediate drop in external Ca^{2+} (signifying Ca^{2+} uptake by zoospores), then a progressive Ca^{2+} release from the cysts, which germinated within 90 minutes. (b,c) In identical experiments, the addition of lanthanum or verapamil (both of which are Ca^{2+} channel blockers) prevented the release or uptake of Ca^{2+}. The vortexed cells were immobilized but did not produce cyst walls. Another calcium-modulator, TMB-8, caused the cells initially to behave like the controls (an early uptake of Ca^{2+} after vortex-treatment) but with no subsequent release of Ca^{2+} and no germination. TMB-8 is known to block the release of Ca^{2+} from intracellular stores. (All data from Warburton & Deacon 1998.)

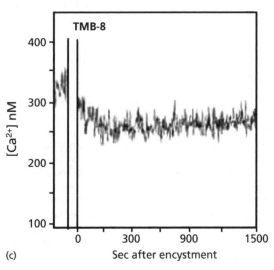

(c)

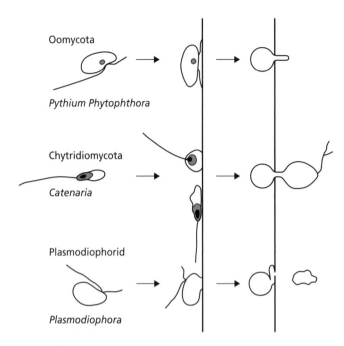

Oomycota

Pythium Phytophthora

Chytridiomycota

Catenaria

Plasmodiophorid

Plasmodiophora

Fig. 10.22 Comparison of the docking and penetration of a host by three types of zoosporic fungus (see text for details). In all three types of organism the zoospore shows precise orientation during encystment on a host surface, related to the fact that these zoospores have a predetermined site from which they penetrate the host.

Collectively, the results in Fig. 10.21 demonstrate a central role of Ca^{2+} uptake and subsequent release in the transition from a zoospore to a germinating cyst.

In normal conditions, when zoospores encyst on a host surface they germinate within 20 or 30 minutes by the emergence of a germ-tube, which usually penetrates the host directly. However, the zoospores of *Pythium* and *Phytophthora* also have a default option. If a zoospore does not locate a suitable host after several hours of swimming, it encysts before its nutrient reserves are depleted and the cyst then releases a further zoospore (Fig. 10.19). This process of repeated emergence can occur two or three times before the zoospore exhausts its nutrient reserves.

Parallel development among zoosporic species

Many of the features described above for Oomycota are also found in other zoosporic organisms, but the details vary (Fig. 10.22). For example, zoospores of Oomycota always dock onto a host or other surface with the flagella and ventral groove located next to the surface. The Chytridiomycota always seem to dock "head-on," with the flagellum orientated away from the host. Plasmodiophorids show yet another variation: their zoospores always seem to dock with the two flagella orientated **away** from the host. But these obligately parasitic organisms also display a unique mode of behavior. When the zoospore settles and encysts on a

host surface, the cyst vacuole enlarges and a small **adhesorium** is produced at the site of contact with the host. Then a pre-formed bullet-like **stylet** is shot through the host wall and the contents of the cyst enter the host cell as a wall-less **plasmodium**. Again, this provides evidence of precisely orientated encystment, because the stylet must be positioned correctly to ensure that the protoplast will be delivered through the host cell wall.

Zoospores as vectors of plant viruses

About 20 plant viruses are currently known to be transmitted by zoospores, and in some cases this is their main or only means of transmission (Table 10.2). These zoospore vectors belong to three genera: *Olpidium* (Chytridiomycota), *Polymyxa* (plasmodiophorids), and *Spongospora* (plasmodiophorids). All are common and usually symptomless parasites of roots. The feature that makes them significant as vectors is that the zoospore encysts on a root and then germinates to release a naked protoplast into the plant. Any virus particles that bind to the surface of the swimming zoospore will therefore be introduced into the host.

There are different degrees of specialization in these virus–vector relationships. *Olpidium* spp. usually transmit isometric viruses such as cucumber necrosis virus and tobacco necrosis virus, but these viruses have additional (and perhaps more important) modes of transmission. The swimming zoospores acquire these viruses from soil when the virus particles adhere to a

Table 10.2 Some important viruses that are vectored by zoosporic fungi.

Virus type and features	Examples	Host	Vector
Furoviruses: always vectored by fungi			
Straight tubular particles,	Soil-borne wheat mosaic	Wheat, barley	*Polymyxa graminis*
250–300 + 100–150 × 20 nm	Beet necrotic yellow vein	Sugar beet, spinach	*Polymyxa betae*
	Potato mop top	*Solanum* spp.	*Spongospora subterranea*
Single-stranded RNA; genome	Peanut clump	Peanut	*Polymyxa graminis*
divided between more than	Oat golden stripe	*Avena* spp.	*Polymyxa graminis*
one particle	Broad bean necrosis	*Vicia faba*	*Polymyxa graminis*
Barley yellow mosaic type: always vectored by fungi			
Filamentous particles,	Barley yellow mosaic	*Hordeum*	*Polymyxa graminis*
350–700 × 13 nm	Wheat yellow mosaic	*Triticum*	*Polymyxa graminis*
	Wheat spindle streak	*Triticum*	*Polymyxa graminis*
	Oat mosaic	*Avena*	*Polymyxa graminis*
	Rice necrosis mosaic	*Oryza*	*Polymyxa graminis*
Tobacco stunt type: characteristically vectored by fungi			
Straight tubular particles,	Tobacco stunt	*Nicotiana* spp.	*Olpidium brassicae*
200–375 × 22 nm	Lettuce big vein	Lettuce	*Olpidium brassicae*
Double-stranded RNA			
Tobacco necrosis type: various means of transmission, including fungi			
Isometric particles, 26–30 nm	Tobacco necrosis	Tulip, *Solanum, Vicia,* many other plants	*Olpidium brassicae*
Single-stranded RNA	Cucumber necrosis	Cucumber	*Olpidium* spp.
	Melon necrotic spot	Melon, cucumber	*Olpidium radicale*

zoospore surface component. The virus presumably remains on the plasma membrane when the zoospore encysts, and will later be carried on the membrane of the protoplast that enters the host. By contrast, the zoospores of *Polymyxa* and *Spongospora* cannot acquire viruses by swimming in virus suspensions; instead, they acquire the virus inside an infected plant. Many of these viruses are of a distinct type, termed **furoviruses** (**fu**ngally-transmitted **ro**d-shaped **viruses**), and they have no other natural means of transmission. They include some of the most economically important soil-borne viruses, such as beet necrotic yellow vein virus, soil-borne wheat mosaic virus, and potato mop-top virus. Other fungus-transmitted viruses, such as the filamentous types (Table 10.2), show some affinities to furoviruses but have a different particle shape.

There is no evidence that any of these viruses multiply within the vectors. But they are carried internally in the resting spores, which can persist in soil for 20 years or more. So, once established in a field site, these viruses are almost impossible to eradicate. Further details of the zoospore-vectored viruses can be found in Adams (1991), Hiruki & Teakle (1987), and Brunt & Richards (1989).

Dispersal of airborne spores

Most terrestrial fungi produce airborne spores that are dispersed by wind or rain-splash. These are the spores of most significance in plant pathology and for allergies and fungal infections of humans. In this section we consider how spores become airborne (take-off), how they remain airborne (flight), and how they are finally deposited in appropriate environments for future development (landing). These are features of fundamental significance in understanding the ecology of airborne fungi. The subject is covered in detail by Ingold (1971) and Gregory (1973).

Spore liberation – take-off

The essential feature of spore liberation is that a spore needs to break free from the **boundary layer** of still air that surrounds all surfaces. Above this boundary layer the air becomes progressively more turbulent in local eddies, until there is net movement of the air mass which can carry spores to a new site. The depth of the boundary layer can vary considerably – from a fraction of a millimeter on a leaf surface on a windy day, to a meter or more on a forest floor on a perfectly calm day. So the fungi that grow in these different types of environment require different strategies for getting their spores airborne. Some of these strategies are

shown in Fig. 10.23. Often they involve adaptations of the spore-bearing structures rather than of the spores themselves.

Fungi that grow on leaf surfaces sometimes produce chains of spores from a basal cell so that the mature spores are pushed upwards through the boundary layer as more spores are produced at the base of the chain (e.g. *Blumeria* (*Erysiphe*) *graminis* and other powdery mildew pathogens). The spores are then removed by wind or, sometimes more effectively, by mist-laden air (e.g. *Cladosporium*). Other types of spore are flung off the spore-bearing structures by hygroscopic (drying) movements that cause the spore-bearing hyphae suddenly to buckle (e.g. *Phytophthora infestans* and downy mildew fungi such as *Peronospora*). Fungi that grow on

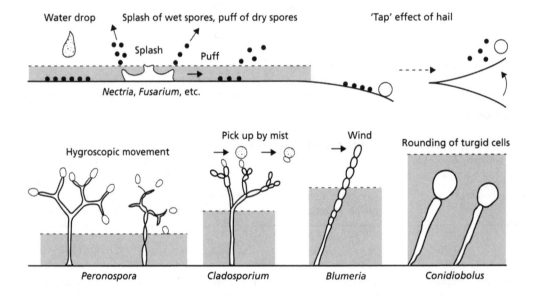

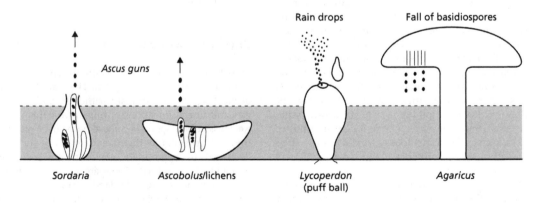

Fig. 10.23 The diversity of mechanisms of spore liberation through a boundary layer of still air (shown by shading).

more rigidly supported surfaces can release the spores by active processes. For example, the asci of many Ascomycota function as small guns, shooting ascospores to 1 or 2 cm distance, to break free of the boundary layer. Spores can also "pop" from a supporting structure when an enclosing wall layer suddenly ruptures, like two balloons that pop apart if compressed and suddenly released (e.g. *Conidiobolus*).

Some other fungi are dispersed by rain-splash. In these cases the spores often are linear or curved (e.g. *Fusarium*, Fig. 10.24) and are produced in mucilage on a pad of tissue (an acervulus) or in a splash cup so that raindrops are caused to fragment on impact and rebound as many tiny droplets which can carry the spores into the air. Splash cups are also found in a group of Basidiomycota that grow on decomposing wood chips, on the surface of organic soils or on dung. A good example is the genus *Cyathus* (see Fig. 2.21).

Raindrops or hail can also release dry spores that are lying on a surface, by "puff" and "tap" mechanisms.

When a raindrop falls on a rigidly supported surface, the drop splodges sideways, and the resulting puff of air disturbs the boundary layer, causing the dry spores to become airborne. Hailstones act differently – they are most effective in redistributing spores from lightly supported surfaces such as leaves. In puffballs (Basidiomycota) such as *Lycoperdon* (see Fig. 2.23) the mature basidiospores are enclosed in a papery fruitbody with an apical pore, so that raindrops "puff" the spores into the air, like bellows.

Toadstools display a different strategy from all those above, made necessary by the deep layer of still air that often exists on a woodland floor. When the stalk (stipe) of a toadstool elongates, the cap (pileus) projects into turbulent air. The basidiospores develop on short stalks (sterigmata) and the spores are popped from the sterigmata when the continuous outer wall surrounding the sterigma and the spore breaks down, so that the spores can drop from the gills or pores and be carried away by wind. Much of the variation in shape and

(a) (b) (c)

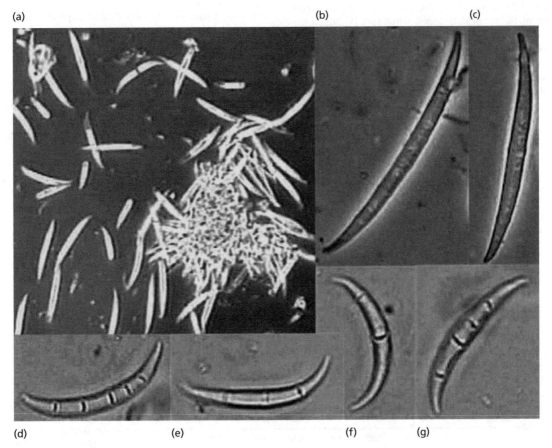

(d) (e) (f) (g)

Fig. 10.24 Curved spores (macroconidia) with several compartments, typical of many species of *Fusarium* and other splash-dispersed fungi.

size of toadstools is related to this strategy – the toadstools with thick, rigid stipes (e.g. *Boletus, Amanita*) and the large brackets produced by tree-rotting fungi (e.g. *Ganoderma*) have very closely spaced and deep gills or pores, just wide enough for spores to be popped from the basidia and then to fall vertically into turbulent air. Toadstools with thin, bendable stipes (e.g. *Marasmius oreades*, the common fairy-ring fungus of grass turf) have widely spaced, shallow gills to ensure that the spores fall free. Many of these strategies are illustrated in Chapter 2 (see Figs 2.24–2.32).

This is more than just a catalogue of examples, because it demonstrates how fungi have an **integrated lifestyle**. The only reason for producing a fruitbody is to disperse the microscopic spores, and the only reason for producing a large or massive fruitbody is to overcome the constraints to spore dispersal imposed by a boundary layer.

Flight

The fate of spores in the air is determined largely by meteorological factors – wind speeds, rain, etc. – but at least two features of spores are significant for long-distance dispersal: their resistance to desiccation, conferred by hydrophobins in the walls, and their resistance to ultraviolet radiation, conferred by wall pigments. Thus, the hyaline (colorless), thin-walled conidia of *Blumeria graminis* (cereal powdery mildew) or the wind-borne sporangia of *Phytophthora infestans* (potato blight) remain viable for only a short time on bright, cloudless days, whereas the pigmented uredospores of rust fungi (e.g. *Puccinia graminis*) and conidia of *Cladosporium* can remain viable for days or even weeks in air.

The use of spore-trapping devices mounted on the outsides of aircraft has provided clear evidence of long-distance dispersal of fungi. Figure 10.25 shows an example where spores were carried on the westerly winds across the North Sea from the English coast to Denmark. The spore clouds were found to be clustered at different altitudes and distances from the English coast. From knowledge of the wind speeds it was possible to distinguish between spores released on different days in England and also to distinguish between spores released in daytime (e.g. *Cladosporium*) and those released at night (the pink yeast *Sporobolomyces*, and various ascospores). Such long-distance dispersal can be highly significant for plant disease epidemiology, especially when new pathogenic races or fungicide-resistant strains develop and are spread across or between continents.

Spore deposition – landing

Spores suspended in the air can be removed in three major ways – by sedimentation, impaction, or washout. The shape, size and surface properties of spores have major effects on these processes – even to the extent that an understanding of a spore's properties enables us to predict the circumstances in which it will be deposited.

Sedimentation

All spores settle out of the air by sedimentation in calm conditions, and the heavier (larger) spores settle faster than lighter (smaller) spores. The sedimentation rates can be measured in closed cylinders and, except for unusually shaped spores for which correction factors are needed, the rates are found to agree closely with Stokes's Law for perfect spheres of unit density (1.0). The relevant equation is:

$$V_t = 0.0121r^2$$

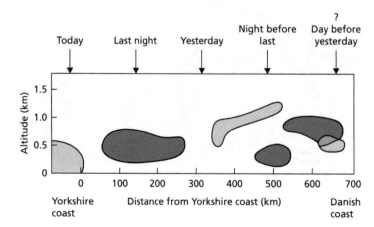

Fig. 10.25 Positions of peak spore concentrations of *Cladosporium* spp. (light shading) and damp-air spore types (dark shading) at different altitudes over the North Sea and at different distances from the English coast. (From Lacey 1988; based on the work of P.H. Gregory and J.L. Monteith.)

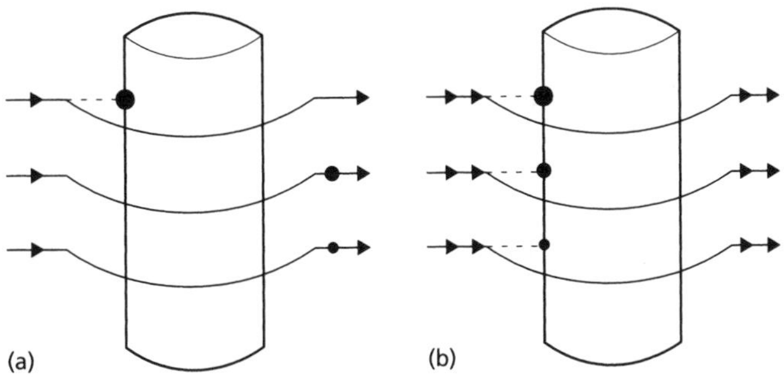

Fig. 10.26 Illustration of the relationship between spore size, wind speed, and impaction of spores onto a cylinder. (a) At low wind speeds only the largest (heaviest) spores impact. (b) At higher wind speeds progressively smaller (lighter) spores impact. (a)   (b)

where V_t is the terminal velocity (cm s^{-1}) and r is the spore radius (μm). For example, spores of the cereal smut pathogen *Tilletia caries* (17 μm diam) had a measured V_t of 1.4 cm s^{-1}, whereas puffball spores (*Bovista plumbea*, 5.6 μm diam) had a V_t of 0.24 cm s^{-1}. Differences of this order probably have little effect in outdoor environments, and so the spores of all types would remain airborne or settle according to the prevailing conditions. However, small differences can be significant in buildings and in the respiratory tract, discussed later.

Impaction

Impaction is one of the major mechanisms by which large spores are removed from the air, and it has special significance for plant pathogens. As shown in Fig. 10.26, when spore-laden air moves towards an object (or vice-versa), the air is deflected around the object and tends to carry spores with it. But the **momentum** (mass × velocity) of a spore will tend to carry it along its existing path for at least some distance.

Three points arise from this:

1 at any given air speed the larger spores (greater mass) have more chance of impacting than do smaller spores;
2 as the air speed (velocity) increases, so progressively smaller spores can impact;
3 as the size of the receiving object increases, so the deflection of air is greater and this reduces the chances of impaction.

These points are directly relevant to spores in nature. All the fungi that infect leaves or that characteristically grow on leaf surfaces (phyllosphere fungi) have large spores, with sufficient mass and therefore sufficient momentum to impact at normal wind speeds (up to 5 m s^{-1}). Examples include *Cladosporium herbarum* (spores 8–15 μm diameter), *Alternaria* spp. (about 30 μm), and the leaf-infecting pathogens *Blumeria graminis* (about 30 μm) and *Puccinia graminis* (about 40 μm). By contrast, the typical soil fungi such as *Penicillium*, *Aspergillus* and

Trichoderma spp. have spores about 4–5 μm diameter, too small to impact at normal wind speeds but they can sediment out of the air in calm conditions.

The receiving object also determines the efficiency of spore impaction. A classic demonstration of this involved placing young branches of apricot trees in a wind tunnel, and exposing the branches to air containing spores of *Eutypa armeniacae* (Ascomycota), a pathogen of apricot trees. This fungus naturally releases its spores as clusters of eight ascospores held together in mucilage – a relatively large propagule with sufficient momentum to impact at relatively low wind speeds. As shown in Fig. 10.27, at all wind speeds the spore clusters impacted best on the narrow leaf stalks (petioles, about 1–2 mm diameter), less well on the thicker young apricot stems, and even less well on the broader leaf blades. And, as the wind speed was increased, so the efficiency of impaction increased. It might be considered that the impaction efficiencies were quite low in all cases – never more than 3%. But in an apricot orchard the spores that do not impact on one shoot system would impact on another. On this basis, it was estimated that most spore clusters would be removed from the air as it travelled through an orchard at 2 m s^{-1} – a typical wind speed that was measured in field conditions. *Eutypa* is a wound pathogen, which commonly infects grape vines and apricots through natural wounds or pruning wounds. After impaction, the fungus relies on secondary spread by rain or irrigation water, which disperses the mucilage and carries the separate spores down to any wound sites.

Washout

Even light, steady rain will remove almost all suspended particles from the air. However, the spore surface properties then come into play. Wettable spores become incorporated within the raindrops and finally come to rest where the water does – spreading as a film across a wettable surface or dripping from a nonwettable one. By contrast, nonwettable spores that are covered with rodlets of hydrophobins remain on the

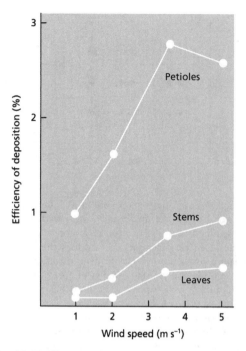

Fig. 10.27 Efficiency of impaction of ascospore tetrads of *Eutypa armeniacae* on leaf blades, petioles, and branch stems of apricot shoots, at different wind speeds. Most impaction occurs on the narrower objects (petioles) and in all cases the efficiency of impaction is increased as the wind speed increases. (From Carter 1965.)

surface of the raindrop, and if this rolls across a non-wettable surface such as a leaf cuticle it will leave the spores behind as a trail. This is easily demonstrated in a laboratory by rolling a water drop over a *Penicillium* colony, when the drop becomes coated with the nonwettable spores. If the spore-laden drop is then rolled across a plastic Petri dish, the spores are left as a trail across the plastic. The nonwettable spores can become airborne again by the "puff and tap" mechanisms mentioned earlier.

Air sampling devices and human health

Spores in the respiratory tract

The respiratory tract of humans and other animals is a natural spore-trapping device, as people with allergies know only too well. The respiratory tract is also the route of entry for spores of several fungal pathogens of humans, discussed in detail in Chapter 16. Three mechanisms of spore deposition in the respiratory tract are important: **impaction**, **sedimentation**, and **boundary layer exchange**.

During each intake of breath the air speed is fastest in the nose, trachea and bronchi (about 100 cm s^{-1}) and it diminishes with successive branching of the bronchioles. Therefore impaction will only occur in the upper respiratory tract, where the air speed is fastest, and only the heaviest (largest) spores will have sufficient momentum to impact. The hairs in the nostrils are narrow and covered with mucilage, making them highly efficient for intercepting the larger fungal spores and pollen grains. Some of these airborne particles cause rhinitis and other typical symptoms of hayfever.

All other particles that are too small to impact will be carried deep into the lungs and reach the terminal bronchioles and alveoli. This includes most particles of 4–5 μm diameter or less, including the spores of many common airborne fungi. Most of these spores are expelled again, but a few will settle onto the mucosal membranes by sedimentation during the brief period (usually less than 1 second) when the air in the alveoli is static between inhalation and exhalation. Particles even smaller than this, including the airborne spores of actinomycetes (1–2 μm diameter), can be trapped by boundary layer exchange. This is a process in which small particles that are positioned very close to the boundary layer (the lining epithelium) can "flip" into the boundary layer by electrostatic or other forces. The airborne spores of potentially pathogenic fungi such as *Aspergillus fumigatus*, *Blastomyces dermatitidis*, *Histoplasma capsulatum*, and *Coccidioides* spp. can settle in the alveoli by sedimentation, as do the spores of several other *Aspergillus* and *Penicillium* species. Some of these fungi, such as *Aspergillus clavatus*, cause acute allergic alveolitis in people who have been repeatedly exposed to spore dusts and have become sensitized. Several occupational diseases are of this type – farmer's lung, malt-worker's lung, etc. Once a spore has been deposited in the alveoli it persists until it is engulfed by a macrophage. By contrast, the upper regions of the respiratory tract are lined with ciliated epithelium which continuously sweeps mucus upwards and removes any particles deposited there.

The importance of airborne spores in relation to crop pathology, human ailments, and air quality in general has led to the design of air-sampling devices for the monitoring of spore loads. We end this chapter by considering the main types of device and the principles on which they operate.

The rotorod sampler

The rotorod sampler is a very simple air-sampling device, used mainly as an experimental tool (Fig. 10.28). It consists of a U-shaped metal rod with two narrow upright arms, attached to a spindle. The upright arms revolve rapidly (about 2000 rev min^{-1}) by a battery or

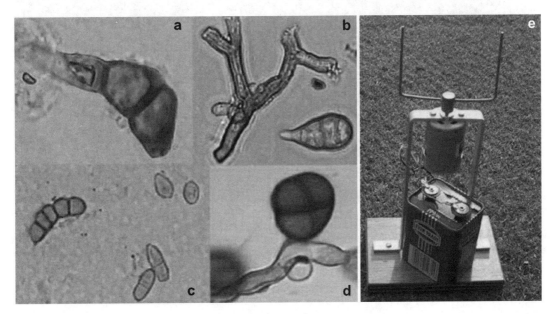

Fig. 10.28 The rotorod air sampler (e), and some representative spores and hyphal fragments commonly seen when rotorod tapes are examined. Images are at various magnifications. (a) Uredospore of *Puccinia graminis*. (b) Darkly pigmented hyphae of *Cladosporium*, with a spore of *Alternaria*. (c) Various pigmented spores, including *Cladosporium* (in the lower part of the frame). (d) An immature spore of *Epicoccum purpurascens* attached to a hyphal fragment.

electric motor. In order to sample spores and other airborne particles, the arms are covered with narrow strips of double-sided, transparent adhesive tape so that spores impact on the tape and can be examined with a microscope. This apparatus is cheap and portable, and is highly efficient at trapping relatively large particles, in the range 10–30 µm diameter, including pollen grains and the spores of most leaf fungi. It is much less efficient at trapping small particles. It can be used to home-in on the source of a particular type of spore, by making successive samplings in a small area. For example, it has been used in a field site in southern Britain to find the source of spores of *Pithomyces chartarum*, a toxigenic fungus that causes the facial eczema condition of sheep (see Fig. 7.20).

The Burkard spore sampler

The Burkard spore sampler (Fig. 10.29) is a continuous monitoring device that works on the same principle as the rotorod sampler. It is used commonly in crop epidemiology and for monitoring allergen levels, including the pollen and spore counts announced on radio and television.

This sampler consists of a sealed drum with a narrow slit orifice (arrowhead in Fig. 10.29) beneath a weather-shield. Spores and other particles entering the orifice impact as a narrow band on a reel (double arrowhead in Fig. 10.29) covered with transparent sticky tape. The reel rotates slowly past the orifice on a daily or weekly cycle, providing a continuous record of the particles present in the air throughout the sampling period. Finally, the tape is removed, cut into sections representing different time periods, and examined microscopically. Like the rotorod sampler, the Burkard sampler is based on the principle of impaction. The air entering the orifice is travelling at relatively low speed, so only the larger (heavier) particles are deposited on the tape. These include many pollen grains, the larger spores of plant-pathogenic fungi, and the spores of many common allergens.

The Anderson spore sampler

The Anderson spore sampler (Fig. 10.30) is perhaps the most ingenious and is claimed, with some justification, to simulate the deposition of airborne particles in the human respiratory tract (Fig. 10.31). It consists of a stack of perforated metal plates which fit together to form an airtight cylinder, with space for open agar-filled Petri dishes to be inserted between them. Each metal plate has the same number of holes in its base, but these holes become progressively smaller down the stack. Air is drawn in at the top, and down through

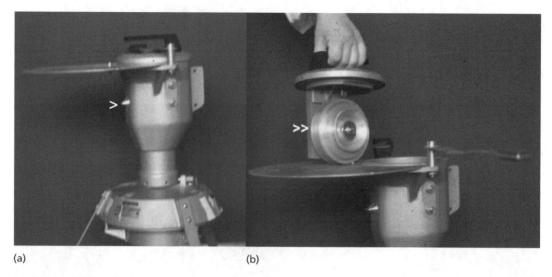

Fig. 10.29 (a,b) The Burkard continuous monitoring sampler in assembled form (a) showing the air intake orifice, arrowhead and (b) showing the reel covered with adhesive tape (double arrowhead) on which the spores are trapped.

Fig. 10.30 (a) The Anderson spore sampler in assembled form, and (b) three of the metal disks with perforations in their bases. (c–e) Examples of colonies from agar plates at different levels in an Anderson sampler.

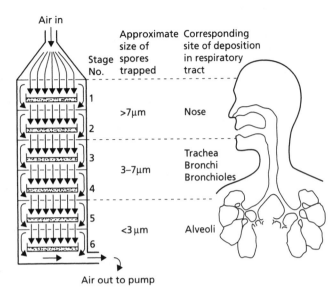

Air in

Stage No.	Approximate size of spores trapped	Corresponding site of deposition in respiratory tract
1	>7µm	Nose
2		
3	3–7µm	Trachea Bronchi Bronchioles
4		
5	<3µm	Alveoli
6		

Air out to pump

Fig. 10.31 Representation of the Anderson sampler, which simulates the deposition of different sized particles in the human respiratory tract.

the apparatus by a motor-driven suction pump at the base.

The air striking the first agar plate is travelling at low speed, so only large particles will impact on this plate. Any particles that do not impact are carried round the agar plate in the airstream and pass through the perforations at the next stage, and this is repeated down the whole stack of plates. Since the same volume of air has to pass through successively smaller perforations (while the number of perforations remains the same), its speed is increased progressively as it exits the perforations, and this causes successively smaller particles to impact on the agar plates. In fact even the smallest spores such as those of the actinomycetes *Thermoactinomyces vulgaris* and *Faenia rectivirgula* (1–2 µm) will impact at the high air speeds on the lower agar plates. These spores cause debilitating respiratory conditions such as "farmer's lung" (acute allergic alveolitis).

It is important to note that all the metal plates in an Anderson sampler have perforations that are large enough for spores of all sizes to pass through them. So the Anderson sampler is not a sieving device. Instead, it sorts the airborne particles according to their momentum (mass × velocity). In this respect it acts on the same principle as the rotorod and Burkard spore traps.

After this sampler has run for an appropriate time, depending on the spore load, it is dismantled and the agar plates are incubated to identify the colonies that grow on them. In the example shown (the bottom row of Fig. 10.30) the agar plates were, from left to right, colonies from one of the uppermost agar plates,

colonies from the center of the stack of agar plates, and colonies from the lowest agar plate, consisting entirely of small actinomycete colonies. The agar plate at the extreme right is an example of a split plate with three different types of agar, designed to detect different types of organism simultaneously.

Cited references

Adams, M.J. (1991) Transmission of plant viruses by fungi. *Annals of Applied Biology* **118**, 479–492.

Brasier, C.M. (1995) Episodic selection as a force in fungal microevolution, with special reference to clonal speciation and hybrid introgression. *Canadian Journal of Botany* **73**, S1213–S1221.

Brunt, A.A. & Richards, K.E. (1989) Biology and molecular biology of furoviruses. *Advances in Virus Research* **36**, 1–32.

Carter, M.V. (1965) Ascospore deposition of *Eutypa armeniacae*. *Australian Journal of Agricultural Research* **16**, 825–836.

Cho, C.W. & Fuller, M.F. (1989) Ultrastructural organization of freeze-substituted zoospores of *Phytophthora palmivora*. *Canadian Journal of Botany* **67**, 1493–1499.

Coley-Smith, J. (1987) Alternative methods of controlling white rot disease of *Allium*. In: *Innovative Approaches to Plant Disease Control* (Chet, I., ed.), pp. 161–177. Wiley, New York.

Deacon, J.W. & Donaldson, S.P. (1993) Molecular recognition in the homing responses of zoosporic fungi, with special reference to *Pythium* and *Phytophthora*. *Mycological Research* **97**, 1153–1171.

Deacon, J.W. & Mitchell, R.T. (1985) Toxicity of oat roots, oat root extracts, and saponins to zoospores of *Pythium* spp. and other fungi. *Transactions of the British Mycological Society* **84**, 479–487.

Deacon, J.W., Donaldson, S.J. & Last, F.T. (1983) Sequences and interactions of mycorrhizal fungi on birch. *Plant and Soil* **71**, 257–262.

Fox, F.M. (1986) Groupings of ectomycorrhizal fungi on birch and pine, based on establishment of mycorrhizas on seedlings from spores in unsterile soils. *Transactions of the British Mycological Society* **87**, 371–380.

Gregory, P.H. (1973) *Microbiology of the Atmosphere*, 2nd edn. Leonard Hill, Aylesbury.

Gubler, F. & Hardham, A.R. (1988) Secretion of adhesive material during encystment of *Phytophthora cinnamomi* zoospores, characterized by immunogold labeling with monoclonal antibodies to components of peripheral vesicles. *Journal of Cell Science* **90**, 225–235.

Hardham, A.R. (1995) Polarity of vesicle distribution in oomycete zoospores: development of polarity and importance for infection. *Canadian Journal of Botany* **73**(suppl.) S400–407.

Hardham, A.R. (2001) Cell biology of fungal infection of plants. In: *The Mycota VIII. Biology of the Fungal Cell* (Howard, R.J. & Gow, N.A.R., eds), pp. 91–123. Springer-Verlag, Berlin.

Hardham, A.R., Cahill, D.M., Cope, M., Gabor, B.K., Gubler, F. & Hyde, G.J. (1994) Cell surface antigens of *Phytophthora* spores: biological and taxonomic characterization. *Protoplasma* **181**, 213–232.

Hiruki, C. & Teakle, D.S. (1987) Soil-borne viruses of plants. In: *Current Topics in Vector Research*, vol. 3 (Harris, K.F., ed.), pp. 177–215. Springer-Verlag, New York.

Hsu, S.C. & Lockwood, J.L. (1973) Soil fungistasis: behavior of nutrient-independent spores and sclerotia in a model system. *Phytopathology* **63**, 334–337.

Ingold, C.T. (1971) *Fungal Spores and their Dispersal*. Clarendon Press, Oxford.

Lacey, J. (1988) Aerial dispersal and the development of microbial communities. In: *Micro-organisms in Action: concepts and applications in microbial ecology* (Lynch, J.M. & Hobbie, J.E., eds), pp. 207–237. Blackwell Scientific, Oxford.

Lockwood, J.L. (1977) Fungistasis in soils. *Biological Reviews* **52**, 1–43.

Mitchell, A.G. & Brasier, C.M. (1994) Contrasting structure of European and North American populations of *Ophiostoma ulmi*. *Mycological Research* **98**, 576–582.

Moss, S. (1986) *The Biology of Marine Fungi*. Cambridge University Press, Cambridge.

Reichle, R.E. & Fuller, M.F. (1967) The fine structure of *Blastocladiella emersonii* zoospores. *American Journal of Botany* **54**, 81–92.

Ron, E.Z. & Rosenberg, E. (2001) Natural roles of biosurfactants. *Environmental Microbiology* **3**, 229–236.

Warburton, A.J. & Deacon, J.W. (1998) Transmembrane Ca^{2+} fluxes associated with zoospore encystment and cyst germination by the phytopathogen *Phytophthora parasitica*. *Fungal Genetics and Biology* **25**, 54–62.

Chapter 11

Fungal ecology: saprotrophs

This chapter is divided into the following major sections:

- a theoretical model: the concept of life-history strategies
- the biochemical and molecular toolbox for fungal ecology
- a "universal" decomposition sequence
- the fungal community of composts
- fungal decomposers in the root zone
- fungal communities in decaying wood

The importance of fungi in ecosystem processes is undeniable. Fungi are the main agents of decomposition in many terrestrial and aquatic environments. They are particularly important in the breakdown and recycling of cellulose and hemicelluloses, which together account for nearly 70% of all the plant wall material that is recycled annually. In addition, fungi have a unique role in degrading woody substrates, which contain cellulose intimately complexed with lignin (**lignocellulose**). And, fungi degrade many other natural and manmade materials, causing serious economic losses.

In previous chapters we dealt with the physiology, growth, genetics, and dispersal of fungi – the basis for understanding fungal ecology. But when we turn to fungi in natural environments we face a major problem, because natural communities are extremely complex: they contain many types of substrate, interacting species, and microhabitats. Therefore, at a practical level we need to find well-defined communities that can be dissected to provide key insights

into fungal behavior. We will do this by focusing on a few natural "model" systems that have been well researched – the leaf zone, leaf litter, the root zone, self-heating composts, and wood decay. The principles derived from these natural model systems apply more generally across the fungal kingdom. We will also explore the **biochemical and molecular toolbox** that enables us to track and identify fungi in complex natural materials.

A theoretical model: the concept of life-history strategies

Ecology lends itself to theoretical models as a basis for synthesizing complex information. The references at the end of this chapter cite some key publications in this field. Here we will briefly discuss one of these models, first developed by animal ecologists, then applied to plants and later to fungi – the concept of **life-history strategies** (Fig. 11.1).

Initially, ecologists developed the concept of **r-selected** and **K-selected** organisms, these being the two extreme expressions of the life history of an organism. The r-selected organisms (**ruderals**) establish themselves and grow rapidly in suitable conditions, but after a short phase of growth they produce many offspring, each of which has only a low chance of reproductive success. Weed plants are classic examples of this strategy, as are many of the fast-growing fungi of the Zygomycota (*Mucor*, *Rhizopus*, etc. – see Chapter 2). At the other extreme, K-selected organisms typically establish and develop slowly, live for a long time, and produce few offspring, but each of these offspring has a significant

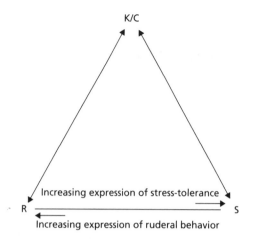

Fig. 11.1 A method of defining the life-history strategy of a fungus according to its degree of expression of the three "primary strategies" – ruderal behavior (R), K-selected behavior, or combative behavior (K/C) and stress-tolerance.

chance of establishing in a new environment. The larger mammals are good examples of this, as (in some respects) are the Basidiomycota that produce perennial fruiting bodies on trees (Fig. 11.2). So the distinction between r- and K-selected organisms is, in essence, a distinction between how these types of organism invest their food resources for producing offspring.

A third factor, **stress-tolerance**, was added to the two original life-history strategies, creating a triangle

(Fig. 11.1) representing the extreme expressions of ruderal, stress-tolerant, or K-selected behavior. In theory, any organism should be able to be placed within this triangle, depending on its degree of expression of the three **primary strategies**, ruderal behavior, stress-tolerance, or K-selection, renamed **combative behavior** in the case of fungi. The value of this and other theoretical models is discussed by Andrews (1992).

The biochemical and molecular toolbox for fungal ecology

Assessing the fungal content of natural materials by biochemical "signature molecules"

Several biochemical methods have been used to estimate the fungal content of natural materials, but all have limitations. One of the early methods developed in the 1970s was to determine the **chitin content** of soils and other materials, by hydrolysing chitin to the monomer N-acetylglucosamine, which can be assayed colorimetrically. This method has also been used to estimate the fungal content of plant tissues colonized by plant pathogens or mycorrhizal fungi. The main drawback is that the chitin content of fungal walls can vary in different fungi, in different growth conditions, or at different stages of a fungal life cycle. Also, it is difficult to distinguish between the chitin of fungal walls and the chitin of insect exoskeletons in natural

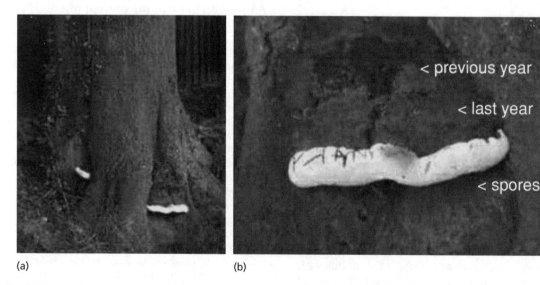

(a)   (b)

Fig. 11.2 *Ganoderma adspersum*, which exhibits predominantly K-selected behavior. This fungus progressively rots the heartwood of standing trees (especially beech, *Fagus sylvatica*) and produces woody, perennial, hoof-like fruitbodies which release millions of spores. The current year's young growth is white.

materials. This limits the value of chitin assays except in well-defined model systems.

An alternative method is to estimate the **ergosterol content** by extracting ergosterol from a sample and assaying it by high performance liquid chromatography (HPLC) and UV detection at 280 nm. Ergosterol is the characteristic sterol of most fungal membranes and is absent from plants and other soil organisms. However, again the ergosterol content can vary markedly between fungal taxa. In a critical analysis of all such methods, including the assay of other "signature molecules" such as phospholipid fatty acids, Olsson *et al.* (2003) concluded that the comparison of biomass between different fungi is very difficult using any currently available biochemical marker.

Antibody techniques

Fluorescent antibody techniques enable fungi to be identified on the basis of their surface or internal antigens. Antibodies are raised to the antigens of a specific fungus by standard techniques, involving the inoculation of rabbits, then the antibodies are used to detect the fungus in a natural sample. For this, a secondary antibody is used which binds to the primary antibody and which is also tagged with a fluorescent compound so that the target fungus can be seen under a fluorescence microscope. Alternatively, the secondary antibody can be linked to an enzyme which releases a dye from a colored substrate. This technique, termed **ELISA (enzyme-linked immunosorbent assay)**, can be used to quantify the amount of specific antigen in a sample. The main problem with both techniques is that the antibodies can cross-react with the antigens of other fungi. **Monoclonal antibodies** offer more specificity because each is derived from a single cell line producing one specific antibody. An example of their use in a different context was discussed in Chapter 10 (see Fig. 10.16).

Fluorescent vital dyes

Fungi can be visualized directly by treatment with fluorescent dyes (**fluorochromes**) that bind to the cell surface or accumulate in specific cellular organelles. The cells are then viewed under a fluorescence microscope (Fig. 11.3). Some of these dyes, when pre-loaded into hyphae, can be translocated for up to 9 mm as the

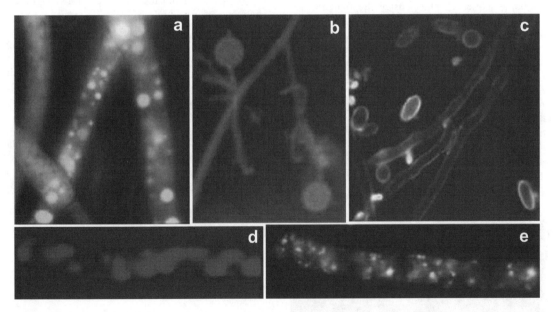

Fig. 11.3 Examples of some fluorescent vital dyes; see color images on the Internet for higher resolution. (a) Hyphae of *Sclerotium cepivorum* treated with **CMFDA**, which accumulates in fungal vacuoles and shows as bright green fluorescence. (b) Hyphae and spores of *Pythium oligandrum*, stained with **Cellufluor** which binds to chitin in fungal walls and fluoresces blue. (c) Hyphae of *Fusarium oxysporum* stained with **DiI** and **DiO**, giving green or red fluorescence of the fungal membranes. (d) Hyphae of *Botrytis cinerea* stained with **CMAC**, giving blue fluorescence of the vacuoles. (e) Hyphae of *Fusarium oxysporum* stained with **Nile red**, showing intense yellow fluorescence of lipid droplets, while the cytoplasm fluoresces orange.

hyphal tips extend. They can potentially be used to trace the growth and interactions of fungi in soils or other natural substrata. They also distinguish between living and dead cells if the dye is accumulated by an active metabolic process. Soils have a strong quenching effect on fluorescence, which limits the usefulness of fluorochromes in this environment, but fluorescent vital dyes can be valuable as tracers in microcosm studies (Stewart & Deacon 1995).

One of the most commonly used fluorochromes is the stilbene dye, **Cellufluor** (or Calcofluor White) which binds to β-1,4 linked polymers in fungal walls and produces an intense blue fluorescence. However, it does not stain the cytoplasm, so it cannot be used to distinguish living from dead hyphae. Another common fluorochrome, with many applications, is the phenoxazine dye, **Nile red**, which is highly specific for neutral lipids. It enters the cell membrane and causes lipid droplets to fluoresce bright yellow, while the cytoplasm fluoresces orange. Another compound, **carboxyfluorescein diacetate** (CFDA), is valuable as an indicator of living fungal hyphae, because it is hydrolysed within the cells to release carboxyfluorescein, a hydrophilic compound that accumulates in vacuoles and fluoresces bright green. Two chloromethyl dyes (**chloromethylfluorescein diacetate**, CMFDA, and **aminochloromethyl coumarin**, CMAC) are also membrane-permeable. Once inside the hyphae they are thought to undergo a reaction mediated by glutathione-*S*-transferase, resulting in a cell-impermeable product. The products of both of these dyes accumulate in vacuoles and fluoresce either green (CMFDA) or blue (CMAC). They are very photostable and can retain their fluorescence for at least 72 hours. Two further types of fluorochrome are widely used: the **carbocyanine dyes** called $DiIC_{18}$ and $DiOC_{18}$, which are lipophilic and bind to membranes, producing red and green fluorescence respectively, and **DAPI** (4′ 6-diamidino-2-phenylindole-2HCl) a blue-fluorescent compound that intercalates in the A-T-rich regions of DNA and enables nuclei to be seen (Fig. 11.4).

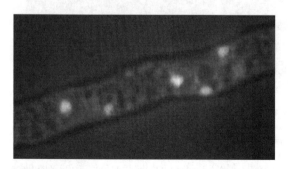

Fig. 11.4 Nuclei fluorescing blue in a hypha stained with the nucleus-specific dye, DAPI.

Identifying fungi by oligonucleotide probes

As described in Chapter 9, an increasing range of molecular tools such as **RFLP**, **RAPD**, **AFLP** (**amplified fragment length polymorphism**), or **fluorescent AFLP** can be used for tracking fungi. All are available in kit form and require no detailed knowledge of molecular biology. They have had a significant impact on studies of fungal population biology.

More recently, the technique of **fluorescence in-situ hybridization** (**FISH**) has been used widely by microbial ecologists (and also for "chromosome painting" in genetics), but only a few fungal ecologists have used this technique to date. It is based on the principle that single-stranded DNA or RNA probes will hybridize with complementary nucleic acid sequences in an environmental sample. Fluorescent labeling of the probes enables these nucleic acid sequences to be visualized even if the organism cannot be cultured. The **oligonucleotide probes** are typically based on the small subunit (18S) ribosomal RNA gene sequences (Chapter 9). Some are universal probes, complementary to a 16S rRNA or 18S rRNA gene sequence found in all known organisms – for example [5′-GAA TTA CCG CGG TAA CTG CTG-3′]. Others are specific for particular groups of organisms, and are termed **signature sequences**. For example:

- the sequence **UUCCCG** is found in more than 95% of bacteria but never in the archaea or eukaryotes;
- the sequence **CACACACCG** is found only in archaea, never in bacteria or eukaryotes;
- the sequence **AAACUUAAAG** is found in all eukaryotes and archae but never in bacteria.

Other sequences can be constructed for particular species, an example being a probe that was developed specifically to study the population dynamics of *Sydowia polyspora* (*Aureobasidium pullulans*) on leaf surfaces (Li *et al.* 1997). A colorimetric variant of this technique, termed **CISH**, can be used when the autofluorescence of fungi or their substrates precludes the use of **FISH** (Schröder *et al.* 2000).

A "generalized" decomposition sequence

> No man is an *Iland*, Intire of it selfe; every man is a peece of the *Continent*, a part of the *maine* . . ." (John Donne 1624)

The decomposition of organic matter is brought about by a wide range of fungi and other organisms (bacteria, nematodes, mites, springtails, etc.) acting in consort. The types of organisms that are active in these communities can vary, but the overall course of decomposition is similar in many situations (Fig. 11.5) because it involves an overlapping sequence of activities of fungi with different patterns of behavior and different abilities to degrade particular types of substrate (Table 11.1).

In this section we will look at the main stages in a generalized decomposition sequence, starting with the organisms that colonize the surfaces of living

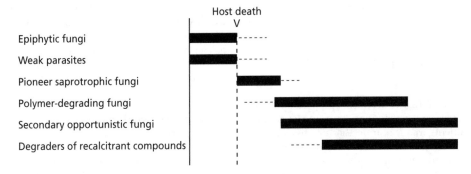

Fig. 11.5 Representation of the overlapping phases of activity of different types of fungi in a decomposition sequence.

Table 11.1 The main behavioral groupings of decomposer fungi.

Group	Features
1. Resident microbes on living tissues	(i) Grow on generally low levels of available substrates (ii) Often tolerate stress conditions (iii) Are usually displaced when the host tissues die
2. Endophytes, weak parasites and pathogens	(i) Grow initially by tolerating host resistance factors or other special conditions (ii) Generally utilize simple soluble substrates or storage compounds but not structural polymers (iii) Generally poor competitors for dead organic matter
3. Pioneer saprotrophic fungi	(i) Generally utilize simple soluble substrates or storage compounds but not structural polymers (ii) Good competitors, with fast growth, etc. and short life cycles (iii) Cannot defend a resource against subsequent invaders
4. Polymer-degrading fungi	(i) Degrade the main structural polymers (cellulose, hemicelluloses, chitin, etc.) (ii) Have an extended growth phase, defending a resource by antibiosis or by sequestering mineral nutrients, etc. (iii) Substrate-specialized, and sometimes tolerant of stress factors (extremes of temperature, pH, etc.)
5. Degraders of recalcitrant materials	(i) Specialized to degrade recalcitrant organic materials (lignin etc.) and gain access to polymers (cellulose etc.) complexed with them (ii) Long growth phase, and defend a resource by antagonism or mutual inhibition (deadlock) (iii) Can gain access to mineral nutrients (nitrogen, etc.) that previous colonizers have exploited
6. Secondary opportunists	(i) Nutritionally opportunistic: grow on dead remains of other fungi, insect exoskeletons, etc., or parasitize other fungi, or grow commensally with polymer-degraders. (ii) Tolerant of the metabolic byproducts of other fungi (iii) Often antagonistic

tissues, through to the final stages of decomposition that result in the production of soil humus – a highly complex mixture of organic compounds that persist for tens or hundreds of years and that play a vital role in soil fertility. For reviews of decomposition sequences the following texts are strongly recommended: Carroll & Wicklow (1992), Harper & Webster (1964), and Hudson (1968).

The resident fungal population of living tissues

The surfaces of all higher organisms support a population of bacteria and yeasts that grow on the simple, soluble nutrients that leak from the host tissues. This is as true of humans as it is of the living leaves of plants – it is estimated that, of all the cells in a human body, fewer than 0.1% are human cells! Normally, these resident microbes cause no harm and can even be beneficial in suppressing invasion by potential pathogens. But they can become invasive as the host tissues begin to senesce, initiating the decomposition process.

The yeasts and yeast-like fungi that colonize the surfaces of living plant leaves have been studied in detail. They grow in an environment termed the **phyllosphere**, where the conditions are relatively hostile because of exposure to UV irradiation and wide fluctuations in temperature, moisture and nutrient levels. Several common yeasts are found in this environment, including both Ascomycota and Basidiomycota, such as species of *Candida*, *Cryptococcus*, *Rhodotorula*, *Torulopsis*, and *Sporobolomyces*. Many of them have carotenoid pigments, which quench the toxic effects of reactive oxygen species (Chapter 8). Several also produce capsules and slime for adhesion to the plant surfaces. One of the basidiomycetous yeasts, *Filobasidiella* (*Cryptococcus*) *neoformans*, is notable because it has become one of the most serious life-threatening pathogens of immunocompromised people in western societies, and a different form of this fungus (variety *gattii*) causes an endemic disease of immunocompetent people in Australia, Papua New Guinea, and parts of Africa, India, south-east Asia and Central and South America (Chapter 16). This variety of the fungus is commonly associated with eucalypt trees, and the extensive plantations of eucalypts that have been established outside of Australia may provide the inoculum source for spread of this human disease.

In stark contrast to this, the yeast *Candida oleophila*, which grows on leaf and fruit surfaces, has been marketed commercially as a biological control agent. It is applied to fruit surfaces, where it competes for nutrients, helping to reduce the incidence of fruit-rot

fungi that invade through minor wounds (Chapter 12).

The distribution of leaf surface yeasts can be studied by pressing leaves against the surface of malt-extract agar, then removing the leaves and incubating the agar plates (Fig. 11.6). The surfaces of young leaves support a relatively low microbial population, but as the leaves start to senesce the population increases and is often dominated by dimorphic fungi such as *Sydowia polyspora* (*Aureobasidium pullulans*; see Fig. 6.7) and a yellow *Candida* species that produces a fringe of fungal hyphae with clusters of yeast cells (Fig. 11.6).

Endophytes, weak parasites, and pathogens

In recent years, many fungi have been discovered to grow as symptomless endophytes within plants that show no obvious sign of infection or distress. Often they occur sparsely as hyphae in the intercellular fluids and wall spaces of their plant hosts. A specialized group of these fungi – the "clavicipitaceous endophytes" – are known to produce toxic alkaloids in grasses (Chapter 14). Other fungal endophytes have attracted interest as potential sources of novel metabolites (bioprospecting), although recent studies of the endophytic *Colletotrichum* isolates from tropical rainforests in Guyana suggest that the diversity of endophytic fungi (and thus of their products) may have been overestimated: most of the endophytes were nonhost-specific and were found repeatedly in different samples (Lu *et al.* 2004).

Pathogens and weak parasites colonize living plant tissues or tissues that are beginning to senesce. They gain an advantage over purely saprotrophic fungi because they tolerate or overcome the host defense factors. They can remain active for a time after the tissues have died, usually by exploiting the more readily utilizable nutrients. Several mitosporic fungi that produce melanized hyphae and spores are often seen at this stage. They include *Alternaria* spp. and *Cladosporium* spp., and a range of similar fungi (Fig. 11.7) which utilize sugars and other soluble nutrients, including the sugar-rich "honeydew" that aphids excrete when they tap into plant phloem. Several of these fungi grow well in laboratory culture when supplied with soluble cellulose sources such as carboxymethylcellulose, but they seldom grow well on "crystalline" cellulose (filter paper, cotton fabric, etc.). This suggests that they might utilize the swollen (hydrated) cellulose and hemicelluloses in moist plant tissues, as explained in Chapter 6.

Some plant pathogens that attack living plant tissues can continue to grow after the plant has died, by slowly using the dead tissues that they colonized

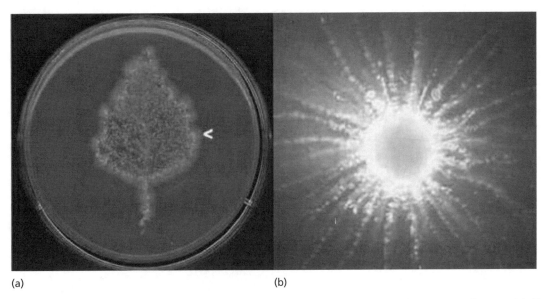

(a)   (b)

Fig. 11.6 (a) Leaf print of a mature birch leaf, incubated on agar, showing a predominance of a yellow *Candida*-like fungus that produced a fringe of hyphae from the leaf margin (arrowhead). (b) A single colony of the *Candida*-like fungus, showing widely spaced radiating hyphae with clusters of yeast-like cells at intervals along the hyphae.

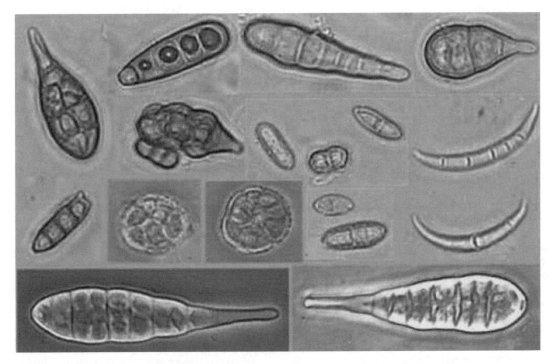

Fig. 11.7 An assortment of (mainly) darkly pigmented spores of fungi that typically colonize the dead or dying tissues of stem bases and the above-ground parts of senescing leaves and stems.

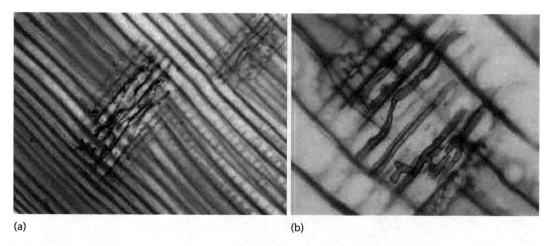

(a)   (b)

Fig. 11.8 (a,b) Darkly pigmented sapstain fungi growing in the medullary rays of felled trees.

while the plant was still alive. Classic examples of this are the stem base fungi such as the take-all fungus (*Gaeumannomyces graminis*, Chapter 12) and the eyespot fungus (*Tapesia yallundae*) of cereal crops. The inoculum surviving from the previous growing season can then initiate infection of a subsequent cereal crop.

Pioneer saprotrophic fungi

Several species of *Mucor, Rhizopus* and other Zygomycota are commonly found in soils or fecal-enriched materials, and also in the rhizosphere (root zone) of plants, where they utilize sugars and other simple soluble nutrients. Several *Pythium* species also fall into this category because they colonize fresh plant residues in soil, although they are best known as pathogens of living root tips and other fleshy plant tissues (Chapter 12).

The spores of these and similar fungi germinate and grow rapidly in response to soluble nutrients, and within a few days they produce a further batch of asexual dispersal spores or sexual resting spores (see Fig. 2.8). They usually cannot degrade the more complex structural polymers (cellulose, etc.). In laboratory culture they are found to be intolerant of the antibiotics or growth metabolites of other fungi, and they do not themselves produce antibiotics. So, these pioneer fungi typically have a short exploitative phase and a high competitive ability.

The **sapstain fungi** that rapidly colonize newly felled trees are typical pioneer saprotrophs. They grow in the nonwoody (parenchymatous) medullary rays and they discolor the wood by their darkly pigmented

hyphae (Fig. 11.8). Although they cause no structural damage to the timber, they can seriously reduce its marketable value. These staining fungi include several species of *Ophiostoma* (related to *Ophiostoma ulmi* which causes Dutch elm disease; Chapter 10). Vanneste *et al.* (2002) describe how natural products and biological control agents are being developed to control the sapstain fungi in New Zealand. However, similar "bluestain" fungi such as *Chlorosplenium aeruginascens* (Ascomycota) cause an attractive turquoise pigmentation of wood. This used to be highly valued for producing inlaid veneers in decorative objects, termed Tunbridge Ware because the industry was based in the Kentish town of Tunbridge Wells.

Polymer-degrading fungi

Several fungi have an extended phase of growth on the major structural polymers such as cellulose, hemicelluloses, or chitin. Once established, these fungi tend to defend the resource against potential invaders, either by sequestering a critically limiting nutrient such as nitrogen or by producing inhibitory metabolites (Chapter 6).

The large number of fungi in this category differ in the types of substrate they utilize, their different environmental requirements, and their growth in different phases of a decomposition sequence. The following examples illustrate this diversity.

When cereal straw and other cellulose-rich materials are buried in soil they are colonized by fungi such as *Fusarium, Chaetomium, Stachybotrys, Humicola,* or *Trichoderma* spp. The *Fusarium* spp. tend to be favored by low water potentials (drought stress) because they can grow at almost undiminished rates when the water potential is lowered to –5 MPa in culture (Chapter 8). *Trichoderma* spp. often are favored by soil acidity (pH stress) because they can grow at about pH 3.0 in culture. A spectrum of salt-tolerant cellulolytic fungi is found on plant remains in estuarine waters (e.g. *Lulworthia* spp., *Halosphaeria hamata*, and *Zalerion varium*) and a different spectrum is found in freshwater streams (some of the fungi with tetraradiate spores, such as *Tetracladium*, *Lemonniera*, and *Alatospora* – Chapter 10). Chitinous materials are colonized by yet another group of fungi, such as *Mortierella* spp. (Zygomycota) in soil or *Chytridium confervae* (Chytridiomycota) in freshwater habitats.

Polymer-degrading fungi can exploit different microhabitats, and thereby coexist in a single substrate resource. Evidence for this has come from microscopical examination of transparent cellulose film ("Cellophane") buried in soil (Tribe 1960). In wet soils the film is colonized by cellulolytic chytrids (e.g. *Rhizophlyctis rosea*; Fig. 11.9) and chytrid-like organisms (e.g. *Hyphochytrium catenoides*). These degrade the cellulose locally by forming finely branched rhizoids between the layers of the film. Other parts of the film can be colonized by mycelial fungi such as *Humicola grisea*, *Fusarium* spp., and *Rhizoctonia*. The *Rhizoctonia* and *Fusarium* colonies form loose networks over the surface of the film, whereas *Humicola* is seen as localized, compact colonies that "root" into the film and produce fans of hyphae within it, like the fans of *Pythium graminicola* shown in Fig. 11.9.

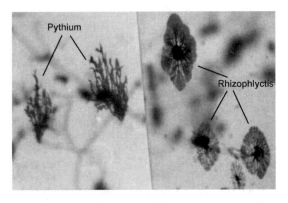

Fig. 11.9 A small piece of transparent cellulose film (about 3 × 2 cm) retrieved from soil and stained with trypan blue to show fungi growing at different levels in the film. *Rhizophlyctis rosea* produces finely branched rhizoidal systems. At a different plane of focus, *Pythium graminicola* produces finger-like branching systems.

Fungi that degrade recalcitrant (resistant) polymers

Fungi that degrade the more resistant polymers such as lignocellulose (cellulose complexed with lignin) often predominate in the later stages of decomposition. Several of them are Basidiomycota, including *Mycena galopus*, a common small toadstool in the leaf litter of woodlands. Other common examples are the **fairy ring fungi**, frequently seen in old grasslands. These fungi grow on the thick mat of accumulated dead leaf sheaths of grasslands and spread progressively outwards year by year. Several fungi produce these rings, including the common edible field mushroom (*Agaricus campestris*). They usually cause no damage because they grow on dead organic matter, and often the only sign of their presence is a ring of more vigorous, lush growth of the grass, resulting from mineral nutrients released from breakdown of the organic matter. But the "fairy ring champignon" *Marasmius oreades* is an interesting exception because it can kill the grass (Fig. 11.10) as the rings spread progressively outwards. The reason is that this fungus produces a mass of fungal hyphae just beneath the grass surface, and these hyphae dry in mid summer, becoming hydrophobic and preventing water from penetrating the soil. Even fungicides are ineffective in preventing death of the grass, but surfactants (including dilute washing-up liquids) can reduce much of this damage.

Keratin-degrading fungi, related to the dermatophytic pathogens of humans and other animals (Chapter 16), are found in the later stages of decomposition of hair or animals' hooves. They might occur earlier, but they become conspicuous only after an initial phase of exploitation by Zygomycota and *Penicillium* spp. which utilize the more readily available proteins and lipids.

The ecological success of fungi that develop later in the decomposition sequence is related to their specialized ability to degrade polymers that most other fungi cannot utilize. However, it does not necessarily follow that they use these complex polymers as their main energy source. In fact, *Mycena galopus*, mentioned above, cannot degrade lignin in culture unless it is also supplied with cellulose or hemicelluloses as more readily utilizable substrates. The chief attribute of these fungi seems to be that they degrade or modify the recalcitrant polymers and so gain access to other substrates that are chemically or physically complexed with the resistant polymers. Much of the cellulose in plant cell walls is intimately associated with lignin, either covalently bonded to it or encrusted by it, and this "lignocellulose" is largely unavailable to fungi that cannot modify the lignin. In one of the

(a)   (b)   (c)

(d)   (e)

Fig. 11.10 Part of a fairy ring caused by *Marasmius oreades* (a), and part of the turf lifted to show the mass of white fungal hyphae just below the soil surface (b). A ring of small toadstools of *Marasmius* (c) develops near the killing zone in autumn. (d) Close-up of two fruitbodies of *Marasmius*, showing the typical widely spaced gills. (e) Toadstools of another fairy-ring fungus – the waxy, scarlet-colored *Hygrocybe* sp. These toadstools form rings of fruitbodies but they do not kill the turf.

best-studied lignin-degrading fungi, *Phanerochaete chrysosporium*, the production of lignin-degrading enzymes is strongly stimulated by a critical shortage of nitrogen. This could explain why these fungi occur in the later stages of decomposition – they might need to wait until other fungi have sequestered most of the available nitrogen.

Secondary (opportunistic) invaders

At many stages in the decomposition of natural materials there are opportunities for secondary invaders to grow. Some of these opportunists might grow in close association with polymer degraders, using some of the breakdown products released by enzyme action. As discussed later, *Thermomyces lanuginosus* seems to behave in this way in composts. Other secondary invaders are known to parasitize other fungi in culture

(e.g. *Pythium oligandrum*, Chapter 12) or they grow on dead hyphal remains; yet others (e.g. *Mortierella* spp.) might grow on the fecal pellets or shed exoskeletons of microarthropods that are abundant in decomposing materials. The range of potential behavior patterns of these fungi makes it difficult to generalize, but their features commonly include: (i) nutritional opportunism, because they can scavenge low levels of nutrients, and (ii) the ability to tolerate the metabolic byproducts of other fungi.

Eventually all the decomposable materials will be utilized, leaving a residue of humic substances (soil humus). These are heterogeneous polymers consisting of a framework of aromatic and aliphatic molecules complexed with proteins and sugars. They are essentially nondegradable and they play an essential role in soil fertility by enhancing soil structure and water retention, and by providing ion-exchange sites for micronutrients in the soil.

The fungal community of composts

Composts can be made from any type of decomposable organic matter, stacked into a heap (or long heaps called windrows in the case of mushroom composts) and supplied with adequate mineral nutrients (especially nitrogen), moisture, and aeration. They are used on a vast scale for the commercial production of mushrooms (*Agaricus*), where the compost normally consists of a mixture of cereal straw and animal manure. They are also used extensively for the processing of horticultural and urban wastes.

Provided that the basic requirements are met, the decomposition process follows an essentially standard pattern, shown in Fig. 11.11 for an experimental compost made from wheat straw supplemented with nitrogen in the form of ammonium. Within a few days the temperature in the bulk of the compost rises to 70–80°C, caused by the heat generated when microorganisms grow on the more readily utilizable nutrients. Bacteria are primarily responsible for this initial heating phase. They include the thermophilic actinomycetes that cause farmer's lung disease (Chapter 10), and members of the bacterial genus *Thermus*, which used to be considered a typical inhabitant of thermal springs but has recently been found to be almost universally present in composts and other self-heating materials. Fungi play only a minor role in the early stages of composting because they are slower to initiate growth, and the maximum temperature for growth of any fungus is about 62–65°C (Chapter 8).

After "peak-heating" the compost starts to cool because most of the microorganisms are killed in the center of the heap, but some *Bacillus* spp. can survive as spores, and other bacteria can recolonize from the surface, giving rise to second and third smaller temperature peaks. Then the temperature gradually declines to ambient temperature over 20–30 days. Bacterial populations remain high throughout this time, but fungi are considered to play the most important role when they recolonize after peak-heating and degrade much of the organic matter as the temperature slowly declines.

Chang & Hudson (1967) recognized several behavioral groups of fungi, occurring at different stages of the composting process. They are modified here for ease of presentation.

1 A mixture of **weak parasites** and **pioneer saprotrophic fungi** are found in the first few days. Many of them were present on the original material, including the leaf-surface fungi (*Cladosporium*, etc.) and *Fusarium* spp. which often grow on cereal straw. A few thermophilic fungi also grow in the first few days, notably *Thermomucor pusillus* and *Mucor* (or *Thermomucor*) *miehei*. These are pioneer saprotrophic fungi, which have maximum growth temperatures of 57–60°C. They are inactivated or killed by the peak-heating and they do not reappear.

2 **Cellulolytic** species of Ascomycota and mitosporic fungi colonize after peak-heating and grow over the next 10–20 days. High temperature might activate the dormant ascospores of some of these fungi (Chapter 10). Their incidence can differ between different types of compost, but these cellulolytic fungi commonly include *Chaetomium thermophile*, *Humicola insolens*, *Thermoascus aurantiacus* (Fig. 11.12), *Scytalidium thermophilum*, and *Aspergillus fumigatus* (see Fig. 8.4). The temperature range of each fungus is likely to determine how quickly it colonizes after peak-heating (see Fig. 8.2). For example, *C.*

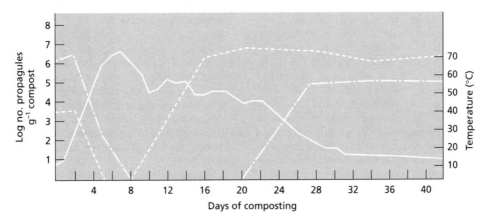

Fig. 11.11 Changes in temperature (thick line), populations of mesophilic fungi (broken line), and populations of thermophilic fungi (narrow line) in a wheat straw compost. (Based on Chang & Hudson 1967)

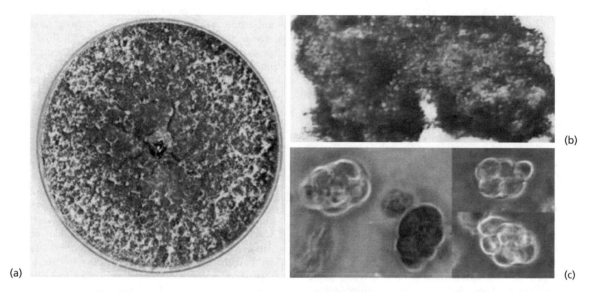

Fig. 11.12 (a) *Thermoascus aurantiacus* produces a thick, crusty, golden-brown colony on agar. (b) The crusty appearance is due to the presence of many cleistothecia (closed ascocarps) which contain asci. (c) Within each ascus is a cluster of eight ascospores, shown in different stages of maturity.

thermophile (maximum about 55°C, optimum about 50°C) might start earlier than *H. insolens* (maximum about 55°C, optimum about 37°C), and these two fungi are known to recolonize before *A. fumigatus* (maximum 52°C, optimum 37°C). During the prolonged warm-temperature phase after peak-heating a compost can lose up to half of its original dry weight, and nearly two-thirds of the main plant wall components such as cellulose and hemicelluloses (mixed polymers of arabinose, xylose, mannose, and glucose). The mitosporic fungus, *Thermomyces lanuginosus* (see Fig. 8.3) has a temperature maximum of about 62°C, and optimum of about 47°C. It grows during the high temperature phase and is one of the commonest fungi in all types of compost. However, it is noncellulolytic, and it seems to grow as a commensal, utilizing some of the sugars released by the enzymes of other fungi (see later).

3 As the temperature falls below 35–40°C, the thermophilic fungi start to decline but *A. fumigatus* remains active. The compost is then colonized progressively by mesophilic fungi, including some mitosporic fungi (e.g. *Fusarium*, *Doratomyces*) and Basidiomycota such as *Coprinus cinereus* (max. about 40°C but its optimum is much lower). *Coprinus* spp. represent the **degraders of recalcitrant polymers**. They utilize lignocellulose and are highly antagonistic to many other fungi, damaging their hyphae on contact by a process termed **hyphal interference** (Chapter 12). Spores of the commercial mushroom,

Agaricus bisporus, can be introduced into the compost once this has cooled to below 30°C. It cannot be added earlier because of its temperature requirements, but it must be added before *Coprinus* becomes established or it will be antagonized. In practice, this problem is overcome (and the whole commercial composting process is accelerated) by pasteurizing commercial composts soon after peak heating so that most of the resident fungi are killed before *Agaricus* is introduced (Chapter 5).

The central role of nitrogen in the composting process

An adequate supply of mineral nutrients is essential for composting and for the decomposition of organic matter in general. A supply of nitrogen is particularly important, and can often be the rate-limiting factor. The availability of nitrogen is often expressed in terms of the **carbon-to-nitrogen (C : N) ratio** – the ratio of elemental nitrogen to elemental carbon, assuming that both the nitrogen and the carbon are available in usable forms.

To appreciate the significance of this, we must make two points:

1 Fungal hyphae typically have a C : N ratio of approximately 10 : 1, although it can vary depending on the age of the hyphae and other factors.

2 During growth on a carbon substrate, fungi convert approximately one-third of the substrate carbon into cellular material (a substrate conversion efficiency of about 33% as explained in Chapter 4). The other two-thirds is respired to CO_2. These values are only approximate, but they serve as guidelines.

It follows that a material with a C : N ratio of about 30 : 1 is a "balanced" substrate – it can be degraded rapidly and completely because:

10 of the carbon "units" (units are unspecified) are incorporated into fungal biomass;
20 of the carbon units are released as CO_2;
1 unit of nitrogen is incorporated into fungal biomass.

Now, consider a material of C : N ratio 100 : 1 (wheat straw is roughly 80 : 1, sawdust ranges from 350 : 1 to 1250 : 1, and newsprint has essentially no nitrogen). The fungus starts to grow on this material, but the available nitrogen is depleted long before all the organic carbon has been used. Viewed in simple terms, decomposition will stop at the point where:

10 C units are combined with 1 N unit in the mycelium;
20 C units have been respired to CO_2;
70 C units are left in the residual substrate.

In effect the fungus is now starved because there is no nitrogen available for growth. At this point several things can happen. Some fungi seem to recycle their cellular nitrogen, perhaps by controlled autolysis of the older hyphae. Other fungi preferentially allocate nitrogen to essential metabolic processes – for example, wood-decay fungi such as *Coriolus versicolor* seem to do this (discussed later). A third possibility is to recruit extra nitrogen from soil, but this is not possible in a "closed" system such as a compost. In practice, many of the fungal hyphae will die when nitrogen becomes limiting, and the nitrogen can then be reused, either by the remaining cells of the same species or by other species. The same points apply to any essential mineral nutrient that is in short supply.

Can nitrogen-depletion *drive* a decomposition sequence?

This question arises from the comments above: if one fungus depletes the available nitrogen and cannot recycle it, then a proportion of the cells will die and their products might be used by another fungus later in the succession.

This possibility has been tested experimentally in laboratory conditions (Fig. 11.13), with flasks containing 7 g sterile filter paper (almost pure cellulose) plus nitrate as the nitrogen source (C : N, 200 : 1) and

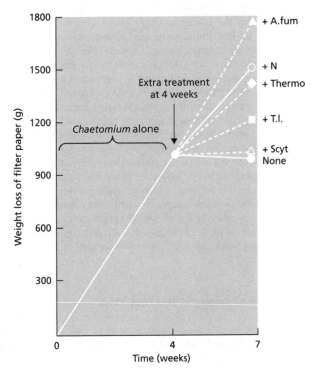

Fig. 11.13 A laboratory experiment simulating a succession of thermophilic fungi growing on cellulose (Deacon 1985; see text for details). A. fum = *Aspergillus fumigatus*; N = nitrogen; Scyt = *Scytalidium thermophilum*; Thermo = *Thermoascus aurantiacus*; T.l. = *Thermomyces lanuginosus*.

other mineral nutrients. All flasks were inoculated with the cellulolytic fungus, *Chaetomium thermophile* and incubated for 4 weeks at 45°C. At that stage *Chaetomium* had degraded 1 g of the original flask contents but its growth had stopped and there was no further breakdown after 7 weeks. This was shown to be due to nitrogen depletion, because the addition of extra nitrogen at 4 weeks led to further cellulose breakdown (Fig. 11.13).

Other treatments were also introduced at 4 weeks (the stage at which *Chaetomium* was nitrogen-depleted) but **no extra nitrogen was supplied.** In all cases these treatments involved placing a small inoculum block of a thermophilic fungus on the original *Chaetomium* colony. When *Scytalidium thermophilum* was added to the *Chaetomium* flasks at 4 weeks there was no further breakdown of the cellulose, although *Scytalidium* can degrade cellulose when grown alone. By contrast, the addition of either *Thermoascus aurantiacus* or *Aspergillus fumigatus* to the 4-week *Chaetomium* flasks led to a further weight loss after 7 weeks. *A. fumigatus* was the most effective in this respect – it almost doubled the original weight loss. These fungi were seen to grow over the original *Chaetomium* colony. Even a noncellulolytic fungus, *Thermomyces lanuginosus*, caused some additional weight loss when it was added after 4 weeks.

The implication of these findings is that the early colonizers in a substrate succession (in this case *Chaetomium* which has the highest temperature optimum for growth) can be displaced by fungi that occur later in the succession (e.g. *Thermoascus* or *A. fumigatus*) in conditions of nitrogen starvation.

Basidiomycota (*Agaricus bisporus, Coprinus cinereus*, etc.) typically occur late in the succession of fungal activities in natural materials, and many of them have been shown to use proteins as nitrogen sources. Indeed, *A. bisporus* can grow in culture media when nitrogen is supplied only in the form of living or heat-killed bacteria. We noted earlier that *Phanerochaete* is induced to synthesize lignin-degrading enzymes in response to nitrogen limitation; and in Chapter 12 we will see that Basidiomycota such as *Coprinus* disrupt the hyphae of other fungi on contact, which perhaps provides a source of organic nitrogen. So it seems that nitrogen availability is a key factor in fungal successions, and that the later colonizers have special abilities to obtain (recycle) the nitrogen that earlier colonizers have utilized.

Fungal decomposers in the root zone

Living roots provide a continuous input of nutrients into soil, evidenced by the fact that motile pseudomonads and zoospores of Oomycota or plasmodiophorids accumulate near the root tips. Amino acids, sugars, and other organic molecules are found frequently in root exudates, and the total microbial population increases progressively with distance behind the tips until it reaches a plateau. At this point the population level is likely to represent the "carrying capacity" of the root – in other words, the rate of continuing nutrient release is matched by the rate at which these nutrients are **utilized by the existing population.**

The zone of soil influenced by the presence of a root is termed the **rhizosphere**. As we saw in Chapter 10, fungal spores can be induced to germinate by the presence of root exudates or even by small signalling molecules that may not serve as nutrients. But this is only the start of a continuous process of root colonization by saprotrophs, parasites, and pathogens until, eventually, the roots die and their nutrients are returned to the soil.

The behavior of roots is difficult to investigate in field conditions, but this can be done in a glasshouse if plants are grown in soil containers with a sloping transparent face so that periodic observations can be made (the roots being blacked out except during periods of observation). Figures 11.14 and 11.15 show the results of one such experiment where groundnut plants (*Arachis hypogea*) were grown over a 14-week period (Krauss & Deacon 1994). The whole visible part of the root system was traced at weekly intervals, and the individual roots were scored as being either alive (white roots) or dead (brown decaying roots or roots that had disappeared).

Figures 11.14 and 11.15 reveal an astonishingly high rate or **rhizodeposition**, i.e. the shedding of root material into the soil where it becomes available to the microbial population of the rhizosphere. Although the total (cumulative) root length increased throughout the period of observation, up to plant maturity at 14 weeks or 20 weeks (depending on the groundnut cultivar), the **maximum length of living (white) roots** on any plant was reached at between 2 and 4 weeks after sowing the seeds. Beyond that time, the taproot continued to grow and produced more root laterals, but the rate of root death (disappearance) exceeded the rate of new root production, leading to a decline in total root length. An early onset of root lateral death, 5 weeks after sowing, was also recorded in a field plot of groundnuts in Malawi. So, at least for this crop, there is a continuous process of "root shedding" which would provide a continuous input of organic matter for fungi and other soil microbes.

Other studies have focused on the roots of cereals and grasses, using vital dyes to follow the progressive death of the root cortex. **Acridine orange** penetrates roots, binds to the DNA in the nuclei, and fluoresces bright green under a fluorescence microscope (Fig. 11.16a,c). Alternatively, roots can be infiltrated with

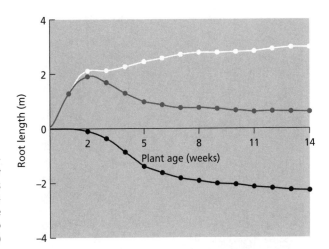

Fig. 11.14 Cumulative root production (white symbols), cumulative root decay (black symbols below the line), and white living roots (black symbols above the line) of groundnut plants grown for 14 weeks (near-maturity). Nearly identical results were obtained for five different groundnut cultivars (two replicates of each). (From Krauss & Deacon 1994.)

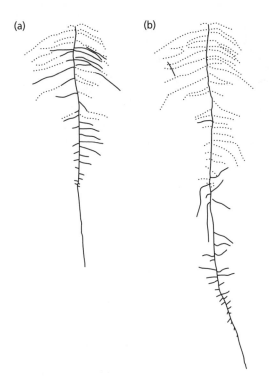

Fig. 11.15 Representative tracings of the root system of a single groundnut plant. Root distribution was traced onto a transparent overlay at weekly intervals, but only the tracings at 6 weeks (a) and 13 weeks (b) are shown. Solid lines represent roots that were white (alive) at the times of these tracings. Broken lines represent roots that were present earlier but had decayed and disappeared. Nearly identical results were found for five groundnut cultivars (two replicates of each). (From Krauss & Deacon 1994.)

neutral red which accumulates in the intact protoplasts, then a solution of strong osmotic potential is used to shrink the protoplasts (Fig. 11.16b,d). In both methods the number and distribution of living cells can be counted. For glasshouse-grown wheat seedlings, the outermost root cortical cell layer dies first – typically in root regions that are 7–10 days old – and then the successively deeper cortical cell layers die at approximately 3–4 day intervals, until only the innermost cortical cell layer (next to the root endodermis) remains alive. This **progressive root cortical death** behind the extending root tips is seen when roots are grown **in soil or in aseptic conditions**. It is faster in some cereals than in others, faster in cereals than in grass roots, and faster in laboratory than in field conditions. But in all cases the root cortex dies, and this does not affect root function, because the root tips continuously extend into new zones of soil, while the older root regions serve a transport function, with a largely redundant, senescent root cortex.

Fungal invasion of cereal and grass roots

The pattern of root cortical cell death, described above, precisely matches the pattern of fungal invasion in the roots of perennial ryegrass, *Lolium perenne* (Fig. 11.17), originally described by Waid (1957). The sequence is summarized below, but essentially it involves the successive invasion of roots by parasites, weak parasites, and saprotrophs, responding to the progressive natural senescence of the root cortex:

1 The growing root tips are virtually free from fungal hyphae, but an increasingly complex fungal community develops with distance (age) behind the tips.

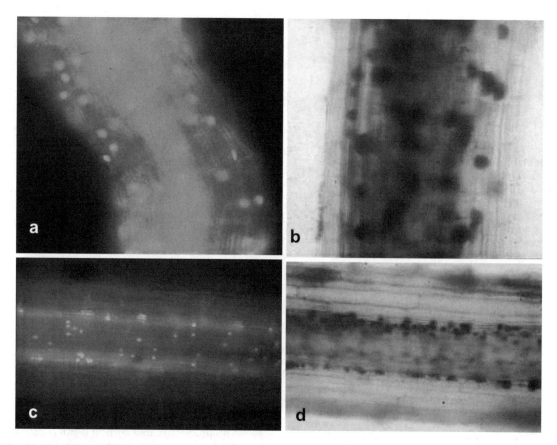

Fig. 11.16 Natural (nonpathogenic) cortical cell death in soil-grown wheat roots stained with acridine orange (a,c) or neutral red (b,d). All the cortical cell layers are alive in (a) and most layers are alive in (b). The outer five (of a total six) cell layers have died in the older root regions (c and d) but the innermost (sixth) layer always remains alive until the root itself dies. (From Lascaris & Deacon 1991.)

2 Even in young root zones, and persisting into the old root zones, a fungus with hyaline (colorless) hyphae is seen in the innermost cortical cells. This is almost certainly an arbuscular mycorrhizal fungus growing in the living host cells (Chapter 13).

3 Behind the root tips, *Mucor* and *Penicillium* species are found on the root surface, and they remain associated with the surface of roots of all ages. These fungi are assumed to grow on the soluble nutrients that leak from root cells, or possibly the root surface mucilage (a pectin-like material). However, their hyphae cannot be seen or identified easily in the root zone, so it is possible that they have an initial phase of growth or periodic phases of growth and then persist as spores on or near the root surface.

4 A fungus with darkly pigmented hyphae is found on the root surface in young root zones, and is seen to occur progressively deeper in the cortex as the roots age. This fungus is *Phialophora graminicola*, one of the most abundant fungi on grass roots. It is a weak parasite that exploits a narrow window of opportunity, invading the root cortical cells as they start to senesce, but ahead of purely saprotrophic species (Chapter 12).

5 As *Phialophora* progresses inwards through the root cortex, it is followed by a sequence of fungi, including an unidentified sterile (nonsporing) fungus that grows as hyaline (colorless) hyphae, and *Fusarium culmorum* which is a characteristic colonizer of dead and dying tissues but also can be pathogenic to cereals and grasses in conditions of water-stress.

6 In the old root zones *Trichoderma* spp. and *Clonostachys rosea* (formerly called *Gliocladium roseum*) colonize the outer cortex. These are characteristic soil and rhizosphere fungi. They are known

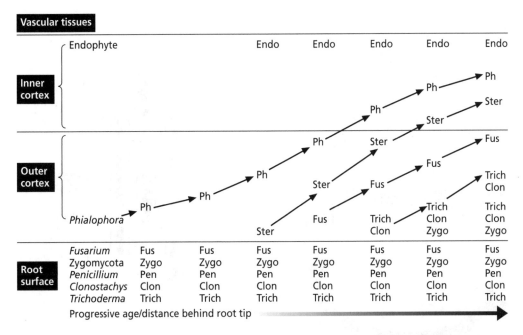

Fig. 11.17 The sequence of fungal invasion of ryegrass roots, with increasing age and distance behind the extending root tips. *Phialophora graminicola* (Ph), a weak parasite, initiates the invasion sequence as the root cortex starts to senesce. Subsequent invaders include a sterile hyaline fungus (Ster), *Fusarium culmorum* (Fus), and then a suite of saprotrophic fungi, including Zygomycota (Zygo), *Penicillium* (Pen), *Clonostachys rosea* (Clon), and *Trichoderma* (Trich). (Based on Waid (1957) but with additional information and interpretation.)

to antagonize and to overgrow other fungi on agar plates (Chapter 12). So they probably grow in the root cortex as secondary (opportunistic) invaders.

Fungal communities in decaying wood

In terms of both its physical and chemical properties, wood is an exceptionally difficult substrate to degrade, so it is largely unavailable to most fungi. Wood consists mainly of cellulose (40–50% dry weight), hemicelluloses (25–40%), and lignin (20–35%). Of these, lignin often presents the main obstacle to wood decay, because it is a complex aromatic polymer that encrusts the cell walls, preventing access of enzymes to the more easily degradable cellulose and hemicelluloses. Lignin is highly resistant to breakdown by conventional enzyme systems because it is chemically complex, variable, nonhydrolysable, and water-insoluble. Wood also has a very low nitrogen content (commonly a C : N ratio of about 500 : 1) and low phosphorus content. And, it contains potentially fungitoxic compounds, which are deposited in the heartwood. In broad-leaved trees the toxic compounds are usually **tannins**, well know for their ability to cross-link proteins, making animal skins resistant to

decay. By contrast, conifers contain a range of phenolic compounds such as terpenes, stilbenes, flavonoids and tropolones. The most toxic of the tropolones are the **thujaplicins** which act as uncouplers of oxidative phosphorylation; they are particularly abundant in cedarwood, making this a naturally decay-resistant wood for high-quality garden furnishings, etc.

Despite this formidable list of obstacles, woody tissues are degraded by fungi, and these fall into three types according to their mode of attack on the woody cell walls – **soft-rot fungi**, **brown-rot fungi**, and **white-rot fungi**.

Soft-rot fungi

Soft-rot fungi grow on wood in damp environments. They are the characteristic decay fungi of fence posts, telegraph poles, wooden window frames, the timbers of cooling towers, and wood in estuarine or marine environments. They have a relatively simple mode of attack on wood, illustrated in Fig. 11.18. Their hyphae grow in the lumen of individual woody cells, usually after entering through a "pit" (depression) in the wall. Then they produce fine penetration branches that grow through the thin, lignin-coated S3 layer of the

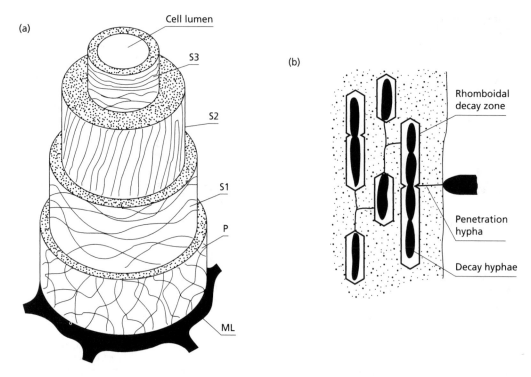

Fig. 11.18 (a) Diagram of the cell wall layers in woody tissue, showing the arrangement of cellulose microfibrils. ML = middle lamella between adjacent woody cells; P = thin primary wall with loosely and irregularly arranged microfibrils; S1–S3 = secondary wall layers. (b) Characteristic decay pattern of a soft-rot fungus in the S2 layer. The fungus penetrates by narrow hyphae, then forms broader hyphae in planes of weakness in the wall, and these hyphae produce rhomboidal cavities where the cellulose has been enzymatically degraded.

wall, to gain access to the thick, cellulose-rich S2 layer. When the penetration hyphae find a longitudinal plane of weakness in the S2 layer, they produce broader T-shaped hyphae which grow along the plane of weakness and secrete cellulase enzymes. The diffusion of these enzymes creates a characteristic pattern of decay, seen as rhomboidal cavities within the cell wall. These persist even when the fungi have died, leaving the characteristic "signature" of a soft-rot fungus. The soft-rot fungi have little or no effect on lignin, which remains more or less intact. All the soft-rot fungi need relatively high nitrogen levels for wood decay, typically about 1% nitrogen content in the wood. If this is unavailable in the wood itself, then nitrogen can be recruited from the environment, such as the soil at the bases of fence posts.

The fungi that cause soft rots include several Ascomycota and mitosporic species, such as *Chaetomium* and *Ceratocystis* in terrestrial environments and species of *Lulworthia*, *Halosphaeria*, and *Pleospora* in marine and estuarine environments.

Brown-rot fungi

Brown-rot fungi are predominantly Basidiomycota, including common species such as *Schizophyllum commune*, *Fomes fomentarius* (the "hoof fungus" of Scottish birch woods; see Fig. 2.24), and the "dry-rot fungus", *Serpula lacrymans* (see Fig. 7.8). Many of the brown-rot fungi produce bracket-shaped fruitbodies (basidiocarps) on the trunks of dead trees, but the characteristic feature of these fungi is that the decaying wood is brown and shows brick-like cracking – a result of the uneven pattern of decay, causing the wood to split along lines of weakness (Fig. 11.19) (Bagley & Richter 2002).

The "dry-rot" fungus, *Serpula lacrymans*, causes major damage to structural timbers in the buildings of Europe. However, the term dry-rot does not indicate a different type of wood decay; instead, it signifies that this fungus, once established, can generate sufficient water from the breakdown of cellulose in the wood to continue growing even when the environmental

Fig. 11.19 Part of a pine stump showing the characteristic brick-like decay by brown-rot fungi.

conditions are quite dry. *Serpula lacrymans* is also notable because it posed an enigma for a long time: it is very common in poorly ventilated buildings in Britain and much of Europe, but nobody had ever found it in a natural environment. Singh *et al.* (1993) then discovered that it occurs rarely in the Himalayan forests of northern India, which seem to be its natural habitat. Dry rot has been recorded in Europe since about 1765, before there was any export of timber from India. So this "rare" fungus seems to have arrived in Europe as air-borne basidiospores and then flourished in buildings where the climatic conditions are similar to those in its natural habitat.

The term "brown rot" refers to the characteristic color of the decayed wood, because most of the cellulose and hemicelluloses are degraded, leaving the lignin more or less intact as a brown, chemically modified framework. However, the decay is typically irregular, with some groups of wood cells being heavily degraded while others are only slightly so. This causes the wood to crack in a brick-like manner.

The hyphae of brown-rot fungi occur very sparsely in the wood, often restricted to the lumen of woody cells, and yet they cause a generalized decay in which the S2 wall layer is almost completely degraded. This type of decay cannot be explained by the diffusion of cellulase enzymes, which are too large to diffuse very far, and too large even to pass through the pores in the S3 layer. In fact, the cellulases of brown-rot fungi have little effect on cellulose *in vitro*, unlike the cellulases of soft-rot fungi. Instead, the brown-rot fungi degrade cellulose by an **oxidative process**, involving the production of hydrogen peroxide during the

breakdown of hemicelluloses. Being a small molecule, H_2O_2 can diffuse through the woody cell walls to cause a generalized decay. In support of this, the characteristic decay pattern of brown-rot fungi can be mimicked experimentally by treating wood with H_2O_2 alone, and at least one of these fungi, *Poria placenta*, has been shown to degrade cellulose only if hemicelluloses also are present, as substrates for generating H_2O_2. This mode of attack is an efficient way of using the scarce nitrogen resources in wood, because it does not require the release of large amounts of extracellular enzymes.

White-rot fungi

White-rot fungi are more numerous than brown-rot fungi. They include both Ascomycota (Fig. 11.20) and Basidiomycota (e.g. *Armillaria mellea*, the boot-lace fungus – see Fig. 5.14, and *Coriolus versicolor*, shown in an unexpected setting, Fig. 11.21). Is nothing sacred?!!

The white-rot fungi seem to use conventional cellulase enzymes for wood decay, but they are extremely efficient in their use of nitrogen. For example, the nitrogen content of *Coriolus versicolor* is about 4% when the fungus is grown on laboratory media of C : N ratio 32 : 1, but only 0.2% when grown on a medium of C : N 1600 : 1. In nitrogen-poor conditions this fungus seems preferentially to allocate nitrogen to the production of extracellular enzymes and essential cell components, and it also efficiently recycles the nitrogen in its mycelia (Levi & Cowling 1969). White-rot

(a)   (b)

Fig. 11.20 Two common Ascomycota that cause white rots. (a) *Xylaria hypoxylon*, the "candle snuff" fungus often seen on rotting stumps. (b) *Xylaria polymorpha* ("dead man's fingers").

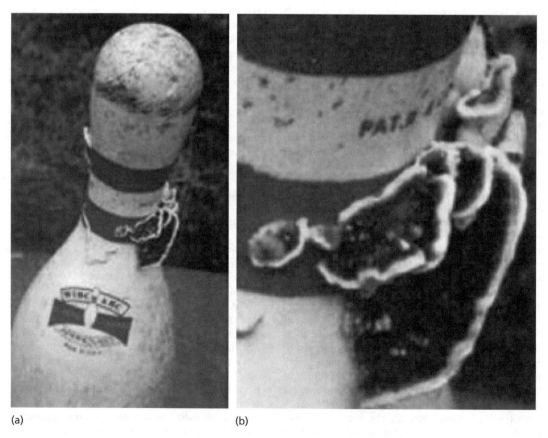

(a)   (b)

Fig. 11.21 Small, leathery, bracket-shaped fruitbodies of *Coriolus versicolor*.

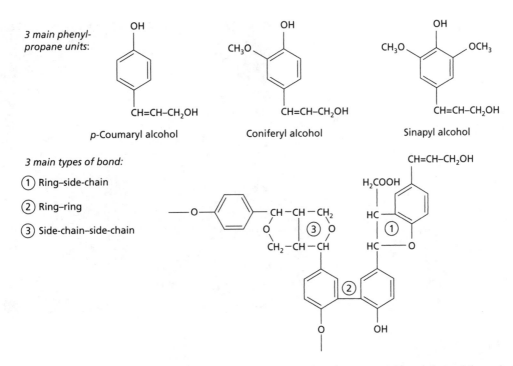

Fig. 11.22 Structural components of lignin, showing the three main phenyl-propane units and three of the main types of bond that link them. Note that only a small part of the lignin molecule is shown.

fungi might also benefit from the growth of nitrogen-fixing bacteria in wood.

The most remarkable feature of white-rot fungi is their ability completely to degrade lignin – they are the only organisms known to do this. As shown in Fig. 11.22, lignin is a complex polymer composed of three types of phenyl-propane unit (six-carbon rings with three-carbon side chains) bonded to one another in at least 12 different ways. If lignin were to be degraded by conventional means it would require a multitude of enzymes. Instead, lignin is degraded by an oxidative process. The details of this are complex, but essentially the white-rot fungi produce only a few enzymes (lignin peroxidase, manganese peroxidase, H_2O_2-generating enzymes, and laccase) and these generate strong oxidants, which virtually "combust" the lignin framework (Kirk & Farrell 1987).

The major enzyme that initiates ring-cleavage is **laccase**, which catalyses the addition of a second hydroxyl group to phenolic compounds. The ring can then be opened between two adjacent carbon atoms that bear the hydroxyl groups (Fig. 11.23). This process occurs while the ring is still attached to the lignin molecule. It is termed *ortho* fission, in contrast to *meta* fission which bacteria employ to cleave the phenolic rings of pesticide molecules (where the ring is opened at a different position – see Fig. 11.23).

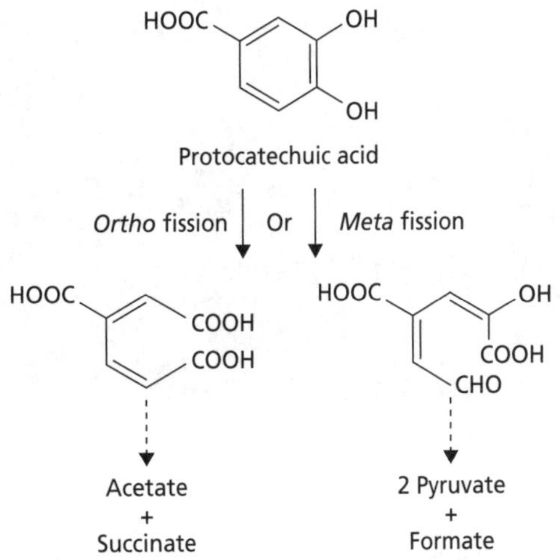

Fig. 11.23 Patterns of opening of aromatic rings. (Left) During lignin breakdown, by the process of *ortho* fission. (Right) During the breakdown of pesticides and other xenobiotics, by *meta* fission. Initially the ring is substituted with two hydroxyl groups on adjacent carbon atoms. Then it is opened either between these two carbons (ortho fission) or adjacent to one of them (meta fission, which is a plasmid-encoded function of bacteria).

(a)

(b)

(c)

Fig. 11.24 (a) Part of a decaying beech stump showing dark zone lines at the junctions between mutually incompatible fungal colonies. (b) A decorative wooden bowl made from spalted (zone-lined) beech wood. (c) Part of an extensively degraded woody stump showing continuous plates of melanized cells within the stump tissues.

The other enzymes are involved mainly in generating or transferring oxidants. They include **glucose oxidase** which generates H_2O_2 from glucose, **manganese peroxidase** which oxidizes Mn (II) to Mn (III), and which can then oxidize organic molecules, and **lignin peroxidase** which catalyses the transfer of singlet oxygen from H_2O_2 to aromatic rings and is one of the main initiators of attack on the lignin framework. These initial oxidations involving single electron transfers generate highly unstable conditions, setting off a chain of chemical oxidations.

Clearly, the degradation of lignin is highly dependent on a supply of oxygen, so it does not occur in waterlogged conditions. Lignin degradation also poses potential hazards to the fungus because some of the oxidative intermediates can be fungitoxic. The white-rot fungi generate such compounds from phenyl-propane units *in vitro*, but detoxify them by polymerization into melanin-like pigments. Evidence of this can often be seen where two colonies of wood-decay fungi meet and produce heavily melanized **zone lines** (actually continuous plates of melanized cells) in the region of contact (Fig. 11.24). When fungi are isolated from either side of a zone line they are found to belong to different species or, more commonly, different strains of a single species that are somatically incompatible with one another. Zone-lined wood (also termed "spalted" wood) has become fashionable for ornaments (Fig. 11.24).

Biotechnology of wood-decay fungi

Lignocellulose is abundant as a byproduct of the wood-processing industries and also in crop residues, so there is the potential to use it as a cheap commercial substrate. For example if the cellulose could be degraded to sugars, these could be used to produce fuel alcohol by microbial fermentation, as an alternative to fossil fuels. This prospect has stimulated research on delignification, especially by the white-rot fungus *Phanerochaete chrysosporium* which grows rapidly in submerged liquid culture and, unusually for Basidiomycota, produces abundant conidia. In near-optimum culture conditions it can degrade as much as 200 mg lignin per gram of mycelial biomass per day. However, it also degrades and utilizes the cellulose component of lignocellulose, defeating the object of the exercise. Genetic engineering offers a potential solution to this, and was made possible by the discovery that *Phanerochaete* produces lignin peroxidase in the early stationary phase of batch culture (Chapter 4). The enzyme production was strongly promoted by nitrogen limitation and further stimulated by addition of lignin. By comparing the mRNA produced in these conditions and in corresponding noninducing conditions, the gene for lignin peroxidase was identified, then sequenced and cloned into *E. coli*. The enzyme produced by the recombinant bacterium acts on a range of lignin "model compounds" *in vitro* so there is a prospect of using high-yielding recombinant microorganisms for delignification processes.

White-rot fungi and their enzyme systems also have potential for bioremediation of land contaminated by aromatic pollutants, and many other processes, discussed by Ralph & Catcheside (2002).

Cited references

Andrews, J.H. (1992) Fungal life-history strategies. In: *The Fungal Community: its organization and role in the ecosystem* (Carroll, G.C. & Wicklow, D.T., eds), pp. 119–145. Marcel Dekker, New York.

Bagley, S.T. & Richter, D.L. (2002) Biodegradation by brown rot fungi. In: *The Mycota X. Industrial Applications* (Osiewicz, H.D., ed.), pp. 327–341. Springer-Verlag, Berlin.

Carroll, G.C. & Wicklow, D.T. (1992) *The Fungal Community: its organization and role in the ecosystem.* Marcel Dekker, New York.

Chang, Y. (1967) The fungi of wheat straw compost: paper II. *Transactions of the British Mycological Society* **50**, 667–677.

Chang, Y. & Hudson, H.J. (1967) The fungi of wheat straw compost: paper I. *Transactions of the British Mycological Society* **50**, 649–666.

Deacon, J.W. (1985) Decomposition of filter paper cellulose by thermophilic fungi acting singly, in combination, and in sequence. *Transactions of the British Mycological Society* **85**, 663–669.

Harper, J.L. & Webster, J. (1964) An experimental analysis of the coprophilous fungus succession. *Transactions of the British Mycological Society* **47**, 511–530.

Hudson, H.J. (1968) The ecology of fungi on plant remains above the soil. *New Phytologist* **67**, 837–874.

Kirk, T.K. & Farrell, R.L. (1987) Enzymatic "combustion": the microbial degradation of lignin. *Annual Review of Microbiology* **41**, 465–505.

Krauss, U. & Deacon, J.W. (1994) Root turnover of groundnut (*Arachis hypogea* L.) in soil tubes. *Plant & Soil* **166**, 259–270.

Lascaris, D. & Deacon, J.W. (1991) Comparison of methods to assess senescence of the cortex of wheat and tomato roots. *Soil Biology and Biochemistry* **23**, 979–986.

Levi, M.P. & Cowling, E.B. (1969) Role of nitrogen in wood deterioration. VII. Physiological adaptation of wood-destroying and other fungi to substrates deficient in nitrogen. *Phytopathology* **59**, 460–468.

Li, S., Spear, R.N. & Andrews, J.H. (1997) Quantitative fluorescence in situ hybridization of *Aureobasidium pullulans* on microscope slides and leaf surfaces. *Applied and Environmental Microbiology* **63**, 3261–3267.

Lu, G., Cannon, P.F., Reid, A. & Simmons, C.M. (2004) Diversity and molecular relationships of endophytic *Colletotrichum* isolates from the Iwokrama Forest Reserve, Guyana. *Mycological Research* **108**, 53–63.

Olsson, P.A., Larsson, L., Bago, B., Wallander, H. & van Aarle, I.M. (2003) Ergosterol and fatty acids for biomass estimation of mycorrhizal fungi. *New Phytologist Letters* **159**, 7–10.

Ralph, J.P. & Catcheside, D.E.A. (2002) Biodegradation by white-rot fiungi. In: *The Mycota X. Industrial Applications* (Osiewacz, H.D., ed.), pp. 303–326. Springer-Verlag, Berlin.

Schroder, S., Hain, M. & Sterflinger, K. (2000) Colorimetric in situ hybridization (CISH) with digoxigenin-labeled oligonucleotide probes in autofluorescent hyphomycetes. *International Microbiology* **3**, 183–186.

Singh, J., Bech-Andersen, J., Elborne, S.A., Singh, S., Walker, B. & Goldie, F. (1993) The search for wild dry rot fungus (*Serpula lacrymans*) in the Himalayas. *The Mycologist* **7**, 124–130.

Stewart, A. & Deacon, J.W. (1995) Vital fluorochromes as tracers for fungal growth studies. *Biotechnic and Histochemistry* **70**, 57–65.

Tribe, H.T. (1960) Decomposition of buried cellulose film, with special reference to the ecology of certain soil fungi. In: *The Ecology of Soil Fungi* (Parkinson, D. & Waid, J.S., eds), pp. 246–256. Liverpool University Press, Liverpool.

Vanneste, J.L., Hill, R.A., Kay, S.J., Farrell, R.L. & Holland, P.T. (2002) Biological control of sapstain fungi with natural products and biological control agents: a review of the work carried out in New Zealand. *Mycological Research* **106**, 228–232.

Waid, J.S. (1957) Distribution of fungi within the decomposing tissues of ryegrass roots. *Transactions of the British Mycological Society* **40**, 391–406.

Chapter 12

Fungal interactions: mechanisms and practical exploitation

This chapter is divided into the following major sections:

- the terminology of species interactions
- antibiotics and their roles in species interactions
- antibiotics and disease control by *Trichoderma* species
- hyphal interference: a classic example of applied biological control
- mycoparasites: fungi that parasitize other fungi
- competitive interactions among fungi
- commensalism and mutualism among fungi

Microorganisms interact in many diverse ways. In this chapter we consider some of the major types of interaction involving fungi, and the potential for exploiting these interactions for practical benefit. We will also discuss the use of several commercially available biological control agents. (See Online resources for this chapter.)

The terminology of species interactions

The terminology of species interactions is difficult, but for most purposes we can distinguish three broad categories of interaction among fungi:

1 The ability of one species to exclude another by **competition** (sometimes called **exploitation competition**), i.e. by being faster or more efficient in exploiting a resource (space, substrate, etc.).
2 The ability of one species to exclude or replace another by **antagonism** (sometimes called **inter-ference competition** or **combat**), i.e. by directly affecting another organism through antibiotic production, parasitism, etc.
3 The ability of two species to coexist (**commensalism**) to the benefit of one or both (**mutualism**).

It must be emphasized that these types of interaction grade into one another, because fungi can behave differently in different situations.

Antibiotics and their roles in species interactions

An antibiotic can be defined somewhat arbitrarily as a diffusible secondary metabolite of one (micro)organism that inhibits another (micro)organism at a concentration of $100 \, \mu g \, ml^{-1}$ or less. This definition serves to exclude general metabolic byproducts such as CO_2 or organic acids. It restricts the term to specific highly active compounds that affect specific cellular targets.

The best-known antibiotics from fungi are the **penicillins** (Chapter 7), **cephalosporins**, and **griseofulvin** (Chapter 15), which are used clinically to control bacteria or, in the case of griseofulvin, other fungi. Some more recently discovered antibiotics include **fusidic acid** from the mitosporic fungus *Fusidium coccineum*, which is active against Gram-positive bacteria, **fumagillin** from *Aspergillus fumigatus*, which is used to control parasitic protozoa in veterinary medicine, and **sordarin**, from a species of *Sordaria* (Ascomycota) to control fungal infections of humans.

These are only a few of the 1000 or more antibiotics known to be produced by fungi. Most of the antibiotics discovered in routine screens by chemical companies are not used commercially because of their nonspecific toxicity or other undesirable side-effects. **Patulin** is one such example (see Fig. 7.13); it was a promising antibiotic but its development was abandoned because, among other things, it was found to be a potent mycotoxin. However, antibiotics are produced naturally in a wide range of environments and they play significant roles in species interactions. Some of the most common fungi that produce antibiotics in natural and agricultural environments are species of Ascomycota and mitosporic fungi, including *Penicillium*, *Aspergillus*, *Fusarium*, and *Trichoderma*. The Basidiomycota also produce several antibiotics, but very few antibiotics have been recorded from Chytridiomycota, Zygomycota, or Oomycota, perhaps because these organisms have a short life cycle and do not need to defend a substrate against invaders.

Antibiotics in natural environments: the control of fungi by fluorescent pseudomonads

The past two decades have brought major advances in the detection of antibiotics in the root zone (rhizosphere) of crops, helping to explain how fluorescent pseudomonads can control fungal pathogens of roots. Fluorescent pseudomonads are found on the roots of many plants, often at high population levels, and can be detected easily by plating soil dilutions onto "King's B agar." This is an iron-deficient medium, so it induces these bacteria to release fluorescent siderophores (iron-chelating compounds) to capture iron (Chapter 6). However, only a small subset of fluorescent pseudomonads (including strains of *Pseudomonas fluorescens* and *P. aureofaciens*) are highly effective in controlling fungi. These strains produce specific antifungal antibiotics such as **phenazine-1-carboxylic acid** (PCA) and **2,4-diacetylphloroglucinol** (DAPG) (Fig. 12.1).

Such strains were first recognized in field conditions, where crops such as tobacco or wheat grown in sites with naturally high populations of PCA- or DAPG-producers grew better than crops with low population levels of these bacteria. The terms **disease-suppressive soil**, and **disease-conducive soil** are used to describe this difference. In experimental studies, disease-suppressive soils can be converted to disease-conducive soils by pasteurization (treatment at about 60°C for 30 minutes). Conversely, the reintroduction of antibiotic-producing pseudomonads, at sufficiently high levels, can render the soils suppressive again.

The most detailed studies on soil suppressiveness have been made for the **take-all fungus**, *Gaeumannomyces graminis* – one of the most important root pathogens of cereal crops. This fungus infects roots from inoculum that persists in soil from a previous cereal crop, and it then grows along the roots as darkly pigmented "runner hyphae" (see Fig. 9.11). From these, it sends infection hyphae into the root, destroying the cortical cells and entering the vascular system, where it destroys the phloem (sugar-conducting cells) and blocks the water-conducting xylem vessels with dark vascular gels. The level of take-all infection increases progressively from one season to the next if cereals are grown repeatedly in a site. But after 3 or 4 years of cereal monoculture the disease reaches a peak and then spontaneously declines to a level at which cereals can be grown continuously without suffering serious yield losses.

This spontaneous decline in the disease level is termed **take-all decline** (Fig. 12.2) and it is always strongly correlated with a high population of antibiotic-producing fluorescent pseudomonads, which seem to be favored by continuous cereal cropping.

A significant advance in understanding this phenomenon was made in the early 1990s, when Thomashow *et al.* (1990) used **high performance liquid chromatography** (HPLC) to detect phenazine-1-carboxylic acid (PCA) in the rhizosphere of young

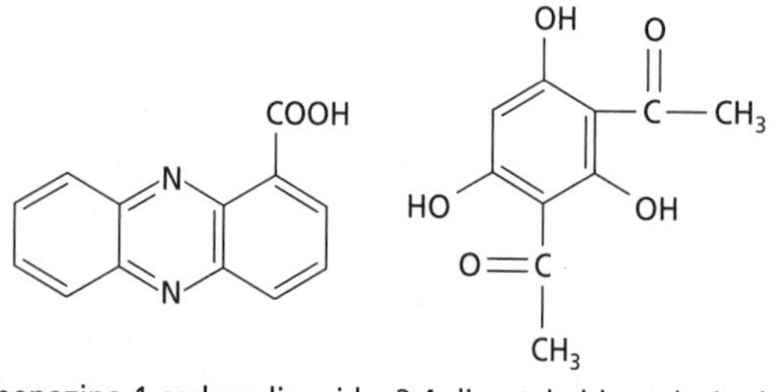

Phenazine-1-carboxylic acid   2,4-diacetyl phloroglucinol

Fig. 12.1 Two antibiotics from fluorescent pseudomonads that are widely implicated in control of plant-pathogenic fungi.

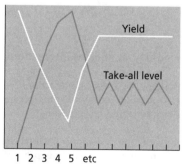

1 2 3 4 5 etc
Years of cereal cropping

Fig. 12.2 Take-all decline when cereals are grown continuously (year after year) in field conditions.

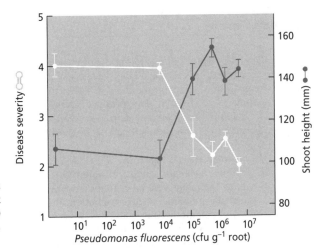

Fig. 12.3 The effects of different population levels of a diacetyl-phloroglucinol producing fluorescent pseudomonad on disease level (or shoot height) of wheat seedlings grown in take-all infested soil. (Reproduced from Raaijmakers & Weller 1998.)

wheat seedlings. These seedlings had been raised from seeds inoculated with phenazine-producing pseudomonads and grown in pots of natural soil containing the take-all fungus. After 4 weeks the seedlings were harvested and the rhizosphere soil was found to have antibiotic levels of 30–40 nanograms per gram of root with adhering soil. The levels of PCA were about 10 times higher when the inoculated seedlings were grown in sterile soil, but this would be expected because of the lack of competition from other rhizosphere bacteria. Compared with controls (untreated seeds) the PCA-treated seeds caused a significant reduction of disease. In contrast to the natural phenazine-producing strains, phenazine-minus strains obtained by transposon mutagenesis gave no reduction of disease and had no detectable antibiotic levels, but back-mutation to a phenazine-plus phenotype restored the ability to reduce disease. Identical results were obtained with a diacetyl-phloroglucinol (DAPG)-producing strain. So, there is clear evidence for a role of antibiotic-producing fluorescent pseudomonads in take-all suppressive soils.

An interesting feature revealed by these studies is that the population of antibiotic-producing pseudomonads builds up progressively on wheat crops, but only (or largely) when the take-all fungus is present. Wheat can be grown repeatedly in the absence of the take-all fungus in glasshouse conditions, and this does not lead to a build-up of the antagonistic pseudomonads. So it seems that the crop has to go through a build up of disease before the disease suppression sets in.

The likely reason for this is shown by the experimental results in Fig. 12.3. Wheat seeds were inoculated with different inoculum densities of DAPG-producing pseudomonads and then sown in a glasshouse in pots of natural soil containing take-all

inoculum. The seedlings were sampled after 4 weeks and assessed for disease severity and plant height. There was no disease control until the population level of DAPG strains on the roots exceeded 10^4 colony-forming units per gram of root. But, at colony levels of 10^5 and above, there was a marked and significant decrease in disease severity and a corresponding increase in plant height. This "all or nothing" effect is characteristic of a bacterial signalling system called **quorum sensing** – a term derived from the meetings of committees, where a certain number of people (a quorum) has to be present before a decision can be taken. Quorum-sensing by a population of Gram-negative bacteria involves the continued release of molecules called N-acyl homoserine lactones (Fig. 12.4). When the concentration of these molecules reaches a certain level (indicating that the population is large enough) the relevant genes are switched on. In this case, it is the genes controlling antibiotic production and therefore the control of take-all disease. The production of **phenazine** antibiotics is known to be under the control of a quorum-sensing system (Chin-A-Woeng et al. 2003), but there is no evidence as yet that **DAPG** production is regulated in a similar way.

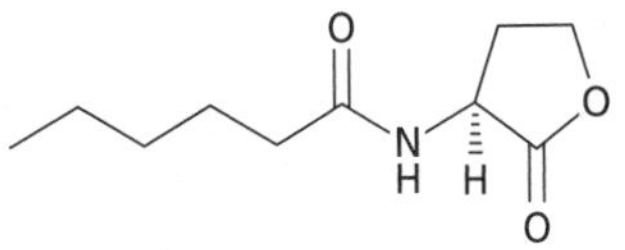

Fig. 12.4 Structure of N-hexanoyl-L-homoserine lactone, a quorum-sensing molecule that regulates the synthesis of phenazine antibiotics by fluorescent pseudomonads. (From Wood et al. 1997.)

Antibiotics and disease control by *Trichoderma* species

Trichoderma species (Fig. 12.5) are well-known for their ability to antagonize other fungi. They were among the first fungi to be shown to produce antibiotics in soil (Weindling 1934). In fact, they produce several antibiotics, including both volatile and nonvolatile compounds, active against fungi, bacteria or both (Fig. 12.6).

Many *Trichoderma* species, such as *T. viride*, *T. harzianum*, and *T. hamatum*, produce **6-pentyl-α-pyrone** (6-PAP) as the major volatile antibiotic, while the major nonvolatile antibiotics include **trichodermin**, **suzukacillin**, and **alamethicine**. But *Trichoderma virens* (as distinct from *T. viride*) produces a different spectrum of antibiotics: some strains produce **viridin**, **viridiol** (a reduction product of viridin), **gliovirin**, and **heptelidic acid**, whereas other strains of this fungus produce **viridin** and **gliotoxin**.

The differences in antibiotic production between strains of *Trichoderma* seem to be important, and have led to the widespread use of *Trichoderma* species as commercial biological control agents against plant-pathogenic fungi (see later). For example, 6-PAP is produced by some (but not all) strains of *T. viride*, *T. harzianum*, and *T. hamatum*, and these 6-PAP producers are more antagonistic than are nonproducing strains when tested against plant pathogens *in vitro* or in seedling bioassays. The strains of *T. virens* that produce gliovirin inhibit the growth of *Pythium ultimum* but not *Rhizoctonia solani in vitro*, and when applied experimentally to cotton seeds only the gliovirin-producing strains controlled seedling diseases caused by *P. ultimum*. Conversely, gliotoxin is more active against *Rhizoctonia* than against *Pythium in vitro*, and gliotoxin-producing strains are better at controlling *Rhizoctonia* on seedlings (Howell *et al.* 1993).

Antibiotics are not the only factors in the antagonistic repertoire of *Trichoderma*. Many strains grow rapidly in culture – up to 25 mm 24 h^{-1} at room temperature. They secrete chitinase and β-1,3-glucanase when grown in the presence of other fungi or on fungal wall components. The hyphae of *Trichoderma* also coil round the hyphae of other fungi (Fig. 12.5) and eventually penetrate the hyphae from these coils. Few other fungi have such a formidable arsenal of antagonistic mechanisms.

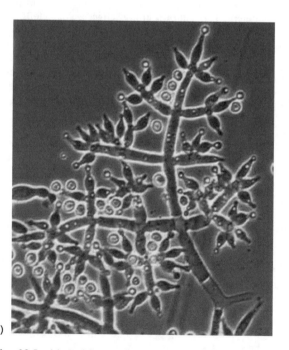

(a)

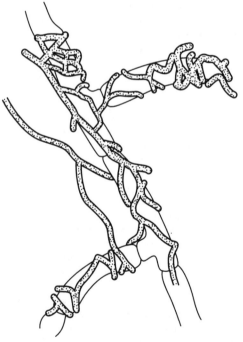

(b)

Fig. 12.5 (a) Aerial spore-bearing structures of *Trichoderma harzianum*: the condiophores branch at right angles and produce conidia at the tips of flask-shaped phialides. (b) *Trichoderma* species frequently coil round the hyphae of other fungi (in this case, *Rhizoctonia solani*) and penetrate or disrupt the host hyphae. ((a) Courtesy of Samuels, G.J, Chaverri, P., Farr, D.F. & McCray, E.B. Trichoderma Online, Systematic Botany and Mycology Laboratory, ARS, USDA; from http://nt.ars-grin.gov/taxadescriptions/keys/TrichodermaIndex.cfm)

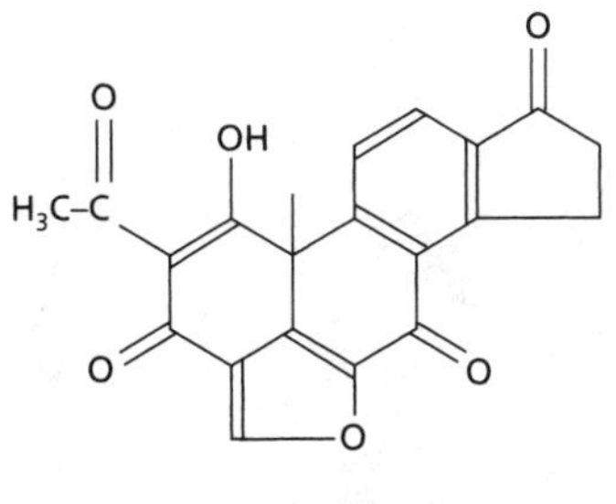

Fig. 12.6 Structures of some antibiotics produced by species of *Trichoderma*.

Viridin

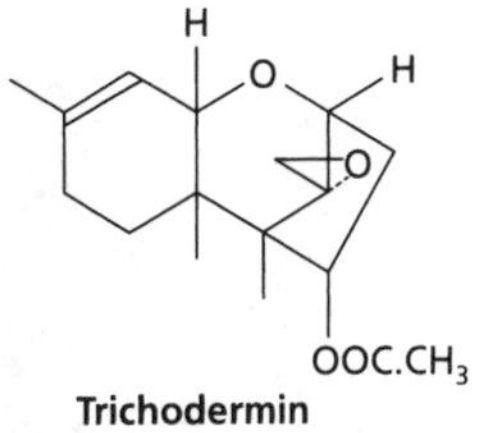

Trichodermin

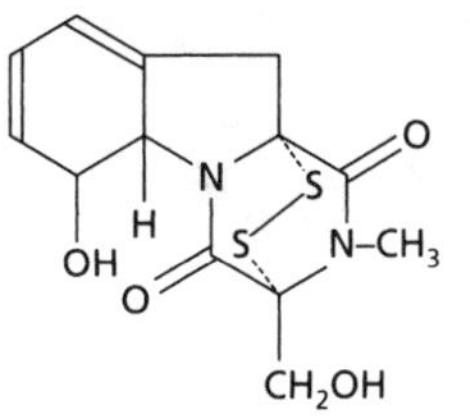

Gliotoxin

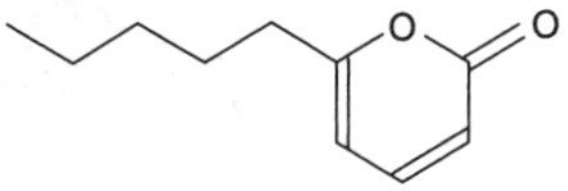

6*n*-Pentyl-2H-pyran-2-one (6-PAP)

Commercial biocontrol formulations of *Trichoderma*

Trichoderma species are easily grown in culture media and have been marketed in many formulations, with varying degrees of success. One example, among many, is the product **Trichodex™** (*T. harzianum*) for control of *Botrytis cinerea* (gray mould) on grapes. Although it was reported not to be as effective as commercial fungicides it can be used as an alternating treatment with fungicides (O'Neill *et al.* 1996), halving the total fungicide input, which is environmentally desirable. *Trichoderma* is also impregnated into dowels, which are hammered into drill-holes in trees to help control wood-rot fungi, and similarly can be inoculated into plum trees to help control silver leaf disease, caused by toxins of the fungus *Chondrostereum purpureum* (Basidiomycota) growing in the wood. A commercial product of this type, called **Trichodowels™**, is marketed through garden centers in New Zealand.

However, one role of *Trichoderma* spp. in biological control exceeds all others – the inoculation of soil-less rooting media in commercial glasshouses where seedlings are raised. Soil-less rooting media composed of peat-based or similar materials have a low resident microbial population, and so have little biological "buffering" capacity. If any seedling pathogens become established in these conditions they can spread rapidly and cause major damage. So, various formulations of *Trichoderma*, with wheat bran or other organic food bases, are incorporated into the rooting medium, to help establish an antagonistic microflora. One of the most successful commercial biocontrol strains, *Trichoderma harzianum* strain T-22, was developed by fusing the protoplasts of two "wild" strains, then allowing the hybrid strain to regenerate a wall and to revert to a stable phenotype. The resulting T-22 strain exhibits very strong "rhizosphere competence" – it colonizes the whole root system and persists throughout the life of a crop. This strain is marketed under the name **Bio-Trek** 22G™ for golf course turf, and under other names for field crops or greenhouse potting mixes. Novel strains generated by protoplast fusion do not involve genetic engineering, so they do not fall under the regulatory requirements for GM (genetically modified) products.

Hyphal interference

Hyphal interference is a specific term describing the behavior of several Basidiomycota that antagonize other fungi at points of contact (Fig. 12.7). This antagonism can be between different species of Basidiomycota, or between Basidiomycota and other fungi. It was discovered during *in vitro* studies on the

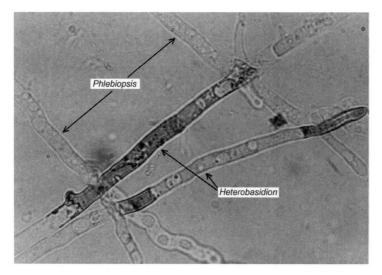

Fig. 12.7 Hyphal interference on an agar plate. Hyphae of *Heterobasidion annosum* (previously termed *Fomes annosus*) have been antagonized where they were contacted by single hyphae of *Phlebiopsis gigantea* (previously termed *Peniophora gigantea*). The agar plate was then flooded with a dilute solution of neutral red, which was not taken up by the undamaged hyphae but entered the damaged hyphae of *Heterobasidion* (seen as darker pigmentation).

interactions of fungi from herbivore dung (see Fig. 10.8) and it helps to explain how the Basidiomycota can ultimately dominate a fungal community, truncating the activities of other fungi (Ikediugwu 1976). It also explains why the Basidiomycota tend to be mutually exclusive. For example, toadstools of both *Coprinus heptemerus* and *Bolbitius vitellinus* occur on dung, but never in the same piece of dung because these two species antagonize one another (Ikediugwu & Webster 1970).

Hyphal interference occurs rapidly, a few minutes after hyphal contact, and often is localized to a single hyphal compartment. The first visible sign is vacuolation and loss of turgor in the affected hyphal compartment. If a dye such as neutral red is added at this stage it is taken into the affected compartment but does not enter the healthy hyphal compartments, indicating that hyphal interference causes loss of normal membrane integrity. In electron micrographs the affected cytoplasm is seen to be degenerate, the mitochondria are swollen, and a wide gap is seen between the retracted plasma membrane and the hyphal wall. The damage is often contained within a hyphal compartment by zones of dense, coagulated cytoplasm on either side of the contact point. The mechanism of hyphal interference is still unclear, but a poorly diffusible factor seems to be involved, because hyphal interference can occur between hyphae separated by a cellophane membrane up to 50 μm wide.

Hyphal interference is not a parasitic phenomenon, because there is no evidence of invasion of the damaged hyphae. Instead, it seems to be an efficient means of inactivating other hyphae that are potential competitors for the same substrates.

Control of pine root rot by hyphal interference

Soon after hyphal interference was discovered in the 1970s, it was recognized as a possible basis for explaining an important biological control system – the control of pine root rot by an antagonistic fungus, *Phlebiopsis gigantea*. Pine root rot is caused by *Heterobasidion annosum* (Basidiomycota), which is the most important pathogen of coniferous trees in the northern hemisphere, responsible for major economic losses in the forestry industry. This fungus grows slowly but progressively along the woody roots, rotting them and spreading from tree to tree by root-to-root contact. Eventually, it can spread into the base of the trunk, causing a **butt rot** which destroys some of the most valuable timber. When established in a site, *Heterobasidion* is almost impossible to eradicate, except by mechanical extraction of all the infected stumps and major roots. So, control measures have focused on preventing it from becoming established, especially in newly afforested sites (Fig. 12.8).

Heterobasidion produces air-borne basidiospores from bracket-shaped fruitbodies at the bases of infected trees. These spores pose little threat in undisturbed forests because they have insufficient food reserves to initiate infection of woody roots when washed into the root zone. However, the situation is different in commercial forestry, where trees are felled for harvest or thinned to create the desired plant density as the plantation develops. The tissues of the exposed stump surfaces can remain alive for several months, but with declining resistance to infection. These exposed stumps provide a highly selective environment for

(a)   (b)   (c)

(d)   (e)   (f)

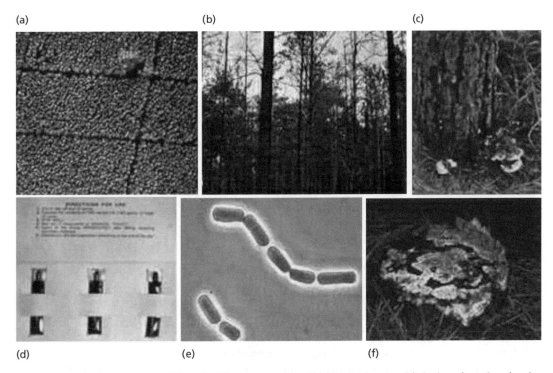

Fig. 12.8 Control of pine root rot (*Heterobasidion annosum*) by *Phlebiopsis gigantea*. (a) A pine plantation showing a large disease gap (near the top) caused by *H. annosum*. (b) Trees within this gap are spindly and dying. (c) Bracket-shaped fruiting bodies of *H. annosum* growing from the base of a heavily diseased tree. (d) Sachets containing spores of *P. gigantea*, which can be diluted with water and inoculated onto freshly exposed stump surfaces. (e) Brick-like spores of *P. gigantea* are produced abundantly in laboratory culture by fragmentation of the hyphae. (f) A pine stump 1 year after treatment with spore suspension of *P. gigantea*, showing abundant growth of *P. gigantea* over the stump surface. (Images courtesy of the late J. Rishbeth.)

pathogens like *Heterobasidion*, which can colonize from basidiospores and then grow down into the dying roots and infect the adjacent healthy trees (Fig. 12.9).

The simplest way to avoid this is to kill the stump surface tissues with phytotoxic chemicals such as urea or boron-containing compounds, enabling saprotrophs to rot the stumps and to exclude *Heterobasidion*. This is common practice in many forests, but is environmentally undesirable, especially if the forests are in catchment areas for domestic water supplies. Rishbeth (1963) developed an alternative control method in which spores of *Phlebiopsis gigantea* were applied to exposed stump surfaces immediately after the trees had been felled. *P. gigantea* is a weakly parasitic fungus that rapidly colonizes the stump tissues of pine trees, but poses no threat to healthy trees. It then grows down into the major roots and prevents *Heterobasidion* from becoming established. This type of biological control is termed **pre-emptive (competitive) niche exclusion.**

We will see other examples of this on leaf and fruit surfaces, later in this chapter.

Hyphal interference also seems to be involved in the control of *Heterobasidion*, because the hyphae of *P. gigantea* antagonize *Heterobasidion* on contact (Fig. 12.7). As shown in Fig. 12.9, *P. gigantea* might have multiple roles in this biocontrol system, helping to protect fresh stump surfaces, helping to contain any existing pockets of infection by *Heterobasidion* in the root zone, and potentially preventing the pathogen from growing up to the stump surface and sporulating there.

In developing this practical biocontrol method, Rishbeth made use of the fact that *P. gigantea* is one of the relatively few Basidiomycota that sporulate readily in laboratory culture. The hyphae fragment behind the colony margin to produce many brick-shaped conidia, which can be used as inoculum to apply to stump surfaces. In the initial biocontrol formulations, *Phlebiopsis* spores were packaged in sachets containing sucrose solution at a water potential that prevented

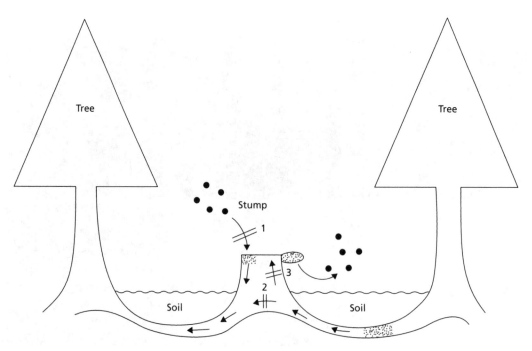

Fig. 12.9 Possible mode of action of *Phlebiopsis gigantea* in controlling *Heterobasidion annosum* in recently felled pine stumps. 1, *P. gigantea* is inoculated onto freshly exposed stump surfaces and prevents colonization of the stumps from airborne basidiospores of *H. annosum* (shown by the double lines). 2, When *P. gigantea* has colonized the dying stump and root tissues it prevents the spread of *H. annosum* from existing foci of infection in the root zone. 3, *P. gigantea* might also prevent *H. annosum* from growing up to the stump surface to sporulate.

spore germination, and also containing a dye (bromocresol purple). For application to stump surfaces, the sachets were diluted with water, allowing the spores to germinate, and the dye enabled foresters to see that the stump surfaces had been treated. More recently, the spores have been produced as dry powder formulations by commercial companies.

This biocontrol system proved highly effective in pine forests over much of Europe and North America. But in Britain it had one significant limitation – *P. gigantea* could be used only to protect **pine stumps** and was not effective in protecting **Sitka spruce**, which is the most common plantation tree in Britain. Now it seems that there is a simple explanation for this: the strains of *P. gigantea* used to control the disease in Britain are endemic strains, specialized to colonize pine stumps, whereas the strains used in Scandinavia have a broader host range and can protect spruce as well as pines. In Finland a commercial preparation of *P. gigantea*, marketed as **Rotstop™**, effectively controls *H. annosum* on spruce. The UK Forestry Commission is currently investigating the possibility of introducing such strains into Britain.

Mycoparasites: fungi that parasitize other fungi

The fungi that parasitize other fungi can be grouped into two broad categories: **necrotrophic mycoparasites**, which invade and destroy other fungal cells and then feed on the dead cell contents, and **biotrophic mycoparasites**, which can establish a specialized feeding relationship, usually by producing haustoria to penetrate and absorb nutrients from living fungal hyphae. These two types of mycoparasite are equivalent to the necrotrophic and biotrophic parasites of plants, discussed in Chapter 14.

Biotrophic mycoparasites

There are several types of biotrophic mycoparasite with different feeding mechanisms (Jeffries & Young 1994), but the most common and distinctive group are the **haustorial biotrophs**. These fungi penetrate living host hyphae to produce a haustorium inside the host wall but separated from the cell contents by a host cell

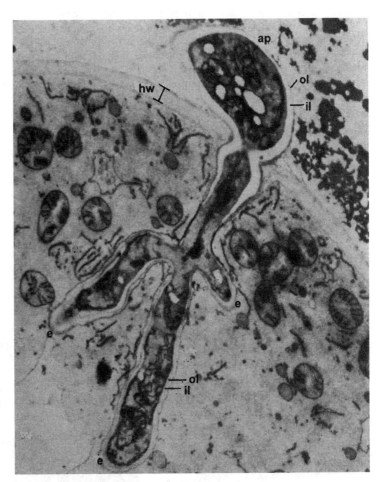

Fig. 12.10 Electron micrograph of an appressorium (ap) and a branched haustorium of the mycoparasite *Piptocephalis unispora* (Zygomycota) in a fungal host, *Cokeromyces recurvatus.* hw = host wall; ol and il = outer layer and inner layer of the *Piptocephalis* wall. The haustorium is surrounded by a continuous membrane (the extra-haustorial membrane, labelled e). (Courtesy of P. Jeffries; from Jeffries & Young 1976.)

membrane (Fig. 12.10). The parasite draws nutrients from the host hyphae, and uses these nutrients to produce sporulating structures on the host colony. Often this type of parasitism causes little damage, as long as the host fungus has an adequate food supply.

Biotrophic mycoparasitism similar to that in Fig. 12.10 is found in several Zygomycota (*Piptocephalis, Dispira,* and *Dimargalis* spp.) that have elongated, few-spored sporangia (merosporangia; see Fig. 2.11). With only few exceptions, these fungi parasitize other Zygomycota such as *Mucor* and *Pilaira* on dung or in soil. Most of these biotrophic mycoparasites can be grown in laboratory media containing extracts of host or nonhost hyphae. The need for hyphal extracts can be replaced by relatively high concentrations of vitamins (especially thiamine) and amino acids, and by providing glycerol instead of glucose as the carbon source. These nutrients could be expected to occur in the host hyphae.

The haustorial biotrophs seem to depend entirely on their fungal hosts in nature. Their spores are triggered to germinate near host hyphae, and the germ-tubes show pronounced tropism towards the host. Then the germ-tube tip produces an appressorium on the host surface and a penetration peg enters the host to form a haustorium. The mycoparasite *Piptocephalis virginiana* shows evidence of specific recognition in the infection process (Manocha & Chen 1990). It infects only some members of the order Mucorales, and the walls of these hosts were found to have two surface-located glycoproteins. Removal of these glycoproteins by treating hyphae with NaOH or proteinase led to impaired attachment and appressorium formation. The treated hyphae (with their glycoproteins removed) were found to bind lectins that recognize fucose, *N*-acetylgalactosamine, and galactose, and this binding pattern of the treated hyphae was identical to that of nonhost hyphae. However the *untreated* host hyphae did not bind these lectins, so it is suggested that the glycoproteins that occur normally on host hyphae cover the fucose, *N*-acetylgalactosamine, and galactose residues (which somehow interfere with infection), and enable the parasite to attach and infect. In support of this, the normal hosts are found to be resistant to parasitism if they are grown in liquid culture media, but susceptible when grown on agar plates. The hyphae from liquid culture were found to have the three sugar residues on their surface, presumably because the glycoproteins that cover them were not produced in liquid culture conditions.

Potential control of potato black scurf by Verticillium biguttatum

The mitosporic fungus *Verticillium biguttatum* is a biotrophic mycoparasite with potential for use as a biocontrol agent of the specific strains of *Rhizoctonia solani* (anastomosis group AG4) that cause black scurf disease of potatoes. This disease is characterized by the familiar black or brown crusts that develop on the surface of potato tubers. The crusts are sclerotial masses of *Rhizoctonia* and they significantly reduce the marketability of the crop even though the damage is only superficial. Like all biotrophs, *V. biguttatum* has relatively little effect on the growth of its fungal host. It penetrates from germinating spores, produces haustoria within the host hyphae (Fig. 12.11), and then forms a limited mycelium outside the host, where it sporulates. The infected host hyphae grow more or less unimpeded, but even a localized infection of a host colony can markedly suppress the production of

sclerotia over the colony. Apparently *V. biguttatum* creates a nutrient sink towards itself within the host mycelial network, counteracting the normal nutrient sink towards developing sclerotia.

V. biguttatum can be grown easily on normal laboratory media, and it produces abundant spores for use in biocontrol. These spores have been shown to reduce black scurf in field experiments, but there is one significant limitation: *V. biguttatum* needs a relatively high temperature (at least 12°C) for growth, whereas *Rhizoctonia* can start to grow at about 4°C so it colonizes potato plants earlier in the growing season, before the mycoparasite can grow. This problem would not be important if potato tubers could be inoculated later in the growing season, because sclerotia are produced only when the potato skins begin to mature. Agronomists in the Netherlands are exploring potential ways of doing this by lifting the potato tubers before the skins mature, inoculating the tubers with *V. biguttatum*, and then returning the potatoes to the soil until the skins mature – a technique termed "green crop harvesting" (van den Boogert *et al.* 1994).

Ampelomyces quisqualis, *a biotrophic mycoparasite of powdery mildew fungi*

Ampelomyces quisqualis is one of several naturally occurring biotrophic mycoparasites that are being

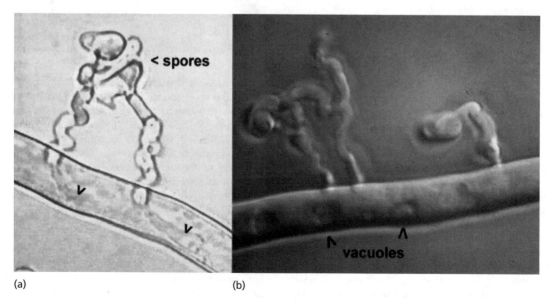

(a)   (b)

Fig. 12.11 Mycoparasitism of *Rhizoctonia solani* (anastomosis group 4) by *Verticillium biguttatum*. (a) Spores of *V. biguttatum* germinate near the host hyphae and grow towards the host in a spiral manner, then penetrate the living hyphae to produce club-shaped haustoria (arrowheads). (b) A similar image, showing that the parasitized host hyphae are still alive; they have intact vacuoles, and videotaped sequences show active protoplasmic streaming. (From van den Boogert & Deacon 1994.)

developed for commercial control of the damaging powdery mildew fungi in glasshouse crops. The host range of *Ampelomyces* is restricted to the powdery mildew fungi (Chapter 14) but it infects many of these and is found in at least 28 countries around the world. It has the potential to control several important diseases, including powdery mildew of grapes, cucumbers, strawberries, tomatoes, and ornamentals. *Ampelomyces* produces flask-shaped pycnidia (see Fig. 2.33) that contain many asexual spores, which are extruded from a pore and are dispersed by water splash. These spores germinate and directly infect the hyphae, conidiophores or immature sexual structures (cleistothecia) of the powdery mildew hosts. *Ampelomyces* does not kill the host tissues initially, but as the powdery mildew colonies grow it produces pycnidia over the colony surface and at this stage the host hyphae are killed. Several cycles of infection can occur in a single growing season, leading to significant control of the powdery mildew hosts.

This mycoparasite survives as pycnidia between crops or growing seasons, because it is unable to grow in the absence of an appropriate host fungus. This is one of the practical limitations of exploiting *Ampelomyces* for disease control, because the powdery mildew fungi must always be present, resulting in at least some crop damage. But this can be minimized by applying spore formulations of *A. quisqualis* early in the growing season. The fungus is marketed commercially, under the trade name AQ10™.

Necrotrophic mycoparasites

Necrotrophic mycoparasites are quite different from the biotrophs discussed above. They have characteristically wide host ranges and in many cases they produce inhibitory toxins or other metabolites as part of the parasitic process. Below, we consider three examples to illustrate these points.

Antagonism by Clonostachys rosea (formerly Gliocladium roseum)

Clonostachys rosea, formerly known by its asexual stage *Gliocladium roseum*, is a very common soil fungus that produces abundant asexual spores. Its sexual stage is *Bionectria ochroleuca*, although this is much less widespread than the asexual stage. *C. rosea* colonizes organic matter in soil and frequently overgrows other fungi on agar plates. In fact, the easiest way to isolate this fungus is to sprinkle crumbs of soil onto the colonies of other fungi: *C. rosea* then produces its typical sporing structures in 8–10 days (Table 12.1). In nutrient-rich conditions, *C. rosea* can kill some fungi by producing diffusible inhibitors, but on water agar it antagonizes the hyphae by causing localized

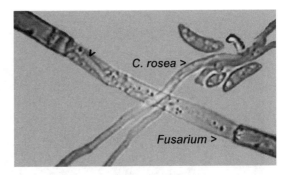

Fig. 12.12 Antagonism of a hyphal compartment of *Fusarium oxysporum* by *Clonostachys rosea*. The damaged compartment has lost its turgor but the adjacent compartments (top left and bottom right) are unaffected. A narrow *Fusarium* hypha (arrowhead) has regrown into the damaged compartment from a septal pore. (Courtesy of V. Krauss.)

vacuolation and loss of turgor about 30–90 minutes after contact (Fig. 12.12). This is slower than hyphal interference by Basidiomycota, which can take only a few minutes, but *C. rosea* also branches and coils round the damaged compartments, so it seems to use the host hyphae as a nutrient source. In nature *C. rosea* is probably a secondary (opportunistic) invader of decomposing organic matter, gaining some of its nutrients by antagonizing living hyphae, some by exploiting dead hyphae, and some from the underlying substratum.

A fungus called *Gliocladium catenulatum* is very closely related to *Clonostachys rosea* (and probably should be reclassified as *Clonostachys*). A strongly antagonistic strain of this fungus is marketed by a company in Finland (Kemira Agro OY) as a biological control agent of several common plant diseases such as damping off, seed rot, root rot, and stem rots.

Antagonism by Talaromyces flavus

Talaromyces flavus (Ascomycota) has a *Penicillium*-like asexual stage. It first attracted attention as a parasite of *Rhizoctonia solani* because it coils round *Rhizoctonia* hyphae on agar plates, penetrating the hyphae and causing localized disruption. However recent interest has focused on its potential to control the vascular wilt pathogen *Verticillium dahliae* (see Chapter 14). It invades the melanized microsclerotia of *Verticillium* on diseased roots, and sporulates on the surface of the microsclerotia. *T. flavus* produces up to four antibiotics in culture, one of them being an antifungal compound, **talaron**. However, it exerts its main effect – at least on the hyphae of *V. dahliae* – by secreting the enzyme **glucose oxidase**, which generates H_2O_2 from glucose (Kim *et al.* 1988). Consistent with this,

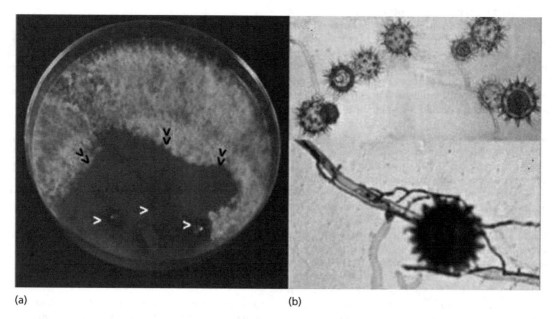

(a)   (b)

Fig. 12.13 (a) A method for detecting *Pythium oligandrum* or other mycoparasites in soil. Three soil crumbs (white arrowheads) were placed on an agar plate previously colonized by a soil fungus, *Phialophora*. Within 4 days, *P. oligandrum* started to grow across the *Phialophora* colony, collapsing the aerial hyphae (black arrowheads) and producing spiny-walled oogonia (top right). (b) Narrow hyphae of *P. oligandrum* attacking and destroying a *Phialophora* hypha, leading to production of a spiny-walled oogonium.

Talaromyces was found to antagonize *V. dahliae* only when glucose was present in culture media, and not when other common sugars were used. A commercial source of glucose oxidase had the same effect on hyphae of *V. dahliae*, but only in the presence of glucose; and these effects could be reproduced by low concentrations of H_2O_2 alone. It seems that *Talaromyces* has several potential mechanisms for antagonizing other fungi; they might act separately or in combination.

Mycoparasitic Pythium *species*

Most *Pythium* species are aggressive plant pathogens that cause economically damaging seedling diseases and root rots (Chapter 14). But six *Pythium* species are non-pathogenic to plants and instead they are aggressive parasites of other fungi. In Britain the most common of these is *Pythium oligandrum* which has distinctive spiny-walled oogonia (Fig. 12.13). Like *Trichoderma* and *Clonostachys*, discussed earlier, *P. oligandrum* can grow from soil crumbs sprinkled onto agar plates previously colonized by other fungi. Using this technique as a method of detection, *P. oligandrum* has been found in nearly half of the soils sampled in Britain, especially from agricultural sites. *P. oligandrum* and similar fungi (e.g. *P. acanthicum* and *P. periplocum*) are equally common in the USA, continental Europe, and New Zealand.

Mycoparasitism by *Pythium oligandrum* has been studied in detail by videomicroscopy (Fig. 12.14). In the sequence shown, a hyphal tip of the plant pathogen *Botrytis cinerea* contacted a hypha of *P. oligandrum* and then lysed within 3 minutes, spilling the contents of the *Botrytis* hypha. *P. oligandrum* then produced many hyphal branches in the lysate and penetrated the *Botrytis* hypha. Many similar sequences, involving different host fungi and different mycoparasitic *Pythium* species, are described in Laing & Deacon (1991).

Some fungi are not invaded by *P. oligandrum* until 1 or 2 hours after contact, and during this time the mycoparasite coils round the host hypha from branches that originate at the initial contact point. *P. oligandrum* then penetrates from beneath the coils. This is commonly seen in interactions with plant-pathogenic *Pythium* spp. and in interactions with mature hyphae of *Rhizoctonia solani* (Fig. 12.15).

Detection of mycoparasites on precolonized agar plates

Necrotrophic mycoparasites are easily detected and isolated from soil using the precolonized plate method (Fig. 12.13). This reveals an interesting feature, because different types of mycoparasite that occur naturally in soil are detected on different types of precolonized agar

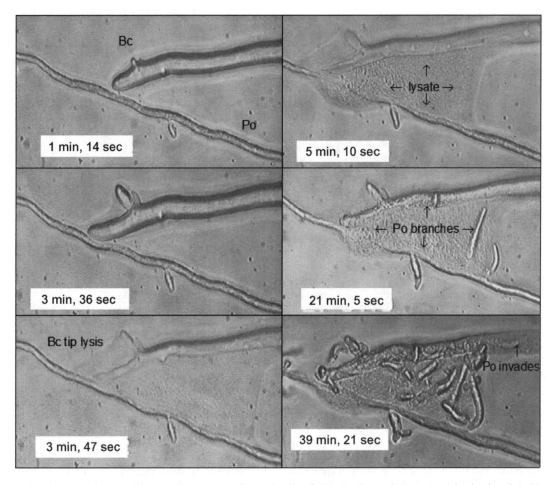

Fig. 12.14 Frames from a videotaped sequence when a hypha of *Botrytis cinerea* (Bc) contacted a hypha of *Pythium oligandrum* (Po) on a film of water agar. The times shown are from the start of recording (0 minutes).

Fig. 12.15 Frame from a videotaped sequence when *Pythium oligandrum* (growing from top right) contacted a hypha of *Rhizoctonia solani*. The *Pythium* hypha grew on, but within 5 minutes it branched at the contact point and coiled round the *Rhizoctonia* hypha. At about 49 minutes after the initial contact, *P. oligandrum* penetrated the host hypha and caused the protoplasm to coagulate.

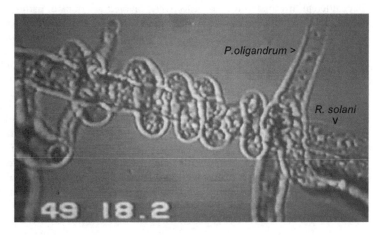

Table 12.1 The ability of different mycoparasites to grow across colonies of other fungi on agar plates. (Data from Mulligan & Deacon 1992.)

Mycoparasite	Precolonizing fungus	Number of soils yielding the mycoparasite (max. 28)
Pythium oligandrum	Fusarium culmorum	18
	Trichoderma aureoviride	3
	Rhizoctonia solani	0
	Botrytis cinerea	12
Clonostachys rosea	F. culmorum	20
	T. aureoviride	27
	R. solani	26
	B.cinerea	25
Trichoderma harzianum	F. culmorum	2
	T. aureoviride	0
	R. solani	22
	B. cinerea	13
Papulaspora sp.	F. culmorum	0
	T. aureoviride	11
	R. solani	3
	B. cinerea	19

Note: Soils were collected from 28 different locations in Britain and were stored air-dried. Twenty-eight plates of potato-dextrose agar were inoculated with *Fusarium culmorum*, and similarly 28 plates were inoculated with *Trichoderma aureoviride, Rhizoctonia solani*, or *Botrytis cinerea*. When these fungi had covered the agar plates a small amount of air-dried soil was added, and the plates were incubated to detect any mycoparasites that grew across the fungal colonies. All the mycoparasites either grew completely across the established fungal colonies or not at all. The results show that four mycoparasites are common in soil (detected in at least 18 of the 28 soils tested). But, with the exception of *Clonostachys rosea* which grew across any fungus, the different mycoparasites are selectively favored by different "host" fungi.

plate (Table 12.1). For example, in the top data line of the table, *Pythium oligandrum* was detected in 18 (of a total 28) different soil samples that were sprinkled onto agar precolonized by *Fusarium culmorum*, but *P. oligandrum* was detected on only three plates precolonized by *Trichoderma*, and on none of the agar plates precolonized by *Rhizoctonia*. These data indicate that different mycoparasites are selectively favored by different "host" fungi – at least in laboratory conditions.

A mycoparasite unique to Britain!

Well, that's not exactly true. There **is** a mycoparasite that has only ever been recorded in Britain, and it is very common. *Pythium mycoparasiticum* (see Fig. 10.2) is a relatively slow-growing parasite that has been recorded frequently on precolonized agar plates, but only if the soil is serially diluted with sand to dilute out the spores of more aggressive mycoparasites, which outcompete *P. mycoparasiticum* in "raw" soil (Deacon *et al.* 1991). It almost certainly occurs in many countries, and this example serves to illustrate

that other fungi await discovery by using novel isolation techniques.

Pythium oligandrum *as a biological control agent*

Pythium oligandrum and the other mycoparasitic *Pythium* spp. have the potential to be used for biocontrol of plant pathogens, especially as seed treatments to control seedling pathogens. In this respect, *P. oligandrum* grows rapidly on cheap commercial substrates such as molasses, and it produces large numbers of oospores in shaken liquid culture. These oospores have been applied to seeds of cress, sugar beet, and other crops, using a commercial seed-coating process, and were reported to give good protection against seedling disease caused by *Pythium ultimum* in experimental conditions (McQuilken *et al.* 1990). A commercial powder formulation of *P. oligandrum* has been developed in the former Czech Republic. However, it is not used widely, if at all, because of technical problems. The major problem is that the oospores of *P. oligandrum* germinate slowly and poorly because they exhibit constitutive dormancy (Chapter 10). Only a maximum

10–40% of oospores freshly harvested from culture broth will germinate readily, and this value drops to 5% or less when the oospores are dried for storage, making commercial production uneconomic. *P. oligandrum* and other mycoparasitic *Pythium* spp. could only be used as commercial biocontrol agents if methods can be developed to ensure consistently high oospore germination. However, this does not preclude a **natural role** for *P. oligandrum* as a biocontrol agent. There is strong evidence that *P. oligandrum* helps to suppress disease caused by *Pythium ultimum* in the irrigated cotton fields of California. It seems to do this by competing with *P. ultimum* for the crop residues at the end of the growing season, preventing *P. ultimum* from using these residues to build up its inoculum level in the soil (Martin & Hancock 1986).

Competitive interactions on plants

Competition, as defined at the start of this chapter, embraces all the types of interaction in which one organism gains ascendency over another because it arrives sooner, grows faster, uses the substrate more efficiently, etc. Competition is probably the most common type of interaction in natural environments, but this is difficult to prove unless all other potential mechanisms can be discounted. Here we consider a few cases where the evidence for competition is strong.

Control of the take-all fungus by *Phialophora graminicola*

As we saw in Chapter 9 (see Fig. 9.11), the take-all fungus *Gaeumannomyces graminis* attacks the roots of cereals and grasses by growing as dark runner hyphae and invading the vascular tissues of the root. There are three forms of this fungus, termed varieties. *G. graminis* var. *tritici* is a major pathogen of wheat and barley roots; *G. graminis* var. *graminis* is often non-pathogenic; and *G. graminis* var. *avenae* attacks oat crops and the fine turf grasses (*Agrostis* spp.) used on golf-course greens, bowling greens, and other fine playing surfaces (Fig. 12.16).

The **take-all patch disease** of fine turf grasses, caused by *G. graminis* var. *avenae*, is rarely seen because it occurs only in specific circumstances:

1 When the existing turf is heavily limed to correct for over-acidity.
2 When new turf is established in sites that have been chemically fumigated or fungicide-treated to destroy pests and diseases.
3 When natural turf on golf course greens is removed and replaced by new grass, laid on a bed of sand.

All three of these circumstances have one factor in common – the natural, **resident microbial antagonists**, which normally prevent take-all patch disease from developing, are reduced or destroyed. For example, over-acidity (case 1 above) is caused by the repeated use of ammonium fertilizers, applied year after year, to maintain the greenness of the turf. The ammonium ion is absorbed by roots, and H$^+$ ions are released in exchange, leading to a progressive lowering of the turf pH. Eventually the turf pH is so low that the grass no longer responds to added fertilizer. The turf is then heavily limed to restore the pH to near-neutrality. This leads to serious outbreaks of take-all patch disease (Fig. 12.16), initiated by air-borne ascospores of

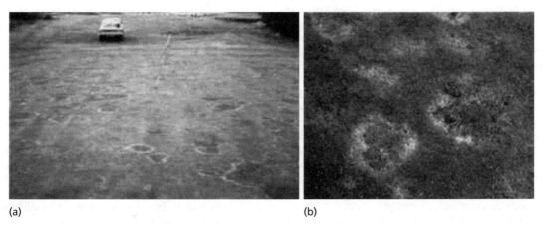

(a)   (b)

Fig. 12.16 (a,b) Take-all patch disease of turf on a golf-course green, caused by *G. graminis* var *avenae*. Note the poor quality of the turf, where the fine turf grasses have been killed by the fungus and have been replaced by weeds and course-leaved grasses.

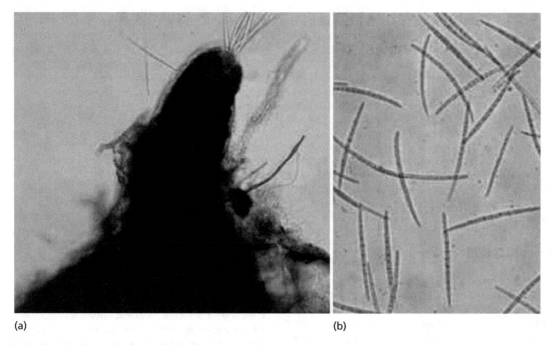

(a)   (b)

Fig. 12.17 (a,b) The neck of a perithecium of *G. graminis* var *avenae*, with long (c. 80 μm) ascospores that are released from the apical ostiole and can infect turf grasses that are not protected by natural antagonists.

G. graminis var. *avenae* (Fig. 12.17). These spores would not have sufficient nutrient reserves to infect if they had to compete with the normal resident microbial antagonists. But these antagonists are pH-sensitive, and have been largely eliminated by the very low turf pH.

Chemical fumigants and fungicides (case 2) act in a similar way, because they progressively reduce the population of resident antagonists. And, sand beds (case 3) have a naturally low resident population of antagonists, because of their low organic matter content. The disease patches then spread rapidly for 2 or 3 years, killing the fine-leaved turf grasses and enabling the coarser grass species and weeds to invade. Then the patches stop spreading, but the quality of the playing surface is seriously damaged.

Two different types of natural antagonist have been implicated in the control of take-all patch disease. First, antibiotic-producing fluorescent pseudomonads, which are known to suppress *G. graminis* in turf of near-neutral pH. Second, the fungus *Phialophora graminicola*, which is a nonpathogenic invader of the grass root cortex (Fig. 12.18) and also is known to suppress *G. graminis* in turf of near-neutral pH (Deacon 1973).

In summary, the key to preventing the take-all patch disease of grass turf lies in the management of turf pH, which must be maintained at a level that supports the natural antagonists, *Phialophora*, and/or fluorescent

pseudomonads. This example illustrates that competition for a resource (**competitive niche exclusion**) can be a highly effective means of disease control.

Frost damage, and postharvest decay of fruit

Leaf and fruit surfaces support a characteristic population of epiphytic bacteria and yeasts. Among these are strains of *Pseudomonas syringae* that have natural surface proteins with ice-nucleation activity. The presence of these bacteria on plant surfaces promotes the formation of ice crystals when the air temperature falls even slightly below 0°, resulting in frost damage. If these bacteria are absent then the temperature can cool to −4° or lower before icing occurs, and this difference can be important for frost-sensitive plants. Lindow and his co-workers (see Lindow 1992) showed that mutant strains (ice⁻) of *P. syringae* lacking the surface protein can be pre-inoculated onto leaves to exclude the wild-type (ice⁺) strains and give significant protection against frost injury – another example of **pre-emptive competitive niche exclusion**.

Other strains of *P. syringae*, and strains of the yeast *Candida oleophila*, have been exploited to protect against postharvest damage to fruit. Even slight

Fig. 12.18 Colonization of cereal or grass roots by *Phialophora graminicola*, a nonpathogenic parasite of roots that can effectively suppress invasion by the aggressive take-all fungus. (a) Dark runner hyphae of *Phialophora* in a grass root cortex. (b) Clusters of darkly pigmented cells of *Phialophora* within the root cortex. (c) Heavy invasion of the cortex of a cereal root by *Phialophora*, but the vascular tissues are not invaded. (d) At higher magnification, *Phialophora* hyphae have colonized the root cortex, but the endodermis (e) prevents invasion of the vascular tissues.

damage to the fruit surface during processing and packing can create portals for entry of decay fungi (*Botrytis cinerea*, *Penicillium* spp., *Mucor* spp., etc). To counteract this, biocontrol strains can be applied to fruit surfaces and are reported to give significant protection by colonizing minor wounds. Like the ice⁻ strains of *P. syringae*, they act by competing for nutrients. Currently, at least two companies market these products in the USA, using different protectant strains. The products based on strains of *P. syringae* are marketed for control of postharvest decay of apples and pears, under the tradename **Biosave 110**, while **Biosave 100** is marketed to control decay of citrus fruits. A product based on a strain of *Candida oleophila* is marketed under the tradename **Aspire**, to control *Penicillium* spp. on citrus fruit, and *B. cinerea* and some *Penicillium* spp. on apples. Fruit treated with these naturally occurring strains is exempt from the normal "tolerance" levels that conventional fungicides are subject to.

Commensalism and mutualism

If we exclude lichens as organisms in their own right (Chapter 13), there are few well-established examples of fungi that live in association with other fungi, to their mutual benefit (**mutualism**) or to the benefit of one and not to the detriment of the other (**commensalism**). These types of association might be common, but they are not well documented. Here we consider one example, from laboratory experiments on the interaction between *Thermomyces lanuginosus* (a noncellulolytic fungus of composts) and the cellulose-

degrading fungus, *Chaetomium thermophile*. It extends the example discussed earlier (see Fig. 11.13).

Both *T. lanuginosus* (T.l.) and *C. thermophile* (C.t.) grow during the prolonged high temperature phase of composts, when much of the cellulose is degraded (Chapter 11). It is assumed that T.l. grows in these conditions by using sugars made available by C.t. or other cellulolytic fungi. But this is difficult to demonstrate in the complex environment of a compost, so most of the evidence has come from *in vitro* studies. The methods used are similar to those described in Chapter 11. Flasks containing sterile filter paper with nitrate and other mineral nutrients were inoculated with C.t. alone, T.l. alone, or C.t. + T.l. The flasks were incubated for up to 7 weeks at 45°, and the loss of dry weight of the flask contents was determined (Fig. 12.19).

Thermomyces could not grow alone, because it cannot degrade cellulose and it cannot use nitrate as a nitrogen source. By contrast, *Chaetomium* grew well when inoculated alone, degrading about 1000 mg of cellulose after 4 weeks, and about 1200 mg after 7 weeks. But the combination of *Chaetomium* and *Thermomyces* gave an even larger weight loss – nearly 1500 mg after 4 weeks and over 1600 mg after 7 weeks. All these differences were statistically significant.

In addition to using a "standard" level of nitrogen (giving a C : N ratio of 174 : 1), this experiment included a double level of nitrogen (C : N of 88 : 1). The effect of this was to allow the rate of decomposition to continue for longer, so that after 7 weeks *Chaetomium* alone had degraded 2000 mg (of the original 7000 mg) of filter paper, while the combination

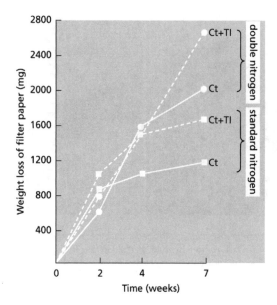

Fig. 12.19 Weight losses of filter paper (originally 7 g dry weight) in flasks inoculated with *Chaetomium thermophile* alone (Ct) or *C. thermophile* with the noncellulolytic fungus *Thermomyces lanuginosus* (Ct+Tl) and incubated at 45°C. Mineral nutrients were supplied in the flasks with either a standard amount of nitrogen (C : N ratio 174 : 1) or double nitrogen (C : N ratio 88 : 1). (Data from Deacon 1985.)

of *Chaetomium* and *Thermomyces* resulted in a 2700 mg weight loss.

The interpretation of even simple experiments like these is difficult, but a number of points can be made:

- The losses in dry weight of the flask contents actually underestimate the amount of cellulose degraded, because some of the breakdown products of cellulose are converted into fungal biomass (which remains in the flasks).
- *Thermomyces* grew well in association with *Chaetomium*. It was seen to grow over the filter paper and produced its characteristic spores. So, in some way *Thermomyces* must have obtained both a carbon source (sugars) and a nitrogen source from the association. Yet there was no evidence of parasitism, and this was confirmed by study of hyphal interactions on agar.
- *Thermomyces* in some way enhanced the breakdown of cellulose by *Chaetomium*. This might be expected from knowledge of the regulation of cellulase enzymes (Chapter 6), because any sugars that accumulate would slow the rate of enzyme action and also

repress the synthesis of further cellulases. So, by using some of these sugars *Thermomyces* might have relieved this negative feedback.
- *Thermomyces* must also have influenced the efficiency of nitrogen usage so that nitrogen was available to maintain the rate of cellulose breakdown for longer than when *Chaetomium* was growing alone. We saw in Chapter 11 that *Chaetomium* soon becomes nitrogen-limited, presumably because it cannot recycle nitrogen efficiently.

It would be interesting to know if *Chaetomium* benefits in some way from the interaction, perhaps by increasing its own biomass or by staving off its replacement by other fungi in composts. However, there is no evidence on this point, so this example must be described as one of commensalism. At the least, this example shows that fungi do not always have negative impacts on one another.

Online resources

Database of Microbial Biopesticides (DMB). http://www.ippc.orst.edu/biocontrol/biopesticides/
Weeden, C.R., Shelton, A.M., Li, Y. & Hoffman, M.P., eds. Biological Control: a guide to natural enemies in North America. Cornell University. http://www.nysaes.cornell.edu/ent/biocontrol/

General texts

Cook, R.J. & Baker, K.F. (1983) *The Nature and Practice of Biological Control of Plant Pathogens.* American Phytopathological Society, St Paul, Minnesota.
Deacon, J.W. (1983) *Microbial Control of Plant Pests and Diseases.* Van Nostrand Reinhold, Wokingham.
Gnanamanickam, S.S., ed. (2002) *Biological Control of Crop Diseases.* Marcel Dekker, New York.

Cited references

Chin-A-Woeng, T.F.C., Bloemberg, G.V. & Lugtenberg, B.J.J. (2003) Phenazines and their role in biocontrol by *Pseudomonas* bacteria. *New Phytologist* **157**, 503–523.
Deacon, J.W. (1973) Factors affecting occurrence of the *Ophiobolus* patch disease of turf and its control by *Phialophora radicicola*. *Plant Pathology* **22**, 149–155. [Note: *Ophiobolus* is now called *Gaeumannomyces*; *P. radicicola* is now called *P. graminicola*.]
Deacon, J.W. (1985) Decomposition of filter paper cellulose by thermophilic fungi acting singly, in combination, and in sequence. *Transactions of the British Mycological Society* **85**, 663–669.
Deacon, J.W., Laing, S.A.K. & Berry, L.A. (1991) *Pythium mycoparasiticum* sp. nov., an aggressive mycoparasite from British soils. *Mycotaxon* **62**, 1–8.

Howell, C.R., Stipanovic, R.D. & Lumsden, R.D. (1993) Antibiotic production by strains of *Gliocladium virens* and its relation to the biocontrol of cotton seedling diseases. *Biocontrol Science and Technology* 3, 435–441.

Ikediugwu, F.E.O. (1976) The interface in hyphal interference by *Peniophora gigantea* against *Heterobasidion annosum. Transactions of the British Mycological Society* 66, 291–296. [See also pp. 281–290; note that *Peniophora* is now called *Phlebiopsis*.]

Ikediugwu, F.E.O. & Webster, J. (1970) Antagonism between *Coprinus heptemerus* and other coprophilous fungi. *Transactions of the British Mycological Society* 54, 181–204.

Jeffries, P. & Young, T.W.K. (1976) Ultrastructure of infection of *Cokeromyces recurvatus* by *Piptocephalis unispora* (Mucorales). *Archives of Microbiology* 109, 277–288.

Jeffries, P. & Young, T.W.K. (1994) *Interfungal Parasitic Relationships*. CAB International, Wallingford.

Kim, K.K., Fravel, D.R. & Papavizas, G.C. (1988) Identification of a metabolite produced by *Talaromyces flavus* as glucose oxidase and its role in the biocontrol of *Verticillium dahliae. Phytopathology* 78, 488–492.

Laing, S.A.K. & Deacon, J.W. (1991) Video microscopical comparison of mycoparasitism by *Pythium oligandrum, P. nunn* and an unnamed *Pythium* species. *Mycological Research* 95, 469–479.

Lindow, S. (1992) Ice⁻ strains of *Pseudomonas syringae* introduced to control ice nucleation active strains on potato. In: *Biological Control of Plant Diseases; progress and challenges for the future* (Tjamos, E.C., Papavizas, G.C. & Cook, R.J., eds), pp. 169–174. Plenum Press, New York.

McQuilken, M.P., Whipps, J.M. & Cooke, R.C. (1990) Control of damping-off in cress and sugar-beet by commercial seed-coating with *Pythium oligandrum. Plant Pathology* 39, 452–462.

Manocha, M.S. & Chen, Y. (1990) Specificity of attachment of fungal parasites to their hosts. *Canadian Journal of Microbiology* 36, 69–76.

Martin, F.M. & Hancock, J.G. (1986) Association of chemical and biological factors in soils suppressive to *Pythium ultimum. Phytopathology* 76, 1221–1231.

Mulligan, D.F.C. & Deacon, J.W. (1992) Detection of presumptive mycoparasites in soil placed on host-colonized agar plates. *Mycological Research* 96, 605–608.

O'Neill, T.M., Elad, Y., Shtienberg, D. & Cohen, A. (1996) Control of grapevine grey mould with *Trichoderma harzianum* T39. *Biocontrol Science and Technology* 6, 139–146.

Raaijmakers, J.M. & Weller, D.M. (1998) Natural plant protection by 2,4-diacetylphloroglucinol-producing *Pseudomonas* spp. in take-all decline soils. *Molecular Plant–Microbe Interactions* 11, 144–152.

Rishbeth, J. (1963) Stump protection against *Fomes annosus*. III. Inoculation with *Peniophora gigantea. Annals of Applied Biology* 52, 63–77. [Note: *Fomes annosus* is now called *Heterobasidion annosum; Peniophora* is now called *Phlebiopsis*.]

Thomashow, L.S., Weller, D.M., Bonsall, R.F. & Pierson III, L.S. (1990) Production of the antibiotic phenazine-1-carboxylic acid by fluorescent *Pseudomonas* species in the rhizosphere of wheat. *Applied and Environmental Microbiology* 56, 908–912.

van den Boogert, P.H.J.F. & Deacon, J.W. (1994) Biotrophic mycoparasitism by *Verticillium biguttatum* on *Rhizoctonia solani. European Journal of Plant Pathology* 100, 137–156.

van den Boogert, P.H.J.F., Kastelein, P. & Luttikholt, A.J.G. (1994) Green-crop harvesting, a mechanical haulm destruction method with potential for disease control of tuber pathogens of potato. In: *Seed Treatment: progress and prospects* (Martin, T., ed.), pp. 237–246. British Crop Protection Council, Farnham.

Weindling, R. (1934) Studies on a lethal principle effective in the parasitic action of *Trichoderma lignorum* on *Rhizoctonia solani* and other soil fungi. *Phytopathology* 24, 1153–1179.

Wood, D.W., Gong, F., Daykin, M.M., Williams, P. & Pierson, L.S. (1997) *N*-acyl-homoserine lactone-mediated regulation of gene expression by *Pseudomonas aureofaciens* 30–84 in the wheat rhizosphere. *Journal of Bacteriology* 179, 7663–7670.

Chapter 13

Fungal symbiosis

This chapter is divided into the following major sections:

- the major types of symbiosis involving fungi
- mycorrhizal associations
- lichens
- *Geosiphon pyriforme* – a remarkable "new" symbiosis
- fungus–insect mutualisms

Fungi are involved in a wide range of intimate symbiotic associations with other organisms. Some of the more important examples are discussed in this chapter, and it would be no exaggeration to say that they have shaped the history of life on land. In several cases the fungi and their partners have become so intimately dependent on one another that they have lost the ability to live alone. In other cases the fungi can be cultured in laboratory media but they are, in effect, ecologically obligate symbionts because they seldom if ever grow as free-living organisms in nature. The many thousands of species of lichen are classic examples of this. They grow in some of the most inhospitable environments on earth, where no other organisms can grow, including cooled lava flows and arid desert sands, where they literally hold the place in place!

Examples of new types of symbiotic association continue to be discovered. In 1996 a unique association between a mycorrhizal fungus and a cyanobacterium was reported for the first time. In this case the fungus engulfs cyanobacteria, which then provide the fungus with its source of sugars (Gehrig *et al.* 1996). This "dual organism," *Geosiphon pyriforme,* is known from only a few natural sites in Germany. Even more recently, a nonphotosynthetic liverwort *Cryptothallus mirabilis* (related to the mosses) was shown to form a partnership with a species of *Tulsanella* (Basidiomycota), a mycorrhizal fungus of birch trees. When birch seedlings were supplied with radiolabeled CO_2 the label was translocated from the birch seedlings to the liverwort, via the mycorrhizal hyphal network, supplying the liverwort with organic carbon nutrients (Bidartondo *et al.* 2003).

As we shall see in this chapter, the below-ground networks of fungal hyphae provide potential links between several different types of organism. There is every reason to believe that further examples remain to be discovered.

Symbiotic associations are also significant in economic terms. The fungal endophytes of pasture grasses produce toxic alkaloids such as lolitrem B and ergovaline, which are now known to be responsible for several diseases of horses, sheep, and other grazing animals (Chapter 14). These fungi have been found in over 60% of pastures in the USA. The many types of mycorrhizal fungi are exploited commercially in forestry and cropping systems. And in a quite different context, fungi form several mutualistic associations with insects, some of which cause serious economic losses.

The major types of symbiosis involving fungi

Type of symbiosis	Fungi	Other partners
Mycorrhizas	Ascomycota, Basidiomycota, or Glomeromycota	Land plants – bryophytes, pteridophytes, gymnosperms, angiosperms
Lichens	Ascomycota or sometimes Basidiomycota	Green algae or cyanobacteria
Geosiphon pyriforme	Glomerales-like fungus	*Nostoc* (cyanobacterium)
Fungal endophytes of grasses (Chapter 14)	*Neotyphodium* and Clavicipitaceous fungi	Several grasses in the family Poaceae
Rumen symbioses (Chapter 8)	Obligately anaerobic Chytridiomycota	Ruminant animals
Bark beetles and ambrosia fungi	Various Basidiomycota, Ascomycota, and mitosporic fungi	Beetles (Scolitidae)
Siricid wood wasps	*Amylostereum areolatum* (Basidiomycota)	Trees
Fungus gardens of leaf-cutting ants, termites, and wood-boring beetles	Basidiomycota (*Termitomyces*; *Leucagaricus*)	Attine ants, termites, beetles

Mycorrhizas

> Under agricultural field conditions, plants do not, strictly speaking, have roots, they have mycorrhizas. (Stephen Wilhelm)

The term **mycorrhiza** means, literally, "fungus root." It refers to the fact that fungi form many types of symbiotic association with the roots or other underground organs of plants. These symbioses are extremely common. They are estimated to occur in at least 80% of all vascular plants, including angiosperms (the flowering plants), gymnosperms (the cone-bearing plants), many pteridophytes (ferns and their allies), and some bryophytes (especially liverworts). Many of these associations are thought to be mutualistic, because the fungus typically absorbs mineral nutrients from the soil and channels these to the plant, while the plant provides the fungus with sugars. However, there are several different types of mycorrhiza, with different properties and features.

We saw in Chapter 1 that fossils from the Rhynie Chert deposits in Scotland (see Fig. 1.2) contain fungal structures similar to those of the most common mycorrhizal fungi today – the **arbuscular mycorrhizal fungi**. So, it seems that some of the earliest land plants had already established mycorrhizal associations, and these might even have been a prerequisite for life on land (Simon *et al.* 1993).

The various types of mycorrhiza

Table 13.1 lists some of the main types of mycorrhiza and their main ecological roles. In addition to the common arbuscular mycorrhizas, there are several other types that have evolved independently of one another (Harley & Smith 1983) and that serve different roles. We will consider the commonest types in terms of the plant–fungus interaction and their frunctional significance.

Arbuscular mycorrhizas (AM)

Arbuscular mycorrhizas are the most common type of mycorrhiza and are found world-wide on crop plants, wild herbaceous plants, trees, many pteridophytes, and some bryophytes. Until recently, the AM fungi were classified as members of the Zygomycota. But analysis of the genes encoding the small subunit (18S) ribosomal RNA shows clearly that AM fungi are not related to Zygomycota and probably share common ancestry with Ascomycota and Basidiomycota. So they have been assigned to a new monophyletic group, the *Glomeromycota* (Scheussler *et al.* 2001; see Fig. 2.4). Seven genera are recognized within this, based primarily on features of the spores: *Acaulospora*, *Entrophospora*, *Archaeospora*, *Glomus*, *Paraglomus*, *Gigaspora*, and *Scutellospora*. None of these fungi can be grown in

Table 13.1 The major types of mycorrhiza and their ecological significance.

Mycorrhizal type	Typical host plants	Fungi involved	Major significance
Arbucular mycorrhizas	Many	Glomeromycota	Phosphorus uptake from soil
Ectomycorrhizas	Forest trees, mainly in temperate and boreal regions	Basidiomycota, Ascomycota	Nitrogen uptake from soil
Ectendomycorrhizas	Mainly pines, spruce, and larch	Ascomycota of the genus *Wilcoxina*	Mineral nutrient uptake from soil
Arbutoid mycorrhizas	*Arctostaphylos, Arbutus, Pyrola*	Basidiomycota, similar to ectomycorrhizal fungi	Mineral nutrient uptake from soil
Monotropoid mycorrhizas	Nonphotosynthetic plants, e.g. *Monotropa*	Basidiomycota such as *Boletus edulis*	Plants obtain sugars from ectomycorrhizal fungi attached to trees
Ericoid mycorrhizas	Heathland plants. *Erica, Calluna*, etc.	Ascomycota and mitosporic fungi; *Hymenoscyphus ericae*	Nitrogen uptake from soil
Orchid mycorrhizas	Orchids	*Rhizoctonia*-like fungi (basidiomycota)	Fungi supply the plant with sugars

culture, away from their host plants, so they are considered to be wholly dependent on plants for their carbon and energy sources.

Roots containing arbuscular mycorrhizal fungi show no outward signs of infection. Instead, they look like normal roots, and the extent of colonization by fungal hyphae can only be seen by special techniques. One method for this is to use differential interference contrast microscopy. But the more common method is to treat roots with strong alkali, which destroys the plant protoplasm, and then to stain the roots with a fungal dye such as trypan blue. Fungal hyphae are seen to colonize the roots extensively by growing between the root cortical cells, often producing large, swollen **vesicles**, and by penetrating individual root cortical cells to form tree-like branching structures termed **arbuscules** (Fig. 13.1).

The AM fungi produce large spores, up to 400 μm diameter, which are easily visible to the naked eye. These spores can be extracted by washing soil through a series of sieves with successively smaller mesh sizes, then floating the spores to retrieve them from the sieves. Single spores can then be used to initiate "pot cultures" with seedlings in sterilized soil – the standard way of maintaining different strains. The spores germinate and infect the roots from an appressorium-like infection structure on the root surface, similar to the infection structures of plant pathogens discussed in Chapter 5. From these initial entry points, the fungi grow extensively between the cells of the root cortex (Fig. 13.2), often producing large, swollen vesicles within the root. These vesicles are thought to have a

storage function. Other hyphae penetrate individual root cells and branch repeatedly within these to form dichotomous tree-like structures, termed **arbuscules** (Fig. 13.3). These are thought to be the main sites of nutrient exchange between the fungus and the root cells. Consistent with this, the invaded cells remain alive because the plant cell membrane invaginates to surround all the individual branches of the arbuscule – the membrane itself is never penetrated. The individual arbuscules live for a relatively short time (2–3 weeks) and then degenerate, being replaced by new arbuscules in other parts of the root.

Ecological significance of arbuscular mycorrhizas

Many studies have shown that the principal role of AM fungi is to provide plants with mineral nutrients, and especially **phosphorus** from the soil, while the plant provides the fungus with sugars. In terms of plant nutrition, phosphorus is second only to nitrogen as the major mineral nutrient that plants require, and yet phosphorus is a highly immobile element. When added to soil in the form of soluble phosphate fertilizers, the phosphate ion readily combines with calcium and other divalent cations, to form insoluble inorganic phosphates, or it combines with organic matter to produce insoluble organic phosphates. The natural rate of release of phosphate is thus extremely slow and is often the limiting factor for plant growth. The AM fungi produce extensive hyphal networks in soil, providing a large surface area for absorption of phosphorus. These fungi also release **acid phosphatases** to cleave phosphate from

(a)

(b)

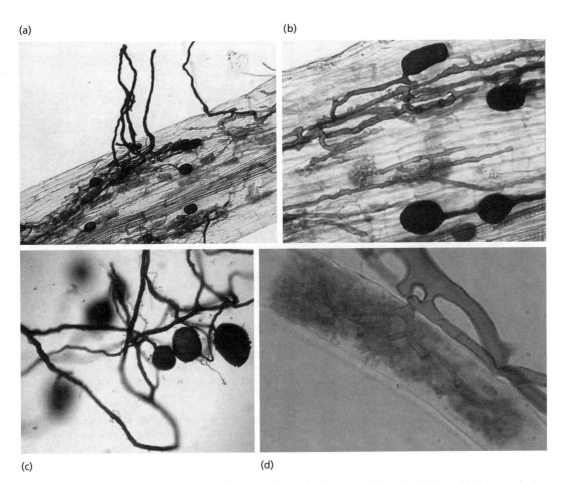

(c)

(d)

Fig. 13.1 The principal features of arbuscular mycorrhizal (AM) fungi, observed by clearing the root tissues with strong alkali and then staining roots with the fungal dye, trypan blue. (a) A root heavily colonized by AM fungi, with hyphae that radiate into the soil. (b) When observed through the depth of the root cortex, AM fungal hyphae are often seen to run parallel to the root axis, growing between the root cortical cells. These hyphae are irregular, with constrictions and bulges, quite unlike the hyphae of most other fungi. They frequently produce large, swollen vesicles within the root tissues. (c) Some of the external hyphae and hyphal aggregates produce clusters of spores in the soil. (d) Some of the root cortical cells are penetrated by hyphae that branch repeatedly to produce intricately branched arbuscules, often completely filling the root cells.

Fig. 13.2 Part of a clover root, partly crushed and cleared of protoplasm by treatment with strong alkali, then stained with trypan blue to show the distribution of fungal structures in the root. The main features shown are root hairs (rh), an **entry point** with a characteristic diamond-shaped swelling equivalent to an appressorium, large swollen **vesicles**, and "fuzzy" **arbuscules**.

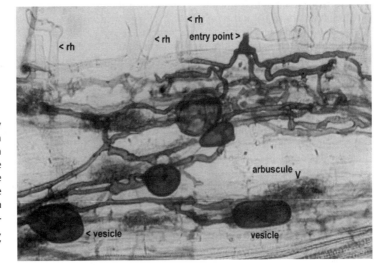

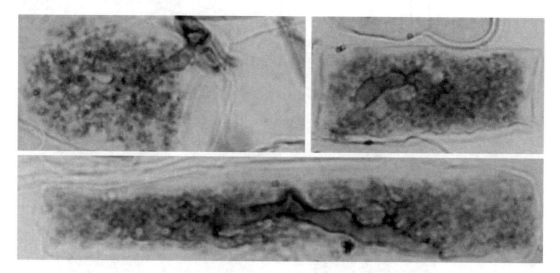

Fig. 13.3 Single arbuscules of AM fungi growing within root cells. The arbuscules enter individual root cells from intercellular hyphae, then branch dichotomously and repeatedly so that most of the cell volume is filled with finely branched AM hyphae, providing a large area of interface for potential nutrient exchange between the fungus and its plant host.

organic matter. They absorb phosphate in excess of requirements and store it in the form of polyphosphates, which can then be released to the plant.

Host ranges and communities of AM fungi

The AM fungi have astonishingly wide host ranges. At least in artificial inoculations, an AM fungus obtained from one type of plant can colonize the roots of many unrelated plants. Nevertheless, a major experimental study by Van den Heijden *et al.* (1998) strongly suggests that plants in natural communities are colonized preferentially by different strains of AM fungi, and that the diversity of AM fungi in a site can influence the plant biodiversity in natural ecosystems.

One line of evidence leading to this conclusion is shown in Fig. 13.4. Small field plots were established with gamma-irradiated soil and sown with 100 seeds of each of 15 plants typical of North American "old field" systems. The plots received either no AM fungi or mycorrhizal inoculum composed of different numbers of AM species (1, 2, 4, 8, or 14). After one season's growth the plots were harvested and assessed by several different parameters. The results clearly show that, as the number of AM species is increased in the plots, so there is a corresponding increase in **total shoot biomass**, total **root biomass**, total **length of AM fungal hyphae** on the roots, an increase in **plant species diversity** and **plant phosphorus concentration**. In other words, with an increase in AM fungal diversity there is an increase in plant productivity and plant biodiversity.

Another part of this study was a greenhouse experiment in containers of sterilized soil, using four strains of AM fungi that had been isolated from a calcareous grassland. There were six treatments: (1) soil with no AM fungi, (2–5) soils with four separate strains of AM fungi (labeled A, B, C and D), and (6) soil with all four strains of AM fungi. Each soil container was sown with a mixture of 70 seeds (of 11 different plant species) collected from the calcareous grassland. Over two growing seasons the above-ground parts of each plant species were harvested separately and the cumulative plant biomass of each species was determined (Fig. 13.5).

The most striking feature in Fig. 13.5 is that different species in the plant community seem to respond differently to different mycorrhizal fungi. For example, the grass *Brachypodium pinnatum* grew poorly in the absence of mycorrhizas and also poorly with mycorrhizal fungus B, but better with mycorrhizal fungus C. The grassland herb *Prunella vulgaris* (commonly known as "self-heal") also grew poorly in the absence of mycorrhizas but better in the presence of any mycorrhizal fungus. The sedge *Carex flacca* grew best in the absence of mycorrhizal fungi, and very poorly in the mixture of all four AM fungi. This is not surprising because some families of flowering plants, including sedges (Cyperaceae), rushes (Juncaceae), crucifers (Cruciferae), and Chenopodiaceae, are typically nonmycorrhizal. They tend to be plants of open habitats and are primary colonizers of bare or frequently disturbed soils. They were probably mycorrhizal at some stage but as typical open-habitat species they probably benefit from being nonmycorrhizal.

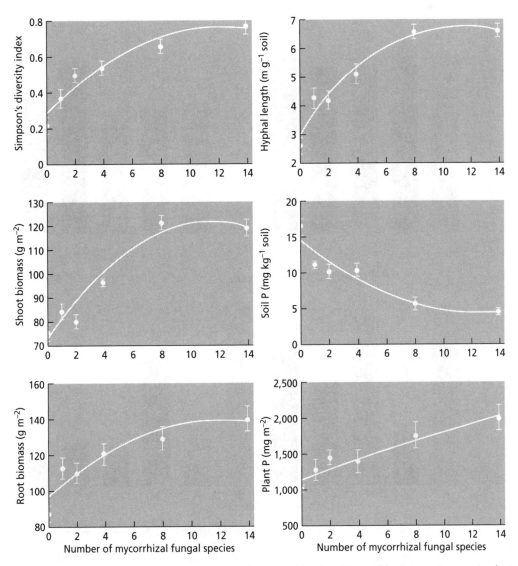

Fig. 13.4 An increase in the number of different arbuscular mycorrhizal fungi in a soil leads to an increase in plant productivity and plant biodiversity. (From van der Heijden *et al.* 1998, with permission from the publisher.)

Ectotrophic mycorrhizas

Ectotrophic mycorrhizas, or ectomycorrhizas, are found mainly on woody plants, including many species of coniferous and broad-leaved trees outside of the tropics. For example, ectomycorrhizas are typically found on trees such as pine, spruce, larch, oak, beech, birch, and eucalypts. But tropical trees and even some temperate trees (sycamore, ash, poplars) have arbuscular mycorrhizas, and some trees (e.g. willows) can have both types. The fungi involved in ectomycorrhizal associations are principally members of the Basidiomycota

that produce many of the common toadstools of the forest floor (e.g. *Amanita, Boletus, Cortinarius, Hebeloma, Lactarius* spp.; Fig. 13.6) but ectomycorrhizas also are formed by some Ascomycota, including the truffle fungi. Given the range of plants and fungi involved in this type of symbiosis, it is thought that ectomycorrhizal associations evolved independently on several occasions in the last 130–180 million years.

The characteristic feature of ectomycorrhizas is the presence of a substantial sheath of fungal tissue that encases the terminal, nutrient-absorbing rootlets (see Fig. 7.10), and the rootlets themselves are often short

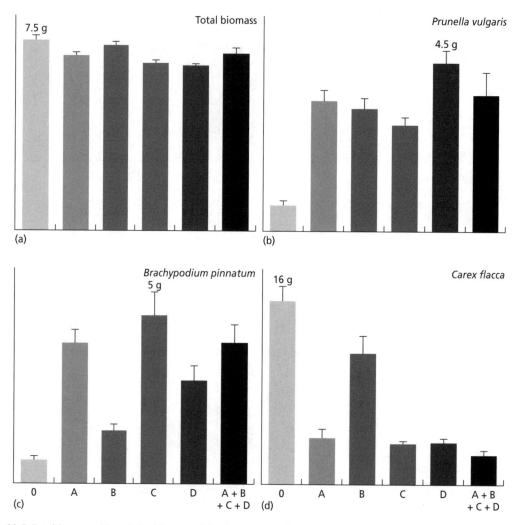

Fig. 13.5 Total biomass (a) and the biomass of three representative plant species (b–d) grown in soil with no mycorrhizal fungus (0) or with four separate AM fungal species (A, B, C, D) or a combination of all four AM species. Note that the vertical scale of each histogram is different but the largest biomass is shown in each case. (Data from van der Heijden *et al.* 1998, with permission from the publisher, but only some of the plant species are shown in this figure.)

and stumpy, with no root hairs. An extensive network of individual hyphae or aggregated **mycelial cords** (see Fig. 5.12) radiates from the surface of the root sheath, while beneath the sheath the fungus invades **between the root cortical cells** to form a "Hartig net." Although the fungus is in close contact with the root cells in this region, there is no penetration of the host cells, hence the name of these mycorrhizas – ectotrophic (outside-feeding) mycorrhizas (Fig. 13.7).

Because of the lack of root hairs and the encasement of the feeder roots by a fungal sheath, virtually all the mineral nutrients that enter the root must be channelled through the fungus. The uptake of mineral nutrients

from soil is facilitated by the mass of fungal hyphae that radiate into the soil and transport nutrients back to the mycorrhizal sheath. The fungus benefits from these associations by obtaining sugars from the plant. Trees invest a considerable amount of photosynthate to support the fungal network – conservatively estimated at 10% or more of the annual photosynthetic production of a tree.

Classic early experiments by Harley demonstrated the transport of sugars from the plant leaves to the roots. In these experiments the leaves of young trees were exposed to ^{14}C-labeled CO_2, and the fate of the label was followed through the plant tissues. Label was

(a)

(b)

Fig. 13.6 The fruitbodies (basidiocarps) of some representative fungi that form ectomycorrhizas with forest trees. (a) *Hebeloma* sp. growing on the roots of an oak seedling in a plantpot. (b) *Amanita muscaria* growing on pine roots in a field site. (c) *Lactarius* sp. (which, typical of the genus, secretes a milky fluid when the gills are damaged) growing on birch roots.

(c)

found in the form of common plant sugars such as sucrose and glucose in the stems and root tissues, but when the label entered the fungal sheath it was found in the form of ^{14}C-labeled sugar alcohols such as mannitol, arabitol, and erythritol. Thus, there is effectively a one-way flow of carbohydrates from the plant to the fungus, because most plants cannot utilize these "fungal sugars" (Harley & Smith 1983; Harley 1989).

Many ectomycorrhizal fungi can grow in laboratory culture on simple organic media, but they have little or no ability to degrade cellulose and lignin, unlike the common decomposer fungi of the Ascomycota

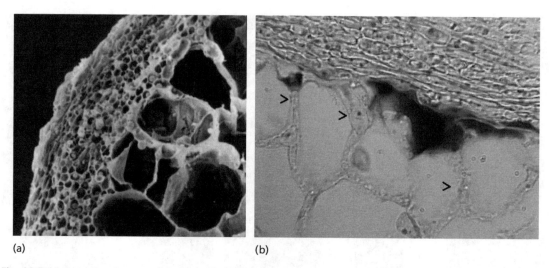

(a)   (b)

Fig. 13.7 (a) Scanning electron micrograph of a cross-section of part of a mycorrhizal root, showing the fungal sheath that surrounds the root. (b) Thin section of part of an ectomycorrhizal root. The arrowheads show hyphae invading between the root cortical cells, forming the Hartig net. Nutrient exchange between the fungus and the root is thought to occur in this region.

and Basidiomycota. So, the ectomycorrhizal fungi seem to be ecologically adapted to grow as symbionts. They seldom show a high degree of host-specificity, so it is quite common to find mycorrhizas of 10 or more different fungi on a single mature tree. However, in broad terms the ectomycorrhizal fungi can be grouped into two types. Some are "generalists" with wide host ranges, especially on young trees in newly afforested sites (e.g. *Laccaria* spp., *Hebeloma* spp., *Thelephora terrestris, Paxillus involutus*); others are more host-restricted and tend to predominate on mature trees (e.g. *Suillus luteus* on pines, *Suillus grevillei* on larch, and the truffles, *Tuber* spp., on oak or beech). These "mature" types have been shown to produce proteolytic enzymes in culture, and this is thought to be significant for their nitrogen nutrition. A few mycorrhizal fungi, such as *Laccaria* and *Hebeloma*, can take up nitrate from soil, but nitrate is readily leached out of soil by rainwater. Most mycorrhizal fungi can take up ammonium or amino acids from soil, but in the cool, wet, acidic conditions of many northern regions of the globe the rates of decomposition of organic matter are very slow, and this can significantly limit the rates of plant growth. The release of proteases by these fungi can therefore be important in providing ectomycorrhizal plants with nitrogen in the form of amino acids.

Fungal networking

The ectomycorrhizal fungi play major roles in ecosystem functioning (Allen 1992). An extensive network of

hyphae and mycelial cords ramifies through the soil from the mycorrhizal sheath (Fig. 13.8), and this network can link many different plants within a habitat – even plants of different species because of the general lack of host specificity of these fungi. For example, it is known that young tree seedlings can be linked to a "mother" tree or "nurse" tree by a common mycorrhizal network, such that $^{14}CO_2$ fed to the leaves of a larger tree can be found as label in the roots and shoot tissues of nearby seedlings. The amounts of nutrients transferred in this way may not be large, but a seedling that has tapped into an existing mycorrhizal network might benefit from this (Smith & Read 1997).

There are at least two other potentially significant functions of this subterranean network. An estimated 70–90% of ectomycorrhizal rootlets die and are replaced each year. If these rootlets were not interconnected they would decompose and at least some of the nutrients would be leached from the soil. The mycelial connections could help to retain mineral nutrients by withdrawing them from the degenerating mycorrhizas to others that are still functioning. Furthermore, as seen in Fig. 13.8, the system of mycelial cords and fans of hyphae extends far beyond the root zone. In observation chambers the peat substrate can be allowed to dry to the point where non-mycorrhizal seedlings die, but the mycorrhizal plants remain healthy because the mycelial cords can transport water from deeper in the container, beyond the reach of the roots themselves. This role can be particularly important in soils with poor water retention, such

Larch
seedling

Fig. 13.8 A young larch seedling, about 3 cm high, growing in a peat-based substrate against a sloping face of an observation chamber. Mycorrhizas can be seen at the base of the stem (arrow) but almost all the visible growth is mycelial cords that explore the soil for nutrients. (Courtesy of D. Read.)

as former mining sites where trees are planted for land reclamation. In fact, this role is not restricted to the cord-forming fungi, because we noted in Chapter 8 that all fungi can grow at water potentials beyond those that plants can tolerate. For this reason even the arbuscular mycorrhizal fungi can be significant in semiarid environments, such as the deserts of southwestern USA. The vast majority of plants in this type of environment (Fig. 13.9), including the giant cacti, have AM fungi on their roots (Bethlenfalvay *et al.* 1984).

Ericoid mycorrhizas

The cold, nutrient-poor, acidic upland soils of the northern hemisphere tend to be dominated by heathland plants of the family Ericaceae, such as *Calluna* (ling), *Erica* (the bell heathers), and *Vaccinium* (bilberry, cranberry, etc.). Equivalent soils in the southern hemisphere support a different family – the Epacridaceae. All these heathland plants have a distinctive type of

mycorrhizal association with Ascomycota that produce coils of hyphae in the thin lateral roots termed "hair roots." The coils develop within the root cells but outside of the host plasma membrane, and nutrient exchange is thought to occur primarily through this interface. The fungi that produce ericoid mycorrhizas are unusual because they seem to be free-living saprotrophs in soil. They grow in laboratory culture, producing septate hyphae with a fragmented, zigzag growth form, but only one of them (*Hymenoscyphus ericae*) has been studied in detail. DNA and RNA profiles indicate that there is considerable genetic diversity between isolates that are similar in colony appearance.

There is strong evidence that a primary role of the ericoid mycorrhizas is to provide the host plants with nitrogen. This was shown initially in laboratory conditions, by supplying plants with ^{15}N-labeled ammonium, when the label was taken up and incorporated into the plants. But when ^{15}N-ammonium was added to natural, acidic heathland soils the mycorrhizal plants actually took up **less label** than the nonmycorrhizal

Fig. 13.9 The giant saguaro cactus, *Carnegia gigantea*, more than 7 meters tall, and other large cacti (chollas and prickly pear cacti) are heavily dependent on arbuscular mycorrhizal fungi because these plants have thick, fleshy roots with few or no root hairs.

control plants, even though the mycorrhizal plants had accumulated more total nitrogen. Evidently, in these conditions the fungus was obtaining nitrogen and supplying it to the host from a different (unlabeled) source, while the uptake of ammonium was simultaneously suppressed. This led to the discovery that the fungus secretes a proteinase with optimum activity at about pH 3, releasing amino acids from the soil organic matter. All members of the Ericaceae are strongly mycorrhizal, consistent with a major role of the mycorrhizal fungi in supplying mineral nutrients in these conditions.

Orchid mycorrhizas

Orchid mycorrhizas are entirely different from any of those above, because orchids are parasitic on a fungus for at least the early part of an orchid's life. The seeds of orchids are extremely small, consisting of an embryo and only a few nutrient reserves. When orchid seeds are triggered to germinate they produce a few root hairs, and these must be colonized by a fungus at an early stage or the seedling will die. The fungi in these cases are species of *Rhizoctonia* (Basidiomycota) or closely related fungi, which grow on soil organic matter, degrading cellulose and other polysaccharides. The fungus penetrates the orchid embryo and produces hyphal coils, called **peletons**, which are surrounded by the host cell membrane (Fig. 13.10). These coils last only a few days, before they degenerate and are replaced by further coils in other cells. Presumably, this repeated production and degeneration of the coils provides the main source of sugars to the developing orchid. These sugars are likely to include trehalose or other fungal carbohydrates (Chapter 7). Consistent with this, orchid seedlings can be raised artificially in commercial conditions by supplying them with trehalose in a culture medium. In natural conditions the mycorrhizal fungi provide orchids with their sole source of carbohydrates during the early years of life. Most orchids do not emerge above ground and produce chlorophyll until they are 3–5 years old, and about 200 species do not produce chlorophyll at all throughout

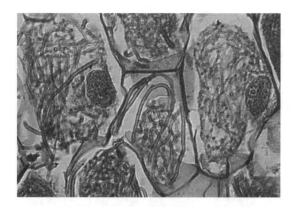

Fig. 13.10 Section through part of the protocorm (basal stem region) of an orchid, *Neottia*, showing coils of hyphae (peletons) within the orchid cells. The cells were alive, as evidenced by the presence of nuclei (darker structures) in two of the orchid cells.

their lives – they remain entirely dependent on the mycorrhizal fungus.

Monotropoid mycorrhizas

Plants of the family Monotropaceae lack chlorophyll throughout their lives and depend on mycorrhizal fungi for all their nutrient requirements. These plants are found in the deep shade beneath forest trees – sometimes broadleaved trees such as oak and beech, but more commonly coniferous trees such as pine, spruce, and fir. The fungi involved in these associations are Basidiomycota, such as *Boletus edulis*, that produce typical ectomycorrhizas attached to the roots of forest trees. From the tree host, these fungi radiate into the soil as hyphal networks or mycelial cords, and they form a hyphal sheath around the roots of *Monotropa* and related genera. So, in effect, the monotropoid plants are simply parasitic on ectomycorrhizal fungi, which in turn draw their sugars from the tree. This transfer of nutrients has been demonstrated by supplying $^{14}CO_2$ to the leaves of trees and following the label as it moves down to the roots as sucrose before entering the mycorrhizal sheath as labeled sugar alcohols or trehalose, and thence through the mycelial cords and ultimately into the flowering spikes of *Monotropa*. The sheath that develops around the roots of *Monotropa* is typical of an ectomycorrhizal sheath, with a Hartig net. But the fungus also produces small pegs that project into the root cells and then expand, surrounded by the plant cell membrane.

This three-membered symbiosis, involving a direct nutritional connection between a tree host, a mycorrhizal fungus, and a parasitic higher plant, provides another example of the roles of mycorrhizal fungi in linking different types of organism in plant communities.

Lichens

Lichens are remarkable organisms, unique in many ways. They represent a symbiosis between at least two separate organisms – a fungus and a photosynthetic partner, which can be either a green alga or a cyanobacterium. When the two (or more) partners come together they produce an entirely different type of organism, with a distinct morphology, leading to a long-term symbiotic relationship. But sooner or later, in many lichens, this relationship breaks down, and then the partners have to re-establish the relationship from the separately dispersed fungal spores and photosynthetic cells.

Lichens are extremely common and can even be the dominant organisms in some environments, such as arctic tundra and semiarid desert regions. Because of their unique symbiotic relationship, lichens are able to grow in conditions that no other organisms can tolerate. In the following sections we will look at the structure and physiology of lichens, and their environmental significance. Nash (1996) provides details of many aspects of lichen biology. But, unfortunately, lichenology has a lexicon of almost impenetrable obscurity, which we will try to avoid as far as possible!

The lichen partners

There are estimated to be between 13,500 and 17,000 species of lichen, but we must begin with a note on taxonomy. Because lichens are "dual organisms" composed of at least two separate species, there has always been a difficulty in naming them. This issue was only recently resolved by formally assigning lichens to the fungal kingdom. So, for example, the common orange-yellow coloured lichen *Xanthoria parietina* (Fig. 13.11), which grows on rocks in coastal areas, is classified as a fungus (*Xanthoria*) that contains a photosynthetic partner – in this case, the green alga *Trebouxia*.

In many lichens the fungus, termed the **mycobiont** (the fungal symbiont), is a member of the Ascomycota (cup fungi), but in a few cases it can be a member of the Basidiomycota. The photosynthetic partner (the **photobiont**) can be either a green alga or a cyanobacterium. However, a few lichens contain both a green alga and a cyanobacterium, representing a **symbiosis of three kingdoms of organisms**.

Almost all of the lichen fungi seem to be ecologically specialized, because they are found only in lichen partnerships and very rarely in a free-living state.

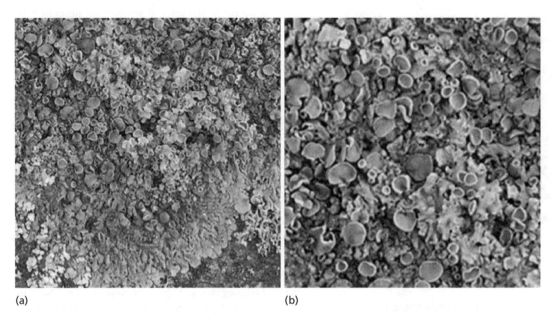

(a)   (b)

Fig. 13.11 The lichen *Xanthoria parietina*, which commonly grows on rocky shores exposed to salt spray. (a) The lichen thallus. (b) Close-up of part of the thallus, showing the disk-shaped apothecia which release ascospores.

Some have been grown in laboratory culture, but they grow very slowly and they lack the enzymes necessary for degrading organic polymers, so they are, effectively, constrained to live in a symbiosis. Only about 100 green algae have been found associated with lichens. The most common are single-celled algae of the genus *Trebouxia*, found in most lichens of temperate and cool regions. *Trebouxia* species also seem to be ecologically specialized because they are found infrequently as free-living species in nature. However, in Mediterranean and tropical regions the green alga *Trentepohlia* (and other genera) is often found instead of *Trebouxia*, and *Trentepolia* can often be found growing independently in nature, so it is not as ecologically specialized as *Trebouxia*.

About 10% of lichens have cyanobacteria as their main or only photosynthetic partner. The most common examples in Northern Europe and Scandinavia are lichens of the genus *Peltigera*, which have *Nostoc* as their photobiont. However, some lichens that contain green algae can also have cyanobacteria in special wart-like structures (**cephalodia**) on the lichen surface. These structures are found in about 3–4% of lichens and their role is probably to exploit the nitrogen-fixing abilities of the cyanobacteria.

The fact that lichens can be formed by more than one type of fungus (Ascomycota or Basidiomycota) and more than one type of photosynthetic partner (green algae or cyanobacteria) suggests that this type of symbiosis has evolved independently on several occasions. It is impossible to trace the evolutionary history of lichens, because they are not single organisms, but the lichen symbiosis is unlikely to have evolved until the Ascomycota developed, more than 300 million years ago.

The range of forms in lichens

Lichens exhibit a variety of forms, ranging from almost casual associations of fungi and photosynthetic cells in some of the desert crusts (discussed later) to highly differentiated structures with a clear zonation of tissues. Germinating ascospores can produce typical fungal colonies on normal culture media, but a suitable photobiont is required to trigger morphogenetic changes leading to several different forms of the lichen **thallus** (the "body" of the lichen) which are commonly grouped into four categories (Figs 13.12–13.15):

1 Foliose lichens, which have a flat, leaf-like structure.
2 Fruticose lichens, which have an erect or pendulous structure.
3 Squamulose lichens, which produce small, scale-like plates.
4 Crustose lichens, which produce flat crusts on rock, soil or tree surfaces.

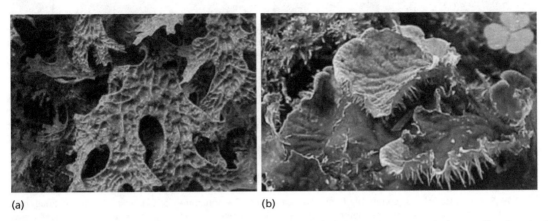

(a)　　　　　　　　　　　　　　　(b)

Fig. 13.12 Two foliose lichens. (a) *Lobaria pulmonaria* (lungwort), which grows on tree trunks in unpolluted parts of Britain. The lobes are bright green, about 1–2 cm diameter, and can have brown fungal fruiting bodies (apothecia) on the ridges. (b) *Peltigera canina* (the "dog lichen") is a common lichen that produces flat, gray lobes about 2–3 cm diameter on mossy banks. (Note the conspicuous root-like projections – **rhizinae** – on the lower surface.)

(a)　　　　　　　　　　　　　　　(b)

Fig. 13.13 Two fruticose lichens. (a) *Cladonia rangiferina* ("reindeer moss") which produces a gray-green, brittle, multiple-branched thallus. There are several *Cladonia* spp. similar to this in upland heathland habitats. They provide a major source of winter food for reindeer in Scandinavia. (b) Long, pendulous, bright green strands of the lichen *Usnea* sp. attached to the branches of trees.

Structural organization of lichens

Many of the larger lichens have a well-defined structure, with a clear zonation, as shown in a cross section of *Xanthoria parietina* (Fig. 13.16) and in the "tissues" of *Peltigera* (Fig. 13.17). Typically, there is an **upper cortex** of tightly packed fungal cells, which progressively merges into a **medulla** of more loosely arranged hyphae. The cells of the photobiont often occur at the junction of the cortex and medulla. In many lichens, including *Xanthoria*, there is often a **lower cortex** which sometimes has anchoring projections termed **rhizinae**. The significant feature of this zonation is that the photosynthetic cells are protected from desiccation and exposure to intense sunlight by the tightly packed cells of the upper cortex. But the algal cells also need gas exchange for photosynthesis, so they are associated with a loose network of fungal hyphae which are probably coated with hydrophobins, providing air spaces for gas exchange (Fig. 13.16).

(a)   (b)

Fig. 13.14 Squamulose lichens, commonly found on peaty soils. The squamules of *Cladonia* spp (a) are small, green, scale-like structures, about 1–2 mm diameter, like flakes of skin (hence the name squamulose). These lichens are very variable because they can grow in different forms, including goblet-shaped and candle-shaped structures (b) with squamules at their base.

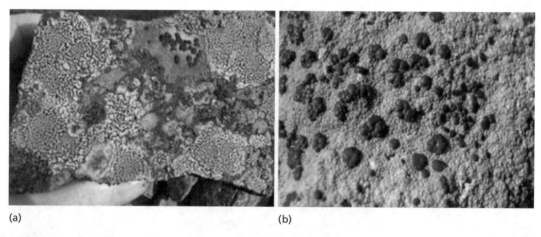

(a)   (b)

Fig. 13.15 Crustose lichens, which usually grow on bare rock surfaces, often produce a patchwork of colonies with apothecia. (a) A rock surface, about 15 × 10 cm, colonized predominantly by a green-coloured lichen, *Rhizocarpon geographicum* (the "map lichen"). (b) A gray, crustose lichen that has covered a rock surface and produced conspicuous red-brown apothecia, up to 5 mm diameter.

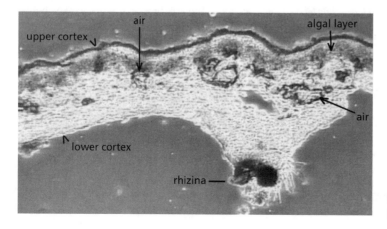

Fig. 13.16 Cross-section of *Xanthoria parietina*, showing the zonation of tissues.

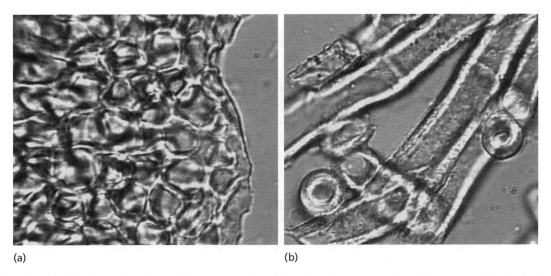

(a) (b)

Fig. 13.17 (a) Tightly packed tissue of the upper cortex of the lichen *Peltigera canina*. (b) Thick-walled, hydrophobic hyphae of the medulla of *Peltigera*.

Physiology of lichens: water relations and nutrient exchange

Many lichens will tolerate prolonged drought and resume activity rapidly after rewetting. Lichens that contain green algae can recover by absorbing water from humid air and then begin to photosynthesize within a short time. However, lichens that contain cyanobacteria can only resume photosynthesis after absorbing free (liquid) water. In all cases the drought tolerance of lichens is likely to be conferred by water-repellent hydrophobins that coat the hyphal walls of the medulla. The hydrophobins seem to be produced only by the fungus, because cells of *Trebouxia*, which have a naturally hydrophilic surface, become covered with a hydrophobic material when grown in the presence of a lichen fungus.

The principal roles of the lichen fungi are to provide physical protection for the photosynthetic cells and to absorb mineral nutrients from the underlying substrate or from rainwater. Lichen fungi are especially adept at accumulating nutrients from trace amounts in the environment. They are so efficient in this respect that they tend to accumulate atmospheric pollutants, particularly sulphur dioxide, to levels that are toxic. This is why many lichens are rarely seen in major towns and cities. Only a few species, such as *Hypogymnia* and *Xanthoria parietina*, are found commonly in towns, and even then they grow poorly, compared with their growth in nonpolluted environments.

The principal role of the photosynthetic partner is to provide sugars for growth of the lichen. Radiolabeling studies have shown that green algae and cyanobacteria can release up to 90% of their photosynthate to the fungal partner. In lichens with *Trebouxia* as the photobiont, and presumably with other green algae, the fungal hyphae can produce short branches that penetrate through the algal wall to act as nutrient-absorbing structures (haustoria; Fig. 13.18). In lichens with cyanobacterial symbionts the hyphae similarly can form protrusions that penetrate the hydrophilic gelatinous sheaths around the cyanobacterial cells (Fig. 13.18).

The major soluble carbohydrates in lichens are **sugar alcohols** (polyols). These are present in the form of mannitol and, to a lesser degree, arabitol in the fungal hyphae. The green algae also produce sugar alcohols as their main photosynthetic products – for example, **ribitol** in the case of *Trebouxia*. However, the cyanobacteria seem to release **glucose** to the fungal partner, apparently through a glucose carrier in the cell membrane after intracellular glucans have been enzymatically degraded. It is notable that the maximum rates of nutrient release from the photobionts occur in optimal moisture conditions, whereas the photobionts retain most of their carbohydrate in conditions of water stress. On this basis it has been suggested that periodic cycles of wetting and drying might be advantageous in maintaining a lichen symbiosis, each partner gaining carbohydrate at different stages of the wetting–drying cycle.

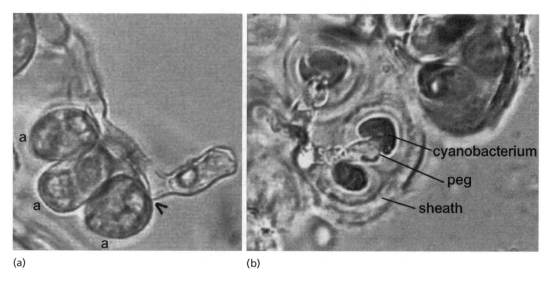

(a)   (b)

Fig. 13.18 Sites of potential nutrient exchange in lichens of the soil crust community of semiarid desert soils. (a) Three green algal cells (labeled a) firmly attached to a hypha in the medulla of the lichen *Peltula* (see Fig. 13.22 for an image of *Peltula*). The arrowhead shows a hyphal projection into an algal cell. (b) A cyanobacterial lichen (*Collema* sp., which is one of the gelatinous lichens); fungal pegs are closely associated with depressions in the cyanobacterial cells, which are surrounded by a gelatinous sheath.

Reproduction and dispersal mechanisms

The only way that a lichen can maintain itself indefinitely is by vegetative propagation, and this can be achieved in several ways. In some lichens, such as *Cladonia* spp., the dry lichen thallus is brittle, so fragments can be broken off and transported to new sites by wind or by animals. Some other lichens produce special stalk-like structures termed **isidia**, which are brittle, so they might be broken off and dispersed (Fig. 13.19). Other lichens produce masses of powdery propagules called **soredia**. These consist of a few photosynthetic cells enveloped in fungal hyphae, and they are readily dispersed by wind (Fig. 13.19).

(a)   (b)

Fig. 13.19 Isidia (a) and soredia (b): two methods of vegetative dispersal of lichens.

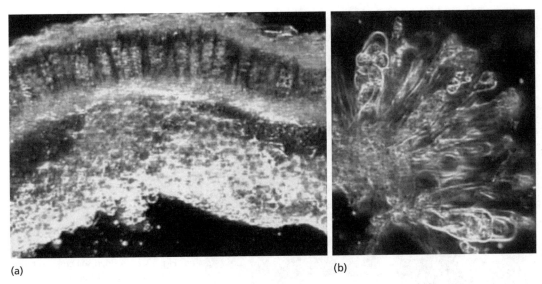

(a)   (b)

Fig. 13.20 Part of a thallus of *Xanthoria parietina*, sectioned through an apothecium. (a) Many asci are seen just beneath the upper surface of the apothecium. (b) Part of a crushed apothecium, composed of packing hyphae (paraphyses) and asci containing ascospores.

A large number of lichens produce apothecia to disperse their sexual spores (ascospores), as shown in Fig. 13.20. When these spores germinate they must contact the cells of a suitable photosynthetic partner to establish a new lichen thallus. This process of "reassembly" can be demonstrated in experimental conditions, and evidence suggests that it also occurs in nature. But its frequency may vary between different types of lichen. For example, the separate dispersal (and subsequent reassociation) of fungal and algal (*Trebouxia*) cells might be the major means of dispersal for the common lichen *Xanthoria parietina* (Fig. 13.20) because this species does not produce soredia or other specialized vegetative propagules.

Lichen ecology and significance

Lichens are classic "pioneer" colonizers in a wide range of environments. They grow on the bark of temperate trees or as epiphytes on the leaves of tropical rain forest trees. Some lichens occupy the most inhospitable environments on earth, growing on cooled lava flows and bare rock surfaces. Other types grow abundantly on tundra soils, providing a winter food source for reindeer and caribou in arctic and subarctic regions. Yet other lichens grow on or in the perennial leaves of economically important tropical plants such as coffee, cacao, and rubber. In all these respects, lichens are significant components of ecosystems. But perhaps their most significant role lies in their contribution to

soil formation, and in many cases this is brought about by inconspicuous lichens that are so rudimentary that they represent little more than a juxtaposition of fungal and photosynthetic cells. For example, in the **soil crusts** that cover many semiarid regions of the world the microbial population consists of mats of cyanobacteria interspersed with fungal hyphae and a few small lichens just visible to the naked eye (Figs 13.21–13.23).

Summary of the lichen symbiosis

The great diversity of lichens and the many environments in which they occur attest to the fact that lichen symbioses are highly successful. Lichens are slow-growing organisms, so they do not necessarily contribute greatly to biomass production. But their unique symbiosis enables them to grow in a range of environments that no other organism can tolerate and, importantly, in conditions that none of the constituent partners could tolerate alone.

Geosiphon pyriforme

Geosiphon pyriforme is a remarkable organism that was first reported in 1996 and is known from only very few sites in Germany. It is a soil fungus, closely related to the arbuscular mycorrhizal fungi, but it incorporates a cyanobacterium (*Nostoc punctiforme*) into its cells as

Fig. 13.21 A dried soil crust community of lichens, cyanobacteria and fungal hyphae that form a thin covering which binds the surfaces of semiarid soils. Such a community would be more than 100 years old and represents a relatively early stage in soil formation.

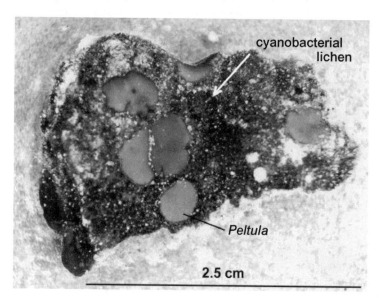

Fig. 13.22 A small fragment of a desert crust community when remoistened, showing several small lichen thalli (*Peltula* sp.) and one small cyanobacterial lichen (*Collema*). Most of the soil surface is covered with filaments of the cyanobacterium *Scytonema*.

an endosymbiont. In the mature stage, *Geosiphon* produces transparent bladder-like structures on the soil surface, each about 1 mm high, with the cyanobacteria located towards the top of the bladders (Fig. 13.24). Hyphae radiate into the soil from the base of the bladders, and it is possible (but as yet unproven) that these hyphae interact with plant roots to form arbus-

cular mycorrhizas. When the partnership is fully established, the cyanobacteria are photosynthetically active, and the cyanobacteria produce heterocysts, which can fix atmospheric nitrogen.

Experimental studies have revealed several stages in the development of this unique symbiosis. The two partners initially live independently on the soil surface,

(a)

(b)

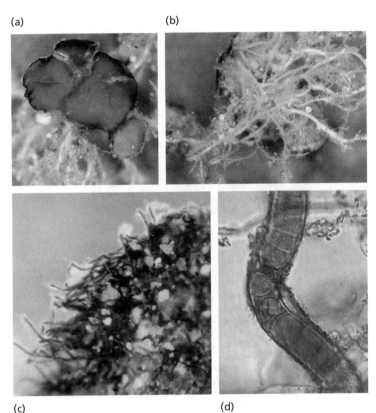

(c)

(d)

Fig. 13.23 (a) The desert crust lichen, *Peltula*, which typically grows as a thallus composed of several squamules. (b) The same lichen seen from below, showing a mass of branched rhizinae that "root" into the desert sand. (c) Part of Fig. 13.22, enlarged to show the mass of cyanobacterial filaments (*Scytonema* sp.). (d) A single filament of *Scytonema* encased in a mucilaginous sheath with soil particles.

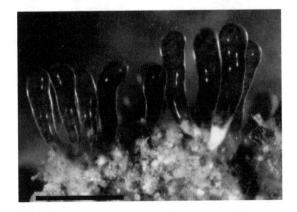

Fig. 13.24 Bladders of *Geosiphon pyriforme* growing on the surface of soil. Scale bar = 1 mm. (Courtesy of A. Schuessler.)

and the partnership is only established when the cyanobacteria are present in a specific stage of their life cycle, termed the primordium stage. When cyanobacteria in this stage make contact with the fungus, the tip of the hypha bulges and surrounds some of the cyanobacterial cells, which are then incorporated into the fungus by **endocytosis**. This leads eventually to the development of a **symbiosome** – a membrane-bound structure containing the cyanobacteria but separating them from the rest of the cell contents (Fig. 13.25). It is notable that each bladder represents the result of a single incorporation of cyanobacteria, so this would have to happen many times in the colony. Once formed, the individual bladders can live for more than half a year.

The cyanobacterial symbiont in these associations (*Nostoc punctiforme*) can be grown quite easily in laboratory conditions, and several strains of this cyanobacterium can be used to establish the *Geosiphon* symbiosis – even cyanobacterial strains from other symbiotic associations such as the liverwort *Blasia* and the higher plant, *Gunnera*.

There is strong evidence that *Geosiphon* represents a symbiotic and **mutualistic** association. The fungus benefits primarily from a supply of photosynthate from the cyanobacterium, while the cyanobacterium probably depends on the fungus for a supply of phosphate. Consistent with this, the *Geosiphon* bladders have been shown to be impermeable to even small organic compounds such as sugars, so the fungus

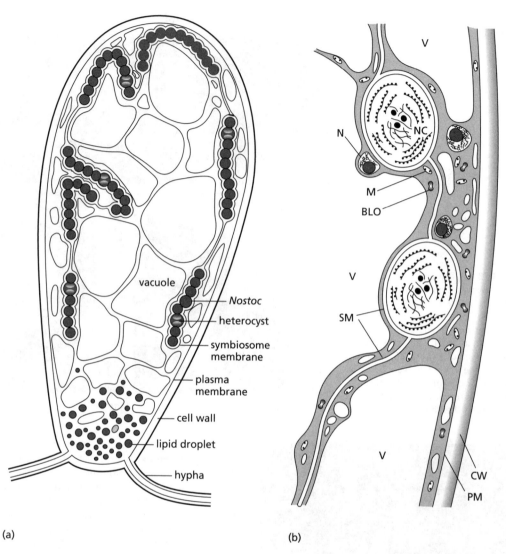

Fig. 13.25 Drawings of the *Geosiphon* bladder compartmentation. (a) Cells of *Nostoc* are located in membrane-bound symbiosomes towards the periphery of the fungal cell. (b) Detail showing a bacteria-like organism (BLO), cell wall (CW), mitochondrion (M), nucleus (N), *Nostoc* cell (NC), plasma membrane (PM), symbiosome membrane (SM), and vacuole (V). (Image courtesy of A. Scheussler & M. Kluge; from Schuessler & Kluge 2001.)

would need to be supplied with these by the photosynthetic partner. This astonishing **newly discovered type of symbiosis** should spur renewed interest in the study of rudimentary associations among the microscopic organisms in soil.

Fungus–insect mutualisms

Several insects have developed mutualistic associations with fungi, to provide the insect with a suitable food source. The likely driving force for this is the inability of insects (or other animals) to degrade cellulose. Instead, the insect has to harness the cellulolytic activities of fungi to obtain food. In the more highly evolved associations of this type, the insect ensures the perpetuation of the mutualism by:

• carrying and transmitting the fungus in specially adapted organs, called **mycangia**;
• inoculating a suitable substrate with spores of the fungus;

Fig. 13.26 The wood wasp, *Sirex noctilio*, boring a hole in a weakened tree to deposit eggs and fungal spores. (Courtesy of M.P. Coutts, J.E. Dolezal and the University of Tasmania; see Madden & Coutts 1979.)

- tending or "gardening" the substrate to promote the growth of the fungus which will be used as a food source by the insects.

The *Sirex* wood wasp associations

Female wood wasps of the genus *Sirex* inject their eggs into the wood of weakened or damaged trees by means of an auger-like ovipositor (Fig. 13.26). Then, using the same entrance hole, the wasp produces two or three further tunnels at different angles, and she deposits further eggs in these. But in the final tunnel the wasp injects spores of a wood-rotting fungus *Amylostereum areolatum* (Basidiomycota) from pouches termed **mycangia** that are located at the base of the ovipositor. The mucus containing the fungal spores is toxic and causes the foliage to wilt, while the growth of the fungus causes the wood to dry out locally and disrupts water movement in the trunk. This combination of factors provides ideal nutritional conditions for the development of the wasp larvae, which feed on the fungus and the cellulolytic breakdown products of the wood. After pupation, the larvae produce a new batch of adults, and as the females emerge they collect spores from the rotting wood and store them in the mycangia, to repeat the cycle of inoculation in other trees.

Several types of wood wasp oviposit in dead trees, but the most significant damage to living trees is caused by *Sirex noctilio*. This wasp can be an economically significant pest in pine plantations and can also attack several other conifers. *Sirex noctilio* occurs naturally in the northern hemisphere, including Britain and several Asian countries, but in the past century it has spread to many countries in the southern hemisphere, including parts of Africa, South America, Australia, and New Zealand.

Slippers *et al.* (2002) analyzed the population structure of *Amylostereum* strains associated with different wood wasp species, as a potential basis for implementing pest control and quarantine measures. Using PCR-RFLP (polymerase chain reaction–restriction fragment length polymorphisms) and analysis of nuclear intergenic spacer DNA, it was possible to match different isolates of the fungus to recent geographical movements of *S. noctilio*. The results indicate that, since the introduction of *S. noctilio* into the southern hemisphere from the northern hemisphere, around 1900, all subsequent movement of this pest has been **between** countries in the southern hemisphere and not by further introductions from the north. This provides the potential basis of a management strategy: local quarantine measures could be applied to prevent further spread of the pest.

Leaf-cutting ants, "gardening" termites, and ambrosia fungi

Several insects and arthropods have co-evolved with fungi in mutualistic associations. Three classic examples of this are exhibited by the leaf-cutting ants of Central and South America, the "garden-tending" termites of Africa and South-east Asia, and wood-boring beetles in several parts of the world (Mueller & Gerardo 2002).

The leaf-cutting ants, commonly termed **attine ants**, produce large nests with many subterranean chambers. The nests are started by a single winged female – the future queen – who carries inoculum of a fungus in a pouch at the back of her mouth. She deposits the fungus on suitable plant material, and as the fungus grows she begins to lay eggs. Worker ants bring leaf pieces back to the nest, lick them, chew them and defecate on them, then inoculate them with tufts of the fungus and pack the inoculated material into chambers. Remarkably, these "fungus gardens" consist almost exclusively of one fungus – often a species of the mushroom-forming genus *Leucoagaricus*. The mycelium of the fungus develops a distinctive growth form within the gardens, producing hyphae with balloon-like swollen tips, called **gongylidia** or **bromatia**, which are fed to the larvae and provide their food source. In addition, the gongylidia are foraged by the ants, to supplement their diet of plant sap. Recent evidence (Ronhede *et al.* 2004) suggests that the fungi in these associations provide many of the enzymes, including proteases, laccases, pectinases, and carboxymethylcellulases, for degrading the leaf material. These enzymes seem to be protected and possibly concentrated during passage through the ant gut, and are deposited in the fecal droplets. By cladistic analysis of both the ants and their fungal associates, Mueller & Gerardo have shown that the ability of ants to cultivate fungi arose only once, about 50–60 million years ago, but has given rise to about 200 species of fungus-growing ants, matched to their specific fungal associates.

An essentially similar type of behavior is found in a subfamily (Macrotermitinae) of termites in Africa and southeast Asia. Unlike most termites, this subfamily cannot digest wood, because they do not have cellulolytic protozoa in their guts. Instead they have evolved a symbiosis with the fungus *Termitomyces* (Basidiomycota) which is cultured in fungus gardens, using plant debris that is continually added to the colony. The termites tend these gardens by weeding out any contaminant fungi. After periods of heavy rainfall the fungus produces large, mushroom fruitbodies with long tapering underground stalks. This association is highly specialized because *Termitomyces* species are only ever found associated with termite nests.

A third association of this type is found in some wood-boring beetles that produce tunnels in damaged or dead trees. The beetles deposit eggs in these tunnels and smear the tunnel walls with spores of fungi, collectively known as **ambrosia fungi**. The spores of these fungi are stored in special sacs (mycangia) on the beetle's body to provide inoculum for establishment in new sites. The developing beetle larvae create extensive galleries beneath the bark and feed predominantly, if not exclusively, on the fungus. These "gardens" typically contain only one type of fungus, because contaminant fungi are weeded out.

There is a remarkable parallel in the evolution of these three types of "fungus-farming" systems, because about 40–60 million years ago the termites, ants, and beetles independently evolved the ability to culture fungi as their source of food. In ants and termites the ability to culture fungi originated only once in each group, but in ambrosia beetles it originated at least seven times. Now all of these fungus-cultivating insects and their fungal partners are mutually dependent on one another. There is no evidence that they have ever reverted to an independent existence.

Cited references

Allen, M.F. (ed.) (1992) *Mycorrhizal Functioning.* Chapman & Hall, New York.

Bethlenfalvay, G.J., Dakessian, S. & Pacovsky, R.S. (1984) Mycorrhizae in a southern California desert: ecological implications. *Canadian Journal of Botany* **62**, 519–524.

Bidartondo, M.I., Bruns, T.D., Weiss, M., Sèrgio, & Read, D.J. (2003) Specialized cheating of the ectomycorrhizal symbiosis by an epiparasitic liverwort. *Proceedings of the Royal Society of London, B* **270**, 835–842.

Gehrig, H., Scheussler, A. & Kluge, M. (1996) *Geosiphon pyriforme,* a fungus forming endocytosymbiosis with *Nostoc* (Cyanobacteria) is an ancestral member of the Glomales: evidence by SSU rRNA analysis. *Journal of Molecular Evolution* **43**, 71–81.

Harley, J.L. (1989) The significance of mycorrhiza. *Mycological Research* **92**, 129–139.

Harley, J.L. & Smith, S.E. (1983) *Mycorrhizal Symbiosis.* Academic Press, London.

Madden, J.L. & Coutts, M.P. (1979) The role of fungi in the biology and ecology of woodwasps (Hymenoptera Siricidae). In: *Insect–Fungus Symbiosis* (Batra, L.R., ed.), p. 165. Allanheld, Osmun & Co., New Jersey.

Mueller, U.G. & Gerardo, N. (2002) Fungus-farming insects: multiple origins and diverse evolutionary histories. *Proceedings of the National Academy of Sciences, USA* **99**, 15247–15249.

Nash, T.H. (1996) *Lichen Symbiosis.* Cambridge University Press, Cambridge.

Ronhede, S., Boomsma, J.J. & Rosendahl S. (2004) Fungal enzymes transferred by leaf-cutting ants in their fungus gardens. *Mycological Research* **108**, 101–106.

Scheussler, A. & Kluge, M. (2001) *Geosiphon pyriforme,* an endocytosymbiosis between fungus and cyanobacteria, and its meaning as a model system for AM research. In: *The Mycota,* vol. IX. *Fungal Associations* (Hoch, B., ed.), pp. 151–161. Springer-Verlag, Berlin.

Scheussler, A., Schwarzott, D. & Walker, C. (2001) A new fungal phylum, the Glomeromycota: phylogeny and evolution. *Mycological Research* **105**, 1413–1421.

Simon, L., Bousquet, J., Levesque, R.C. & Lalonde, M. (1993) Origin and diversification of endomycorrhizal fungi and coincidence with vascular land plants. *Nature* **363**, 67–69.

Slippers, B., Wingfield, M.J., Coutinho, T.A. & Wingfield, B.D. (2002) Population structure and possible origin of *Amylostereum areolatum* in South Africa. *Plant Pathology* **50**, 206–210.

Smith, S.E. & Read, D.J. (1997) *Mycorrhizal Symbiosis,* 2nd edn. Academic Press, San Diego.

van der Heijden, M.G.A., Wiemken, A. & Sanders, I.R. (1998) Mycorrhizal fungal diversity determines plant diversity, ecosystem variability and productivity. *Nature* **396**, 69–72.

Chapter 14

Fungi as plant pathogens

This chapter is divided into the following major sections:

- the major types of plant-pathogenic fungi
- necrotrophic pathogens of immature or compromised hosts
- pathogens of fruits: the roles of pectic enzymes
- host-specialized necrotrophic pathogens
- vascular wilt diseases
- the smut fungi
- fungal endophytes and their toxins
- *Phytophthora* diseases
- biotrophic pathogens

Fungi are pre-eminent as plant pathogens. Roughly 70% of all the major crop diseases are caused by fungi, or the fungus-like Oomycota. One of the most notorious examples is **potato late blight** caused by *Phytophthora infestans* (Oomycota), which devastated potato crops in Ireland in the 1840s leading to widespread famine. It is estimated that more than half a million people died of starvation in that period, and a similar number emigrated to the rest of Europe and North America. Potato blight is still a potentially serious disease wherever potatoes are grown, despite many attempts to control it by plant breeding and fungicides (Chapter 17). A more recent example is the **Great Bengal Famine** of 1943, which is estimated to have cost the lives of some 2 million people due to failure of the rice crops caused by a leaf-spot pathogen, *Helminthosporium oryzae* (now variously known as *Bipolaris oryzae* or *Cochliobolus miyabeanus*).

Fungi also cause many serious diseases of landscape and amenity trees. **Dutch elm disease**, caused by *Ophiostoma ulmi* and the related species *O. novo-ulmi*, swept repeatedly across North America, Britain, and continental Europe over the last century, decimating the elm populations. Similarly, **chestnut blight**, caused by *Cryphonectria parasitica*, has devastated the native American chestnut tree, *Castanea dentata*. This disease swept across much of the eastern part of the USA, after it was first recorded in the New York Zoological Garden in 1904. The magnificent American chestnut forests have now virtually disappeared and are represented only by a shrub-like understorey layer. **Cinnamomi root rot**, caused by *Phytophthora cinnamomi*, is another example of an introduced pathogen, currently threatening large tracts of natural eucalypt forest in Australia. And, at the time of writing, **sudden oak death** caused by *Phytophthora ramorum* is ravaging the natural oak woodlands in the coastal fog belt of northern California and southern Oregon. This "new disease," first recorded in 1996, has now spread to many parts of Europe and is causing serious alarm. It has the potential to destroy many of Britain's ancient woodlands. These are just a few examples of many that could be cited. Taken together, the epidemics in crop plants, plantations, and natural woodlands have caused inestimable damage. In this chapter we focus on the activities of plant-pathogenic fungi and their several adaptations for causing disease. In Chapter 17 we will consider how these diseases can be controlled.

The major types of plant-pathogenic fungi: setting the scene

It is helpful to begin with some definitions:

1 A **parasite** can be defined as an organism that gains all or part of its nutritional requirements from the living tissues of another organism – the **host**. In other words, parasitism describes a **nutritional relationship**.
2 A **pathogen** can be defined as an organism that causes disease. It is almost always a parasite, but the distinction between a parasite and a pathogen is important because parasites do not always cause obvious or serious disease.

There are many thousands of plant-pathogenic fungi, so we need to group them in some way that reflects their fundamental biology. The system we will use is based on two principal features: (i) the type of nutritional relationship between the pathogen and the host, and (ii) whether the pathogen has a broad host range or is host-specialized.

In terms of the nutritional relationship, we distinguish between **necrotrophic** pathogens and **biotrophic** pathogens. Nectrotrophs kill the host tissues, usually by directly invading them, or by producing toxins or degradative enzymes, and then feed on the tissues that they kill (Greek: *nekros* = dead; *trophos* = a feeder). By contrast, biotrophic pathogens feed on living host tissues (Gr. *bios* = life), often by producing special nutrient-absorbing structures that tap into the host's tissues. A third group – **hemibiotrophic** pathogens – initially grow as biotrophs but then invade and kill the tissues. Some of the *Phytophthora* species such as *P. infestans* behave in this way.

The secondary distinction is between pathogens of **broad host range** and those that are host-specialized, with a **narrow host range**. Pathogens of broad host range usually attack immature or senescing tissues, or plants whose resistance is compromised by environmental factors. By contrast, host-specialized pathogens are adapted to overcome the specific defense mechanisms of plants, such as the fungitoxic compounds that many plants produce to ward off infection. The biotrophic pathogens are necessarily host-specialized.

So, we can categorize almost all pathogens in the following simple scheme, which will form the basis for the rest of this chapter. We will discuss representative examples of the major types of plant disease, to build up a picture of the diversity of plant-pathogenic fungi and their activities.

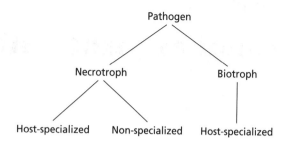

Necrotrophic pathogens of immature or compromised hosts

A large number of fungi attack the tissues of young plants, causing seed rots or seedling diseases. Other fungi typically attack the older, senescing tissues of plants, and yet others attack plants that are compromised by drought or other stress factors. Although these fungi do not have a very sophisticated mode of parasitism, they can cause serious economic damage. Several examples are discussed below.

Seed rots and seedling pathogens

A characteristic range of fungi cause seed rots or seedling diseases – sometimes called **damping-off** diseases because they are often associated with wet soil conditions. For example, many *Pythium* spp. (Oomycota) attack the root tips of seedlings, and strains of *Rhizoctonia solani* often attack the young shoot base, just above or below the soil level. Other fungi with similar behavior include *Fusarium* spp. which cause seed-rots and seedling diseases of cereals. All these fungi respond rapidly to the presence of seed or seedling exudates, and invade the young tissues that have little inherent host resistance. To help counteract this problem, seeds are usually coated with broad-spectrum fungicides such as maneb, thiram, or dinocap (Chapter 17) which interfere with several basic metabolic processes of fungi. But there is an increasing trend to replace fungicidal seed coatings with biological control agents. Several of these are now available commercially. They include antibiotic-producing bacteria such as *Bacillus subtilis*, used widely in the USA for cotton and several other field crops, and *Pseudomonas chlororaphis*, used widely in Scandinavia to protect against seedling diseases of cereals. Several fungi are marketed for biocontrol of seedling diseases, including *Trichoderma* spp. and *Clonostachys rosea* (under its former name *Gliocladium catenulatum*).

All these products work well in appropriate conditions; but seedling pathogens can infect rapidly, before a biological control agent can become established. For example, soil-borne sporangia of the common seedling pathogen *Pythium ultimum* can germinate within 1–2 hours in response to volatile metabolites released from germinating seeds (probably ethanol but possibly also acetaldehyde), and the seed tissues can be heavily colonized within 6–12 hours of planting in *Pythium*-infested soil. Biocontrol agents are unlikely to act fast enough to control such pathogens (Nelson 1987, 1992).

Athelia rolfsii

The sclerotium-forming fungus *Athelia* (*Sclerotium*) *rolfsii* (Basidiomycota) is one of the most devastating seedling pathogens in the warmer parts of the world where rainfall is seasonal. The sclerotia of *A. rolfsii* (Fig. 14.1) survive in soil during the dry season and are triggered to germinate after the first rains. Then the fungus grows on the re-wetted crop residues from the previous season and uses these residues as a food base to attack the newly sown crop, typically rotting the basal stem tissues within 3–4 weeks. Large areas of cereals, and other crops, can be destroyed by this fungus.

The key to understanding *A. rolfsii* lies in the behavior of its sclerotia, as shown by the experiment in Fig. 14.2 and Table 14.1. If sclerotia are harvested from laboratory cultures they exhibit **hyphal germination**, whereby only a few hyphae emerge and these can infect plants at only a short distance from the sclerotia (0.5 cm). But sclerotia that have been air-dried show **eruptive germination** when remoistened – all the sclerotial reserves are used to produce a mass of hyphae, which are organized into mycelial cords and can infect a plant at up to 3.5 cm distance. The sclerotia of this fungus are even more effective in the presence of freshly decomposing organic matter (e.g. crop residues from the previous season, which start to decompose after the first rains). Volatile compounds, especially methanol, are released from these freshly decomposing residues, due to the actions of pectic enzymes that persisted in the residues. These volatile compounds trigger the sclerotia to germinate eruptively and also serve as nutrients, orientating growth towards the source of methanol. In these conditions, the sclerotia can infect seedlings even 6 cm away. These experiments seem to parallel closely the behavior of the fungus in cropping systems, where *A. rolfsii* characteristically colonizes the crop residues and uses these as a food base to support infection of the basal stem tissues of young plants. It has an extremely

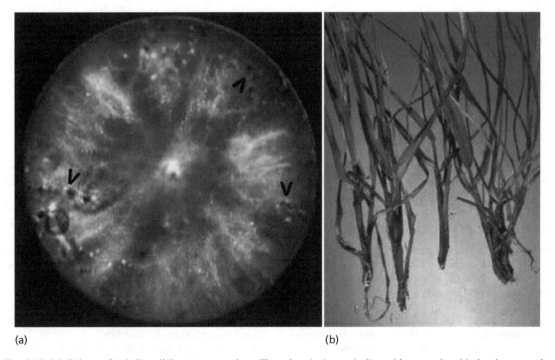

(a)   (b)

Fig. 14.1 (a) Colony of *Athelia rolfsii* on an agar plate. The sclerotia (some indicated by arrowheads) develop near the colony margin. The white mycelium is aggregated into mycelial cords. (b) Young wheat plants rotted by *A. rolfsii* just below soil level.

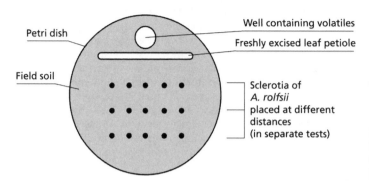

Fig. 14.2 Experimental system to study infection of leaf petioles (living plant tissue) from sclerotia of *Athelia rolfsii* placed at different distances on natural soil.

Table 14.1 Germination of sclerotia of *Athelia rolfsii* and infection of leaf petioles when sclerotia were placed at different distances from the petioles in the experiment shown in Fig. 14.2. (Data from Punja & Grogan 1981.)

	Distance (cm) of sclerotia from leaf petiole						
	0	1	2	3	4	5	6
Nondried sclerotia							
Volatiles absent							
Per cent germination	19	16	11	11	19	14	15
Per cent infection	12	0	0	0	0	0	0
Volatiles present							
Per cent germination	90	73	67	51	17	14	9
Per cent infection	90	71	54	39	0	0	0
Dried sclerotia							
Volatiles absent							
Per cent germination	78	75	71	63	74	68	73
Per cent infection	78	75	60	56	0	0	0
Volatiles present							
Per cent germination	100	100	96	95	100	98	87
Per cent infection	100	100	91	87	100	82	16

aggressive mode of attack, producing pectic enzymes to degrade the plant tissues, aided by the production of copious amounts of oxalic acid (see later).

Rhizoctonia solani

Rhizoctonia solani (sexual stage *Thanatephorus cucumeris*; Basidiomycota) is another common seedling pathogen that attacks the basal stem tissues of plants. It is a taxonomically complex fungus – a species-aggregate consisting of strains that are differentiated by molecular fingerprinting, by host range, and by their ability to anastomose (fuse) with one another when colonies are opposed on agar plates (see Fig. 9.6). Some of these strains cause stunting diseases of cereals, such as **crater disease** of wheat on the heavy, black clay soils of Northern Transvaal, South Africa (Fig. 14.3). In this disease the plants grow normally at first, but large

patches of stunted plants start to appear about 3–4 weeks after sowing, and from this time onwards the stunted plants make little further growth, while the rest of the crop grows normally. The roots of the stunted plants bear bead-like masses of hyphae at about 7–10 cm below the soil surface, where the loose cultivated soil meets the underlying clay soil that has not been ploughed. From these hyphal beads the fungus penetrates and kills the roots, leading to the stunting syndrome. Similar stunting diseases of cereals are found in the compacted sandy soils of parts of Australia, where the roots have brown, spear-pointed ends, caused by rotting of the root tips.

Diseases of this type have a common cause: the compacted soil layer impedes root growth, which is likely to increase nutrient exudation from the root tips, promoting invasion by *Rhizoctonia*. In addition to this, *Rhizoctonia* grows as an extensive mycelial

(a)   (b)

(c)   (d)   (e)

Fig. 14.3 Crater disease of wheat caused by *Rhizoctonia solani* in Northern Transvaal, South Africa (Deacon & Scott 1985). (a) Aerial view showing extensive areas of stunted plants. (b) A disease crater, showing the sharp boundary between the healthy crop and a stunted patch. (c) The boundary between the loose surface soil and the heavily compacted clay soil occurs at about 5–8 cm depth. (d) Beads of *Rhizoctonia* on the first-formed seminal roots; these roots are important because they would normally penetrate deep into a soil profile, so that water can be tapped from a depth when the surface soil layers have dried. (e) Comparison of normal healthy plants and plants within a crater.

network in the lower, undisturbed soil layer. The disease patches represent the areas where this fungal network is present beneath the soil surface, and these patches are spread progressively along the plough lines (Fig. 14.3), increasing in extent from year to year. The most effective way of controlling these diseases is by deep-ploughing, although this is seldom economic on heavy clay soils. However, on the sandy soils of Australia the use of a tine that rips the soil about 10–12 cm below the level at which the seed is sown can allow the roots to penetrate deeply and greatly improve crop yields.

Decline and replant diseases

Several perennial fruit crops, including strawberry, apple, cherry, and avocado, show a progressive decline in yield as the plants age, until production is no longer economic. If the declining plants are removed and replaced by others the new plants often grow poorly or die. These **decline** and **replant diseases** are closely related, indicating a common cause. As we saw in Chapter 10, the young "feeder roots" of plants often have a short life span before they senesce and are replaced by others (see Figs 11.14, 11.15). In the decline or replant diseases this balance between the rates of root production and root decay is altered, so that root tip decay exceeds the rate of root tip production, and eventually the crop dies. Species of *Pythium* (e.g. *P. sylvaticum*) are strongly implicated in the decline of orchard crops such as apples. Similarly, *Phytophthora cinnamomi* (Oomycota) is one of the principal causes of decline in avocado orchards and it also causes a **dieback disease** of native eucalypt vegetation in Australia, raising serious concerns for the future of this unique assemblage of plant species.

One of the major lines of evidence that implicates *Pythium* and *Phytophthora* spp. as the cause of decline and replant diseases is that the acylalanine fungicides (Chapter 17) act specifically against Oomycota, and when these fungicides are applied to field crops experimentally (but at uneconomic levels) they can cause spectacular improvements in plant growth. Similar studies suggest that *Pythium* spp. cause up to 10% yield reduction in wheat crops in the Pacific Northwest of the USA, even though the crops are apparently healthy.

These examples illustrate two important points. First, the root tips and young tissues remain susceptible to infection throughout the life of a plant. So, even mature trees can decline if the rate of pathogen-induced root decay exceeds the rate of root-tip production. Second, many of these pathogens show some degree of host-adaptation. For example, *Pythium graminicola* and *P. arrhenomanes* are characteristically, but not exclusively, associated with graminaceous hosts (the grass family), and are implicated in the progressive decline of sugar-cane crops. Experimental studies comparing "host" with "nonhost" (nongrass) plants strongly suggest that the host adaptation of these two fungi is linked to their efficient **zoospore encystment** on host plants. Similarly, *Phytophthora sojae* is a host-adapted pathogen of soybean crops, and its zoospores show strong attraction to the **flavonoids** released from soybean roots (Chapter 10).

Diseases of senescence: the stalk-rot pathogens

Some fungi that damage seedlings can also grow on senescing plant tissues, exploiting the declining host resistance. Important examples include the stalk-rot pathogens of maize, some of which (e.g. *Macrophomina phaseolina*) attack many crops whereas others are quite host-specialized towards maize (*Diplodia maydis*, *Gibberella zeae*, *Colletotrichum graminicola*). In any case, their similar behavior with respect to senescence allows them to be considered as a group.

Stalk-rot pathogens are extremely common where maize and sorghum are grown, but they cause serious disease in only some sites or growing seasons, whereas in other sites or seasons they cause little or no damage. The reason is that they respond to the plant's physiology, and particularly to any stress conditions that predispose the plants to infection. These stress factors are many and varied. They include temporary drought stress, poor light conditions (overcast weather), mineral nutrient deficiency, or insect (stem-borer) damage. Even an exceptionally high level of grain-setting can lead to stalk-rotting by creating a heavy demand on the plant's nutrient reserves. If any of these conditions occurs at a critical stage in the crop's growth, during grain-filling, then the whole crop can be infected by stalk-rot pathogens, resulting in premature death and collapse, with heavy loss of yield (Fig. 14.4).

A unifying hypothesis to explain this type of disease was developed early in the 1900s and was refined by Dodd (1980). The filling of the grain in maize and sorghum places a heavy demand on plant sugars, and only about 80% of this sugar demand can be supplied by "current" photosynthesis in the leaves. The other 20% is supplied from sugar storage reserves, principally in the base of the stem. As the plant approaches maturity its reserve sugars are progressively depleted, and any stress conditions that temporarily reduce the rate of photosynthesis at this stage cause more sugars to be removed from the stem tissues, leading to premature senescence. Then, the weakly parasitic stalk-rot fungi that were already present but were held in check are able to invade and destroy the stalk base. This is easy to demonstrate experimentally, because stalk rot never occurs if the developing grain is removed to eliminate the sugar stress, but this is hardly a practical solution! Instead, growers must use their experience to predict the likelihood of stress conditions on their farms and adjust the fertilizer and plant spacing accordingly. All the fungi that cause diseases of this type are specialized, natural invaders of senescing plant tissues.

Pathogens of fruits: the roles of pectic enzymes

Ripening fruits and other sugar-rich or starch-rich plant tissues are often rotted by fungi (Figs 14.5, 14.6). One common example is *Botryotinia fuckeliana* (better known by its asexual stage, *Botrytis cinerea*) which causes gray mould of soft fruits such as strawberries, raspberries, and grapes (see Fig. 5.20). This fungus often initiates infection by growing on the senescing flower remains (it is also a problem in the commercial cut flower industry) and then invades the living tissues of the ripening fruit. By contrast, several other fruit-rot fungi infect through minor wounds in the fruit surface, caused by insects or by damage during harvesting and packing. Examples of these fungi include the "green

(a)

(b)   (c)

Fig. 14.4 Stalk rot of maize caused by *Phialophora zeicola* in South Africa (Deacon & Scott 1983). (a) Premature senescence caused by drought stress, typical of thousands of hectares in the year when this photograph was taken. (b) Comparison of a healthy maize stalk (upper) with a shredded, dead maize stalk (lower). (c) Invasion and rotting of the stalk base and roots.

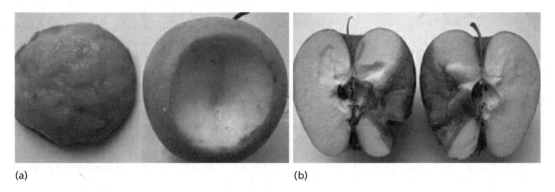

(a)   (b)

Fig. 14.5 (a) Apple inoculated with *Penicillium expansum*, which causes a soft, watery, pale-colored rot (photographed 7 days after wound inoculation). When the apple skin was cut with a scalpel at the margin of the rot, the whole rotted area fell away, leaving uninfected tissue. This type of rot is indicative of polygalacturonase activity. (b) Apple inoculated with *Sclerotinia fructigena*, which produces a firm, irregular, dark brown rot that cannot be separated from the underlying healthy tissue. This might be explained by the "browning reaction" in which phenolic compounds in plant tissues are oxidized when exposed to air. Oxidized phenolics are known to be enzyme inhibitors.

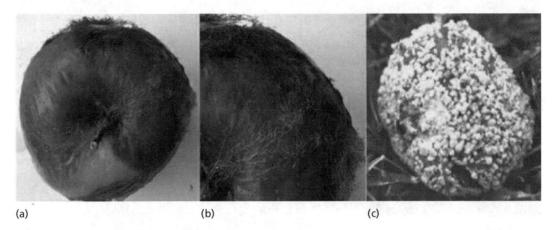

Fig. 14.6 (a,b) Wounded apple naturally contaminated by airborne spores of *Rhizopus* (mucorales), producing a fast-spreading, watery rot that collapses and liquefies the tissues. Dark, spreading hyphae and sporangiophores are seen on the fruit surface. (c) Apple infected by *Sclerotinia fructigena* in field conditions, showing pustules of spores on the fruit surface. The fruit progressively dries out and mummifies.

moulds" *Penicillium italicum* and *P. digitatum* which commonly cause the rotting of oranges, *P. expansum* which commonly rots apples in storage (Fig. 14.5), *Sclerotinia fructigena* which rots apples, peaches, and some other fruits, often through wounds caused by birds and wasps before the fruit is harvested (Figs 14.5, 14.6), and species of *Rhizopus* which rot many types of fruit, including pears (Fig. 14.6).

The interesting feature of these fungi is that they cause different types of rot, ranging from fast-spreading watery rots to firm, dry rots. In all cases they are unable to rot the fruit until it passes through the phase termed **climacteric**, when the fruit begins to ripen. This is the stage at which the acid content declines and sugar levels start to rise.

The roles of pectic enzymes in pathogenesis

The common feature of all fruit-rot diseases is that the tissues are rotted by **pectic enzymes** that degrade the **middle lamella** (the cementing layer) between adjacent plant cell walls (Fig. 14.7). The pectin of the middle lamella consists of chains of α-linked galacturonic acid residues, some of which are methylated, and mixed polymers of galacturonic acid, mannose, and lesser amounts of other sugars. The **pectic enzymes** are of three major types.

1 **Pectin methyl esterase** (PME) is often produced by the plant itself and it demethylates pectin. PME seems to play little role in the breakdown of pectin,

but the methanol released by this enzyme can act as a germination trigger and potential carbon source for *Athelia rolfsii* (see earlier).

2 **Pectic lyase** (PL) is a chain-splitting enzyme with both endo- and exo-acting forms. It cleaves the bonds between the sugar residues in a characteristic way, eliminating water during the process.

3 **Polygalacturonase** (PG) is another chain-splitting enzyme with both endo- and exo-forms. Unlike PL, this enzyme is a hydrolase – it uses water to add H^+ to one sugar residue and OH^- to the adjacent residue during cleavage of the bond.

Fungi are induced to synthesize pectic enzymes in the presence of pectic compounds, and the enzymes probably act in concert. This was shown when the genes encoding the equivalent enzymes of the bacterium *Erwinia carotovora* (the cause of potato black-leg disease) were engineered separately into *E. coli*. The recipient cells required a minimum of one exo-PL, two endo-PLs, and one endo-PG in order to rot the potato tissues. Pectic enzymes of fungi exist as a range of isomers which separate according to size and net charge during gel electrophoresis. The resulting pectic zymograms can be used to distinguish population subgroups of fungi (see Fig. 9.8), but the isomers may not differ significantly in function.

Initially it was thought that the principal role of pectic enzymes in pathogenesis was to separate cells, causing disruption of the fine protoplasmic bridges (plasmodesmata) that link cells through their walls. More recent studies suggest other roles. Pectic enzymes can be directly toxic to plant protoplasts *in vitro*, and

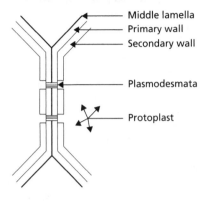

Junctions between plant cells

Middle lamella (composed mainly of pectic compounds)
Primary wall (composed mainly of hemicellulose (mixed polymers))
Secondary wall (composed mainly of cellulose)

Plasmodesmata (protoplasmic connections through the walls)

Protoplast

Pectic compounds

Composed mainly of **pectic acids** and **pectinic acids**.

Pectic acids: straight chains of α-1,4-galacturonic acid.

Pectinic acids: like pectic acids but some residues are methylated.

Pectic enzymes

1 Pectinmethylesterase hydrolyses methyl esters of pectinic acid to produce pectic acid.
2 Polygalacturonases (*exo*- and *endo*- types) hydrolyse pectic acid chains to galacturonic acid.
3 Pectin lyase (*exo*- and *endo*- types). Cleaves pectic acid chains to monomers by eliminating water in the process.

Fig. 14.7 Organization of the middle lamella (cementing layer) between plant cell walls, and summary of the pectic compounds and pectic enzymes that degrade them.

some of the partial breakdown products of pectin – the oligomers composed of 4 or 5 galacturonic acid residues that are generated by endo-PG – are especially toxic, causing rapid cell death.

Plant cells also contain a PG-inhibiting protein, perhaps as a general resistance mechanism. As the plant tissues age, the pectic compounds are increasingly cross-linked with calcium, which renders them more resistant to degradation, perhaps by restricting access by the endo-acting enzymes. For example, the hypocotyls (basal stem tissues) of bean seedlings (*Phaseolus vulgaris*) show a marked increase in resistance to infection by *Rhizoctonia solani* about 2 weeks after seed germination, coinciding with calcification of the middle lamella. The pathogens that rot older plant tissues (*Athelia rolfsii*, and *Sclerotinia sclerotiorum* which

rots the seed heads of crops such as sunflower) overcome this by producing large amounts of oxalic acid, which they also produce in laboratory culture. Oxalic acid can combine with calcium, removing it from association with the pectic compounds and rendering these susceptible to enzymatic attack.

Host-specialized necrotrophic pathogens

In contrast to the pathogens of immature or compromised hosts, many necrotrophic pathogens are host-specialized. They have evolved mechanisms for invading the normal healthy tissues of plants by overcoming the specific defense mechanisms of their hosts. These defense mechanisms can include physical barriers to penetration or fungitoxic chemicals, and they can either be preformed or formed in response to infection. In the following sections we discuss the characteristic features of these host-specialized pathogens.

The role of inoculum: a comparison of *Botrytis fabae* and *B. cinerea*

Botrytis cinerea and *B. fabae* are closely related fungi, but *B. cinerea* invades the compromised tissues of many plants, whereas *B. fabae* is a host-specialized pathogen of broad bean (*Vicia faba*). Both fungi produce ovoid conidia, but those of *B. fabae* are larger (20–25 μm long) than those of *B. cinerea* (10–15 μm), representing a roughly sixfold difference in spore volume. As we saw in Chapter 10, larger spores have a higher efficiency of impaction onto leaves, especially onto wide leaves such as those of broad bean plants. Large spores also have more endogenous food reserves to support infection.

The spores of both *B. fabae* and *B. cinerea* germinate when placed in water drops on the surface of broad bean leaves, but *B. cinerea* often fails to penetrate the leaf cuticle whereas *B. fabae* penetrates and causes a rapid host response – the death of the penetrated cells and their immediate neighbors, giving a large necrotic spot (Fig. 14.8). This reaction to infection is termed the **hypersensitive response**, and it is associated with an accumulation of fungitoxic metabolites in the dead cells, often sufficient to prevent further infection.

The difference in initial penetration by the *Botrytis* species is largely explained by their nutrient reserves. *B. cinerea* will penetrate and cause a necrotic spot if sugars are added to the inoculum drop, or sometimes even if the normal leaf surface microorganisms are killed by antibiotics so that *B. cinerea* can use the leaf-surface

exudates. Conversely, aged spores of *B. fabae*, taken from old parts of agar colonies, can fail to penetrate because their endogenous reserves have been depleted, but they penetrate if supplied with exogenous sugars. Thus, *B. cinerea* has evolved a strategy as a general invader of senescent tissues: it produces many more spores from a given amount of resource but these spores depend on nutrients released from compromised host tissues. *B. fabae* produces fewer spores from the same amount of resource, but each of them has a significant chance of initiating infection of the healthy host.

Such differences in the infectivity of pathogens can be quantified, to predict the likelihood of infection in different conditions. Fig. 14.9 shows this for *B. cinerea* and *B. fabae* when water drops containing different numbers of spores were placed on broad bean leaves. Infection was scored as the presence of absence of a necrotic spot beneath each water drop. The results are plotted as the logarithm of spore dose (spore number) against the percentage response (infection, expressed as a probit value, where the probit of 5 = 50%). This type of plot converts a skewed sigmoid curve into a straight line, and the **estimated dose** of spores that gives 50% probability of infection (the ED_{50} value) can be read from the graphs. The same type of plot is used widely in plant and animal experiments to obtain LD_{50} values for *lethal* toxicity of chemicals, etc. The host-specialized fungus *B. fabae* was found to have an ED_{50} of 4 in water drops, corresponding to a 16% chance of any single spore being able to initiate infection. By contrast, *B. cinerea* had an ED_{50} of about 500, corresponding to a 0.14% chance of a single spore initiating infection.

Experiments of this type provide information relevant to field conditions. For example, several different *Colletotrichum* spp. produce leaf spots on rubber leaves in Malaysia. By inoculating spores in droplets it was possible to predict: (i) the species most likely to infect from low spore numbers and (ii) the age at which the leaves are most infectible, before the development of a thick leaf cuticle prevents any further infection.

The roles of tissue-degrading enzymes

Many necrotrophic fungi produce tissue-degrading enzymes as a primary mechanism of pathogenesis, like the pectic enzymes discussed earlier. But the grass family, Gramineae, seems to be unusual because grasses have different wall matrix components compared with many other plants. Instead of pectin in the middle lamella they have mixed polymers of β-1,4-xylan and α-1,3-arabinose, with lesser amounts of noncellulosic β-1,4- and β-1,3-linked glucans. Cooper *et al.* (1988) investigated the enzymes produced by three stem-base pathogens of cereals – *Rhizoctonia cerealis* which causes

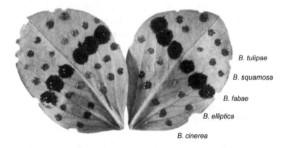

B. tulipae
B. squamosa
B. fabae
B. elliptica
B. cinerea

Fig. 14.8 Broad bean leaflets inoculated with water drops containing spores of different *Botrytis* spp. Only *B. fabae* is a host-specialized pathogen of broad bean plants; the other *Botrytis* spp. have different hosts, but *B. cinerea* is a weak pathogen. (Courtesy of J. Mansfield.)

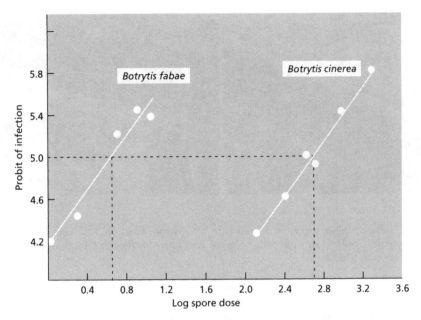

Fig. 14.9 Log dose/probit response plots of infection of broad bean leaves by *Botrytis fabae* and *B. cinerea*. See text for details. (Based on Wastie 1960.)

sharp eyespot disease, *Fusarium culmorum* which causes fusarium foot rot, and *Tapesia yallundae* which causes "true" eyespot disease. When these fungi were grown on cereal cell walls in laboratory culture, all three fungi produced arabinase, xylanase, and β-1,3-glucanase, but only small amounts of pectic enzymes. By contrast, when pathogens of nongrass plants (*Fusarium oxysporum, Verticillium albo-atrum*) were grown on walls of their hosts they produced large amounts of pectic enzymes. So, the cereal pathogens seem to be adapted to degrade the typical wall components of their hosts. But the surprising feature in this study was that the cereal pathogens produced arabinase and glucanase **constitutively**, in the absence of host wall material, and the production of these enzymes was not repressed by sugars. These are unusual features because polymer-degrading enzymes in general are both inducible and repressible. The likely explanation is that these host-specialized fungi grow only on cereals, so they have adapted to produce the necessary wall-degrading enzymes constitutively.

Plant defense against host-specialized necrotrophic pathogens

Plants have several potential lines of defense against invading pathogens. These defenses include both physical barriers and fungitoxic compounds in the plant tissues.

The main physical barrier to infection of many plants is the cuticle on the leaf or stem surface, and in this regard the enzyme **cutinase** has been thought to be important for fungal infection. There is circumstantial and correlative evidence for this. However, when targeted gene disruption was used to block cutinase production by three leaf pathogens, there was no reduction in pathogenicity of two of the fungi and only a partial reduction in a third fungus (VanEtten *et al.* 1995). A possible explanation is that some of the pathogenicity genes might only be expressed during infection and not in laboratory culture. Alternatively, there might be multiple forms of these genes, some of which are expressed *in vitro* and others only in the host environment. This example illustrates the difficulty of unravelling the often-complex host defense systems of plants.

In addition to preformed physical barriers, plants can respond to infection by developing a **papilla** – a localized thickening of the cell wall at the point of attempted invasion. In extreme cases the papilla continues to develop as the hypha grows (Fig. 14.10), so that it encases the penetrating hypha. In electron micrographs, papillae are seen as massive localized accumulations of wall material in which membranes and other degenerate cytoplasmic components have been trapped. Cytochemical stains usually reveal the presence of callose in papillae, but in the Gramineae the papillae often show staining reactions for lignin and suberin.

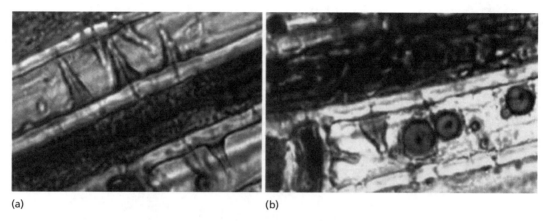

(a)   (b)

Fig. 14.10 (a,b) Lignified papillae produced in response to invasion of wheat root cells by the take-all fungus, *Gaeumannomyces graminis*. Narrow penetration hyphae are seen growing through the cell walls, but they have been stopped by the developing papillae.

The significance of papillae in plant resistance might be to delay invading hyphae while other defenses are activated. The lignin-like precursors in papillae of the Gramineae may be particularly important, because lignin can form complexes with the wall polymers, preventing their enzymatic degradation. Ride & Pearce (1979) demonstrated this for cereal leaves inoculated with *Botrytis cinerea* (not a pathogen of cereals) and several other fungi that induced the production of lignified papillae during attempted penetration. The normal leaf cell walls were digested easily by commercial wall-degrading enzymes, but the papillae and the surrounding "haloes" of lignified wall material were not digested. Thus, lignification and the development of papillae could have three resistance-related functions: to increase the wall thickness that a fungus must penetrate while depending on its endogenous energy reserves; to render the wall resistant to digestion by the fungal enzymes; and to confer a locally toxic environment caused by the phenolic precursors of lignin and suberin.

The oxidative burst

One of the earliest events in the attempted penetration of leaves and other above-ground parts of plants is an oxidative burst, equivalent to the oxidative burst in phagocytes. It has been studied mainly in relation to the hypersensitive response – the rapid death of a group of cells in response to localized invasion (e.g. Fig. 14.8) – but it seems to be a general reaction of plant cells to trauma caused by attempted parasitic invasion or chemical factors. Lamb *et al.* (1994)

have shown that it involves an extremely rapid (2–3 minute) production of H_2O_2 at or near the plant cell surface, by the reaction of oxygen with NADPH to produce superoxide:

$$O_2 + NADPH \rightarrow O_2^- + NADP^+$$

Then O_2^- is transmuted to H_2O_2 by a plasma membrane oxidase. H_2O_2 causes cross-linking of the plant cell wall proteins, which could strengthen the wall against attack by wall-degrading enzymes.

Chemical defenses

Plants produce a wide range of secondary metabolites for defense against invading organisms. Some of these are preformed compounds, collectively termed **phytoanticipins**. They are present either as active compounds or as precursors that are rapidly converted to active compounds in response to infection. A second category of defense compounds are termed **phytoalexins**. Again, they are secondary metabolites but they are synthesized *de novo* in response to infection. However, there is not always a clear difference between these two groups of compounds.

The phytoanticipins and phytoalexins are widely believed to be significant components of plant defense systems – at least in the sense that they can prevent most potential invaders from causing progressive disease of plants. But host-specialized pathogens can sometimes overcome the defenses of their specific hosts, and we consider some representative examples of this below.

Phytoanticipins

Some of the best evidence for a role of phytoantici-pins has come from studies of the take-all fungus of cereals *Gaeumannomyces graminis*, and related studies on the tomato pathogen *Septoria lycopersici*. As discussed in Chapter 9 (see Fig. 9.12), these host-specialized fungi can overcome the phytoanticipins of their host plants by producing enzymes (avenacinase and α-tomatinase) that detoxify the phytoanticipins, whereas fungi that are not specifically adapted to these host plants are killed by these compounds.

Phytoanticipins may be involved in several other plant–fungus interactions. A classic example is the resistance of red- and yellow-skinned onions to *Colletotrichum circinans*, the cause of "onion smudge" disease, whereas white-skinned onions are susceptible. The pigments themselves are not important but they are associated with high levels of the phenolic phyto-anticipins, **catechol** and **protocatechuic acid**.

Phytoalexins (warding-off compounds)

Phytoalexins are distinct from phytoanticipins because they are low-molecular-weight compounds produced **in response to infection**. More than 350 phytoalexins have been reported across a range of crop plants, especially dicotyledonous plants such as beans, pota-toes, peas, and cotton. Most of them are flavonoids, phenolics, or terpenoids, synthesized by three major secondary metabolic pathways: the acetate-malonate route, the acetate-mevalonate route, and the shikimic acid route, similar to the secondary metabolic path-ways of fungi (Chapter 7). The structures of four phytoalexins are shown in Fig. 14.11.

A substantial body of evidence suggests that phyto-alexins act as defense compounds – at least in protecting plants against general attack by fungi. These compounds accumulate rapidly during the hyper-sensitive response, and at growth-inhibitory levels in or around the dead or dying tissues. However, they can also be detoxified by pathogens that are specific to indi-vidual plant species, so there is a dynamic interaction between the speed of host response and the ability of a fungus to detoxify the defense compounds.

A similar localized accumulation of phytoalexins can often be induced by artificial wounding or appli-cation of a toxic chemical to the plant surface. It is always preceded by an oxidative burst, and H_2O_2 is thought to be one of the initial signals that activates the defense genes involved in phytoalexin production. The sequence of events leading to phytoalexin accu-mulation has been studied by exploiting the discovery that low-molecular-weight compounds consisting of a few β-linked sugar residues can act as powerful **elicitors** of the hypersensitive response. These oligosaccharides are released from the extracellular polysaccharides of fungal hyphae, probably by the actions of β-glucanase enzymes that reside in plant cell walls. When applied to plant cells or tissues *in vitro*, the elicitors cause an early oxidative burst and the induction of plant enzymes of the phenylpropanoid pathway, including **phenylalanine ammonia lyase** which is involved in phytoalexin production, and synthesis by the plant

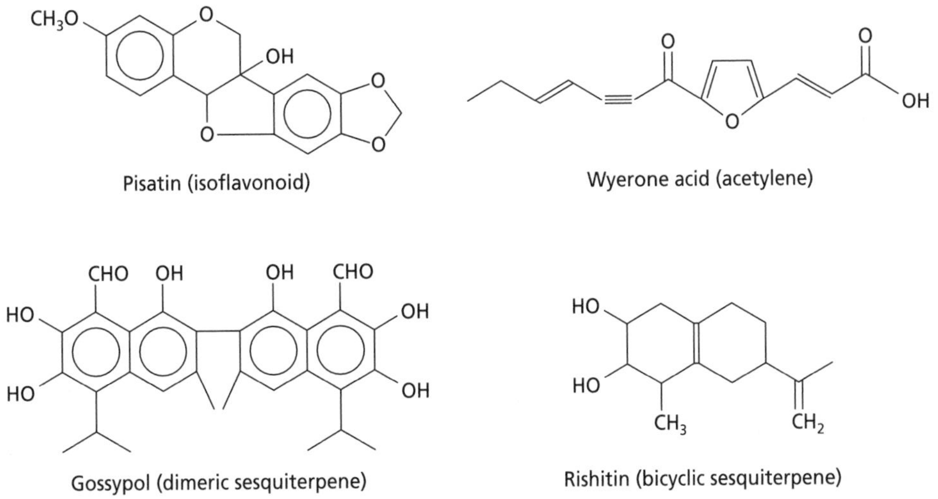

Fig. 14.11 Structures of four phytoalexins: **pisatin** from peas; **wyerone acid** from broad bean; **gossypol** from cotton; **rishitin** from potato.

of **chitinase** and **β-1,3-glucanase** which can degrade fungal walls.

Clearly, this cascade of cellular responses creates a hostile environment for a potential pathogen. So how do host-adapted necrotrophic pathogens overcome or avoid these defenses? Studies on the *Botrytis*–broad bean interaction have shown that both *B. fabae* and *B. cinerea* can cause a hypersensitive response in appropriate conditions (see earlier) and the phytoalexin **wyerone acid** accumulates in the dead cells. But *B. fabae* is more tolerant of wyerone acid than is *B. cinerea in vitro*, and *B. fabae* also causes less wyerone acid to accumulate in the lesion, perhaps by detoxifying it. Thus, *B. cinerea* never spreads further from the necrotic spot, whereas *B. fabae* is checked temporarily but then can spread through the surrounding tissues, killing them progressively. The resulting large brown lesions give rise to the name for this disease – chocolate spot (see Fig. 14.8).

The mechanisms of detoxification of phytoalexins have been reported for some other fungi. The pathogen of pea plants, *Nectria haematococca*, detoxifies **pisatin** by demethylation (Fig. 14.12). Other strains of this pathogen that infect bean plants detoxify one of the bean phytoalexins **kievitone** by hydration (Fig. 14.12). In both cases the ability of the pathogen to infect the plant is correlated with the ability to detoxify the phytoalexin. Moreover, when the pisatin demethylase gene from *N. haematococca* was transformed into *Cochliobolus heterostrophus*, a maize pathogen, this fungus became a pathogen of pea leaves. So, there is strong evidence that host-adapted pathogens have evolved mechanisms for overcoming the inhibitory effects of their hosts' phytoalexins, and even the acquisition of a single phytoalexin-detoxifying gene can change the host range of a fungus. There is, however, a twist to this story. In the **Nectria–bean interaction**, all strains of *N. haematococca* were found to produce the hydrating enzyme, kievitone hydratase, but some strains did not secrete it from the hyphae. These strains were nonpathogenic, in contrast to strains that secreted the enzyme. In the **Nectria–pea interaction** the relationship between pathogenicity and production of the detoxifying enzyme was supported by many lines of evidence, but targeted disruption of the gene for pisatin demethylase led to strains that were still pathogenic. Reviewing all the work on detoxification of phytoanticipins and phytoalexins, VanEtten *et al.* (1995) suggested that pathogens may have evolved several mechanisms for overcoming the host defenses, so that some of these mechanisms, although still expressed, may be redundant.

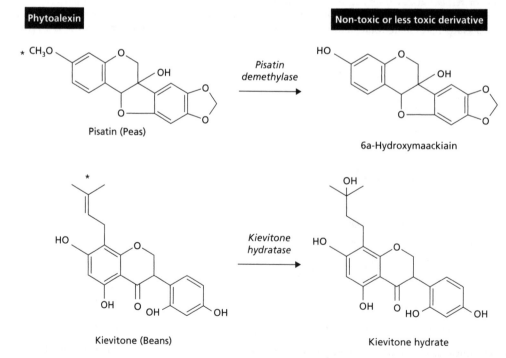

Fig. 14.12 Two examples of detoxification of phytoalexins (at the position marked *).

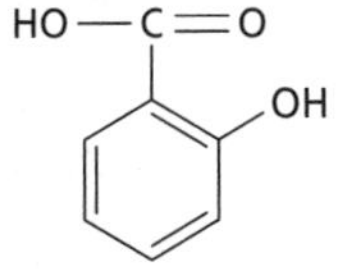

Fig. 14.13 Salicylic acid and jasmonic acid – two key signalling compounds that enhance the resistance of plants to pathogenic attack.

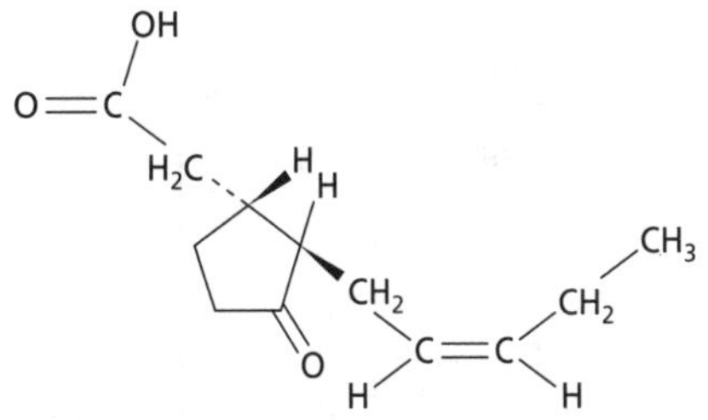

Salicylic acid

Jasmonic acid

Systemic acquired resistance and induced systemic resistance

Earlier in this chapter, we noted that plants can respond to attack from necrotrophic pathogens by mounting a **hypersensitive response**, in which the infected cells and their immediate neighbors die rapidly and accumulate fungitoxic compounds. Although this response is localized, it leads to the production of signalling molecules that are mobilized throughout the plant, so that the plant is, in effect, immunized. Any subsequent infection will lead to increased resistance or reduced expression of disease. This phenomenon is termed **systemic acquired resistance** (SAR). It involves the activation of a set of genes that synthesize pathogenesis-related proteins, including β-glucanases and chitinase which can attack the cell walls of invading fungi. The aspirin-like compound **salicylic acid** (Fig. 14.13) has been identified as the signal molecule involved in SAR, and this is supported by the finding that salicylic acid can be applied exogenously and will induce the synthesis of pathogenesis-related proteins, leading to enhanced plant resistance. In fact, salicylic acid is marketed commercially for this purpose. The remarkable point is that SAR confers a generalized resistance to a wide range of pathogens, including fungi, bacteria, and viruses.

A second type of induced resistance develops in response to colonization of plant roots by specific nonpathogenic strains of rhizosphere bacteria, such as *Pseudomonas fluorescens*. But this defense system is mediated by a different set of signalling compounds, including **jasmonic acid** and **ethylene**. This type of induced resistance is termed rhizobacteria-mediated **induced systemic resistance** (ISR). Like SAR, induced systemic resistance acts against several different types of pathogen, including strains of *Fusarium oxysporum*, the downy mildew pathogen *Peronospora parasitica* (Oomycota), and the bacterial pathogens *Xanthomonas campestris* and *Pseudomonas syringae*. Thus, there are two distinct signalling pathways, one mediated by salicylic acid and one by jasmonic acid/ethylene, and

both confer resistance to pathogenic attack. Recent studies suggest that both of these pathways can act simultaneously, because there is no significant crosstalk between them. Therefore, the combination of SAR and ISR could confer an enhanced level of plant resistance.

Vascular wilt diseases

Vascular wilt pathogens cause some of the most devastating plant diseases. Their characteristic mode of infection is to enter the water-conducting xylem vessels, either by means of vectors such as bark beetles in the case of Dutch elm disease (*Ophiostoma ulmi* and *O. novo-ulmi* – see Chapter 10) or via wounds, or by invasion of the young root tips. Then these fungi proliferate in the xylem as spores or yeast-like budding cells and are carried upwards in the water flow. The spores become trapped on the perforated end walls of the xylem vessels, where they germinate, grow through the pores, and produce further spores that are carried progressively upwards (Fig. 14.14).

Plants respond to invasion by the vascular wilt fungi in several ways (Fig. 14.14):

- Balloon-like swellings (called tyloses) can bulge into the xylem from parenchyma cells adjacent to the xylem vessels; this seems to be an initial response before other defense mechanisms come into play.
- Pectic gels are extruded into the xylem, through bordered pits where the vessel wall is thin because it consists only of the primary wall, not overlaid by the secondary cellulosic wall.
- Phenolic compounds are released into the vessels from phenol store cells. The phenolic compounds are then oxidized and polymerize, helping to stabilize the gels and to create a fungitoxic environment.
- Phytoalexins accumulate in the vessels, presumably as part of a general stress response.

If the host responds fast enough to invasion of the xylem vessels, the infection will be contained and the

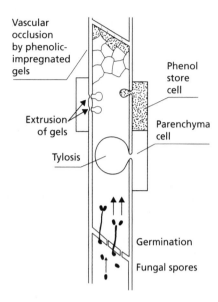

Fig. 14.14 Diagrammatic representation of pathogen spread and occlusion of the xylem vessels in vascular wilt diseases. Spores or yeast-like cells are carried upwards in the water flow and become trapped on the perforated vessel end walls. They germinate, grow through the pores, and produce further cells that are carried upwards to the next vessel end wall, and so on. (Based on Beckman & Talboys 1981.)

plant will suffer no significant damage. But if the host response is slow the fungus can progressively colonize the vessels, causing rapid wilting and plant death (Figs. 14.15, 14.16).

The interesting feature of vascular wilt fungi is that **they remain confined to the nonliving xylem vessels** until the plant is heavily diseased, and only then do they invade the living parenchyma cells, producing thick-walled resting structures which eventually return to the soil when the tissues decompose.

Vascular wilt fungi

The three main fungi that cause vascular wilt diseases are *Fusarium oxysporum*, *Verticillium albo-atrum*, and *V. dahliae*. Of these, *F. oxysporum* is the most important because it consists of more than 80 strains that show specific pathogenicity to particular crop species (Armstrong & Armstrong 1981). These strains are termed "special forms" (*formae speciales*), abbreviated to ff. ssp (plural) or f. sp (singular). For example, the vascular wilt pathogen of bananas is termed *F. oxysporum* f. sp. *cubense*, and the vascular wilt pathogen of tomato is *F. oxysporum* f. sp. *lycopersici*. All these

strains look identical, and all survive by producing **chlamydospores** (spores derived from individual hyphal compartments by accumulating nutrients and developing a thick, resistant cell wall). These spores can survive in soil for many years, but probably also germinate and maintain their population by growing in a saprotrophic mode in the rhizosphere. *Verticillium albo-atrum* and *V. dahliae* also cause serious wilt diseases, such as verticillium wilt of hops and many other plants. However, they seem to be less host-specific than *F. oxysporum*, because strains isolated from one host will often infect several other unrelated hosts in artificial inoculations.

Pathogenicity determinants

The typical symptoms of vascular wilt diseases usually include the progressive yellowing and death of leaves, accompanied by early collapse of the petioles (leaf stalks) so that the leaves hang down – a feature termed epinasty. These symptoms strongly suggest a role of toxins in the disease syndrome. **Fusaric acid** was one of the first potential toxins to be identified. It is known to be produced by *F. oxysporum* in laboratory culture. It has also been detected in diseased plants, and when applied to plant protoplasts *in vitro* it increases their permeability to ions. However, there is still doubt about the role of fusaric acid *in vivo* because its concentrations in xylem fluid are often low. A more convincing role has been established for a toxin termed **cerato-ulmin**, produced by *Ophiostoma ulmi* (the cause of Dutch elm disease). When the bases of cut elm shoots are immersed in solutions of cerato-ulmin they develop the typical leaf symptoms of Dutch elm disease. Cerato-ulmin is a hydrophobin (see Chapter 5), and hydrophobin molecules spontaneously assemble into water-repellent layers at an air–water interface. Therefore they could have a significant role in restricting water flow in xylem vessels.

Role of environmental factors

Environmental factors play a major role in vascular wilt diseases. Early evidence of this was found in the major banana plantations of Central America, where soils could be categorized as either "long-life" or "short-life." In long-life soils, banana plantations could remain productive for many years, with little evidence of disease, whereas short-life soils became nonproductive within a few years of establishing the crop. Chemical analyses revealed a general correlation between the different types of clay mineral and the life of banana plantations. Long-life soils tend to have a higher proportion of montmorillonite clays, as opposed to the kaolinite

Fig. 14.15 Panama disease of banana plants, caused by *Fusarium oxysporum* **forma specialis** ("special form") *cubense*. (a) A young banana plant in an advanced stage of disease; most of the leaves have died and collapsed as a "skirt" around the stem base, and the erect leaves show progressive yellowing and necrosis. (b) Older, fruit-bearing banana plants infected by *F. oxysporum*, showing extensive collapse of the leaves.

clays of short-life soils. This has been linked to a higher population of antagonistic bacteria, because montmorillonite clays hold more nutrients and create a more favorable pH for bacteria.

At one stage the progressive decline of banana plantations in Central America threatened the entire future of banana production in that region, which was based on a single cultivar, "Gros Michel," that is highly susceptible to Panama disease. The threat was averted by the chance discovery of a more wilt-resistant cultivar of the "Cavendish" type. Owing to its value as a disease-resistant cultivar, "Cavendish" bananas were only planted on the more favorable sites in Central America, and they have remained disease-free for more than 50 years. However, the situation is different in the subtropics (South Africa, Taiwan, Canary Islands, and Northeast Queensland) where the Cavendish cultivars have been introduced. Many of these plants have now died from Panama disease, coinciding with the emergence of a new pathotype (race 4) of *F. oxysporum* f. sp. *cubense*. The conditions in these countries are marginal for banana production because the winter temperatures drop to 12°C or lower. Bananas stop growing at this temperature (evidenced by their failure to make any further leaf growth), but *F. oxysporum* f. sp. *cubense* can still grow at 12°C *in vitro*. This could have facilitated the emergence of the new pathotype.

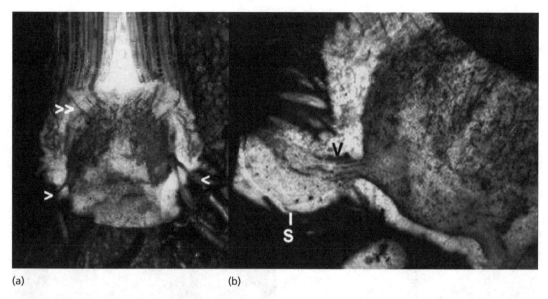

Fig. 14.16 Sections through the base of a diseased banana plant. (a) Heavily discolored vascular strands (single arrowheads) where the fungus entered the base of the corm from roots, and (double arrowhead) where the infection is spreading up the vascular strands in the leaf sheaths. (b) An older corm producing "suckers" (S) that will produce the next banana plant. Vascular disease (V) is already spreading into the new plant.

Fusarium *wilt-suppressive soils*

The terms "long-life" and "short-life" for *Fusarium* wilt-suppressive soils have now been replaced by the terms **disease-suppressive** and **disease-conducive soils**. Many examples of disease-suppressive soils have now been reported for vascular wilt diseases and other pathogens across the world (Schneider 1982). In almost all cases there is clear evidence that microorganisms contribute to suppressiveness, because suppressive soils can be made conducive by pasteurization, and conversely the introduction of a relatively small amount of suppressive soil (10% or less) to pasteurized soil or even to normal field soil can make a soil suppressive.

Some of the most interesting examples of disease suppression are found in sites such as the Chateaurenard region of France, where vegetables have been grown for centuries with little or no disease. In these sites there is strong evidence that nonpathogenic strains of *Fusarium oxysporum* compete with the pathogenic strains for organic substrates. Also, in these suppressive soils the chlamydospores germinate poorly, the germ-tubes grow poorly, the pathogen population in the soil declines more rapidly, and a much higher inoculum level is needed to cause disease. These soils are found to be suppressive to a wide range of *formae speciales* of *F. oxysporum*, but are not suppressive to other, unrelated pathogens such as *Verticillium dahliae*. Attempts

are now being made to exploit nonpathogenic strains of *F. oxysporum* as commercial biological control agents of the pathogenic fusarium wilt fungi. This is not without its dangers, because we still know little about the factors that govern the host ranges – or the pathogenicity – of vascular wilt fusaria (Fravel *et al.* 2003).

The enigma of vascular wilt fusaria

The vascular wilt fusaria are enigmatic because, on the one hand, they cause some of the most devastating plant diseases and show a very high degree of host-specificity, yet on the other hand they grow only in the nonliving xylem tissues (after entry through the immature root tips or through wounds) and only invade the living plant tissues when the plant begins to senesce. In effect, these fungi behave as weak parasites which, for most of their life in the plant, depend on the very low levels of organic nutrients that leak into the xylem vessels. In some respects this behavior is similar to the mode of growth of the nonpathogenic fungal **endophytes** which grow sparsely in the wall spaces and intercellular fluids of plants without causing obvious harm to the host (discussed below). Endophytism is characterized by a basic compatibility between the colonizing fungus and the host cells. Consistent with this, some of the interactions between vascular wilt fusaria and their host plants – for example, *F. oxysporum* f. sp. *pisi* on peas – are governed by a

gene-for-gene relationship typical of biotrophic plant pathogens, discussed later in this chapter.

Smut fungi

The smut fungi are members of the Basidiomycota and gain their name from the fact that they produce millions of black, sooty spores, known as **teliospores**. Over 1000 species of smut fungi are known to parasitize plants, but the species of most economic importance are those that attack cereal crops, including loose smut of wheat (*Ustilago nuda*; Fig. 14.17), maize smut (*Ustilago maydis*), and stinking smut of wheat (*Tilletia caries*), so-named because the spores have a fishy smell due to the presence of trimethylamine. The characteristic feature of smut fungi is that they develop slowly within the plant tissues, often producing only small amounts of hyphal growth, but later in the growing season they produce large numbers of diploid teliospores, which develop in place of the cereal grain or other reproductive structures. For example, in loose smut of cereals (Fig. 14.17) the whole flowering spike is replaced by a mass of spores, which are exposed when a thin sheath which contains these spores breaks down at maturity. Stinking smut is similar, but the grains appear to be normal because they are surrounded by the seed coat (pericarp) and the spores are only released when the grain is threshed.

Plants become infected by loose smut when germinating spores penetrate the developing ovary, so that the resulting seeds already contain the mycelium. Then, when the seeds germinate the fungus progressively colonizes the stem tissues and eventually enters the flowering spike where it replaces the grain with a spore mass. Infection by stinking smut is somewhat different: the spores can survive for up to 15 years, and they germinate at the same time as the grain, penetrate the base of the shoot (coleoptile), and then progressively colonize the tissues, leading finally to the production of a spore mass. In other smuts the process of development is different again. For example, in *Uromyces anemones* (a smut fungus that grows on anemone and buttercup) the spore masses develop beneath the surface of the stem or leaf before they mature and erupt through the plant epidermis.

When the teliospores germinate they undergo meiosis to produce a haploid phase termed the promycelium. This produces uninucleate sporidia, equivalent to basidiospores. Then plasmogamy (cell fusion) of sporidia of compatible mating types gives rise to a dikaryon, and this will eventually produce the mass of teliospores. An interesting feature of smut fungi is that the dikaryotic phase seems to be obligately parasitic on plants, but the monokaryotic, uninucleate sporidia multiply as yeast-like budding cells. These are nonpathogenic and can be cultured on standard laboratory media.

(a)   (b)   (c)

Fig. 14.17 (a,b) Loose smut of wheat caused by *Ustilago nuda*. At maturity the flowering spikes produce a mass of black spores in place of the seeds. The image in the center compares a smutted flowering spike with a normal spike. (c) Warty, ornamented spores (smut spores) about 8 μm diameter.

Fungal endophytes and their toxins

In recent years the leaves and roots of many types of plant have been shown to harbor fungi that grow as **symptomless endophytes**. These fungi do not penetrate the plant cells; instead they grow as relatively sparse hyphae in the spaces between the cells or within the walls of the plant cells, utilizing low levels of nutrients in the intercellular fluids and causing no obvious damage. In many ways, this behavior is similar to the way in which the vascular wilt fusaria colonize the low levels of nutrients in the xylem fluids, discussed earlier.

The significance of fungal endophytes lies in the fact that they produce a wide range of secondary metabolites of potential commercial importance. In addition, some of the endophytes confer abiotic stress tolerance on their plant hosts, or act as insect antifeedants. In the discussion that follows we will focus on one particular group of fungal endophytes – those that grow in some pasture grasses and have repeatedly been shown to reduce insect damage (Bony *et al.* 2001).

The grass-associated endophytes grow very slowly in laboratory culture and do not produce sporing stages. In fact, some of them have probably lost this ability altogether because they always grow within a plant and are transferred from generation to generation by growing into the seed coat. This makes it easy to compare the behavior of endophyte-infected and endophyte-free populations of grasses. The few isolates that have been induced to sporulate in culture produce a simple conidial stage classified as *Neotyphodium* (previously called *Acremonium*) and resembling the conidial stage of the ergot fungus *Claviceps purpurea*. These "clavicipitaceous endophytes" (so-named because they are related to the ergot fungus) are found in several important pasture grasses such as *Lolium* (ryegrass), *Festuca* (fescue), and *Dactylis glomerata* (cocksfoot) in the USA, Europe, and New Zealand. They grow within the leaf tissues but their growth seems to be tightly regulated because it is synchronized with growth of the leaves, and often ceases when the leaves mature (Clay 1989).

Endophyte-infested perennial ryegrass has been shown to contain several mycotoxins, including three that are particularly important:

1 The saturated aminopyrrolizidine alkaloids such as **lolitrem B**, a compound that has been shown to be responsible for the damaging "ryegrass staggers" disease of grazing sheep and cattle in New Zealand. These alkaloids can reach levels of up to 2% in the grass biomass.
2 **Ergovaline**, which causes diseases such as "fescue foot" and "fescue toxicosis" in grazing animals.
3 **Peramine**, a tripeptide that is repellant and toxic to insects, but not to mammals.

The production of these compounds is probably favored by the slow, substrate-limited growth of endophytes within plant tissues, because secondary metabolites characteristically accumulate in conditions that restrict normal growth (Chapter 7). The multiple effects of these mycotoxins raise dilemmas. On the one hand, endophyte-infected grass varieties can pose serious threats to grazing livestock. On the other hand, there are many reports that endophytes increase the stress-tolerance of grasses, and confer a competitive advantage over endophyte-free grasses by deterring insect damage.

The endophyte *Neotyphodium coenophialum* is found in tall fescue grassland and produces **ergovaline** in the plant tissues. Ergovaline causes a range of symptoms in grazing animals, including hyperthermia, weight loss, reduced rates of pregnancy, decreased milk production, and (in horses) birth defects and abortion. The endophyte *N. lolii* in perennial ryegrass produces **lolitrem B** as its major toxin. This neurotoxin is responsible for a condition called ryegrass staggers, where animals experience tremors and loss of co-ordination. The potential extent of these problems in the USA was revealed by a survey of horse pastures (Fig. 14.18; USDA 2000). Nationally, 61.6% of sampled pastures tested positive for endophytes, and a total 28.5% tested positive for toxin. Without intervention, the infection rates would increase with time because infected grasses can outcompete noninfected grasses.

Because the clavicipitaceous endophytes never leave the host plant, pasture and turf managers have options. For grazing animals, a pasture can be reestablished with certified endophyte-free seed, whereas managers of ornamental or recreational turf can elect to sow endophyte-infected seed for improved grass vigor and insect resistance. This has spawned a niche market in the USA, where "endophyte-enhanced grasses" are available through garden centers and seed stores for use in amenity grasslands, golf-courses, sports turf, etc.

Phytophthora diseases

The genus *Phytophthora* contains more than 50 described species, most or all of which are plant pathogens (Erwin & Ribeiro 1996). In fact the name *Phytophthora*, derived from Greek, means literally **plant destroyer**. Among the many species of *Phytophthora*, the potato blight pathogen, *P. infestans*, has become notorious, even though it has a narrow host range – essentially potato plants and, to a lesser extent, tomato (another member of the potato family). But several other species have very broad host ranges and cause various types of disease, including:

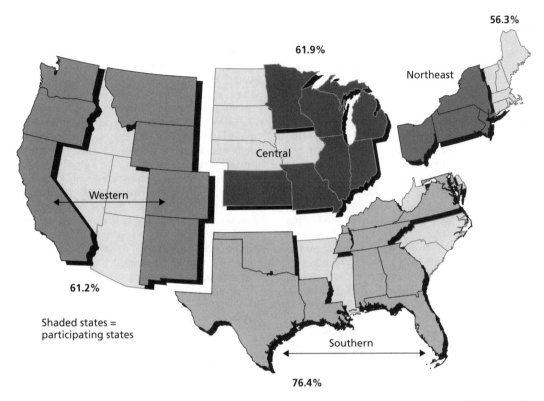

Fig. 14.18 Percentage of pastures testing positive for *Neotyphodium* endophytes in a survey conducted by United States Department of Agriculture in 1998. (Source: http://www.aphis.usda.gov/vs/ceah/Equine/eq98endoph.htm)

- **root rots** (e.g. *P. cinnamomi*, a major pathogen of the root tips of avocado trees, and of *Eucalyptus* vegetation in Australia);
- **cankers**, which develop near the base of trees (e.g. *P. ramorum* which is currently devastating the native oak woodlands of the coastal mist belt of California and southern Oregon, causing **sudden oak death**);
- **collar rots** and **crown rots**, which develop at the base of the stems of woody and herbaceous plants (e.g. *P. capsici*, which attacks tomatoes, cucumber, watermelon, squash, and pumpkins);
- **aerial blights** (e.g. *P. nicotianae* on poinsettias and many other plants);
- **fruit rots** of many plants (e.g. *P. palmivora* on *Capsicum* peppers, papaya, citrus, etc.).

Apart from the economic damage caused by these organisms, *Phytophthora* spp. are notable because they are not fungi at all. Instead, as we have noted repeatedly, they are members of the Oomycota, within the Kingdom Straminipila (which includes the orange-colored algae) and they have several distinctive features, such as cellulosic cell walls, diploid vegetative stages,

motile flagellate spores, and biochemical features that make them insensitive to many conventional fungicides.

Here we will consider two species of *Phytophthora* that cause very different types of disease – *P. infestans* and *P. ramorum*. However, they also have several features in common, including the fact that they are spreading globally and causing serious damage to wild plants and crop plants. They demonstrate the continuing threat posed by members of this genus.

Phytophthora infestans

The centre of origin of wild potatoes is central Mexico, and this is almost certainly the centre of origin of *P. infestans*, because this fungus is heterothallic, requiring two mating types for sexual reproduction (mating types A1 and A2) and both of these are found commonly in central Mexico. The fungus is known to have reached Europe, including Britain, before the 1840s when it caused devastating losses of potato crops. But until the early 1980s only the A1 mating type was found commonly outside of Mexico. Then the A2 mating type began to spread across most potato-growing regions of the world, presumably as a result of international trade, and this has led to an increased diversity of the *P. infestans* population, which was already diverse as evidenced by the existence of many different isozyme profiles.

In field conditions, potato blight is first seen as black, spreading lesions on the foliage, and in cool humid conditions these lesions produce masses of sporangiophores that emerge from the leaf stomata. Potato tubers become infected later in the growing season, perhaps by the spread of motile zoospores (Chapter 10), then the tubers start to rot and are destroyed by secondary bacterial invaders. The main way in which *P. infestans* is disseminated is by the production of detachable sporangia which can be wind-borne or splashed onto foliage (see Fig. 5.17). These sporangia can germinate in two ways, depending on the climatic conditions, and this can be mimicked in laboratory conditions. At temperatures of around 20°C or higher, the detached sporangia germinate by forming a hyphal outgrowth (Fig. 14.19), whereas at temperatures of 12°C or lower the sporangia undergo cytoplasmic cleavage to produce motile zoospores. The production of zoospores at low temperatures could be especially significant in early-season spread of the disease, whereas "direct" germination of sporangia could be more important later in the season – for example, in the infection of tubers. This dual strategy,

also shared by *P. erythroseptica* which causes pink rot of potatoes, is part of the formidable arsenal of *P. infestans*.

However, the major weapon of this pathogen is, undoubtedly, its ability to overcome the "major" race-specific resistance genes that have been bred into potato cultivars over the years. *P. infestans* is a hemibiotrophic plant pathogen: in its early stages of colonizing a leaf it feeds by means of haustoria which tap into the host cells and draw nutrients from them, but at a later stage it switches to a necrotrophic mode and causes generalized tissue breakdown. Every new major resistance gene introduced into potatoes was soon overcome by a mutation in the pathogen, leading to a new "race" of *P. infestans* that could not be controlled by the existing single major genes. Now a new breeding strategy is in place, to try to exploit general resistance (also termed horizontal, rate-limiting or multigenic resistance) coupled with the use of fungicides and cultural practices.

Phytophthora ramorum: sudden oak death

A new species of *Phytophthora* was discovered in Germany and the Netherlands from 1993 onwards, and was described formally in 2001 as *P. ramorum*. It causes damage to the branches of *Rhododendron* and, less often, *Viburnum* bushes, sometimes leading to

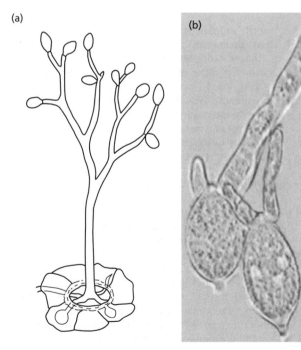

(a)

(b)

Fig. 14.19 *Phytophthora infestans.* (a) A sporangiophore, with attached sporangia, emerging from a stoma on the underside of a potato leaf. (b) Two detached sporangia, producing germ-tube outgrowths from near the apical papilla. Note the characteristic broken "stalk" at the base of the sporangia.

death of these shrubs. This disease is now spreading on ornamental rhododendrons and viburnums in tree nurseries across Europe. Meanwhile, in the coastal fog belt of northern California and southern Oregon, a sudden death of oak trees was occurring, and this disease was also found to be caused by *P. ramorum*. It attacks several types of oak, but mainly the "live oaks" and "tan oaks" which are part of the natural vegetation of the coastal shrub/tree community. The fungus attacks the plants in two ways – either by producing lesions on the leaves and terminal shoots of a wide range of shrubs, leading to shoot dieback, or by producing **cankers** near the bases of the trees, where the cambium (which lies just beneath the bark and produces the annual rings of new wood) is progressively destroyed. The symptoms of this include cracking of the bark and seepage of dark, viscous sap from the bark near the base of the tree – a symptom termed gummosis (Fig. 14.20). In advanced stages of the disease the trees die suddenly and dramatically – hence the name, **sudden oak**

death. However, although the death of the trees occurs suddenly it is preceded by a slow, progressive build up of infection beneath the bark. This can often go unnoticed until the trees suddenly develop advanced symptoms.

There is a close parallel between *P. infestans* and *P. ramorum*, because both have been introduced into parts of the world where they did not occur before and both have caused widespread damage. Current evidence suggests that the European population of *P. ramorum* consists only of the A1 mating type, whereas the North American population is of the A2 mating type. This strongly indicates separate sources of origin of the two populations, but the natural geographical origin of the fungus remains unknown. Here there is a parallel with the spread of another aggressive *Phytophthora* species, *P. cinnamomi*, which attacks many plants across the world, but the geographical origin of this fungus is unknown. It is causing serious damage to the eucalypt vegetation of Australia.

(a) (b) (c) (d) (e)

Fig. 14.20 Sudden oak death caused by *Phytophthora ramorum*. (a) Dead and dying trees of coast live oak (*Quercus agrifolia*) in the mixed oak community of the coastal fog belt of southwestern USA. Dead or dying trees are indicated. (b) Dark viscous sap is exuded from the bark at the base of a heavily infected coast live oak. (c) Removal of the bark reveals the presence of dark zone lines. (d) Terminal die-back and wilting of the shoot tip of tan oak (*Lithocarpus densiflorus*) is one of the characteristic features of the disease. (e) Necrotic, spreading leaf spots are a further symptom of sudden oak death – in this case on leaves of *Azalea* or *Rhododendron*. (Images courtesy of Joseph O'Brien, USDA Forest Service, www.invasive.org; accessed 22 March 2004.)

Whenever a fungus invades an area of natural vegetation it is likely that the fungus has been introduced from elsewhere, because plants in natural communities co-evolve with the resident pathogens.

Biotrophic plant pathogens

Biotrophic plant pathogens are characterized by the fact that they have an extended nutritional relationship with **living host cells**, in contrast to necrotrophic pathogens which kill the plant tissues. The success of this type of parasitism depends on two essential features: (i) the ability to avoid eliciting host cell death and (ii) a means of securing a continuous nutrient supply from the living host tissues. Thus, biotrophs are in many ways the epitome of successful parasites – they feed from the host cells without killing them.

There are many types of biotrophic fungus and fungus-like organisms, because this mode of parasitism has evolved independently on several occasions. The major groups of biotrophs include the **rust fungi** (Basidiomycota), the **powdery mildew fungi** (Ascomycota), the **downy mildews** (Oomycota), and the intracellular **plasmodiophorids** (protists of uncertain taxonomic affinity, but clearly distinct from fungi). In addition to these, the endophytic fungi grow essentially as biotrophs. A classic example is the fungus *Fulvia* (*Cladosporium*) *fulvum* which causes blue mould disease of tomato plants. Unlike the ubiquitous *Cladosporium* spp. that grow as epiphytes on leaves (Chapter 11), *F. fulvum* enters the host stomata from a germinating spore and forms an extensive intercellular network in the leaf, before emerging from the stomata as conidiophores which release further spores. It has no obvious means of feeding from the living host cells except by scavenging leachates; but it might enhance the rate of leakage and also use some of the cell wall polymers. A strain of *F. fulvum* can only feed from the host if it avoids inducing a hypersensitive response, and this is governed by "avirulence" genes of the pathogen and corresponding resistance genes in the host. This **gene-for-gene relationship** with the host plant (Honee *et al.* 1994) is commonly found in biotrophic plant pathogens, and is discussed later.

In the following sections we consider several examples of biotrophic plant pathogens.

Haustorial biotrophs

Many biotrophic fungi initiate infection from spores that land on the host surface and then germinate and enter through the stomata. The initial events involve the precise orientation of germ-tube growth, using cues such as plant surface topography and nutrient gradients, discussed in Chapter 5. The hyphae then grow between the host cells, and attach to individual cells by producing a **haustorial mother cell**. From this, a narrow penetration hypha grows through the host cell wall but it does not penetrate the membrane. Instead, the host cell membrane invaginates to accommodate the invading fungus, which develops into a **haustorium** that is always separated from the host cytoplasm by an **extrahaustorial membrane** (see Fig. 5.7).

As shown in Fig. 14.21, the mature haustorium is surrounded by a fluid matrix, and there is a tight "seal" in the neck region to prevent the fluid from escaping. One of the consequences of this in experimental work is that the whole haustorial complex (enveloping membrane, fluid matrix, and the haustorium itself) can be isolated by digesting the host cell wall with enzymes. From studies on these isolated haustorial complexes of rust and powdery mildew fungi it has been shown that the extrahaustorial membrane allows the free passage of sugars and amino acids. The wall and membrane of the plant cell can also be expected to allow the passage of nutrients through appropriate membrane transporters. So, the crucial interface for nutrient uptake by the fungus is likely to be the haustorial membrane (Fig. 14.21).

In a model proposed by Voegele & Mendgen (2003), sugars and amino acids are believed to be taken into the haustorium in association with H^+ ions by membrane symporters, using the proton motive force generated by ATPase. The sugars are then used for biosynthetic reactions within the haustorium. Fructose (derived by hydrolysis of sucrose to glucose and fructose in the extrahaustorial matrix) is converted to **mannitol**, which is translocated into the fungal hyphae and ultimately into the developing spores. The net result is that nutrients are continuously withdrawn from the living plant cells, to support repeated cycles of sporulation on the host surface. Much of the economic importance of rust and powdery mildew fungi stems from this **continuous withdrawal of nutrients** from the plant to support sporulation of the fungus.

Rust fungi

Rust fungi gain their name from the distinctive rust-colored spores, termed **uredospores**, which are produced in abundance for dispersal during the summer months. Over 4000 rust species have been described on crop or wild plants. Many rust fungi need two separate hosts to complete their life cycle. For example, *Puccinia graminis* (black stem rust of wheat) requires both a cereal and an alternate host (barberry). But other rust fungi, such as mint rust (*Puccinia menthae*), complete their life cycle on a single host. Below, we will consider the life cycle of *P. graminis* because it is one of the most

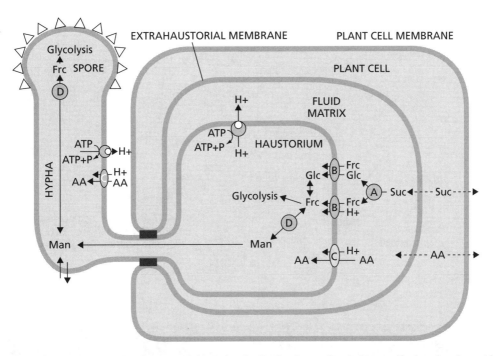

Fig. 14.21 Model for amino acid and sugar uptake and redistribution in rust fungi. Sucrose (**Suc**) and amino acids (**AA**) pass between plant cells and through the extrahaustorial membrane, into the fluid-filled matrix that surrounds the haustorium. Invertase (shown as **A**) cleaves sucrose to glucose (**Glc**) and fructose (**Frc**) in the matrix. Then these sugars, and amino acids cross the haustorial membrane by symport proteins (shown as **B**) in the membrane. Energy for nutrient uptake is supplied by the H$^+$ gradient generated across the membrane by conversion of ATP to ADP + P$_i$ (shown as **C**). Within the haustorium, fructose is converted to mannitol (**Man**) by the enzyme, major alcohol dehydrogenase (shown as **D**). Mannitol is then translocated into the fungal hyphae and provides the nutrients for fungal sporulation. (Based on a diagram in Voegele & Mendgen 2003.)

economically important species, especially in North America. It also displays all of the typical sporing stages of rust fungi.

*Black stem rust of wheat (*Puccinia graminis*)*

The disease cycle of black stem rust is shown in Fig. 14.22. We will begin this cycle at the dispersal phase, when rust pustules burst through the epidermis of cereal or grass leaves in early spring and release large numbers of the characteristic rust-colored **uredospores**, from structures called **uredinia** (see Fig. 14.23b).

Each uredospore is single-celled but contains two nuclei, one of each mating compatibility type, so the spores are dikaryotic as explained in Chapter 1. The uredospores are dispersed by wind and can undergo several cycles of infection in the course of a season, resulting in major epidemics. Towards the end of the season the production of uredospores ceases, and instead the pustules (now termed **telia**) produce two-

celled **teliospores** (Fig. 14.23c). These also are dikaryotic, with two nuclei in each cell, but the nuclei fuse to form diploid cells, and the fungus overwinters in this form. Early in the spring each cell of the teliospore germinates to produce a short mycelium. The diploid nucleus migrates into this and undergoes meiosis, leading to the production of four monokaryotic **basidiospores** (spores with only one haploid nucleus). These spores will only develop further if they land on a barberry leaf. Then they penetrate the leaf and, a few days later, produce flask-shaped **spermogonia**. These have receptive hyphae and also release many small **spermatia** (Fig. 14.23,d,f).

Each spermogonium produces spermatia of a single mating type and they are exuded from the neck of the spermogonium in a sugary fluid. Flies and other insects are attracted to this, and when they visit a spermogonium of a different mating type they transfer the spermatia. The spermatia of one mating type then fuse with receptive hyphae of the opposite mating

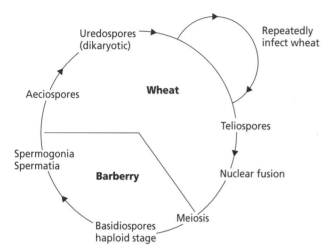

Fig. 14.22 Outline of the disease cycle of *Puccinia graminis*, which alternates between phases of growth on wheat and on barberry.

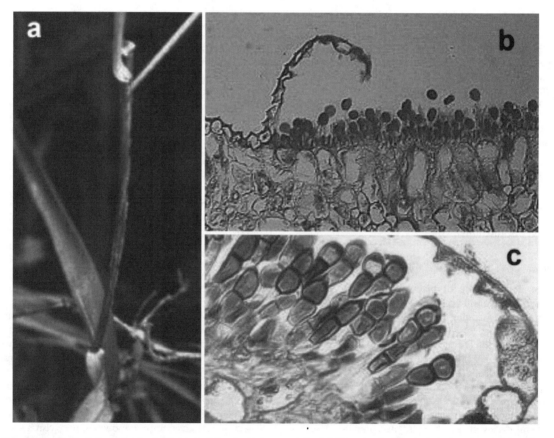

Fig. 14.23 Black stem rust of wheat caused by *Puccinia graminis*. (a) Elongated stem lesions bearing uredospores. (b) Stained section of a wheat stem with a pustule of uredospores breaking though the plant epidermis. (c) Section of a pustule with teliospores.

type and transfer the nucleus through a pore in the wall. This nucleus then divides and migrates through the monokaryotic hyphae that grew from the basidiospores, producing a **dikaryotic mycelium**. By this time, the mycelium has grown through the thickness of the leaf and produces a further sporing structure that erupts through the lower epidermis – an **aecium** containing **aeciospores** (Fig. 14.23e,g). The dikaryotic aeciospores can only infect wheat, and they give rise to infections on which uredospores will develop, thereby completing the cycle.

A potential solution to breaking the cycle of black stem rust would be to eradicate barberry bushes so that the fungus cannot infect its alternate host. A major barberry eradication program was attempted in North America in the last century, but was unsuccessful because the epidemic spreads progressively northwards from the early-sown crops in Mexico to successively later-sown crops in the more northerly regions of the USA.

Powdery mildew fungi

The powdery mildew fungi (Ascomycota) cause significant epidemics in most seasons if they are not treated with fungicides, and can be particularly serious in dry, hot summers. Several crop and wild plants are infected by these fungi, which in general are host-specific. Common examples include powdery mildew of roses (*Sphaerotheca pannosa*), of gooseberries (*S. mors-uvae*), and of hawthorn bushes (*Podosphaera clandestina*). Most people will have seen one or more of these fungi because they produce white, powdery disease pustules on the surfaces of many leaves and fruits. But the most economically damaging species is *Blumeria graminis* (previously called *Erysiphe graminis*), which causes powdery mildew of cereals (Fig. 14.24). Like the rust fungi, *B. graminis* undergoes multiple cycles of infection in a single season and can cause serious

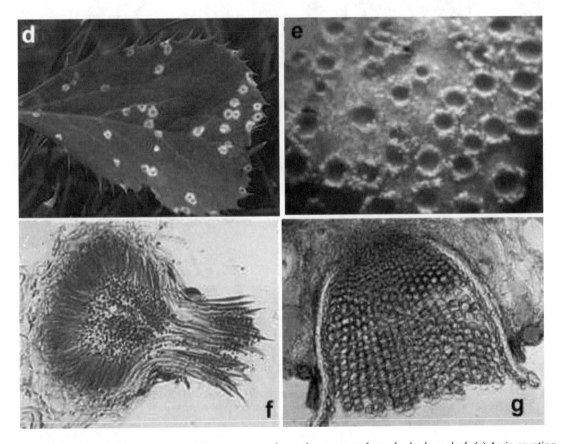

Fig. 14.23 (*continued*) (d) Lesions containing spermogonia on the upper surface of a barberry leaf. (e) Aecia erupting through the lower epidermis of a barberry leaf. (f) A spermogonium, showing the tiny spermatia and receptive hyphae. (g) Cross-section of an aecium.

damage to a range of cereal crops. A single "stategically" positioned lesion at the base of a cereal leaf can effectively prevent that leaf from being a net exporter of nutrients. This can affect root growth as well as shoot growth, because roots are relatively poor competitors for plant assimilates, and so the effects of this fungus are exacerbated in dry soil conditions. Yet, interestingly, the spores of powdery mildew fungi cannot infect in wet leaf conditions, only in drier conditions. Some other features of these fungi are shown in the legend to Fig. 14.24.

Downy mildews (Oomycota)

The downy mildew pathogens are haustorial biotrophs with similar behavior to the rust and powdery mildew fungi although they require more humid conditions for infection and sporulation. Examples include *Bremia lactucae* on lettuce, *Plasmopara viticola* on grape vine, and *Pseudoperonospora humuli* on hops. However, there is a gradation of behavior in the Oomycota, because some *Phytophthora* spp. which are closely related to the downy mildews begin their parasitic phase as

(a) (b)

(c) (d)

Fig. 14.24 Powdery mildew of cereals, caused by *Blumeria graminis* (Ascomycota). (a) Wheat leaves with typical mildew symptoms. (b) Spore chains which develop from a bulbous basal cell and mature progressively from the base to the top. (c) Haustorium with multiple finger-like projections within a host cell, typical of many powdery mildew fungi. Note the host nucleus in the cell, indicating that the cell is alive despite the presence of a haustorium. (d) Two cleistothecia (closed, sexual fruiting bodies containing asci) ornamented with dark projections. The cleistothecia commonly develop on mildewed leaves towards the end of the growing season. They break down to release the ascospores, which serve for dormant survival.

haustorial biotrophs but then kill the host tissues and spread within them as necrotrophs (e.g. *P. infestans* on potato and *P. sojae* on soybean).

Pathogen recognition: the gene-for-gene hypothesis

In many cropping systems, large areas of land are sown with a single crop variety (cultivar) that has been bred for resistance to all known races of the pathogen. Selection pressure then operates on the pathogen to overcome this resistance by mutation. Since many billions of spores are released by the major pathogens each year – even in a single field – there is a strong likelihood of the resistance breaking down. This is especially true for crops that have been bred for single **major gene resistance** (R gene resistance) as opposed to "field resistance" that is based on the combined activities of several "minor" genes. The eminent plant pathologist H.H. Flor, working with flax rust (*Melampsora lini*) in the 1940s and 1950s, proposed a simple scheme to explain the relationship between major gene resistance and the occurrence of disease – the **gene-for-gene hypothesis**. Based on extensive research of the genetics of both the host plant and the pathogen, he showed that for every gene that confers **resistance** (**R**, a genetically *dominant* trait) in flax plants, there is a complementary gene that confers **avirulence** (**AVR**, again a genetically *dominant* trait) in the pathogen. So, the outcome of a host–pathogen interaction can be summarized in the following table:

		HOST GENOTYPE	
		RR or Rr	rr
PATHOGEN GENOTYPE	AVR avr or AVR AVR	Disease-resistant	Susceptible to disease
	avr avr	Susceptible to disease	Susceptible to disease

where **RR** = homozygous resistant, **Rr** = heterozygous resistant, **rr** = homozygous susceptible, **AVR AVR** = homozygous avirulent, **AVR avr** = heterozygous avirulent, **avr avr** = homozygous virulent. From this table we can see that resistance occurs **only** in combinations involving the dominant R allele **and** the dominant AVR allele. The simplest explanation would be that the protein product of the **R** gene interacts with the protein product of the **AVR** gene, leading to the hypersensitive response – a rapid cell death that would

prevent a biotroph from developing. In effect, the R gene product of the host would be like a receptor, interacting with the AVR gene product (an **elicitor**) of the fungus.

Recent research has provided a deeper understanding of this system, and shown that gene-for-gene relationships are common across a range of host–parasite combinations, including interactions of plants with pathogenic fungi, bacteria, viruses, and nematodes, and even in some plant–insect interactions. A **direct protein–protein interaction** governed by R genes and AVR genes, and leading to the hypersensitive response, has been demonstrated in two pathosystems: infection of tomato by *Pseudomonas syringae* (pathovar *syringae*) and infection of rice by *Magnaporthe grisea* (rice blast disease). In these cases the interaction probably involves the recognition of surface-located proteins. However, a **direct** protein–protein interaction has **not been found** in any other host–pathogen systems to date. Instead, there is mounting evidence that at least a third (plant) protein is involved in most gene-for gene interactions, and this third protein (or further proteins) mediates the defense response. There are several lines of evidence for this (reviewed by Luderer & Joosten 2001; Bogdanova 2002). For example, the proteins involved in these interactions seem to be cytoplasmic proteins, not cell surface receptors involved directly in pathogen recognition, as was once thought. The cytoplasmic proteins constitute a family, or families, of related proteins with similar properties. For example, the "model" plant *Arabidopsis thaliana* (which is widely used for molecular genetic studies) contains a gene that codes for resistance to two bacterial pathogens. The product of this gene is similar to the resistance gene products of several other organisms – a gene from tobacco conferring resistance to a virus, a gene from flax conferring resistance to flax rust, a gene from tomato conferring resistance to the fungus *Fulvia fulvum*, and a gene from sugar beet conferring resistance to a nematode. All these resistance gene products have a region of leucine-rich repeats, and all have a nucleotide-binding site that could initiate a signalling cascade leading to activation of the plant's defense response.

Online resources

Endophytes in US Horse Pastures (Aphis Info Sheet; Veterinary Services; April 2000). http://www.aphis.usda.gov/vs/ceah/Equine/eq98endoph.htm
Sudden Oak Death. USDA Forest Service. www.invasive.org

General texts

Agrios, G.N. (1998) *Plant Pathology*, 4th edn. Academic Press, San Diego.

Lucas, J.A. (1998) *Plant Pathology and Plant Pathogens*, 3rd edn. Blackwell Science, Oxford.

Manners, J.G. (1993) *Principles of Plant Pathology*, 2nd edn. Cambridge University Press, Cambridge.

Cited references

Armstrong, G.M. & Armstrong, J.K. (1981) Formae speciales and races of *Fusarium oxysporum* causing wilt diseases. In: *Fusarium: diseases, biology and taxonomy* (Nelson, P.E., Toussoun, T.A. & Cook, R.J., eds), pp. 391–399. Pennsylvania State University Press, University Park, PA.

Beckman, C.H. & Talboys, P.W. (1981) Anatomy of resistance. In: *Fungal Wilt Diseases of Plants* (Mace, C.E., Bell, A.A. & Beckman, C.H., eds), pp. 487–521. Academic Press, New York.

Bogdanove, A.J. (2002) Protein–protein interactions in pathogen recognition by plants. *Plant Molecular Biology* **50**, 981–989.

Bony, S., Pichon, N., Ravel, C., Durix, A., Balfourier, F. & Guillaumin, J.-J. (2001) The relationship between mycotoxin synthesis and isolate morphology in fungal endophytes of *Lolium perenne*. *New Phytologist* **152**, 125.

Clay, K. (1989) Clavicipitaceous endophytes of grasses: their potential as biocontrol agents. *Mycological Research* **92**, 1–12.

Cooper, R.M., Longman, D., Campbell, A., Henry, M. & Lees, P.E. (1988) Enzymatic adaptations of cereal pathogens to the monocotyledonous primary wall. *Physiological and Molecular Plant Pathology* **32**, 33–47.

Deacon, J.W. & Scott, D.B. (1983) *Phialophora zeicola* sp. nov., and its role in the root rot-stalk rot complex of maize. *Transactions of the British Mycological Society* **81**, 247–262.

Deacon, J.W. & Scott, D.B. (1985) *Rhizoctonia solani* associated with crater disease (stunting) of wheat in South Africa. *Transactions of the British Mycological Society* **85**, 319–327.

Dodd, J.L. (1980) The role of plant stresses in development of corn stalk rots. *Plant Disease* **64**, 533–537.

Erwin, D.C. & Ribeiro, O.K. (1996) *Phytophthora Diseases Worldwide*. APS Press, St Paul, Minnesota.

Fravel, D., Olivain, C. & Alabouvette, C. (2003) *Fusarium oxysporum* and its biocontrol. *New Phytologist* **157**, 493–502.

Honee, G. and 10 others (1994) Molecular characterization of the interaction between the fungal pathogen *Cladosporium fulvum* and tomato. In: *Advances in Molecular Genetics of Plant–Microbe Interactions* (Daniels, M.J., Downie, J.A. & Osbourne, A.E., eds), pp. 199–206. Kluwer Academic, Dordrecht.

Lamb, C.J., Brisson, L.F., Levine, A. & Tenhaken, R. (1994) H_2O_2-mediated oxidative cross-linking of cell wall structural proteins. In: *Advances in Molecular Genetics of Plant–Microbe Interactions* (Daniels, M.J., Downie, J.A. & Osbourne, A.E., eds), pp. 355–360. Kluwer Academic, Dordrecht.

Luderer, R. & Joosten, M.H.A.J. (2001) Avirulence proteins of plant pathogens: determinants of victory and defeat. *Molecular Plant Pathology* **2**, 355–364.

Nelson, E.B. (1987) Rapid germination of sporangia of *Pythium* species in response to volatiles from germinating seeds. *Phytopathology* **77**, 1108–1112.

Nelson, E.B. (1992) Bacterial metabolism of propagule germination stimulants as an important trait in the biological control of *Pythium* seed infections. In: *Biological Control of Plant Diseases* (Tjamos, E.C., Papavizas, G.C. & Cook, R.J., eds), pp. 353–357. Plenum Press, New York.

Punja, Z.K. & Grogan, R.G. (1981) Mycelial growth and infection without a food base by eruptively germinating sclerotia of *Sclerotium rolfsii*. *Phytopathology* **71**, 1099–1103.

Ride, J.P. & Pearce, R.B. (1979) Lignification and papilla formation at sites of attempted penetration of wheat leaves by nonpathogenic fungi. *Physiological Plant Pathology* **15**, 79–92.

Schneider, R.W. (1982) *Suppressive Soils and Plant Disease*. American Phytopathological Society, St Paul, Minnesota.

VanEtten, H.D., Sandrock, R.W., Wasmann, C.C., Soby, S.D., McCluskey, K. & Wang P. (1995) Detoxification of phytoanticipins and phytoalexins by phytopathogenic fungi. *Canadian Journal of Botany* **73**, S518–S525.

Voegele, R.T. & Mendgen, K. (2003) Rust haustoria: nutrient uptake and beyond. *New Phytologist* **159**, 93–100.

Wastie, R.L. (1960) Mechanism of action of an infective dose of *Botrytis* spores on bean leaves. *Transactions of the British Mycological Society* **45**, 465–473.

Chapter 15

Fungal parasites of insects and nematodes

This chapter is divided into the following major sections:

- insect-pathogenic fungi
- nematode-destroying fungi

Fungi commonly attack insects, nematodes and other invertebrates in natural environments. In doing so they act as natural population regulators, helping to keep insect and nematode pests in check. Some insect-pathogenic and nematode-destroying fungi can also be exploited as biocontrol agents, and some are available commercially as alternatives to chemical pesticides.

In this chapter we consider the specific adaptations of fungi for this mode of parasitism. It is an important topic not only because it extends our coverage of parasitic interactions, but also because the control of insect pests and nematodes is currently achieved by highly toxic chemicals, with known or potential adverse effects on humans and the environment. For example, **aldicarb** is a systemic insecticide, nematicide and acaricide (mite-control agent) of the carbamate type, currently registered for restricted use on selected crops in the USA. It is one of the most acutely toxic pesticides to mammals, aquatic invertebrates, fish, and birds; it accumulates in groundwater, and it poisons through either oral or dermal contact. The antidote to aldicarb is **atropine** (from the plant *Atropa belladonna*), traditionally used on the poison darts of African bushmen! The development of effective biocontrol agents could provide at least a partial solution to some of these environmental problems.

The insect-pathogenic fungi

Examples of some common insect-pathogenic fungi (**entomopathogens**) are listed in Table 15.1. All these fungi are specifically adapted to parasitize insects, and depend on insects for their survival in nature. In the following sections we will deal first with the general aspects of their mode of parasitism and then focus on specific issues, including the potential for exploiting insect-pathogenic fungi as biocontrol agents. Many aspects of this subject are covered in Butt *et al.* (2001) and Butt (2002).

Among the many insect-pathogenic fungi, the species of *Beauveria* and *Metarhizium* (Fig. 15.1) are commonly found in natural environments and are considered to have strong potential for practical control of insect pests, especially in glasshouses and other protected cropping systems. These two genera produce abundant conidia in laboratory culture and on insect hosts, but are not known to have a sexual stage. *Beauveria* spp. produce their conidia in a **sympodial** fashion, by first producing a terminal spore and then the conidiophore elongates and produces further spores below this spore, arranged in a zig-zag manner (Figs 15.1a, 15.2). The conidia are white, and so these fungal infections are colloquially termed "white muscardine." By contrast, the *Metarhizium* spp. produce chains of green conidia from phialides (Fig. 15.1b) and these fungal infections are termed "green muscardine."

Another common insect-pathogenic fungus, *Lecanicillium lecanii* (formerly known as *Verticillium lecanii*), produces clusters of moist conidia at the ends of long phialides (Fig. 15.1c). This fungus occurs commonly as a parasite of scale insects in subtropical and tropical

Table 15.1 Some common fungi that parasitize insects and other arthropods.

Parasitic fungus	Hosts
Metarhizium anisopliae	Many: Lepidoptera, Coleoptera, Orthoptera, Hemiptera, Hymenoptera
Beauveria bassiana	Most/all
Hirsutella thompsonii	Arachnida (mites)
Cordyceps militaris	Many larvae and pupae of Lepidoptera, some Coleoptera and Hymenoptera
Nomuraea rileyi	Larvae and pupae of Lepidoptera, Coleoptera
Paecilomyces farinosus	Many (Lepidoptera, Diptera, Homoptera, Coleoptera, Hymenoptera, Arachnida)
Lecanicillium lecanii	Several, especially scale insects, thrips, and aphids
Entomophthora, Erynia and similar zygomycota	Various, often host-specific, e.g. *Entomophthora muscae* on flies, *Erynia neoaphidis* on aphids
Coelomomyces spp.	Mosquitoes and midges; often host-specific

Lepidoptera (butterflies and moths); Diptera (flies); Homoptera (bugs); Coleoptera (beetles); Hymenoptera (wasps and bees); Orthoptera (grasshoppers and locusts); Hemiptera (sucking bugs); Arachnida (spiders and mites).

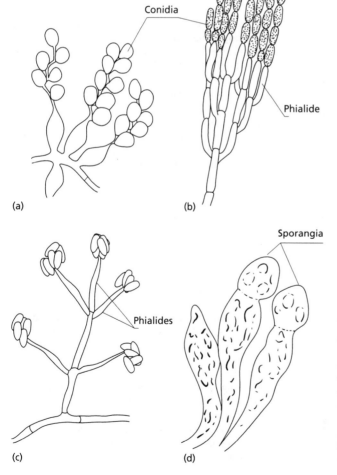

Fig. 15.1 Spore-bearing structures of some common insect-pathogenic fungi. (a) *Beauveria bassiana*, which produces cream-white conidia alternately on an extending tip of a conidiophore. (b) *Metarhizium anisopliae*, which produces green conidia in chains from phialides. (c) *Lecanicillium lecanii*, which produces clusters of conidia in moisture drops at the tips of phialides. (d) *Entomophthora* spp. (Zygomycota), which produce single terminal sporangia that are released at maturity and function as spores.

Fig. 15.2 *Beauveria bassiana* and examples of the diseases that it causes. (a) Spore-bearing structures of *Beauveria bassiana* in laboratory culture. The conidia typically develop in zigzag-like chains on long conidiophores. (b) An adult cicada beetle densely covered with white sporulating pustules of *B. bassiana* that have emerged through the intersegmental plates of the insect cuticle. (c) Sporulation of *Beauveria* from between the cuticular plates of a naturally infected green cockchafer beetle. (d) Heavy infection of an adult pecan weevil by *Beauveria*. (e) Grubs of pecan weevil at different stages of infection by *Beauveria*; healthy grubs are shown at left. ((a),(b) Courtesy of G.L. Barron. (c) Courtesy of Shirley Kerr; http://www.kaimaibush.co.nz/. (d),(e) Courtesy of Louis Tedders (photographer) and USDA, Agricultural Research Service.)

environments, but it has a relatively high temperature optimum for growth, so its use as a biological control agent is restricted to glasshouse environments in the cooler parts of the world. In addition to these conidial fungi, several species of Zygomycota commonly cause insect diseases in natural environments and also in field crops. These fungi belong to a subgroup of Zygomycota termed the **Entomophthorales** (see Fig. 2.11). They include genera such as *Entomophthora* and *Pandora* (previously called *Erynia*). They produce large sporangia at the tips of their hyphae, and the sporangia are released intact, functioning as dispersal spores (Fig. 15.1d). The Entomophthorales often cause natural epizootics in humid conditions, and attempts are being made to develop them as commercial biocontrol agents.

The infection cycle

The general infection cycle of insect-pathogenic fungi is summarized in Fig. 15.3. In almost all cases these fungi initiate infection from spores that land on, and adhere to, the insect cuticle. If the relative humidity is high enough, the spore then germinates and usually forms an appressorium, equivalent to the appressoria produced by many plant-pathogenic fungi (see Fig. 5.2). The appressorium often develops over an intersegmental region of the cuticle, and in one of the common insect pathogens, *Metarhizium anisopliae*, this behavior closely parallels the behavior of the rust fungi (Chapter 5), because the germ-tube recognizes the host surface topography, or artificial surfaces with appropriate

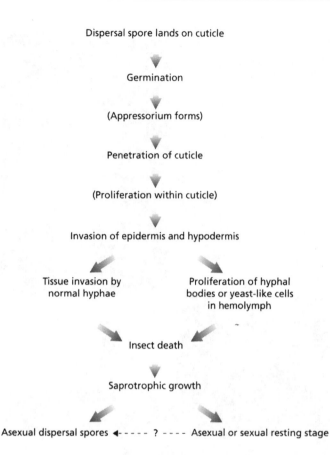

Dispersal spore lands on cuticle

Germination

(Appressorium forms)

Penetration of cuticle

(Proliferation within cuticle)

Invasion of epidermis and hypodermis

Tissue invasion by normal hyphae

Proliferation of hyphal bodies or yeast-like cells in hemolymph

Insect death

Saprotrophic growth

Asexual dispersal spores ◄- - - - - ? - - - - Asexual or sexual resting stage

Fig. 15.3 General infection sequence of an insect-pathogenic fungus. (Based on Charnley 1989.)

spacing of ridges and grooves. Penetration of the cuticle is achieved by means of a narrow penetration peg beneath the appressorium, and this involves the actions of cuticle-degrading enzymes, such as lipases, proteases, and chitinase, all of which are known to be produced by the insect-pathogenic fungi in laboratory culture. The penetration peg either penetrates through both layers of the cuticle – the epicuticle and pro- cuticle – or it penetrates only the hard epicuticle and then forms plates of hyphae between the lamellae of the procuticle, exploiting zones of mechanical weak- ness. Further penetration hyphae develop from these fungal plates.

Up to this stage, the infection will be aborted if the insect moults. Otherwise, the fungus invades the epidermis and hypodermis, causing localized defense reactions. If these are overcome, then the hyphae either ramify in the insect tissues or, most frequently, produce swollen **blastospores** (yeast-like budded cells), **hyphal bodies** (short lengths of hypha), or proto- plasts (*Entomophthora* and related Zygomycota) that proliferate in the haemolymph (insect blood). This "unicellular" phase of growth and dissemination

usually leads to insect death, either by depletion of the blood sugar levels or by production of toxins (see below). Then the fungus reverts to a mycelial, sapro- trophic phase and extensively colonizes the body tissues. Usually, at least some of the tissues are colo- nized before the insect dies – the fat body in particu- lar. Finally, the fungus converts to either a resting stage in the cadaver or, in humid conditions, grows out through the intersegmental regions of the cuticle to produce conidiophores that bear numerous asexual conidia for dispersal to new insect hosts.

In aphids infected by *Lecanicillium lecanii* the coni- diophores can develop on many parts of the body while the insect is still moving. In other host–parasite interactions the insect is killed more rapidly, before the body tissues are extensively invaded, indicating the involvement of toxins. Both *Beauveria bassiana* and *M. anisopliae* produce depsipeptide toxins in laboratory culture, and these are active on injection into insects. The toxins of *M. anisopliae* are termed **destruxins** (Fig. 15.4) and are thought to be significant in patho- genesis because infected insects die rapidly before there is extensive tissue invasion. The role of the toxin

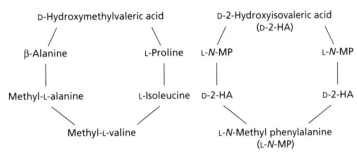

Fig. 15.4 Destruxin B and beauvericin, two cyclic peptides produced by insect-pathogenic fungi.

Destruxin B (*Metarhizium anisopliae*)   Beauvericin (*Beauveria bassiana*)

of *B. bassiana* (termed **beauvericin**, Fig. 15.4) is less clear because this fungus invades the tissues more extensively before the host dies, and the pathogenicity of strains is not always correlated with their *in vitro* toxin production. Several other potential toxins are also produced by *Beauveria* spp. Another insect-pathogenic fungus, *Hirsutella thompsonii*, produces an extracellular insecticidal protein, hirsutellin A, which is lethal to the adult citrus rust mite – the natural host of *H. thompsonii*. These are among a wide range of insecticidal and nematicidal metabolites produced by fungi (Anke & Sterner 2002).

In addition to these toxins and potential toxins, *Beauveria* spp. produce an antibiotic, **oosporein**, after the insects have died. This compound has no effect on fungi but is active against Gram-positive bacteria and could help to suppress bacterial invasion of the cadaver, enabling the fungus to exploit the dead host tissues.

Insect host ranges

Several common species of *Beauveria* and *Metarhizium* have very wide host ranges, including hundreds of insects in the Orthoptera (grasshoppers), Coleoptera (beetles), Lepidoptera (butterflies and moths), Hemiptera (bugs), and Hymenoptera (wasps). These fungi can be grown easily in laboratory culture and they produce large numbers of asexual spores, making them attractive candidates for applied biological control of insect pests. Similarly, *Lecanicillium lecanii* can be grown easily in laboratory culture and it is used commercially to control aphids and whitefly in greenhouse cropping systems (see later).

In contrast to these examples, the entomopathogenic Zygomycota can be either host-specific (e.g. *Entomophthora muscae* on houseflies) or can have broad host ranges. All of these species depend on insect hosts, because they do not grow naturally in the absence

of a host. However, techniques have been developed to culture several of these fungi in artificial conditions, providing inoculum for applied insect-control programs. We return to this subject later.

One other insect-pathogenic organism is of considerable interest because of its unusual biology. *Coelomomyces psorophiae* is a member of the Lagenidales (Oomycota) and it has an unusual life cycle, shown in Fig. 15.5. In its diploid phase it parasitizes mosquito larvae, and releases thick-walled resting spores when the larval host dies. The resting spores germinate and undergo meiosis, releasing motile, haploid gametes, and these can only infect a copepod host, such as *Cyclops*. After passage through this host, the motile spores fuse in pairs and then initiate infection of another mosquito larva. So, *Coelomomyces* displays an obligate alternation of generations.

Coelomomyces cannot be grown in culture, so this limits its potential use as a biological control agent. But there could be a possibility of manipulating the *Coelomomyces* population indirectly, by promoting the populations of copepods so that there is an abundant source of inoculum for infection of mosquito larvae.

Natural epizootics caused by entomopathogenic fungi

Insect-pathogenic fungi can cause natural and spectacular population crashes of their hosts. An example is shown in Figure 15.6, where the population of the broad bean aphid, *Aphis fabae*, in a field crop reached a peak in mid-July but was then dramatically reduced by a natural epizootic caused by the fungi *Pandora neoaphidis* and *Neozygites fresenii* (both obligate parasites in the Zygomycota). This pattern is typical of obligate parasites: their population level always lags behind that of the host because the relationship is host-density-dependent. This means that there is always some degree of crop damage.

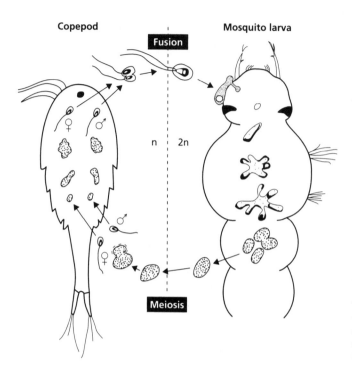

Fig. 15.5 Proposed life cycle of *Coelomomyces psorophorae*, which has an obligate alternation of generations. **Diploid phase in a mosquito larva:** a motile zygote attaches to the larva, encysts, produces an appressorium, then penetrates the host and produces a weakly branched mycelium. This gives rise to thick-walled resting spores that are released when the mosquito host dies. **Haploid phase in a copepod (*Cyclops vernalis*):** meiosis occurs in the resting sporangia and haploid zoospores of different mating types are released. These infect the copepod and produce a thallus that eventually gives rise to gametangia. The gametangia release motile gametes that fuse either inside or outside the host. The resulting diploid zygote can only infect a mosquito larva. (Adapted from Whisler *et al.* 1975.)

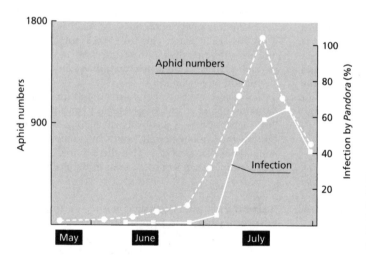

Fig. 15.6 Numbers of aphids (*Aphis fabae*) in a broad bean crop during a growing season in Britain (broken line) and percentage infection by the pathogenic fungus *Pandora neoaphidis* (solid line). (Based on data in Wilding & Perry 1980.)

Even more important is the fact that epizootics caused by insect-pathogenic fungi are strongly influenced by environmental factors. Insect-pathogenic fungi in general require a high humidity for spore germination and also for sporulation on the dead insect hosts. Thus, natural epizootics are unpredictable, and even commercially applied inocula of these fungi are prone to fail unless the environmental conditions are favorable.

Developments in practical biocontrol

Some of the limitations in natural biocontrol can be overcome by using commercially produced inocula if these are formulated to high standards. Feng *et al.* (1994) reviewed the technology for producing *B. bassiana*, which is commercially attractive as a biocontrol agent because of its wide host range. These authors reported that about 10,000 tons of spore powder containing

B. bassiana are produced annually in China for treatment of 0.8–1.3 million hectares of forest and agricultural land. Several other medium- to large-scale field trials or field releases have been conducted with selected species/strains of *Metarhizium*. Unfortunately, many of the reports of successful biocontrol in these field programs are anecdotal; the experimental data are not published, nor supported by statistical analysis.

One of the most comprehensive, field-scale, insect-biocontrol programs to date was termed LUBILOSA – a French acronym, **Lu**tte **Bi**ologique contre les **Lo**custes et **Sa**uteriaux (biological control of locusts and grasshoppers). This program was funded by international development agencies, with the aim of developing a biocontrol agent that could be used in Central African countries to control grasshoppers and locusts (see On-line resouces). The strain used was characterized as *M. anisopliae* var *acridum*, based on ribosomal DNA sequence analysis, and it was developed as an oil-based formulation that could be dried to less than 5% moisture content and sealed hermetically for long-term storage below 20°C.

The results of field trials with this strain are reported to have been successful, but – at least for the LUBILOSA strain – the only economical production method seems to be large-scale solid-state fermentation, which requires the expertise of commercial production companies rather than medium-technology facilities that could operate locally. The LUBILOSA product has now been transferred to private sector companies under trade names that include "**Green Muscle.**"

Several interesting features emerged from the field trials of the LUBILOSA program:

1 The initial small-scale field trials proved to be unsatisfactory because too many treated insects left the trial area, and untreated insects moved in. But when large-scale field plots of 800 hectares were used there was clear evidence of grasshopper mortality within 10 days of spraying, and grasshoppers continued to die until the end of the season.

2 A comparison between plots treated with *Metarhizium* and a commercial insecticide, fenitrothion, showed that the insecticide killed the grasshoppers rapidly but the insects soon recolonized the plots. By contrast, grasshoppers **did not recolonize** the *Metarhizium*-treated plots. The consequence of this was that, 10 days after spraying, the grasshopper populations in the two types of plot were roughly the same, and from that time onwards the *Metarhizium* treatment provided better control than did the insecticide.

3 Grasshoppers and locusts can raise their body temperature in response to fungal infection – a "fever" response that can help the insect to overcome the infection. Locusts do this naturally, but grasshoppers need to expose themselves to sun in order to raise their body temperature. This knowledge could be incorporated into a geographical information system (GIS) that could predict the best conditions for using *Metarhizium* in field situations.

The development of Lecanicillium lecanii *as a commercial biocontrol agent*

Lecanicillium lecanii is one of the most successful commercial biocontrol agents of insects. It was developed in the late 1970s and is used primarily in glasshouses for the control of aphids on potted chrysanthemums and (using a different strain) for control of whitefly on cucumbers and other indoor-grown crops. For these purposes the fungus is produced as fermenter-grown conidia because it is one of the relatively few fungi that produce conidia readily in submerged liquid culture.

L. lecanii occurs naturally as a parasite of aphids and scale insects in the subtropics, but requires relatively warm conditions (>15°C) for infection. Like all the entomopathogenic fungi, it needs a high relative humidity during the germination and penetration phases, but then the humidity can be reduced without affecting its parasitism. All these factors make this fungus an ideal control agent for use on potted chrysanthemums – one of the most important year-round horticultural crops – because chrysanthemums have to be "blacked-out" with polythene sheeting for part of each day to produce the short daylengths that are necessary to initiate flowering. The blackout sheeting raises the humidity for infection by *L. lecanii*, so that a single spray of conidia just before blacking-out can be sufficient to give season-long control of the important aphid pest, *Myzus persicae*. However, experimental trials showed that treatment with *L. lecanii* (under blackout conditions) was less effective for two minor aphid pests of chrysanthemum, *Macrosiphoniella sanborni* and *Brachycaudus helichrysi*. Hall & Burges (1979) showed that this was not related to inherent differences in susceptibility of the three aphid species. Instead, it is explained by their behavioral differences. *M. persicae* tends to feed on the undersides of leaves where the humidity is higher, and it is more mobile than the other aphids on chrysanthemum, feeding for short times and then moving on, because chrysanthemum is not its preferred host plant. *L. lecanii* often sporulates on the body while an aphid is still alive, so the infected individuals of *M. persicae* can spread the infection to other individuals of this species feeding in the same locations on the crop.

L. lecanii is currently marketed as two products, **Mycotal®** and **Vertalec®**, by the Dutch-based company, Koppert BV. The product Mycotal is used primarily to control whitefly and thrips in protected crops such as cucumbers, tomatoes, sweet peppers, beans,

Table 15.2 Some examples of currently registered mycoinsecticides.

Country	Registered product name	Fungus	Target pest	Crop
USA	Mycotrol, Botanigard	*Beauveria bassiana*	Whitefly, aphids, thrips	Glasshouse tomatoes and ornamentals
USA	Naturalis	*B. bassiana*	Sucking insects	Cotton, glasshouse crops
USA	BioBlast	*Metarhizium anisopliae*	Termites	Domestic houses
USA/Europe	PFR-97™	*Paecilomyces fumosoroseus*	Whitefly, thrips	Glasshouse crops
UK, Europe	Vertalec	*Lecanicillium lecanii*	Aphids	Glasshouse crops
UK, Europe	Mycotal	*L. lecanii*	Whitefly, thrips	Glasshouse crops
South Africa	Green Muscle	*M. anisopliae*	Locusts	Natural bushland
Reunion	Betel	*B. bassiana*	Scarab beetle larvae	Sugar cane
Switzerland	Engerlingspilz	*Beauveria brogniartii*	Scarab beetle larvae	Pasture
Switzerland	Beauveria Schweizer	*B. brogniartii*	Scarab beetle larvae	Pasture
France	Ostrinol	*B. bassiana*	Corn borer	Maize
Australia	BioGreen	*Metarhizium flavoviride*	Cockchafer	Pasture, turf

aubergine, lettuce, ornamentals, and cut flowers. The Vertalec formulation contains a different strain of *L. lecanii* that controls many aphid species. Both products have a long history of successful usage, but according to the manufacturer's technical data sheets, the best results require a temperature of 18–28°C and a minimum relative humidity of 80% for 10–12 hours a day.

Several other mycoinsecticides have been registered commercially, and some of these are shown in Table 15.2. Many are based on different strains or species of *Beauveria* or *Metarhizium*. In this respect it is important to note that different strains or species can have quite different host ranges, so the activity of any one strain

cannot be used to predict the activities of other stains within this highly variable group of fungi.

It is interesting to note that *B. bassiana* is reported to colonize some maize cultivars in the USA. It does so by growing as an endophyte in the maize tissues, similar to the endophytes of plants discussed in Chapter 14. When granular formulations of *B. bassiana* conidia were sprayed on the maize foliage the fungus grew into the plant and persisted throughout the growing season, significantly suppressing the damage caused by larvae of the European corn borer, *Ostrinia nubilaris* – the most important insect pest of maize (Fig. 15.7).

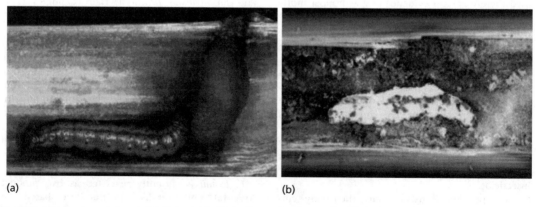

(a)   (b)

Fig. 15.7 European corn borer (*Ostrinia nubilaris*) in the stalk of a maize plant. (a) Healthy insect. (b) Insect killed by *Beauveria bassiana*, showing profuse development of spores and hyphae on the insect cadaver. (Courtesy of Marlin. E. Rice, University of Ohio.)

The nematode-destroying fungi

Nematodes (eelworms or roundworms) are small animals, usually 1–2 mm long. They are extremely common in soil, animal dung, and decomposing organic matter. For example, the nematode population in European grasslands is estimated to range from 1.8 to 120 million per square meter. Most nematodes are saprotrophs that feed on bacteria or other small organic particles, but some are parasites of animals, including humans (e.g. *Trichinella spiralis* which invades human muscle tissue), and some are parasites of crop plants. The important nematodes of crop plants include the root-knot nematodes (*Meloidogyne* spp.), the cyst nematodes (e.g. *Heterodera* and *Globodera* spp.), and various ectoparasitic and burrowing nematodes. The chemicals that can be used to control parasitic nematodes in living plants or organic matter are extremely toxic and thus environmentally undesirable. For this reason

interest has focused on **nematophagous** (nematode-eating) fungi and other parasites of nematodes that might be exploited as biological control agents.

Nematophagous fungi are common in organic-rich environments, and they include representatives of almost all the major fungal groups (Table 15.3). Here we will consider the three major types that have different adaptations for feeding on nematodes – the **nematode-trapping fungi**, the **endoparasitic fungi**, and the **parasites of nematode eggs or cysts**. Barron (1977) gives an extensive account of all these fungi.

The nematode-trapping fungi

The nematode-trapping fungi are predatory species which capture nematodes by specialized devices of various types: adhesive hyphae (*Stylopage* and *Cystopage*; Zygomycota), adhesive nets (e.g. *Arthrobotrys oligospora*,

Table 15.3 Examples of the major types of nematophagous fungi.

Fungus	Behavioral group	Infective unit
Chytridiomycota		
Catenaria anguillulae	Endoparasite	Zoospore
Oomycota		
Nematophthora gynophila	Endoparasite	Zoospore
Myzocytium humicola	Endoparasite	Adhesive zoospore cyst
Zygomycota		
Stylopage and *Cystopage* spp.	Predator	Adhesive hyphae
Mitosporic fungi (but some of these have sexual stages in the ascomycota – see below)		
Arthrobotrys oligospora	Predator	Adhesive nets
Monacrosporium cionopagum	Predator	Adhesive branches
Dactylella brochopage	Predator	Constricting rings
Drechmeria coniospora	Endoparasite	Adhesive conidia
Hirsutella rhossiliensis	Endoparasite	Adhesive conidia
Verticillium chlamydosporium	Egg parasite	Hyphal invasion
Dactylaria candida	Predator	Adhesive knobs and nonconstricting rings
Ascomycota		
Atricordyceps (sexual stage of *Harposporium oxycoracum*)	Endoparasite	Nonadhesive conidia
Orbilia spp (sexual stages of some *Dactylella*, *Arthrobotrys*, and *Monacrosporium* spp.)		
Basidiomycota		
Hohenbuehelia (gilled mushroom – the sexual stage of several *Nematoctonus* species)	Predator	Adhesive conidia
Pleurotus ostreatus	Predator and toxin producer	Adhesive traps and toxic droplets

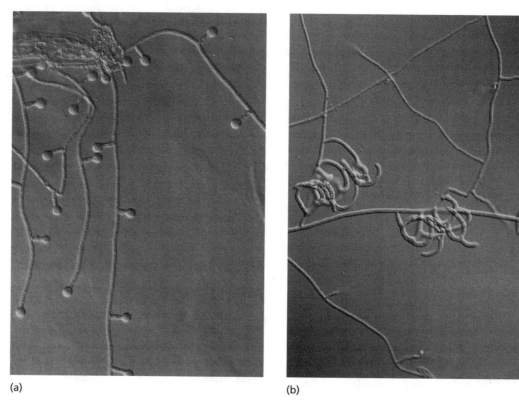

(a)   (b)

Fig. 15.8 (a) Adhesive network of *Arthrobotrys oligospora*. (b) Adhesive knobs of *Monacrosporium ellipsosporum* on hyphae growing from a parasitized nematode (seen at the top left of the image). (Courtesy of B.A. Jeffee; from Jaffee 1992.)

Fig. 15.8), short adhesive branches (e.g. *Monacrosporium cionopagum*), adhesive knobs (e.g. *M. ellipsosporum*, Fig. 15.8), nonconstricting rings, and constricting rings that are triggered to contract when a nematode enters them (e.g. *Dactylella brochopaga*). More than one type of mechanism can be found in different species of a genus.

All these fungi are considered to be essentially saprotrophic because they grow on a range of organic substrates, including cellulose, in laboratory culture, and some are wood-degrading members of the Basidiomycota. Also, some of them (e.g. *A. oligospora*) coil round the hyphae of other fungi in culture, indicative of mycoparasitic behavior (Chapter 12). However, their specialized trapping devices clearly indicate that they are adapted to exploit nematodes. In some cases they produce the traps during normal growth in culture, but in other cases (e.g. *A. oligospora*) the traps are produced only in the presence of nematodes or nematode diffusates. This can be mimicked by supplying small peptides or combinations of amino acids such as phenylalanine and valine. It seems likely that the nematode-trapping fungi exploit nematodes mainly as a source of nitrogen, which could often be

in short supply in the habitats where the nematode-trapping fungi grow, especially in woody materials.

Several details of the fungus–nematode interactions have been established in recent years. For example, the initial adhesion is almost instantaneous and effectively irreversible. Even when the nematode thrashes to free itself from the fungus, the trapping organ will break from the hyphae and remain attached to the nematode, then initiate infection. Yet, the trapping organs are not "sticky" in the general sense, because they do not accumulate soil debris, etc. Instead, the adhesive could be a lectin-like material that binds to specific sugar components of the nematode surface. This has been studied in the interaction between *A. oligospora* and the saprotrophic nematode *Panagrellus redivivus*, where the ability of the fungus to bind to the nematode was annulled in the presence of *N*-acetylgalactosamine (Tunlid *et al.* 1992).

Presumably, this sugar derivative binds to the fungal lectin, blocking the adhesion process. *Panagrellus* is known to have *N*-acetylgalactosamine components on its surface, because it binds to commercially available lectins (e.g. wheat-germ agglutinin) that recognize this sugar derivative. A glycoprotein with this binding

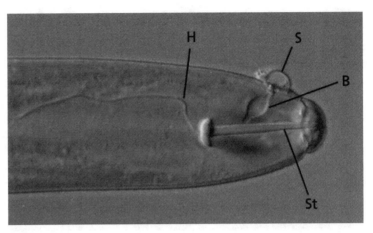

Fig. 15.9 *Hirsutella rhossiliensis*, an endoparasitic fungus on a nematode host. Infection occurred from a spore (S) that adhered near the nematode's mouth and then germinated and penetrated the cuticle to produce a bulb (B). A thin infection hypha (H) has started to colonize the host. (The nematode's piercing stylet (St) is shown; this is used for feeding.) (Courtesy of B.A. Jaffee; from Jaffee 1992.)

property has been isolated from *A. oligospora*, but it is found only on the surface of the traps, not on the normal hyphae, so presumably it is the product of a differentiation-specific gene. Other fungus–nematode combinations have also been investigated for binding specificity by using sugars to try to block adhesion. These studies indicate that *Arthrobotrys conioides* has a lectin that recognizes α-D-glucose or α-D-mannose residues, *Monacrosporium eudermatum* has a lectin that binds to L-fucose, and *Drechmeria coniospora* has a lectin that binds to sialic acid. So the adhesion can be to some degree nematode-specific. For example, *Monacrosporium ellipsosporum* does not capture *Xiphinema* spp. (which feed on the root tips of many plants) but these nematodes are captured by other species of *Monacrosporium*, *Arthrobotrys*, and *Dactylaria*. However, Tunlid *et al.* (1992) cautioned against simple interpretations based on lectin-binding, because ultrastructural studies suggest that the adhesive on fungal traps might change when nematodes become attached to it: more adhesive material might be released or there might be a rearrangement of the adhesive so that different binding sites are exposed in it.

The trapping of nematodes by some wood-decay fungi has focused attention on the role that nematodes might have as supplementary nitrogen sources, overcoming the critically low nitrogen content of wood (Chapter 11). The wood-rotting "oyster fungus" *Pleurotus ostreatus* (Basidiomycota; see Fig. 2.30) not only forms adhesive traps but also produces droplets of toxin from specialized cells. Nematodes immobilized by this toxin are then invaded and digested by the fungus.

The endoparasitic fungi

In contrast to the nematode-trapping fungi, endoparasitic fungi initiate infections from **spores** that adhere to a nematode surface, and then germinate to infect the host. Again, lectins seem to be involved in the initial adhesion process, but the endoparasitic fungi only produce detachable, adhesive spores, and when they have colonized and absorbed the host contents they grow out into soil and produce further spores to repeat the infection cycle. *Hirsutella rhossiliensis* is a good example of this (Figs 15.9, 15.10).

The endoparasitic fungi differ from the trapping fungi because they seem to depend on nematodes as their main or only food source in nature, even though many of them can be grown in laboratory culture media. Consistent with this, they show a strong density dependence on their hosts (Jaffee 1992); in other words, the population density of these fungi is dependent on the population density of the nematode. In contrast to this, the zoosporic fungus *Catenaria anguillulae* (Chytridiomycota), which also can attack nematodes, is one of the least specialized examples of a nematode-control agent because it grows on several types of organic material in nature, including liver fluke eggs. Also, its zoospores do not settle easily on moving nematodes in water films, and instead it accumulates at the body orifices of immobilized or dying nematodes (see Fig. 2.2). This contrasts with *Myzocytium humicola* (Oomycota) which also produces zoospores but these spores encyst soon after release and then germinate to produce an adhesive bud which attaches to a passing nematode.

Some endoparasitic fungi – for example, *Hirsutella rhossiliensis* (Fig. 15.10) and *Drechmeria coniospora* – seem to attract nematodes by means of a chemical gradient, helping to ensure that attachment occurs. Then the adhered spores germinate rapidly and the hyphae fill the host, killing it within a few days. Finally, the hyphae grow out through the host wall and produce a further batch of spores. From a single parasitized nematode, *Hirsutella* can produce up to 700 spores, while

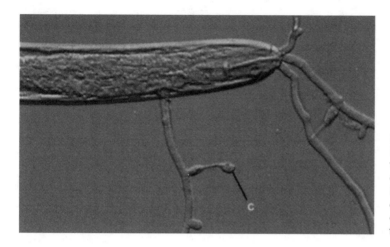

Fig. 15.10 *Hirsutella rhossiliensis* at a later stage than that in Fig. 15.9, when the body contents have been completely colonized by the fungus. Hyphae have grown out from the dead nematode and produced adhesive conidia (C). (Courtesy of B.A. Jaffee; from Jaffee 1992.)

Drechmeria is reported to produce up to 10,000 spores from a single parasitized nematode.

Parasites of nematode eggs and cysts

Cyst nematodes are important pests of several crops, including cereals, potato and sugar beet in Europe. They are characterized by the fact that the female nematode penetrates the root just behind the root tip and lodges with her head inside. The host cells respond by swelling into nutrient-rich "giant cells" from which the nematode taps the host nutrients. As the female grows her body distends into a lemon shape which ruptures the root cortex so that her rear protrudes from the root. Then she is fertilized by wandering males, and her uterus fills with eggs which develop into larvae (Fig. 15.11a). At this stage the larval development is arrested, the female dies and her body wall is transformed into a tough, leathery cyst which can persist in soil for many years, making these nematodes difficult to eradicate.

A classic example of biocontrol of plant-parasitic nematodes was reported by Kerry & Crump (1980). When oat crops were grown repeatedly on field sites in Britain, the population of cereal cyst nematodes (*Heterodera avenae*) was found to increase progressively, but then spontaneously declined to a level at which it no longer caused economic damage. Investigation of these **cyst-nematode decline sites** revealed a high incidence of parasitism of the females by a zoosporic fungus, *Nematophthora gynophila* (Oomycota), coupled with parasitism of the eggs (i.e. the sacs containing the individual arrested larvae) by another fungus, *Verticillium chlamydosporium* (a mitosporic fungus). *Nematophthora* infects the females and fills most of the body cavity with thick-walled resting spores (oospores) so that the cyst, if formed at all, contains relatively few

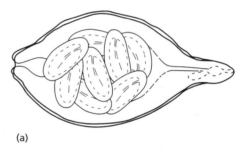

(a)

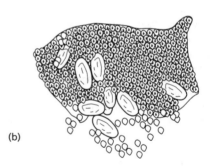

(b)

Fig. 15.11 A female cereal cyst nematode, *Heterodera avenae*, and the cyst-parasitic fungus *Nematophthora gynophila* (Oomycota). (a) Mature healthy nematode containing embryonated eggs. The body wall of the female will subsequently develop a leathery cuticle and become a cyst that persists in soil. (b) A ruptured cyst filled with oospores of *N. gynophila*. Most of the eggs are not infected, but the absence of a cyst wall enables egg parasites such as *Verticillium chlamydosporium* to infect and destroy many of the eggs. (Drawn from photographs in Kerry & Crump 1980.)

eggs but a large number of fungal spores (Fig. 15.11b). *Verticillium* has a different role from this – it is a facultative parasite of nematode eggs, destroying the eggs after they have been released into soil.

The parasitic efficiency of these two fungi – the cyst parasite and the egg parasite – is so high that they bring the level of damage to oat crops below the "economic threshold" at which fungicides or other control agents would be justified. However, although this natural control is highly effective, it has not led to the widespread use of *N. gynophila* for biological control. Part of the reason is that *N. gynophila* cannot be grown in culture – it is an obligate parasite of nematodes. And in any case it is effective only in soils wet enough to favor zoospore activity. The egg parasite *V. chlamydosporum* can be cultured easily but is less effective as a biocontrol agent, especially when acting alone.

Online resources

LUBILOSA (biological control of locusts and grasshoppers). http://www.lubilosa.org/exsumm.htm

Cited references

Anke, H. & Sterner, O. (2002) Insecticidal and nematicidal metabolites from fungi. In: *The Mycota X. Industrial Applications* (Osiewacz, H.D., ed.), pp. 109–127. Springer-Verlag, Berlin.

Barron, G.L. (1977) *The Nematode-destroying Fungi*. Canadian Biological Publications, Guelph, Ontario.

Butt, T.M. (2002) Use of entomogenous fungi for the control of insect pests. In: *The Mycota XI. Agricultural Applications* (Kempken, F., ed.), pp. 111–134. Springer-Verlag, Berlin.

Butt, T.M., Jackson, C.W. & Magan, N., eds (2001) *Fungi as Biocontrol Agents, Problems, Progress and Potential*. CABI Publishing, Wallingford, Oxon.

Charnley, A.K. (1989) Mechanisms of fungal pathogenesis in insects. In: *Biotechnology of Fungi for Improving Plant Growth* (Whipps, J.M. & Lumsden, R.D., eds), pp. 85–125. Cambridge University Press, Cambridge.

Feng, M.G., Poprawski, T.J. & Khachatourians, G.G. (1994) Production, formulation and application of the entomopathogenic fungus *Beauveria bassiana* for insect control: current status. *Biocontrol Science and Technology* **4**, 3–34.

Hall, R.A. & Burges, H.D. (1979) Control of aphids in glasshouses with the fungus *Verticillium lecanii*. *Annals of Applied Biology* **102**, 455–466.

Jaffee, B.A. (1992) Population biology and biological control of nematodes. *Canadian Journal of Microbiology* **38**, 359–364.

Kerry, B.R. & Crump, D.H. (1980) Two fungi parasitic on females of cyst-nematodes (*Heterodera* spp.). *Transactions of the British Mycological Society* **74**, 119–125.

Kerry, B.R., Crump, D.H. & Mullen, L.A. (1982) Studies of the cereal cyst nematode, *Heterodera avenae*, under continuous cereals, 1975–1978. II. Fungal parasitism of female nematodes and eggs. *Annals of Applied Biology* **100**, 489–499.

Tunlid, A., Jansson, H.-B. & Nordbring-Hertz, B. (1992) Fungal attachment to nematodes. *Mycological Research* **96**, 401–412.

Vey, A., Hoagland, R.E. & Butt, T.M. (2001) Toxic metabolites of fungal biocontrol agents. In: *Fungi as Biocontrol Agents* (Butt, T.M., Jackson, C. & Magan, N., eds), pp. 311–346. CAB International, Wallingford, Oxon.

Whisler, H.C., Zebold, S.L. & Shemanchuk, J.A. (1975) Life history of *Coelomomyces psorophorae*. *Proceedings of the National Academy of Sciences, USA* **72**, 693–696.

Wilding, N. & Perry, J.N. (1980) Studies on *Entomophthora* in populations of *Aphis fabae* on field beans. *Annals of Applied Biology* **94**, 367–378.

Chapter 16

"The moulds of man"

This chapter is divided into the following major sections:

- the major fungal pathogens of humans and other mammals
- the dermatophytic fungi
- *Candida albicans* and other *Candida* species
- opportunistic and incidental pathogens: Aspergillosis
- endemic dimorphic fungi: *Coccidioides, Blastomyces, Histoplasma,* and *Paracoccidioides*
- *Cryptococcus neoformans*
- *Pneumocystis* species

In contrast to the many thousands of fungi that infect plants, only about 200–300 fungi are reported to cause diseases of humans and other warm-blooded animals – diseases that are collectively termed **mycoses**. We can be thankful for this, although it remains to be explained satisfactorily in evolutionary terms. Even the fungi that do infect humans and other warm-blooded animals are, for the most part, opportunistic or cause only mild symptoms in normal, healthy individuals. But the situation has changed drastically in recent years, with the increasing use of immunosuppressant drugs in transplant surgery and cancer therapy, and the advance of HIV/AIDS. Fungal infections can be life-threatening in these situations, and we shall see in Chapter 17 that there are few really satisfactory drugs to control them without causing adverse side effects.

In addition to the invasive mycoses, fungi pose a threat to health by producing mycotoxins in foodstuffs and animal feeds (Chapter 7), and airborne fungal spores can be significant causes of asthma, hay fever, and more serious occupational diseases, discussed in Chapter 10. Taking all these factors together, fungi can have a significant impact on human and animal health.

In this chapter we consider the major fungi that infect humans and some other warm-blooded animals.

Major fungal pathogens of humans and other mammals

The human-pathogenic fungi can be grouped into five categories based on features such as their primary route of entry into the host, the type of disease that they cause, and their natural sources of inoculum (Table 16.1). An outline of these groups is given below and will provide the basis for more detailed treatment in later sections of this chapter:

1 The **dermatophytes**, also known as ringworm fungi, grow in the dead, keratinized tissues of the skin, nails, and hair. They are very common, and infect large sections of the human and animal populations. The diseases that they cause are superficial, being confined to the dead tissues, but they can cause severe irritation of the underlying living tissues, leading to secondary invasion by bacteria. Many of these fungi show a degree of specialization for particular hosts (humans, cats, cattle, etc.) but can cross-infect other hosts. The source of inoculum is usually the shed keratinized tissues (flakes of skin, hairs, etc.) in which the fungus can persist in a dormant phase. As a group, the dermatophytes are successful parasites with a clearly defined niche.

Table 16.1 The major types of fungus that cause mycoses of humans.

Primary route of entry	Fungus	Sexual stage	Disease	Natural distribution
Skin	Trichophyton (22 species) Microsporum (19 species) but only 9 are involved in infections Epidermophyton (2 species)	Arthroderma (Ascomycota)	Dermatomycosis: ringworm, tinea, athlete's foot, etc.	Keratinized tissues, humans and wild or domesticated animals
Mucosa	Candida albicans Some other Candida spp.	Recently reported (see text)	Candidosis: thrush, vulvovaginitis, stomatitis	Commensal on mucosa
Lungs	Aspergillus fumigatus	None	Aspergillosis: invasive (systemic) or aspergillomas of lungs	Saprotrophic in soil or organic matter (composts)
	Blastomyces dermatitidis	Ajellomyces (Ascomycota)	Blastomycosis: lungs, skin lesions, bones, brain	Saprotrophic
	Coccidioides immitis	None	Coccidioidomycosis: lungs, systemic	Saprotrophic in soil
	Cryptococcus neoformans	Filobasidiella (Basidiomycota)	Cryptococcosis: lungs, brain, meninges	Bird excreta, vegetation (eucalypt trees)
	Histoplasma capsulatum	Ajellomyces (Ascomycota)	Histoplasmosis: lungs, rarely systemic	Bird and bat droppings
	Paracoccidioides brasiliensis	None	Paracoccidioidomycosis: lungs, cutaneous, lymph nodes	Soil ?
Wounds/ lesions	Phialophora Cladosporium Sporothrix, etc.	Often none	Subcutaneous mycoses: chromomycosis, sporotrichosis, etc.	Saprotrophic in soil, dead plant material
	Rhizopus, Absidia, etc.	Zygomycota	Zygomycosis	Saprotrophic
Lungs	Pneumocystis species	None	Virulent pneumonia	Humans, other mammals

2 Several *Candida* species grow as normal resident **commensals** on the mucosal membranes of healthy individuals, but can become invasive in appropriate conditions. The diploid yeast *Candida albicans* is the prime example of this. It is found commonly on the mucosa of the gut, the mouth, and the vulvo-vaginal tract. But in response to environmental triggers the yeast cells can produce hyphal outgrowths that invade the mucosa, leading to clinical or subclinical conditions. For example, "thrush" is a white, invasive speckling of the throat, which is common in neonates and immunocompromised individuals, including AIDS and cancer patients, and

people in the advanced stages of diabetes. *Candida* can also cause intense irritation of the vulvo-uterine tract of women during stages of menstruation or during pregnancy, and it commonly causes stomatitis in people who wear dentures.

3 A diverse group of fungi that normally grow as saprotrophs in soil or on plant or animal remains can establish infections in the lungs from inhaled spores. These infections invariably stem from an environmental source of spores rather than from direct patient-to-patient transmission. Almost any fungus that produces spores small enough (c. 3–4 μm) to reach the alveoli, that can grow at 37°C, and

that can withstand the host's cellular defenses is a potential threat to health. But these infections are normally restricted to immunocompromised or immunosuppressed individuals, including patients with advanced diabetes or cancers. A notable feature of many of these infections is that they are geographically confined to specific regions – in other words, they are **endemic**. For example, *Coccidioides immitis* (a fungus known to have killed a laboratory worker) is largely confined to the arid desert regions of southwestern USA and northern Mexico.

4 A few fungi with melanized hyphal walls, such as species of *Phialophora*, *Cladosporium*, and *Sporothrix*, can invade traumatized tissues such as deep wounds and cause tissue-degrading lesions. Although they can be important causes of infection, especially among agricultural workers in poorer countries, they are essentially **pathogens of wounds or traumatized tissues**. They will not be considered further in this chapter. Similarly, some common members of the Zygomycota cause traumatic infections of people who suffer from diabetes and ketoacidosis, as discussed in Chapter 2.

5 A specialized group of primitive fungus-like organisms, now classified as different species of *Pneumocystis*, are found in the lungs of humans and a wide range of mammalian hosts. They seem to be host-specific because DNA sequence comparisons of *Pneumocystis* strains from different hosts show wide genetic diversity, but close similarity between strains isolated from any one type of host. The fungus that infects humans, *Pneumocystis jirovici* (formerly

known as *P. carinii*), seems to be particularly common in young children, because often more than 80% of children show an antibody response to the pneumocystis antigen in skin tests. But the fungus seems to disappear as children age, and is found again in AIDS patients, where it causes a virulent pneumonia that can be transmitted from patient to patient by airborne cells. Pneumonia caused by *P. jirovici* in HIV patients is often regarded as one of the first "AIDS-defining" illnesses.

Dermatophytic fungi

The dermatophytes are a clearly defined group of about 40 species, traditionally assigned to three genera of mitosporic fungi – *Trichophyton*, *Microsporum*, and *Epidermophyton*, based on features of their conidial stages (Figs 16.1, 16.2a). Some species of *Trichophyton* (and similarly of *Microsporum*) have been found to produce a sexual stage when strains of different mating types are paired in laboratory culture. The sexual stage, named *Arthroderma*, is a cleistothecium containing asci and ascospores. However its role, if any, in epidemiology is not known, and so most medical mycologists use the familiar "asexual" names because these are associated with specific diseases and geographical distributions.

The characteristic feature of dermatophytes is their ability to grow in the dead, keratinized tissues of the skin, nails, and hair (Figs 16.2b, 16.3), where their metabolic products can induce an inflammatory

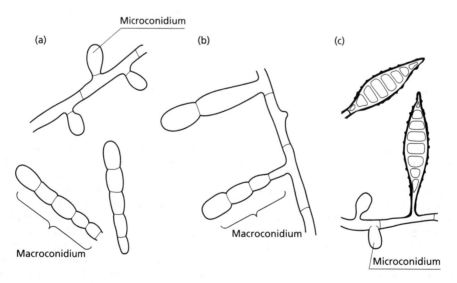

Fig. 16.1 Spores of the three common genera of dermatophytic fungi. (a) Macroconidia (about 50 μm long) and microconidia (about 4 μm) of *Trichophyton* spp. (b) Macroconidia of *Epidermophyton* spp, which do not produce microconidia. (c) Spindle-shaped macroconidia, and microconidia of *Microsporum* spp.

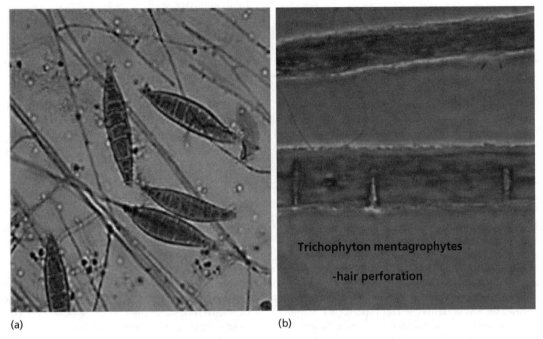

(a)   (b)

Fig. 16.2 (a) Spindle-shaped macroconidia, and microconidia, of *Microsporum* spp. (b) Enzyme-mediated penetration of a hair by hyphae of *Trichophyton mentagrophytes*. (Reproduced by courtesy of the Canadian National Centre for Mycology; http://www2.provlab.ab.ca/bugs/webbug/mycology/dermhome.htm)

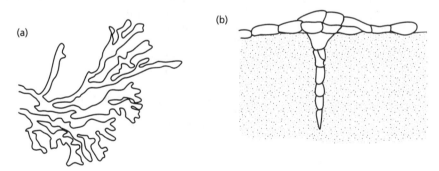

Fig. 16.3 (a) Flattened fronds of hyphae of a dermatophyte, growing in planes of weakness within a stratified substrate such as skin flakes. (b) Diagram of a perforating organ, similar to that in Fig. 16.2b.

response. The living tissues are not invaded because dermatophytes cannot grow at 37°C. Nevertheless, the irritation caused by dermatophytes often leads to scratching and further damage to the tissues, enabling bacteria to invade. Scratching also leads to the shedding of dermatophyte-infested skin and hair, so the shared use of "scratching posts" by animals can lead to transmission of the infection. Similarly, in human populations communal facilities such as bathing areas,

changing rooms, and domestic carpets can provide a source of inoculum. This indirect transmission is thought to be more important than direct host-to-host transmission. Even the role of the conidial stages is debatable, because these are not always found on infected tissues while they are attached to the host, but pigmented hyphae and vegetative arthrospores (thick-walled spores formed by hyphal fragmentation) can be found in these tissues.

The pathogenicity and virulence determinants of dermatophytes

Pathogenicity is the ability to cause disease, while virulence factors determine the severity of a disease. Although this distinction is clear in principle, it is not always easy in practice without knowledge of specific genes and gene products that can be manipulated to establish a direct link between disease and the roles of pathogenicity/virulence determinants. However, two key features are thought to contribute to diseases caused by dermatophytes: (i) their ability to grow on substrates rich in keratin, and (ii) their significant, although not absolute, degree of host specialization.

Keratin is a major protein found in skin, nails, and hair. It is a difficult substrate to degrade, especially when found in the form of "hard keratin" (e.g. nails, hoofs) which contains numerous disulfide bonds between the sulphur-containing amino acids. In laboratory culture the dermatophytes can grow on keratin as their sole carbon and energy source, by producing keratinolytic proteases termed **keratinases**. This is perhaps a unique feature of these fungi, because no other fungus is reported to grow exclusively on proteins. Keratinases could therefore be virulence determinants of the dermatophytes, and an understanding of these enzymes could lead to the development of vaccines. In a recent report (Brouta *et al.* 2002) a metalloprotease (MEP) genome sequence of the unrelated fungus, *Aspergillus fumigatus*, was used as a probe to identify three *MEP* genes in a genomic library of *Microsporum canis*. One of these genes, *MEP3*, was found to encode a 43.5-kDa keratinolytic metalloprotease, and both this protease and another (termed MEP2) were found to be produced *in vivo* during experimental infection studies on guinea pigs. This work represents the first report of a gene family encoding potential virulence determinants in dermatophytes.

Dermatophytes show a quite significant degree of host- or habitat-specialization, so they are commonly grouped into three categories (Box 16.1):

- **Anthropophilic** (human-loving) species are primarily parasites of humans and only rarely infect other species. They typically cause persistent but relatively mild infections. Examples of these fungi include:
  - *Epidermophyton floccosum*, with a worldwide distribution, causing infection of the groin, the body, and feet (athlete's foot);
  - *Trichophyton mentagrophytes* and *T. rubrum* – probably the commonest species, with worldwide distributions, infecting many body parts.

- **Zoophilic** (animal-loving) species typically infect wild or domesticated animals, such as cattle, cats, dogs, and horses. They can also be transmitted to humans who have very close contact with animals, but these infections usually induce a strong inflammatory response in the human host and then resolve spontaneously. A classic example is *Microsporum canis*, commonly acquired by humans who have close contact with domestic pets such as cats and dogs.

- **Geophilic** species live in soil and usually degrade keratin-containing materials such as feathers and hair, but they can occasionally infect human hosts. An example is *Microsporum gypseum* which is found worldwide but commonly in South America, where it is usually acquired from soil but occasionally from animals.

The treatments used to control dermatophyte infections are discussed in Chapter 17. Briefly, because these fungi cause superficial infections they can be treated by topical application of oxidizing agents or antibiotics (e.g. nystatin, amphotericin B). If the infections are persistent they can be treated by oral administration

Box 16.1 Some of the main dermatophytic fungi that infect humans.

Anthropophilic	Zoophilic	Geophilic
Epidermophyton floccosum	*Microsporum canis* (cats, dogs)	*Microsporum gypseum* (commonly infects humans)
Microsporum audouinii	*Microsporum equinum* (horses)	*Trichophyton terrestre*
Microsporum ferrugineum	*Microsporum nanum* (soil/pigs)	
Trichophyton mentagrophytes var. *interdigitale*	*Microsporum persicolor* (rodents)	
Trichophyton rubrum	*Trichophyton equinum* (horses)	
Trichophyton tonsurans	*Trichophyton mentagrophytes* var. *mentagrophytes* (mice, rodents)	
	Trichophyton verrucosum (cattle)	

of the antibiotic griseofulvin or by newer, less toxic drugs such as terebinafine (Chapter 17).

Candida albicans and other *Candida* species

Candida albicans occurs naturally as a **diploid budding yeast** on the mucosal membranes of humans and other warm-blooded animals. It is a common commensal of humans, being found on the mucosa of the mouth, gut, or vagina of more than 50% of healthy individuals. Usually it causes no harm, but a wide range of predisposing factors can cause it to become invasive, leading to conditions collectively known as **candidosis**. A few other *Candida* species, such as *C. glabrata* and *C. tropicalis*, cause similar conditions, but less frequently. Several recent reviews cover significant aspects of *Candida* biology – Brown & Gow (2001), Calderone & Fonzi (2001), and Douglas (2003).

The list of clinical manifestations of *C. albicans* is extensive. This fungus can cause "thrush" in newborn babies, when the fungus invades the mucosa of the mouth and throat, producing speckly white pustules from which the disease gains its name. This is often associated with delivery through an infected birth canal, and the fungus proliferates on the mucosa before a normal, balanced microbial population has developed. *Candida* also causes cystitis, and it can cause inflammation of the mouth (stomatitis) of many people who wear dentures. This is associated with several factors, including abrasion caused by the dentures, adhesion of *Candida* cells to the dental plastic, and probably the closed environment with lack of "flushing" underneath the denture plate.

People whose hands are frequently exposed to water can develop *Candida* infections of the skin and fingernails. In addition, *Candida* can be introduced into the blood through catheters and other surgical procedures, but the yeast population soon declines when the catheter is removed. *Candida* infection of the gut is often associated with prolonged antibacterial therapy, especially with tetracycline antibiotics which suppress the bacterial population. Stress can be an additional factor in *Candida* infections; for example, astronauts have been found to develop high populations of *C. albicans* during space flights. And, in extreme cases of predisposition, such as advanced diabetes, neutrophil or macrophage disorders, immune disorders, and malignancies, *Candida* can grow systemically and become life-threatening. This catalog of examples shows that *Candida* is potentially an ever-present threat to human wellbeing. An understanding of its virulence determinants could pave the way for new approaches to controlling these infections.

The virulence determinants of *Candida albicans*

A large body of evidence suggests that there are two obvious virulence determinants of *C. albicans* – its **ability to adhere** strongly to epithelia and several other surfaces, and its ability to undergo a **dimorphic switch** between a yeast phase and a hyphal or pseudo-hyphal phase in response to environmental factors.

Adhesins

Despite many years of study on the adhesion of *Candida* cells, both *in vitro* and *in vivo*, it is still not possible to define precisely the role that adhesion plays in *Candida* infections, because there seem to be many interacting factors. Adhesion studies *in vitro* show that the yeast cells of *C. albicans* adhere strongly to shed epithelial cells of the vagina, and that adhesion is stronger to cells obtained from pregnant women. The two commonest cell types in the vaginal epithelium are the intermediate cells and the superficial cells. *Candida* binds most strongly to the intermediate cells *in vitro*, and these cells predominate when there are high levels of progesterone (for example, during pregnancy or in women who use oral contraceptives). Progesterone seems to have a direct effect, because *Candida* binds more strongly to the epithelial cells of nonpregnant women if progesterone is added *in vitro*. Additionally, strains of *C. albicans* from active vaginal infections are found to adhere more strongly in *in vitro* studies than do isolates from healthy people. This raises the possibility that venereal spread of vaginal candidosis might involve the transmission of strongly adherent strains.

These general correlations between *in vitro* adhesion and the course of relatively mild clinical infections are supported by other studies. For example, *C. albicans* adheres to cells of the buccal cavity, to the methyl acrylate resin of dentures, and to the surfaces of catheters. By using buccal epithelia or denture resin as model systems *in vitro*, the adhesion of *Candida* was found to be strongly enhanced if the fungus had been grown on high levels of galactose, maltose, or sucrose, rather than on glucose. This is associated with the presence of **mannoprotein adhesins** on the surface of the yeast cells. Nevertheless, the differential effects of sugars seem to be strain-related, because *Candida* strains from active infections show it frequently, whereas strains from asymptomatic carriers can show it much less often. Also, adhesion may be a specific feature of *C. albicans* rather than of *Candida* species in general, because the adhesion of nonpathogenic *Candida* spp. or of the nonpathogenic *Saccharomyces cerevisiae* is not markedly altered when they are grown on different sugars.

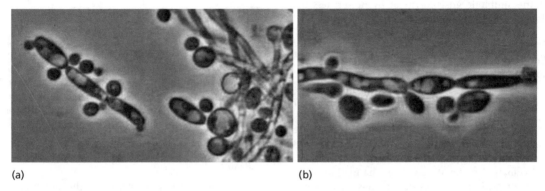

(a)   (b)

Fig. 16.4 (a,b) Pseudomycelium of *Candida albicans* in nutrient-limiting conditions. Yeast cells are produced instead of fungal branches at the septa.

The dimorphic switch

Candida albicans is a **dimorphic fungus**. It normally grows as a budding yeast, but in response to nutrient limitation the yeast cells can produce a hyphal outgrowth (see Fig. 1.4). This dimorphic switch is central to the pathogenicity of *C. albicans*, because invasion of the mucosa is always achieved by hyphal growth. In fact, this distinguishes *C. albicans* from all other (non-pathogenic) *Candida* spp. in clinical samples, and it is used in a simple diagnostic test: the yeast is cultured on a selective medium, then a loopful of inoculum is transferred to a vial containing horse serum and incubated at 37°C for 4–5 hours. Of all the *Candida* strains that might occur in clinical samples only *C. albicans* will sprout germ-tubes in these conditions, although several other *Candida* spp. from nonhuman sources can do so. The hyphal stage is only transitory, and incubation for longer times results in the production of a **pseudomycelium** in which the hyphae consist of strings of sausage-shaped cells that produce budding yeast cells at the septa (Fig. 16.4).

Progress in understanding the behavior of *C. albicans* has been hampered by the fact that *C. albicans* is constitutively diploid, so mutants are difficult to obtain. Some homozygous recessive strains have been generated by repeated treatment with mutagens, especially UV-irradiation. Also, parasexual genetics (Chapter 9) has helped to generate mutants. For this, strains with auxotrophic mutations are combined by protoplast fusion, then allowed to regenerate walls. Some of the resulting strains are tetraploid hybrids or aneuploids which, when grown on minimal medium and subjected to heat shock or antimicrotubule agents, can revert to an altered diploid state by loss of chromosomes. The development of molecular genetic methods for *C. albicans* has also been hampered. Many genes from *C. albicans* can be transformed into, and expressed in, *Saccharomyces cerevisiae*, but not vice versa because

C. albicans has a nonstandard codon usage: the codon CUG encodes serine, whereas it encodes leucine in most other organisms. This is a significant limitation because many of the putative virulence determinants of *C. albicans* cannot be studied easily in the background of the pathogen itself.

The discovery of mating-type genes

As explained in Chapter 9, the mating system of the ascomycetous yeast *Saccharomyces cerevisiae* involves a cassette of three mating-type genes (*Mata1*, *Matα1* and *Mat α2*) and the mating type of a strain can change, depending on whether the *MATa* or *MATα* gene is moved into the expression locus by a specific transformation event. By contrast, nobody had ever identified the equivalent mating-type genes of *Candida albicans*, so this fungus had always been assumed to be a clonal organism, with no capacity for sexual recombination. However, in 1998 a single mating type-like (*MTL*) locus was discovered in a laboratory strain of *C. albicans*. Since there is only one *MTL* locus, the cell must become homozygous for **a** or α by mitotic recombination or gene conversion. An **a** strain (a/a) cannot convert to α (α/α) because each strain has lost the alternate *MTL* allele. This means that normal strains of *C. albicans* cannot mate. But by deleting one of the two *MTL* alleles it is possible to produce mating strains that are either a/– or α/– and that, in appropriate conditions, can fuse with one another. This was found to be rare event, often leading to apparent tetraploids that did not develop further. But a few a/– and α/– strains were found to switch their phenotype from the normal white colony appearance to an opaque colony form, and these opaque strains have an increased mating efficiency of about a million fold. Therefore, it is proposed that *C. albicans* can undergo mating but does so only in the opaque colony form. The opaque phase cells are stable ay 25°C but not at 37°C, and they are far more efficient in colonizing skin surfaces. Therefore it is suggested that mating in this fungus must occur outside the body – on the skin surface or in environmental reservoirs.

Systemic infection by *C. albicans*

The rare cases of systemic spread of *C. albicans* in the blood and lymph tissues are always associated with severely predisposing factors such as leukemia, advanced diabetes, prolonged corticosteroid therapy, etc. But, strangely, *C. albicans* seldom grows **systemically** in AIDS patients, who tend to develop other systemic fungal infections instead. *Candida* grows systemically as a yeast in the body fluids, but dimorphism plays an important role in one respect. If cells of *C. albicans* are mixed with white blood cells (macrophages and polymorphonuclear leucocytes) *in vitro* the yeast cells can be engulfed and destroyed, but some of them break out of the phagocyes by converting to hyphae and then produce a further population of yeasts. In these conditions the vigor of the host defense system will be crucial in containing an infection. In Chapter 17 we will see that the major drugs used to treat systemic candidosis (modern derivatives of ketoconazole and related compounds) act synergistically with the host defenses, because at even low concentrations they suppress the transition from yeast to hyphal growth.

Opportunistic and incidental pathogens

Theoretically, any fungus that can grow at 37°C could be a potential pathogen of humans, but in practice the spectrum is much narrower than this (Table 16.1). A few common saprotrophic species of *Phialophora*, *Sporothrix*, *Cladosporium*, and *Acremonium* can infect wounds and cause damaging subcutaneous mycoses. Many examples of this are described by Kwon-Chung & Bennett (1992). A different spectrum of fungi characteristically infect through the lungs because their airborne spores are small enough to reach the alveoli (Chapter 10). Of all these fungi, the most significant threat is posed by *Aspergillus* species, especially *A. fumigatus* but, to a lesser extent, *A. flavus*. We will focus on these and related fungi in this section, because *Aspergillus* infections can be serious and often life-threatening.

Aspergillosis

Aspergillus fumigatus, *A. flavus* and *A. niger* are very common saprotrophs on a range of organic materials and produce abundant airborne conidia that are small enough to enter the lungs (Chapter 10). In people with impaired respiratory function these spores can germinate to produce dense, localized saprotrophic colonies termed **aspergillomas** (see Fig. 8.4). Usually these remain noninvasive, surrounded by fibrous tissue of the host. Aspergillomas are quite common in poultry that are fed on moulded grain, and can occur in farm workers who regularly handle such materials. Infection of the lungs is entirely incidental so far as the fungi are concerned because they grow as saprotrophs on plant organic matter (Chapter 10) and they have no natural means of disseminating from the lungs to spread the infection to other hosts.

Aspergillus fumigatus is by far the most damaging airborne fungal pathogen of humans, and is responsible for about 90% of aspergillosis cases. Its spores are frequently inhaled and enter the lungs, but normally are destroyed by the body's innate defenses. However, *A. fumigatus* can become invasive and grow systemically in the body, either from infections in the respiratory tract of immunocompromised patients or after entry through surgical wounds. It can be a potential problem in transplant surgery, when the patient's immune system is artificially suppressed.

There are thought to be two principal routes of invasion by *A. fumigatus* – through the ciliated epithelium that lines the upper regions of the respiratory tract or through the alveoli. The cells of the ciliated epithelium can engulf conidia of *A. fumigatus*, and some of these engulfed spores can survive within the host cells. Corticosteroid treatments are also known to be a risk factor in **invasive aspergillosis**, by reducing the release or efficacy of antifungal peptides produced by the lining epithelium. The spores that are not engulfed by the lining epithelium enter the alveoli, where they can germinate, but are destroyed quite rapidly by polymorphonuclear neutrophils. However, low neutrophil numbers resulting from chemotherapy can reduce the efficiency of this defense system. The spores can also be engulfed quickly by alveolar macrophages where they are killed by reactive oxygen species. But the rate of killing is slow, and it can take 2–3 days to clear a respiratory challenge. These factors can be important in determining the outcome of an infection.

To date there is no evidence of a specific virulence factor associated with *A. fumigatus*, and almost any strain of this fungus seems to be able to infect compromised individuals. The model system most commonly used for this is inoculation of mice (the so-called murine, i.e. mouse, model system). But this may not be the most appropriate model for aspergillosis of humans. The complete genome sequence of *A. fumigatus* is now available (http://www.tigr.org/tdb/e2k1/afu1). This should enable comparisons to be made of either the transcriptome (messenger RNA) or the proteome (protein profile) of strains grown *in vitro* compared with *in vivo*, to identify any potential virulence determinants that are expressed specifically in the host environment. Details of the roles of *Aspergillus* spp. as human pathogens are covered by several contributors in Domer & Kobayashi (2004).

Endemic dimorphic fungi

In addition to *C. albicans*, a small group of fungi can cause systemic infections of healthy or immunocompromised individuals. These fungi are particularly interesting because they are geographically localized (endemic) and they are dimorphic, switching from one growth form to another in response to temperature shifts. All of them initiate infection from airborne spores that enter the lungs. People with impaired respiratory function can be particularly prone to lung infection, and people in the advanced stages of diabetes, leukemia, and immunosuppressive disorders can often develop the systemic disease. Yet, skin tests with antigens of some of these fungi indicate that a substantial proportion of the population in the endemic areas has been exposed to infection at some stage and perhaps suffered only mild, flu-like symptoms before the infection spontaneously resolved. Thus, these fungi pose a significant and perpetual threat to a sector of the population. The four main pathogens that cause these diseases are discussed below.

Coccidioides immitis

Coccidioides immitis and a related species, *C. posadasii*, are fungi that grow in the alkaline soils of arid desert regions of California, Arizona, and Texas, extending into parts of Central and South America. The hyphae of these fungi fragment to produce small, thick-walled spores (arthroconidia or arthrospores) which can be dispersed in dry, wind-blown dust and thereby enter the lungs. Most of these cells are engulfed and destroyed by macrophages in the lungs, and cause a flu-like or pneumonia-like infection which resolves spontaneously. The incidence of this is quite common among people who live in these desert areas, because up to 50% of healthy people show a positive response to a skin test with *Coccidioides* antigen, obtained from laboratory cultures of the fungus.

However, in some people – especially those who are immunocompromised – the fungus is not contained by macrophages and it can spread within the lungs and enter the bloodstream, causing a serious generalized infection of the bones, subcutaneous tissues, meninges, and major organs. This disease is called **coccidioidomycosis**.

As shown in Figs 16.5 and 16.6, in the aggressive, generalized phase of coccidiodomycosis, the arthroconidia swell and undergo multiple nuclear divisions, leading to the production of multinucleate, swollen structures termed **spherules** in the lungs. The large

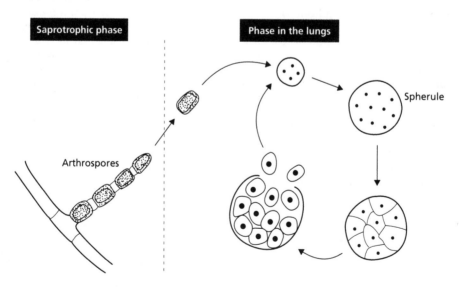

Fig. 16.5 Infection cycle of *Coccidioides immitis*, which grows in desert soils. In its saprotrophic phase, the fungus produces hyphal branches that undergo multiple septation to form a chain of cells. Then the nutrients are withdrawn from every alternate cell, to supply nutrients for the remaining cells (arthrospores) which develop thick walls. Arthrospores carried in wind-blown soil enter the lungs, germinate and produce a multinucleate spherule. At maturity, the cytoplasm cleaves around the individual nuclei, and the spherule releases uninucleate cells that can develop into further spherules.

(a)

(b)

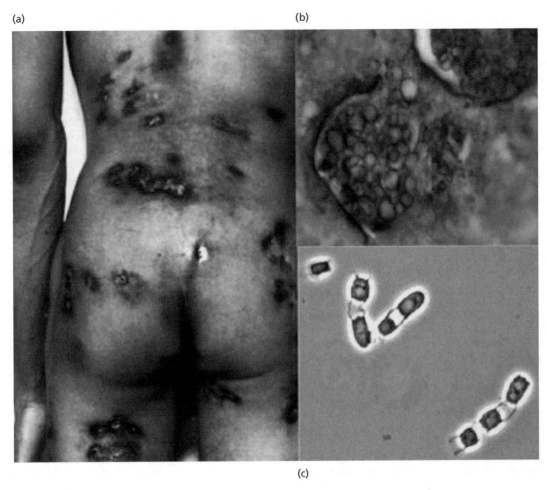

(c)

Fig. 16.6 (a) A patient showing the disseminated stage of disease (coccidioidomycosis). (b) Spherules. (c) Chains of arthrospores interspersed with empty cellular compartments. (Reproduced by courtesy of the DoctorFungus website; http://www.doctorfungus.org/)

spherules then undergo cytoplasmic cleavage to produce smaller bodies (subspherules), which can repeat the infection cycle within the lung tissues.

Coccidioides immitis has several interesting features. First, at common environmental temperatures such as 25°C it can grow as normal hyphae which produce arthrospores for dissemination, whereas it produces spherules at higher temperatures (37–40°C) typical of the host environment. Even so, a change of temperature alone is not sufficient to produce spherules, because *in vitro* studies show that these are only produced in the presence of high CO_2 concentrations and in specific growth media. *C. immitis* and the related species *C. posadasii* (see later) are the only dimorphic fungi that produce spherules; all other endemic fungal pathogens of humans grow in a yeast-like budding phase

in the host tissues. In advanced stages of infection, *C. immitis* becomes disseminated to many of the body tissues, as shown in Fig. 16.6.

The geographical distribution of coccidioidomycosis is not fully known, and even within the endemic regions it shows a patchy occurrence. There are reports that it is most common near the burrows of soil-dwelling rodents. In controlled experiments, dogs, cats, rodents, and other small animals have been shown to be susceptible to infection (but so are bottle-nose dolphins and horses!). It has been suggested that wild desert animals might provide a natural reservoir for infection of humans. Most recently, a molecular comparison of microsatellite DNA sequences from strains isolated from California/Arizona compared with Texas, Central and South America revealed that these represent two distinct species of *Coccidioides* (*immitis* and *posadasii*), both of which cause the same disease. Perhaps the most disturbing feature is that *Coccidioides* has the potential to be an inherently virulent pathogen. It is known to have killed a healthy laboratory technician, and the Occupational Safety and Health Administration of the US Government classifies it as "a biological agent with the potential to pose a severe threat to public health and safety that could potentially be used by terrorists." Put in simple terms, *C. immitis* is the only known fungus that is considered to represent a significant bioterrorism threat. Further information on this fungus can be found at the "Doctor Fungus" website (see Online resources).

Histoplasma capsulatum

Histoplasma is a fungus with a "typical" mould-yeast dimorphism: it grows as hyphae on laboratory media at 25–30°C, but as a budding yeast at 37°C on cysteine-rich media. The hyphae produce single conidia on the ends of short branches. Some of these conidia are large, warty **macroconidia** (8–15 μm) whereas others are **microconidia** (2–4 μm) that are likely to be the main infective agents in the lungs. The spores germinate to form a germ-tube which rapidly gives rise to a budding yeast phase in appropriate conditions. This yeast phase is almost invariably found throughout the course of infection, whereas the hyphae are involved in the saprotrophic phase in dead body tissues or in natural substrates.

Histoplasma capsulatum has a wide geographical distribution, including much of the eastern USA, most of Latin America, and parts of Southeast Asia, Africa and Europe (e.g. Italy). It occurs naturally as a saprotroph in fecal-enriched soil around poultry houses, in bat droppings in caves, and also in starlings' droppings in towns and cities. The birds themselves are not infected, and the fungus is not found in their fresh droppings. However, bats have been found to be naturally infected by *H. capsulatum*, with intestinal lesions. Spores produced from fecal materials are the most likely source of infection of humans, because there are recorded instances of speliologists having developed the disease (**histoplasmosis**) after visiting caves. Skin tests using an antigen, histoplasmin, obtained from culture filtrates suggest that a high proportion of the human population has been exposed to infection in the endemic areas; one such study estimated that 20% of all the people in the USA had been infected at some time from airborne spores.

Even in mild infections this fungus has been detected in the urine, suggesting that it might commonly be generalized, although in the vast majority of cases the infection resolves spontaneously. Like *Coccidioides*, *Histoplasma* can be acquired in the laboratory because there is a recorded instance of students having tested positive after examining cultures in the classroom. The airborne spores can readily enter the lungs and cause acute pulmonary histoplasmosis. However, chronic pulmonary infection is confined to people with lung dysfunction from other causes, and it can lead to dissemination to the other body tissues. This dissemination occurs by carriage of the yeast cells in the lymph system, and is most common in immunodeficient people or in the very young (less than 1 year) or older (50 years or more) sectors of the population. Magrini & Goldman (2001) describe the increasing range of molecular genetic tools that have been used to investigate the pathobiology of *H. capsulatum*.

Blastomyces dermatitidis

Blastomyces dermatitidis is an endemic dimorphic fungus found in the southeastern and south central regions of the USA, along the Mississippi and Ohio rivers, where it causes a disease termed North American blastomycosis, or Chicago disease. It is also reported to be isolated from humans in parts of Africa and the Indian subcontinent, but these strains lack the characteristic antigens of *B. dermatitidis* and might represent a different species. In North America infection is associated with moist soils enriched with organic matter or rotting wood, but little is known about the basic ecology of this fungus. When grown in laboratory culture, it has narrow hyphae (about 2 μm diameter) that produce single globose conidia, 2–5 μm diameter, at the ends of short hyphal branches, similar to (but smaller than) those of *Thermomyces lanuginosus* (see Fig. 5.18). It infects humans from airborne conidia that enter the lungs and transform into a **thick-walled**, yeast-like budding phase. About 50% of infections are asymptomatic, but other infections can progress to an acute pulmonary phase after an incubation period of 30–45 days, producing symptoms in the lungs that strongly resemble a bacterial pneumonia. The infection can also progress to a chronic phase that affects the lungs, skin, bones, genitourinary tract, and other organs – for example, ulcerative lesions of the skin, granulomatous inflammation of the lungs, and spread to many other tissues and organs.

Brandhorst *et al.* (2002) reported that thermally regulated dimorphism is the single most defining trait of *Blastomyces*, and this is linked to a phase-specific gene of *B. dermatitidis* termed *BAD1* (*Blastomyces* adhesin), found only on the surface of the Y-phase and not on the hyphae that grow at lower temperatures. The cell wall glucan of the yeast-phase cells consists of 90% α-glucan, whereas mycelial phase cells have roughly equal content of α-glucan and β-glucan. This is also true for *Paracoccidioides brasiliensis* (see below) and strains of *Histoplasma capsulatum* – in all three cases the reduced content of α-glucan correlates with loss of virulence in animal models. In *B. dermatitidis* this correlation between α-glucan and virulence was confirmed by deleting the gene encoding BAD1. The deleted strain had a diminished ability to bind to macrophages or to mouse lung tissue, and it was highly attenuated in experimental mice. Restoration of the *BAD1* gene by transformation restored the virulence of *B. dermatitidis* and restored the adhesive properties of the fungus. The BAD1 protein has been found to self-associate in the presence of calcium ions, and it is suggested that, in the conditions of the lung, chitin-bound BAD1 molecules could begin to capture further BAD1 molecules, adding further α-glucan layers to the yeast cell wall.

Paracoccidioides brasiliensis

Paracoccidioides brasiliensis is another endemic fungus, found mainly in subtropical forest regions of Central and South America, especially Brazil, Venezuela, and Columbia. It has been isolated from the digestive tract of some animals (e.g. armadillos; Silva-Vergara *et al.* 2000) and from moist soil enriched with proteins, but the natural ecology of the fungus is poorly understood. It is a mitosporic fungus with no known sexual stage, and it grows very slowly in agar culture. *P. brasiliensis* is a dimorphic fungus that grows as mycelia at lower temperatures but as a budding yeast at 37°C or in body tissues. The yeast cells are oval or spherical and range in size from 2 to 10 μm or up to 30 μm or more. Often, they grow as a single large central cell with multiple buds attached to the central cell by narrow necks, resembling a mariner's wheel. The buds remain attached to the mother cell until they, themselves, bud (Fig. 16.7).

P. brasiliensis causes the disease termed paracoccidioidomycosis, or South American blastomycosis. Infection is assumed to occur from airborne spores that enter the lungs, but most infections are asymptomatic, as evidenced by a skin test to the fungal antigen. In other cases the infection can develop several years after initial exposure, indicating a long latent period in which the fungus has remained dormant in the lymph nodes. Such infections are probably associated with immunodeficiency. The characteristic form of the disease involves severe ulcerative lesions of the mouth, nose, larynx, and subcutaneous tissues, causing serious facial disfigurement. In extreme cases the fungus can also affect other organs such as the spleen, liver, bones, and central nervous system. A notable feature of paracoccidioidomycosis is that it occurs primarily in men – the ratio of infection is about 15 : 1 in men compared with women, but can be up

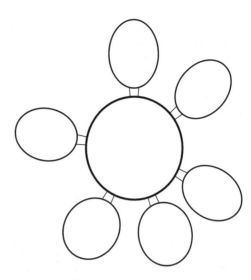

Fig. 16.7 Typical appearance of the budding phase *of Paracoccidioides brasiliensis*: the large parent yeast-like cell produces multiple buds that remain attached to the mother cell.

to 78 : 1, whereas equal numbers of men and women show a positive skin test reaction to the paracoccidioides antigen. The reason is that estrogens inhibit the transition of cells from the conidial or mycelial form to the yeast phase which is necessary for colonization of the tissues (Borges-Walmsley *et al.* 2002).

Cryptococcus neoformans

Cryptococcus neoformans differs from the fungi discussed above because it is not endemic but has a **wide global distribution**. It is common as a saprotroph and has been isolated frequently from old, weathered pigeon droppings in cities and from soils enriched with bird excreta. However, it does not compete well in wet droppings, where bacteria can raise the pH to growth-inhibitory levels, and it does not infect the birds themselves. Studies in New York City indicate that a high proportion of young children have been exposed to the fungus, because they show a reaction when skin-tested with the *Cryptococcus* antigen, but apparently do not develop any symptoms. There are over 30 species of the yeast *Cryptococcus*, but only one species, *C. neoformans*, is pathogenic to humans, causing the disease called **cryptococcosis**.

Cryptococcosis used to be a rare disease, found mainly in people who were naturally immunocompromised or who had undergone transplant surgery involving immunosuppressive drugs. But the increasing incidence of HIV/AIDS in the 1980s and 1990s led to a major surge in cryptococcosis, affecting 7–10% of AIDS patients worldwide. Although the disease can be treated with routine administration of antifungal agents, this is seldom possible in developing countries. The typical route of infection is via the lungs, when spores or yeast cells are inhaled. After an initial subclinical pulmonary infection, the fungus can cause chronic lung infection and then disseminate to the central nervous system, where it shows a predilection for growth in the cerebral cortex, brain stem, cerebellum, and meninges. This is invariably fatal if not treated, and it is common in AIDS patients, perhaps because a primary lesion in the lung has remained quiescent for many years and leads to secondary spread when the immune system is impaired.

There are several features of *C. neoformans* that have yet to be resolved and that could assist in preventing or treating infections. Normally the fungus is isolated from environmental samples as a haploid yeast, and it was commonly assumed that dehydrated yeast cells are the primary source of infection. With a diameter ranging from 2.5 to 10 µm, some of these cells would be small enough to reach the alveoli, rehydrate, and initiate infection. However, more recently it was discovered that *C. neoformans* has a sexual stage in the genus *Filobasidiella* (Basidiomycota; Fig. 16.8),

leading to the production of airborne basidiospores about 1.8–3.0 µm diameter. These would be an ideal size to be deposited in the alveoli (Chapter 10). The basidiospores can be produced in laboratory culture by pairing of the two mating types, termed "a" and "α". But α strains alone can be induced to produce basidiospores (with a single haploid nucleus in each cell) when cultured on media that lack nitrogen and in conditions of water-stress. The α mating-type locus (*MATα*) has been cloned and shown to encode a pheromone-like peptide. The α strains are found much more frequently than are the a mating types in clinical specimens, and in inoculations of mice they cause much more rapid death than do the a-mating-type strains. Thus it seems that the gene product of α-mating-type strains may contribute to virulence.

In addition to these points, there are two distinct forms of *C. neoformans*, classified as varieties – *C. neoformans* var. *neoformans* and *C. neoformans* var. *gattii*. The first of these is commonly associated with AIDS infection in Europe and North America, and presumably originates from bird excreta. By contrast, the variety *gattii* is seldom found in AIDS patients; instead it is associated with endemic disease of nonimmuno-compromised people in Australia, Papua New Guinea, and parts of Africa, India, south-east Asia and Central and South America. The environmental source of

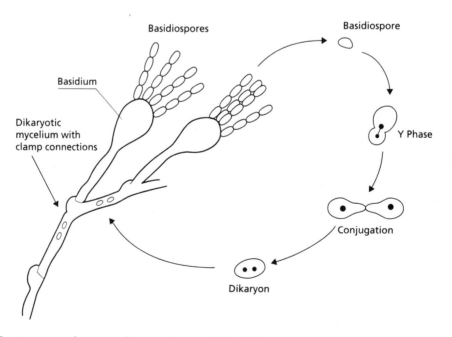

Fig. 16.8 *Cryptococcus neoformans* and its sexual stage, *Filobasidiella neoformans*. The budding yeast cells (Y-phase) of opposite mating types conjugate and form a dikaryotic cell. This forms a mycelium on which the inflated basidia develop. Meiosis in the basidia leads to the production of four haploid nuclei, which migrate to the tips of the basidia. Then chains of four haploid basidiospores are produced, and these germinate to give the haploid Y-phase cells.

var. *gattii* seems to be eucalypt trees and hollows in decaying wood.

The pathogenicity and virulence determinants of *C. neoformans*

As noted earlier in this chapter, pathogenicity determinants include all those factors that enable an invasive organism to live in a host environment. Virulence determinants include the factors that determine the severity of disease. For *C. neoformans* the major **pathogenicity determinants** are the ability to grow at 37°C in an atmosphere of about 5% CO_2 and at a pH of about 7.3. The key to this seems to be a gene that codes for **calcineurin A**, a specific type of protein phosphatase that is activated by Ca^{2+}-calmodulin and that is involved in stress responses in yeasts. Mutants that are defective in calcineurin A can grow at 24°C but not at 37°C, and they cannot survive in 5% CO_2 or alkaline pH.

There are several **virulence determinants** of *C. neoformans*. The most conspicuous of these – both in laboratory culture and in clinical specimens – is the presence of a thick, rigid polysaccharide capsule around the yeast cells, similar to the capsules seen in Fig. 16.9. The capsule consists of a high molecular weight polysaccharide, with a backbone of α-1,3-linked D-mannose units and single units of D-xylose and D-glucuronic acid. It seems to be a key virulence determinant of *C. neoformans*, because naturally noncapsulate strains, or mutants that lack the capsule, are nonvirulent. The encapsulated cells are not as easily phagocytized or killed by neutrophils, monocytes or macrophages, compared with noncapsular cells, and their high net negative charge can reduce cell–cell interactions required for clearance of the cryptococci.

Pathogenic strains of *C. neoformans* also produce brown or black pigments containing diphenolic compounds. Through the action of phenoloxidase, these compounds are oxidized and polymerized to produce **melanin**, which is deposited in the yeast walls, perhaps helping to protect against reactive oxidants in the host tissues. A similar protective role has been proposed for D-mannitol, which is produced by *Cryptococcus* and is especially effective as a scavenger of hydroxyl radicals. Thus, overall there seem to be several potential virulence determinants in *C. neoformans*, even if the precise roles of some of these are not fully understood at present. The virulence determinants of *Cryptococcus* are reviewed by Perfect (2004).

Pneumocystis species

So far in this chapter, we have seen some quite astonishing examples of fungi that cause serious diseases of humans, often as opportunistic or incidental pathogens. They include fungi whose natural habitat is eucalypt trees, or pigeon excreta, or the excreta of cave-dwelling bats, and soil fungi that are endemic to specific regions of the globe. We have also seen fungi that disseminate "silently" in the tissues of humans and that show predilections for certain organs of the body. And yet many of these aggressive fungi have no obvious natural means of transmission from one host to another. For many of them, the infection of a human host seems to be purely incidental.

Now we turn to perhaps the most astonishing fungi of all – the many *Pneumocystis* species that cause virulent pneumonia-like symptoms associated with AIDS and immunosuppressive therapies. These organisms were first described as protozoa (protists), under the name *Pneumocystis carinii*, over 100 years ago, and have a life cycle that more closely resembles that of a protist than a typical fungus. Phylogenetic analysis (based on 18S ribosomal DNA sequence) places them closest to the early ascomycota, along with the fission yeast *Schizosaccharomyces pombe*. But they have one unique feature: they are the only fungi that have cholesterol in the cell membrane, whereas most other fungi (with the exception of some zygomycota) have ergosterol as their characteristic membrane sterol. One consequence of this is that *Pneumocystis* is not susceptible to the common antifungal agents that target ergosterol; instead *Pneumocystis* is controlled by some of the antiprotozoal agents.

The natural occurrence of *Pneumocystis* species

Pneumocystis species are found worldwide in human hosts and in the lungs of a wide range of wild mammals. However, before the widespread development

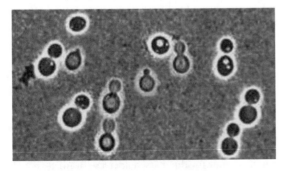

Fig. 16.9 Cells of a *Cryptococcus* species (*C. albidus*) viewed by phase-contrast microscopy in the presence of India ink, which clearly shows the rigid polysaccharide capsules.

of AIDS in the 1980s *Pneumocystis* was only found infrequently as a cause of pneumonitis, called ***Pneumocystis carinii* pneumonia** (PCP). It was found first in malnourished children and then in patients undergoing immunosuppressive therapies. Now PCP has become one of the commonest causes of death of AIDS patients, and there is a direct correlation between the risk of developing PCP and a low number of circulating $CD4^+$ T lymphocytes. A cell count of 200 or less per microliter of blood is the critical lower limit.

It is notable that a very high proportion of children in Europe and the USA show a positive serum test for *Pneumocystis* antigens by the age of 4 years, indicating exposure to the fungus. This is probably also true for most other countries. However, *Pneumocystis* then seems to disappear from the lungs and only returns as people age, or in response to immunosuppression. Thus *Pneumocystis* infection is strongly age-linked, and re-infections later in life probably result from widespread and constant re-exposure to the fungus.

Soon after *Pneumocystis carinii* had been formally transferred to the fungal kingdom, DNA sequence analysis showed that strains obtained from different animal hosts were very diverse and each was specific for the individual host species. Using the polymerase chain reaction, primers were developed to amplify DNA from all known species of *Pneumocystis*, and each of these was found to be distinct and host-specific, with no evidence of cross-infection from one host species to another. Confirmation of this in several studies has led to the proposal that all the host-specific forms of *Pneumocystis* should be regarded as separate species. Thus, the human-host-specific fungus has been named *P. jiroveci*, and the original name *P. carinii* has been retained for one of the two known *Pneumocystis* species found in rats. *Pneumocystis* organisms obtained from rats, mice, ferrets, pigs, and monkeys all show host-species specificity and new names for these are likely to be proposed. Consistent with the co-evolution of parasites with their hosts, the *Pneumocystis* species from humans are most similar to (although distinct from) those of other primates.

Modes of transmission, pathology, and the life cycle of *Pneumocystis*

Pneumocystis cannot be maintained for long in laboratory culture, so it seems to be essentially an obligate parasite. Small amounts of DNA of *P. jiroveci* have been detected in environmental samples, including samples of airborne spores and pond water, but the most likely source of infection is either the activation of a pre-existing latent infection or re-infection from inhaled spores. The fungus escapes the defenses of the upper respiratory tract, and establishes infections in the

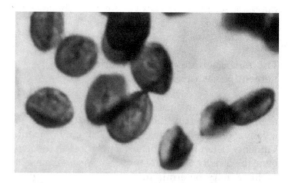

Fig. 16.10 Saucer-shaped cysts of *Pneumocystis jiroveci* (Reproduced with permission from Centers for Disease Control and Prevention, Image Library; http://www.dpd.cdc.gov/dpdx/HTML/ImageLibrary/Pneumocystis_il.htm)

alveoli, apparently by binding to the type 1 epithelial cells.

The life cycle of *Pneumocystis* is complicated, and still incompletely understood. Asexual trophic (feeding) forms proliferate by mitosis in the lungs. There is also a sexual stage in which haploid cells conjugate to produce a diploid pre-cyst. This undergoes meiosis, followed by mitosis, to produce eight haploid nuclei in the "late phase cyst." Mature cysts are variable in shape but often have a saucer-like appearance (Fig. 16.10). They are thought to rupture, releasing haploid vegetative cells.

In immunocompromised people the *Pneumocystis* cells proliferate and can result in tiers of cells up to four layers thick in the alveolar lumen, creating oxygen deficiency. A foamy alveolar exudate is produced, and in advanced stages of disease the fungus can be detected in all major organs of the body, especially the lymph nodes, bone marrow, liver, and spleen.

DNA sequence polymorphisms have often been found in isolates of the human pathogen *P. jiroveci*, indicating that many strains exist within this species. For example, polymorphisms have been found in the mitochondrial 18S ribosomal RNA gene, the mitochondrial small subunit rRNA gene, the internal transcribed spacer (ITS) regions of the nuclear rRNA gene, and the dihydropteroate synthase (DHPS) gene. These polymorphisms provide the basis for investigating the epidemiology of *Pneumocystis* infections in humans. Comparisons of strains collected before (1968–1981) and after (1982 to present) the beginning of the AIDS pandemic indicate that there has been no substantial change in the *Pneumocystis* population that could account for the massive increase in AIDS-related pneumonia. DNA sequence polymorphisms have also been used to investigate whether PCP (pneumonia) exists in a long-term latent form in humans or is reacquired during continuous or periodic exposure to the pathogen. The reactivation of long-term latent infection cannot be ruled out, because theoretically even a single latent *Pneumocystis* cell could give rise to infection when a host is immunocompromised. But studies of population genetics have helped to address this issue, by comparing the strains found in adult patients who have remained close to their birthplace against the strains found in patients who moved far away from their birthplace. The data indicate that adult patients acquire strains that are more similar to those in their current place of residence than in their birthplace, strongly suggesting that infection is reacquired rather than being carried since early childhood.

The following points can be made in summary of the current knowledge of *Pneumocystis*:

- *Pneumocystis* is a genus of worldwide distribution. It exists as many different host-specific species, with no evidence of cross-infection between different hosts.
- *Pneumocystis* seems to be a primitive fungus-like organism that represents one of the earliest basal lineages of fungi and has diversified as an obligate parasite that infects many species of mammals, including humans.
- In human populations, young children are frequently infected but typically exhibit only a transient infection and subsequently develop immunity. Children may be the primary reservoir of infection.
- Adults seem to acquire infection from environmental sources close to their areas of residence, but they only develop disease if they are immunosuppressed or immunocompromised. In these circumstances the infection can progress rapidly, and is closely correlated with a low count of circulating CD4[+] T lymphocytes.

We return to this and other fungal pathogens of humans in Chapter 17, where we consider the methods available for treating human mycoses.

Online resources

Aspergillus fumigatus genome sequence. http://www.tigr.org/tdb/e2k1/afu1

Centers for Disease Control and Prevention, USA, Image library. http://www.dpd.cdc.gov/dpdx/HTML/ImageLibrary/Pneumocystis_il.htm

Doctor Fungus website. http://www.doctorfungus.org. [An excellent, comprehensive source of up-to-date information on all the human mycoses]

General texts

Evans, E.G.V. & Richardson, M.D., eds (1989) *Medical Mycology: a practical approach*. Oxford University Press, Oxford.

Kwon-Chung, K.J. & Bennett, J.E. (1992) *Medical Mycology*. Lea & Febinger, Philadelphia.

Sutton, D.A., Fothergill, A.W. & Rinaldi, M.G., eds (1998) *Guide to Clinically Significant Fungi*. Williams & Wilkins, Baltimore.

Cited references

Borges-Walmsley, M.I., Chen, D., Shu, X. & Walmsley, A.R. (2002) The pathobiology of *Paracoccidioides brasiliensis*. *Trends in Microbiology* **10**, 80–87.

Brandhorst, T.T., Rooney, P.J., Sullivan, T.D. & Klein, B.S. (2002) Using new genetic tools to study the pathogenesis of *Blastomyces dermatitidis*. *Trends in Microbiology* **10**, 25–30.

Brouta, F., Descamps, F., Monod, M., Vermout, S., Losson, B. & Mignon, B. (2002) Secreted metalloprotease gene family of *Microsporum canis*. *Infection & Immunity* **70**, 5676–5683.

Brown, A.J.P. & Gow, N.A.R. (2001) Signal transduction and morphogenesis in *Candida albicans*. In: *The Mycota VIII. Biology of the Fungal Cell* (Howard, R.J. & Gow, N.A.R., eds), pp. 55–71. Springer-Verlag, Berlin.

Calderone, R.A. & Fonzi, W.A. (2001) Virulence factors of *Candida albicans*. *Trends in Microbiology* **9**, 327–335.

Domer, J.E. & Kobayashi, G.S., eds (2004) *The Mycota XII. Human Fungal Pathogens*. Springer-Verlag, Berlin.

Douglas, L.J. (2003) *Candida* biofilms and their role in infection. *Trends in Microbiology* **11**, 30–36.

Kwon-Chung, K.J. & Bennett, J.E. (1992) *Medical Mycology*. Lea & Febinger, Philadelphia.

Magrini, V. & Goldman, W.E. (2001) Molecular mycology: a genetic toolbox for *Histoplasma capsulatum*. *Trends in Microbiology* **9**, 541–546.

Perfect, J.R. (2004) Genetic requirements for virulence in *Cryptococcus neoformans*. In: *The Mycota XII. Human Fungal Pathogens* (Domer, J.E. & Kobayashi, G.S., eds), pp. 89–112. Springer-Verlag, Berlin.

Silva-Vergara, M.L., Martinez, R., Camargo, Z.P., Malta, M.H., Maffei, C. M. & Chadu, J.B. (2000) Isolation of *Paracoccidioides brasiliensis* from armadillos (*Dasypus novemcinctus*) in an area where the fungus was recently isolated from soil. *Medical Mycology* **38**, 193–199.

Chapter 17

Principles and practice of controlling fungal growth

In this chapter we deal with the major practical methods of controlling fungi, under the following headings:

- control by management of environmental and biological factors
- biological and integrated control
- chemical control of fungi
- the principal cellular targets of antifungal agents
- the use of fungicides for plant disease control
- antifungal antibiotics used for plant disease control
- control of fungal infections of humans

Control through the management of environmental and biological factors

We have seen many good examples of how fungi can be controlled through the management of physical and environmental factors. For example, the management of water potential is essential for grain storage, and the combination of temperature and water potential can be used to predict the safe limits for preventing growth of mycotoxin-producing fungi in stored food products (see Fig. 8.10). The storage and shipment of fresh produce, including many fruits, is often achieved by **controlled atmosphere storage** – a combination of cool temperature and elevated levels of carbon dioxide. This delays ripening and the onset of senescence, and can be cheaper than either of the single treatments alone. Similarly, bananas are routinely shipped from Central America to the European market as unripe, green fruits and are then ripened artificially by exposure to ethylene-generating chemicals. The green fruits survive long-distance shipping (2–3 weeks), and while they are green the fruits will not develop lesions of banana anthracnose (*Colletotrichum musae*) which is present as latent infections on the skin (see Fig. 5.3).

Wounding of fruits leads to the release of ethylene, a wound hormone, and can lead to premature ripening and invasion by fruit-rotting fungi. To counteract this, several types of fruit are treated with fungicides or with wax coatings, and held in cool storage. We saw in Chapter 12 that several bacteria and yeasts are now marketed commercially as fruit protectants, as an alternative to the use of fungicides. It may not be necessary to use living organisms at all, because the cell wall fractions of many yeasts, including *S. cerevisiae*, can act as powerful elicitors of plant defense mechanisms. Similarly, there is increasing interest in the use of **chitosan** to protect plants from pathogenic attack. At present, chitosan is obtained commercially by chemical deacetylation of crab shells. It disrupts the growth of several fungi *in vitro*, causing excessive branching, wall alteration, and cytoplasmic disorganization. Chitin is a characteristic wall component of fungi, and some fungi – especially Zygomycota – naturally deacetylate chitin by the enzyme chitin deacetylase. They could be used as an alternative, renewable source of chitosan for triggering plant defense reactions.

Sanitation, quarantine, and similar biologically based strategies

Crop rotation is a traditional, highly effective method of controlling many plant diseases if the pathogen does not survive for long in the absence of a host crop. This is true of the take-all fungus *Gaeumannomyces* *graminis* (see Fig. 9.11), which usually declines to non-damaging levels within 1 year if a different (noncereal) crop is sown. As we saw in Chapter 12 (see Fig. 12.2), the alternative is to grow cereal crops continuously, year after year, until the soil becomes naturally suppressive to take-all.

Crop management practices can be altered to avoid disease. In many countries stinking smut of wheat (caused by *Tilletia caries*, Chapter 14) is seed-transmitted and must be controlled by seed-applied fungicides. But in the Pacific northwest of the USA it is mainly soil-borne, and if winter wheat is sown early it reaches the most susceptible stage while the soil moisture and temperature are unsuitable for germination of the smut spores. Other simple and effective disease avoidance practices include the liming of soil to prevent serious clubroot disease of cruciferous crops caused by the protist *Plasmodiophora brassicae*. **Phosphate deficiency** can be a major cause of yield losses. The browning root rot disease of wheat, caused by *Pythium graminicola* (Oomycota), was one of the most damaging diseases in the North American prairies until the 1940s, but was virtually eliminated when phosphate fertilizers became widely available.

Meteorological forecasting has been used for many years as a disease-management tool. It was developed initially for **control of potato blight** (Chapter 14) by Beaumont in 1947. By studying the relationship between weather and blight epidemics, he established that, after a certain date which varies from region to region, a blight epidemic will develop within 2–3 weeks following a 2-day period in which the temperature is 10°C or more and the relative humidity is more than 75%. This 2-day period became known as a **Beaumont period**, and it enabled growers to time the application of fungicides so that major epidemics could be avoided. This has now been refined further, and similar forecasting methods are widely used in other cropping systems. One example is the facial eczema warning system for *Pithomyces* (see Fig. 7.20).

Sanitation can be highly effective for controlling or avoiding disease. Two **mycoparasites**, *Trichoderma harzianum* and *Pythium oligandrum*, are best known for their potential to control plant pathogens (Chapter 12) but can also be a problem in commercial mushroom production because they attack and destroy the mycelium of *Agaricus bisporus*. A distinctive strain (or strains) of *T. harzianum* has become a problem in some British mushroom sheds, when spores contaminate the trays of compost. *P. oligandrum* also has caused serious cropping losses in some mushroom-production units. A recent example was described from New Zealand, where internal transcribed spacer (ITS) ribosomal DNA (Chapter 9) was used to identify a specific strain of *P. oligandrum* responsible for a serious outbreak of mushroom disease (Godfrey *et al.* 2003).

The growth of fungi such as *Amorphotheca resinae* and *Paecilomyces varioti* in aviation fuel storage tanks has been particularly difficult to control. These fungi grow on the long-chain n-alkanes of **aviation kerosene** (Chapter 6) and cause problems by blocking filters and corroding the walls of fuel-storage tanks when they produce acids as metabolic byproducts. At least one aircrash has been attributed to *A. resinae* in the 1960s, before the seriousness of this problem was recognized. Theoretically, these fungi should not grow in aviation fuel because they require water. But it is almost impossible to prevent water from seeping into fuel storage tanks or condensing during changes of air temperature. Then the fungi grow at the fuel–water interface. This problem is now controlled by a combination of measures – lining the tank walls with rubberized materials to prevent corrosion, regular checking and cleaning of filters to prevent blockages caused by fungal growth in the fuel lines, and the addition of biocides to the fuel – often mixtures of organoborates, or ethylene glycol monoethyl ether.

Quarantine is one of the principal methods for preventing the spread of pathogens, but is becoming increasingly difficult to enforce with the volume of international trade. Most major epidemics are caused by fungi that were introduced inadvertently from other countries. Typical examples include **potato blight** (Chapter 14) and the several major epidemics of **Dutch elm disease** in North America and Western Europe, which have been linked to introductions of elm timber that was not de-barked to remove the beetle vectors (see Fig. 10.9). The devastation caused by **chestnut blight** (*Cryphonectria parasitica*) in North America seems to be traceable to the introduction of an ornamental Asian chestnut tree to the New York Zoological Garden in 1904 (see Fig. 9.13), although recently it has been suggested that the fungus must have been present earlier than this to have caused such extensive devastation.

A unique strain of the **human-pathogenic fungus** *Cryptococcus neoformans* is increasingly being isolated from AIDS patients and has been traced to tropical origins. The same strain was discovered to grow on certain types of eucalypt tree in Australia, and the regions where the clinical strain occurred corresponded to regions where these trees had been imported – an association that could not have been foreseen for quarantine purposes (Chapter 16). The spread of Panama disease of bananas (*Fusarium oxysporum* forma specialis *cubense*, Chapter 14) to most banana-growing regions of the world was almost certainly caused by transport of planting material – bananas are established from pieces of corm at the base of the stem, because commercial bananas are sterile triploids, unable to establish from seeds. This fungus is thought to have its center of origin in the southeast Asian peninsula and to have been disseminated by traders in the nineteenth century. The development of technologies for raising sterile "tissue-culture" plantlets (actually derived by **meristem culture**) should overcome such problems in the future.

More recent examples can be cited. A serious root-rot disease of raspberry canes caused by *Phytophthora fragariae* var. *rubi* is now found in most raspberry production areas of the world. It seems to have been spread in vegetative propagating material (canes) of new, improved cultivars. Consistent with the spread of this pathogen from a single source in recent times is the finding that restriction fragment length polymorphisms (RFLPs, Chapter 9) are virtually identical in pathogenic strains across the world (Stammler *et al.* 1993). The same was true in the past for the spread of *P. fragariae* var. *fragariae*, the cause of red core disease of strawberry plants; and for a particular form (biovar. 3) of the crown gall bacterium, *Agrobacterium tumefaciens*, on grape vines. This pathogen has been discovered to grow as a systemic, symptomless endophyte, so in the past it would have escaped detection in "visibly clean" vegetative propagation material.

This catalogue of examples illustrates an important point: many diseases can be controlled easily and cheaply, without the need for toxic chemicals, but simply by "good housekeeping."

Biological and integrated control

Several examples of biological control have been discussed in this book, so here we review the subject in broader terms. **Biological control (biocontrol) can be defined as the practice in which, or process whereby, an undesirable organism is controlled by the activities of other organisms.** This definition covers both naturally occurring biocontrol and the practical methods used to achieve control, such as:

• the purposeful introduction of one organism to control another;

• the purposeful exploitation and management of naturally occurring biocontrol systems;

• the purposeful change of an environmental factor to promote the activities of natural biocontrol agents;

• any combination of these, because often it is necessary to change some environmental factor in order to favor an introduced biocontrol agent.

Examples of the purposeful introduction of a control agent include the use of *Phlebiopsis gigantea* to control root rot of pine (see Fig. 12.9); the use of hypovirulence to control chestnut blight (see Fig. 9.15); the use of mycoparasites such as *Trichoderma* spp. as seedling protectants (see Fig. 12.5), and the use of various fungi to control insect pests (Chapter 15).

The purposeful exploitation of naturally occurring biocontrol was seen in the decline of cereal cyst nematode (Chapter 15), the exploitation of soils naturally suppressive to *Pythium* diseases of cotton (Chapter 12), and the manipulation of turf pH to control the take-all patch disease of turf grasses (see Figs 12.16–12.18).

The third approach to biocontrol involves changing an environmental factor to promote the activities of naturally occurring biocontrol agents. A classic example of this is shown in Table 17.1, from the work of Olsen & Baker (1968) on **pasteurization of soil** in glasshouse cropping systems. Plant-pathogenic fungi tend to build up during the course of a cropping season in glasshouses, so the soil needs to be treated to destroy pathogens before a new crop is sown. Traditionally this has been done by raising the soil temperature to 65–70°C for 30 minutes, using steam–air mixtures (**aerated steam**) rather than by treatment with steam alone (100°C). The use of steam–air mixtures is not only cheaper but also more effective, as shown by the experimental results in Table 17.1 and Fig. 17.1. In this experiment, trays of soil naturally contaminated with the seedling pathogen *Pythium ultimum* were treated at different temperatures then sown with a thick crop of bell pepper seedlings. A small inoculum of another pathogen, *Rhizoctonia solani*, was placed in one corner of each tray, to simulate a surviving pocket of inoculum that might be found just

	Disease caused by Rhizoctonia solani		Area of disease (cm²) caused by resident Pythium species
Temperature	Area (cm²)	Linear spread (cm)	
No treatment	None	<1	103
100°C	253	18	0
71°C	65	9	0
60°C	3	2	0

Table 17.1 Disease caused by *Rhizoctonia solani* when introduced into one corner of seedling trays containing soil treated at different temperatures to eradicate a natural resident soil population of *Pythium*. (Data from Olsen & Baker 1968.)

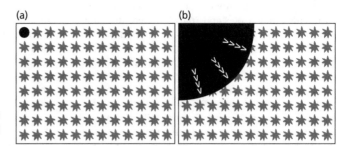

Fig. 17.1 (a,b) Diagram of the spread of disease caused by *Rhizoctonia* from a pocket of surviving inoculum.

beyond a treated area in a commercial glasshouse. The area of diseased plants in each box was then assessed. Treatment of the soil for 30 minutes at any temperature (61, 70 or 100°C) eliminated the resident *Pythium* population. But after treatment at 100°C the soil was sterile and highly conducive to the spread of *Rhizoctonia*, which can grow rapidly through sterilized soil. By contrast, soil treated at 61°C was free from any disease. This was explained by the fact that *Bacillus* spp., which occur naturally as dormant spores in soil, can not only survive pasteurization but also are triggered to germinate at this temperature. These naturally occurring biocontrol agents grow rapidly to fill the biological "vacuum" and they produce several antifungal antibiotics that protect the newly sown crop.

The use of steam–air mixtures in glasshouse crops has now been replaced by the use of *Trichoderma* spp. and similar antagonists. But solar heating of soil (**solarization**) has become common practice for raising valuable **field crops** in many of the warmer parts of the world. It is achieved by covering the soil with thin polyethylene sheeting for a 20- to 30-day period. This creates a "greenhouse effect," raising the temperature at the soil surface to 40°C or more, and often to 30°C or more at even 50 cm soil depth. The spores of most soil-borne pathogens can survive these temperatures *in vitro*, but in moist, solar-heated soil the pathogens are progressively eliminated by the activities of competing and antagonistic microorganisms that flourish at the higher temperatures (Katan 1987).

Chemical control of fungi

Fungitoxic or fungicidal chemicals are often used to control fungal growth, but a distinction must be made between general toxicants and selectively toxic agents.

General toxicants

Fungitoxic chemicals play a major role in **preventing decay** of a range of products including cosmetics,

starch-based materials, cotton fabrics, and structural timbers. In each case the choice of preservative depends on specific needs. For example, the preservatives used in cosmetics and ointments must be nontoxic, nonirritant, and compatible with the product formulation. Often, low-molecular-weight alcohols and esters of *p*-hydroxybenzoic acid (**parabens**) are used in these cases. At the other extreme, marine timbers, fence posts, and telegraph poles need to be treated with persistent compounds that will protect against decay and insect attack for several decades. The three most commonly used wood preservatives are **coal-tar creosote, pentachlorophenol**, and **inorganic arsenicals** such as chromated copper arsenate. Timbers are often pressure-treated with these compounds to ensure high penetration and retention in the wood. Clearly, many of these wood-treatment products pose potential health hazards. Pentachlorophenol is a known carcinogen that is absorbed by the lungs and skin, and it can contain trace amounts of **dioxin**. Arsenic-based wood treatments are widely known to be poisonous, and the Environmental Protection Agency of the USA recently announced (2003) that arsenic will be withdrawn from use. The extreme toxicity of many of these wood preservatives has spurred interest in the possible use of biological control agents for wood protection. Among the potential candidates are strains of *Trichoderma* that frequently colonize the cut surfaces of logs and that antagonize wood-decay fungi in laboratory tests (Vanneste *et al.* 2002).

Control of human and crop pathogens: the principle of selective toxicity

In contrast to general toxicants, chemicals used to control fungi in the tissues of another living organism must show **selective toxicity**. The rest of this chapter will be devoted to this topic, dealing first with the control of plant diseases and then with the control of human mycoses. These areas are more closely related than one might think, because the same types of chemical are used to control both plant- and human-pathogenic fungi. For example, the **azole fungicides** (imidazoles and triazoles) were first developed to control plant diseases, but in modified forms they are now widely used to control human mycoses. Similarly, the naturally occurring antifungal antibiotic **griseofulvin** was first discovered as a "curling factor" that caused the germ-tubes of a plant-pathogenic fungus, *Botrytis allii*, to grow in a distorted spiral fashion, but it was developed commercially as an orally administered antibiotic to control infections caused by the dematophytic fungi. It acts by disrupting fungal microtubules, and this explains its morphogenetic effect because microtubules are involved in the delivery of cellular components to the growing hyphal tip (Chapter 4).

Principal cellular targets of antifungal agents

The main **cellular targets** currently used to control **plant or human diseases** are shown in Fig. 17.2. At first sight it might seem that there are a large number of cellular targets that could be exploited for disease control. But in practice the range is limited. Many of the compounds shown in Fig. 17.2 have a very restricted usage (shown as "R") and are used mainly in Japanese agriculture. If we exclude these compounds then we are left with just **five main types of antifungal target**:

1 the **cell membrane**, because fungi are unique in having **ergosterol** as their characteristic membrane sterol;
2 the **microtubules** and **microtubule-associated proteins**, which are disrupted by the antibiotic griseofulvin, and by benzimidazole fungicides (which have now been withdrawn);
3 **mitochondrial respiration**, which is targeted by some plant fungicides;
4 **fungal cell wall components**, especially β1-3 glucans, for which a new group of drugs, the echinocandins, has recently come into use (2002);
5 various aspects of **general metabolism**.

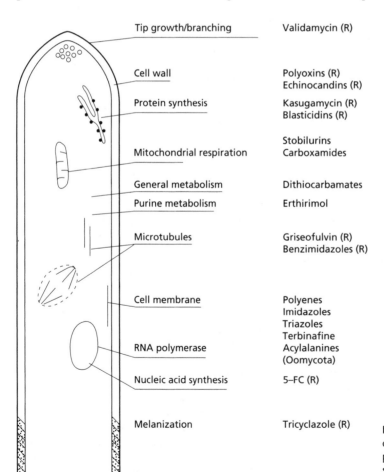

Target	Compound
Tip growth/branching	Validamycin (R)
Cell wall	Polyoxins (R) / Echinocandins (R)
Protein synthesis	Kasugamycin (R) / Blasticidins (R)
Mitochondrial respiration	Stobilurins / Carboxamides
General metabolism	Dithiocarbamates
Purine metabolism	Erthirimol
Microtubules	Griseofulvin (R) / Benzimidazoles (R)
Cell membrane	Polyenes / Imidazoles / Triazoles / Terbinafine / Acylalanines (Oomycota)
RNA polymerase	
Nucleic acid synthesis	5–FC (R)
Melanization	Tricyclazole (R)

Fig. 17.2 The main cellular targets for chemical control of fungi that cause plant and human diseases. Compounds shown as "(R)" have limited usage.

There is an urgent need to find new chemicals (with novel modes of action) and new cellular targets, to provide a greater range of options for controlling fungal diseases.

Fungicides used for plant disease control

The term "fungicide" is used in a broad sense for any compound that kills or inactivates fungi. Most of these compounds are chemically synthesized, but some are modified derivatives of naturally occurring compounds.

All fungicides have to be officially approved for use by the relevant national regulatory body, which also specifies the crops on which the fungicide can be used and any "withholding period" between the last application of a fungicide and the harvesting of a crop. Currently, there are about 80 different chemical groups of fungicides (where a group is defined as a specific type of chemical such as the pyrimidines, the imidazoles, the triazoles, etc.). These groups can be combined into 12 categories according to their primary **cellular target** or **mode of action**, as shown in Table 17.2.

In practice, fungicides can be grouped into three broader categories, reflecting their different roles in plant disease control – the **inorganic fungicides**, the **organic contact (protectant) fungicides**, and the **systemic fungicides**.

Inorganic fungicides

Inorganic fungicides were the first to be used to control plant diseases, beginning with **elemental sulfur** in 1846, **copper salts** in 1882, and **mercury** around 1920 (although its use has been banned since 1992 because of its extreme toxicity, and Britain holds the distinction of being the last industrialized country to accept this ban!).

Inorganic sulfur always had a limited use because of its insolubility, but it can be dusted onto leaves to control powdery mildew fungi, because the hyphae and spores of these fungi develop on or near the leaf surface. Sulfur is the only fungicide available to organic growers because it is an entirely natural product. Fungicides based on various copper salts have been used extensively for their broad antifungal and antibacterial spectrum. They were spectacularly successful against downy mildew of grapevines (*Plasmopara viticola*, Oomycota) in the late 1800s, when "Burgundy mixture" and "Bordeaux mixture" were developed. Usually, the copper fungicides are complexes of **copper sulfate with lime** (Bordeaux mixture) or they are based on **copper hydroxide** or **copper oxychloride**. They control many leaf and fruit diseases, but are potentially phytotoxic. Their mode of action is based on the disruption of several basic metabolic processes, so fungi do not develop resistance to them easily.

Table 17.2 Classification of fungicides according to their main site or mode of action.

Site/mode of action	Comments
Nucleic acid synthesis	A few fungicides, including the **acylalanines** that control diseases caused by Oomycota; e.g. *Phytophthora infestans*
Microtubules (mitosis and cell division)	A few fungicides, including the **benzimidazoles** that act systemically
Respiration	Many fungicides that block steps in the mitochondrial electron transport chain or that inhibit ATP synthesis
Amino acid and protein synthesis	A few fungicides and antibiotics such as blasticidin-S and kasugamycin
Signal transduction	A few fungicides that affect G proteins in cellular signalling or MAP protein kinase
Lipids and membrane synthesis	Several fungicides that affect lipid peroxidation, phospholipid biosynthesis, or cell membrane permeability
Sterol biosynthesis	Many important fungicides that target different steps in the sterol synthesis pathway
Glucan and cell wall synthesis	A few fungicides and antibiotics (polyoxin, validamycin)
Melanin synthesis	A few fungicides that block melanin biosynthesis
Multisite activity	Several inorganic fungicides (sulfur, copper) and protectant (contact) fungicides that disrupt basic metabolic processes
Induced host plant defense	A few compounds (e.g. salicylic acid, chitosan) that activate plant-defense mechanisms
Various; unknown mode of action	A few compounds, including phosphorous acid and fosetyl-aluminum, for controlling *Phytophthora* root rot

Organic contact (protectant) fungicides

Organic fungicides were developed in the 1930s and quickly replaced the inorganic fungicides for most purposes. They are termed "protectant" or "contact" fungicides because they act only near the site where they are applied – to protect the plant surface or to control an established infection. Thus, they need to be applied over the whole plant surface, and must be re-applied to protect any new growth. They are, however, extremely durable fungicides and many of them are still in use today. Fungi do not develop resistance to them easily because they interfere with basic metabolic processes and often have multiple sites of action in fungal cells.

Examples of these protectant fungicides include the **dithiocarbamates** such as **maneb** (Fig. 17.3) and **thiram** (Fig. 17.3). These contain one or more thiol (sulfur-containing) groups and they inactivate sulfydryl (-SH) groups in amino acids, proteins, and enzymes. They are used to control many leaf diseases. Other protectant fungicides include the dinitrophenols such as **dinocap** (Fig. 17.3), which uncouples respiration from ATP synthesis.

Systemic fungicides

Systemic fungicides are absorbed by plants and are then distributed internally, usually by upward movement in the xylem, where they can help to protect new growth. The development of these compounds in the 1960s revolutionized crop protection because these fungicides can eradicate existing infections and protect against subsequent infections. They are now used widely, although most of them do not move downwards in the phloem, so they provide little protection against the major soil-borne pathogens. Their use has also revealed another problem: the active ingredients are not general toxicants but instead have highly specific sites or modes of action – for example, on a single step in an enzymic pathway or by binding to a specific component of a fungal cell. Because of this, fungi can develop resistance to them quite easily, often by a single-gene mutation. When this happens a fungus often shows cross-resistance to all other members of that fungicide group. For this reason, the systemic fungicides need to be used in conjunction with a **resistance-management strategy**. This can involve the alternating use of fungicides with different modes of action, or the combination of systemic and broad-spectrum protectant fungicides. We consider some representative examples below.

Benzimidazole fungicides

The benzimidazole compounds were the first systemic fungicides to be introduced – thiabendazole in 1964, benomyl in 1967, and then others such as carbendazim (1970) and thiophanate-methyl (Fig. 17.4). Ironically, thiabendazole was developed and released in 1962 as an antihelminthic, and only later was its fungicidal role discovered. This empirical approach to fungicide development is still common. It involves the screening of thousands of potentially active compounds against a few major pathogens, typically including a

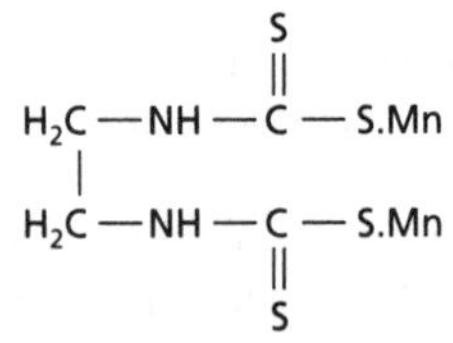

Maneb (dithiocarbamate)

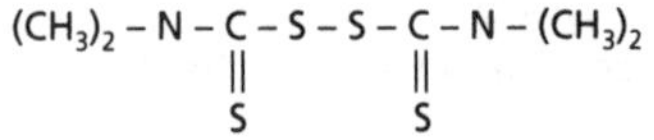

Thiram (dimethyldithiocarbamate)

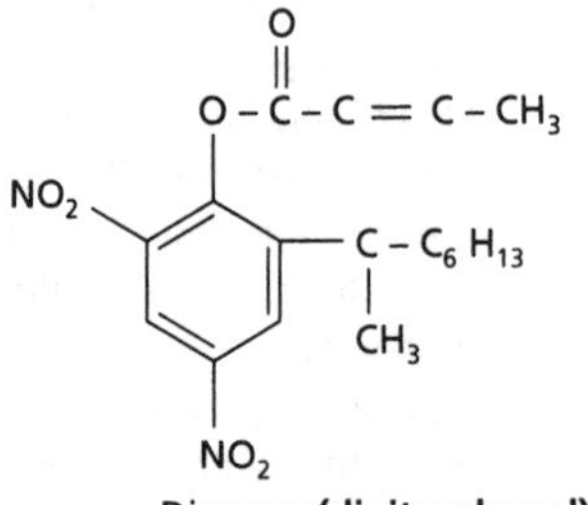

Dinocap (dinitrophenol)

Fig. 17.3 Examples of protectant fungicides: maneb, thiram, and dinocap.

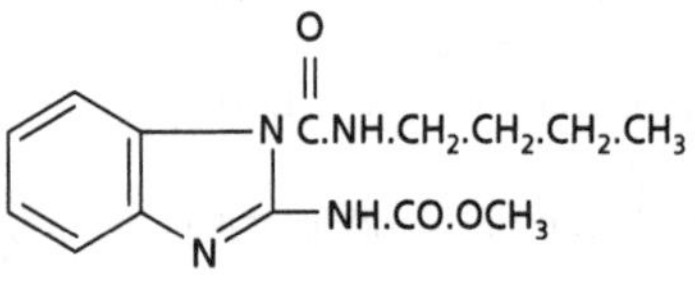

Benomyl (benzimidazole)

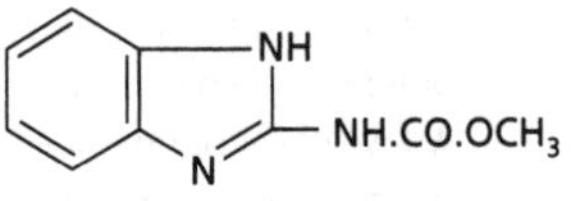

Carbendazim (MBC)

Fig. 17.4 Examples of two systemic benzimidazole fungicides.

cereal rust fungus, a powdery mildew fungus, a downy mildew pathogen (e.g. *Phytophthora infestans*), and a rice pathogen (*Magnaporthe grisea*). Compounds that show promise in this primary screen are chemically modified to produce a range of isomers for further testing, to enhance the fungicidal activity, and to minimize undesirable side effects. This type of approach led to the separate patenting of benomyl and carbendazim from the initial compound, thiabendazole. It is now believed that they have similar or identical modes of action: they are converted to methyl benzimidazole-2-yl-carbamate (MBC) which binds to spindle microtubules and blocks mitosis. They also affect fungal tip growth, by binding to the cytoplasmic microtubules which deliver components to the growing hyphal tip (Chapter 4).

Benzimidazole fungicides are selectively antifungal, with little or no effect on plant and animal cells, despite the fact that plants and animals also have microtubules. The reason for this was shown by *in vitro* studies. Microtubules are formed when β-tubulin forms a dimer with α-tubulin, and the microtubule grows by successive additions of these dimers. MBC binds strongly to the β-tubulin of fungi, preventing its association with α-tubulin, and therefore preventing the self-assembly of microtubules. MBC does not bind strongly to the tubulins of higher animal or plant cells, nor to the tubulins of Oomycota. So it is a selective antifungal agent (Davidse 1986).

The benzimidazoles have been used successfully to control several plant pathogens, but fungi can develop tolerance (resistance) to them if they are used repeatedly. This seems to be caused by point mutations at various sites in the β-tubulin gene, sometimes involving reduced binding of the fungicide to β-tubulin and sometimes affecting the interaction of β-tubulin with α-tubulin or with the tubulin-associated proteins, all of which are necessary for functional microtubules.

Despite the use of benomyl (Benlate) as a fungicide for more than 30 years, it has been linked to a birth defect in which mothers exposed to benomyl in the very early stages of pregnancy can give birth to children with empty eye sockets. According to *The Observer* newspaper (21 December 2003), "more than 40% of pregnant rats fed high levels of benomyl produced foetuses with severe eye defects." The manufacturer recently withdrew Benlate from the global market.

Sterol synthesis inhibitors

Sterols are found in the cell membranes of all eukaryotes and of some archaea (the methanotrophs), but are absent from bacterial membranes. Sterols insert into the phospholipid bilayer and help to maintain membrane stability and fluidity. Different groups of organisms have different types of membrane sterol. Fungi characteristically have **ergosterol** as their membrane sterol, whereas cholesterol is the characteristic sterol of animals, and sitosterol and similar phytosterols are found in plants and Oomycota.

All sterols are synthesized by a complex, multistep pathway, described in Chapter 7 (see Fig. 7.15). In outline, they are derived from the condensation of three molecules of acetyl coenzyme A, to form *isoprene units* (5-carbon compounds), then three isoprene units condense to form the 15-carbon compound, *farnesyl pyrophosphate*. Two molecules of this combine to form the 30-carbon compound, *squalene*, which then undergoes a series of cyclization reactions and ring closures, resulting in the sterol, *lanosterol*. This is the precursor sterol from which all other sterols are produced. There are several intermediate steps from lanosterol to the final sterol, but the key step leading to *ergosterol* is the removal of a methyl group from the C-14 position of the molecule. This step is catalysed by the enzyme **lanosterol 14 α-demethylase**, which has an iron-containing cytochrome P-450 as its coenzyme. This demethylation step occurs only during synthesis of the fungal sterol, not during the synthesis of plant or animal sterols, so it provides an ideal target for antifungal agents.

Several systemic fungicides act by inhibiting sterol demethylation, with the result that the fungus cannot synthesize its normal sterols, and instead other sterols such as lanosterol are incorporated in the fungal membrane. This leads to membrane leakage and ultimately to cell death. The **imidazoles** (e.g. **imazalil**, Fig. 17.5) and **triazoles** (e.g. **propiconazole**, Fig. 17.5) are important examples of fungicides that act in this way. Their common feature is the possession of a five-membered heterocyclic ring containing either two (**imidazoles**) or three (**triazoles**) nitrogen atoms. These azole fungicides are thought to act in the same way as the azole drugs used to treat fungal infections of humans (see below). The lipophilic part of the fungicide is thought to bind to the demethylase enzyme, while nitrogen in the heterocyclic ring associates with an iron-containing coenzyme, blocking demethylation.

Group-specific systemic fungicides

Several systemic fungicides act more or less specifically on particular fungal groups. For example, the **carboxamide** fungicides such as **carboxin** (Fig. 17.6) act against Basidiomycota (rusts, smuts, *Rhizoctonia solani*) by interfering with respiration; they inhibit the step in the TCA cycle where succinate is dehydrogenated to fumarate (see Fig. 7.2). The **2-aminopyrimidine** fungicides such as **ethirimol** (Fig. 17.6) act specifically

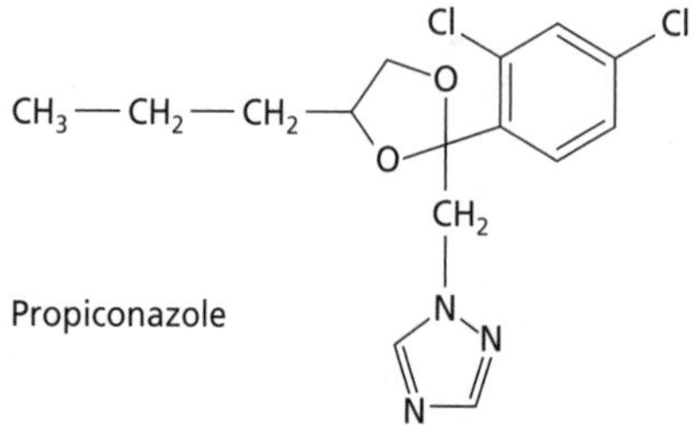

Fig. 17.5 Examples of azole fungicides: propiconazole is a triazole with three nitrogen atoms in the heterocyclic ring; imazalil is an imidazole, with two nitrogen atoms in the heterocyclic ring.

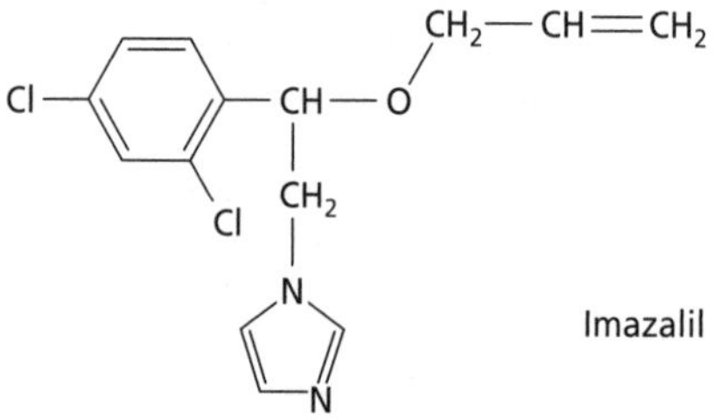

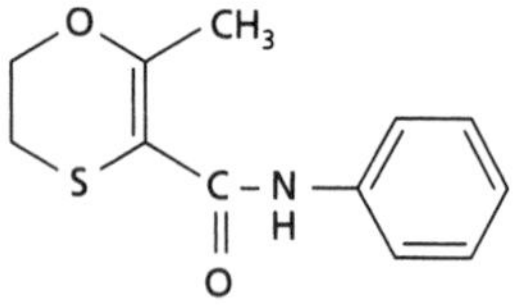

Carboxin (carboxamide) 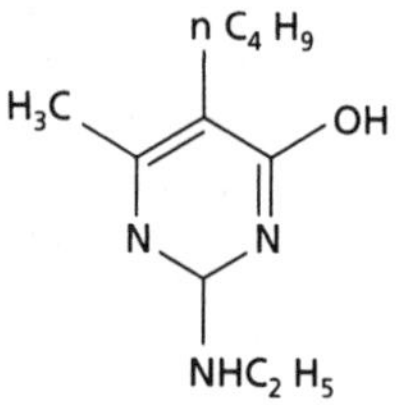   Ethirimol (2-aminopyrimidine)

Fig. 17.6 Systemic fungicides of the carbox-amide and 2-aminopyrimidine classes.

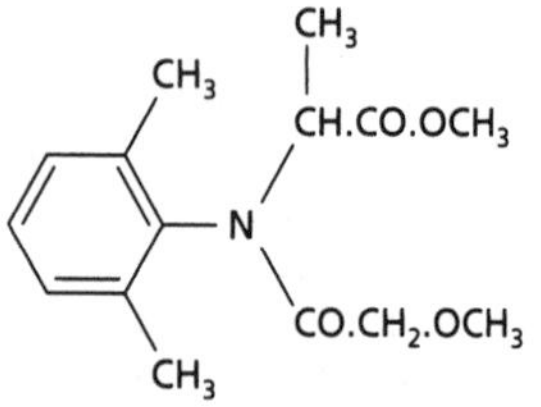

Metalaxyl (acylalanine) 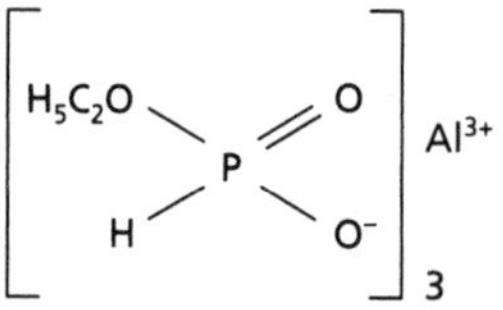   Fosetyl aluminum

Fig. 17.7 Metalaxyl and fosetyl-aluminum, two fungicides that act specifically against Oomycota.

against powdery mildew fungi by inhibiting an enzyme (adenosine deaminase) involved in purine metabolism. In this case, fungicide-tolerant mutants often show no alteration of the enzyme but are thought to circumvent the inhibition by obtaining purines from the host plant.

The **acylalanine** fungicides (e.g. **metalaxyl**, Fig. 17.7) act specifically against Oomycota by inhibiting the nuclear RNA polymerase, thereby blocking RNA synthesis. They have no effect on the true fungi but are highly effective in controlling *Pythium* and *Phytophthora* spp., including *Phytophthora infestans*. Resistance to the acylalanines can develop rapidly in field conditions, so these fungicides are used in combination with protectant fungicides such as the dithiocarbamates. **Fosetyl-aluminium** (Fig. 17.7) is another compound that acts specifically against Oomycota, and it has the remarkable property of being fully mobile in phloem. So it can be applied to the shoots and it moves into the roots, making it extremely valuable for control of root-infecting

Phytophthora spp., such as *Phytophthora cinnamomi*. The mode of action of fosetyl-Al is still unclear. It shows little or no activity against Oomycota in laboratory culture, and this led initially to the view that it might act by inducing host resistance. Subsequent work showed that fosetyl-Al decomposes readily in plants to yield phosphorous acid (H_3PO_3) and this simple mineral acid can suppress the growth of Oomycota in culture media of low phosphate content.

A few fungicides act as **melanin synthesis inhibitors**. They have no effect on fungal growth in culture, but interfere with the infection process, especially when fungi infect from melanized appressoria (see Fig. 5.2). These fungicides have found a limited application in the control of *Colletotrichum* spp. and especially of *Magnaporthe grisea* (the cause of rice blast). Appressoria attach strongly to a host surface, then become melanized, enabling them to build-up an astonishing osmotic pressure equivalent to about 8 atmospheres (8 megaPascals). This is essential for "focusing" of the narrow penetration peg which develops from a small nonmelanized region of the wall in contact with the host surface. If the fungus cannot synthesize melanin then there is a broad zone of contact with the host surface, and the attempted penetration fails. Melanin-deficient mutants of these pathogens show a similar inability to penetrate the host. The main target pathogens, including rice blast, can rapidly develop resistance to these fungicides, whose mode of action is based on a specific enzyme – either a reductase or a dehydratase in the melanin biosynthetic pathway (Wolkow *et al.* 1983).

Strobilurin fungicides

Strobilurin A and **Oudemansin A** are natural products found in the toadstools of two wood-rotting Basidiomycota, *Strobilurus tenacellus* and *Oudemansiella mucida* (Fig. 17.8). They were discovered independently but later were found to be chemically identical, with a common mode of action – they inhibit mitochondrial respiration in fungi, by blocking the oxidation of ubiquinol at a specific site of the cytochrome bc1 complex in the electron-transport chain. Thus, they inhibit the generation of ATP. Although these compounds had been known for some time, they only started to be developed commercially in the early 1980s when agrochemical companies produced synthetic analogues, known as the **strobilurins**, which were photochemically stable and had other desirable properties such as low mammalian toxicity, appropriate mobility within plants and acceptable crop safety. The first commercial strobilurins – **azoxystrobin** and **kresoxim methyl** (Fig. 17.9) – were launched in 1996,

Fig. 17.9 Structures of two strobilurin fungicides.

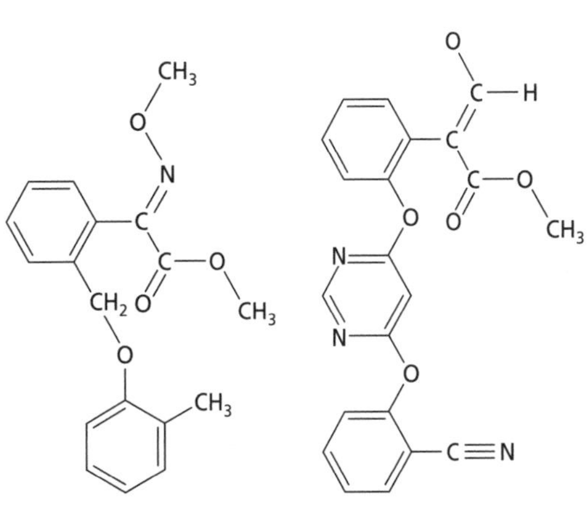

Kresoxim-methyl   Azoxistrobin

and there are now over 700 patents filed on these compounds by the major agrochemical companies.

The strobilurin fungicides show an astonishingly wide range of activity against all the major taxonomic groups of plant pathogens. For example, azoxystrobin is registered for use against more than 400 plant pathogens. It strongly inhibits spore germination, shows excellent preventative activity, and has eradicant and antisporulation properties. Specific formulations are marketed for many different types of crop, including all the major diseases of cereals. The question arises as to why these compounds are toxic to fungi but have little effect on plant hosts, which have the same mitochondrial target sites. The proposed explanation is that there could be differential penetration and degradation of these fungicides in plants compared with in fungi. Like all fungicides that act on specific metabolic targets, there is strong evidence that fungi can develop resistance to the strobilurins. So they need to be used in mixtures or as alternating treatments with other fungicides, as part of a resistance-management strategy.

Antifungal antibiotics used for plant disease control

There are several antifungal antibiotics (Table 17.3) but only a few are selective enough to be used for disease control. For example, **cycloheximide** is a broad-spectrum antifungal agent and is widely used as an experimental tool in laboratory studies, but it acts by blocking protein synthesis on 80S ribosomes and is therefore toxic to all eukaryotes. It is **not used for disease control**. The few antibiotics that have been used

Fig. 17.8 *Oudemansiella mucida*, the "porcelain fungus" that grows on rotting wood and is one of the original sources of strobilurin-type fungicides. (Courtesy of Marek Snowarski, Fungi of Poland; www.grzyby.pl)

Table 17.3 Some antifungal antibiotics used for the control of plant or human mycoses.

Antibiotic	Produced by	Fungi affected	Site/mode of action
Griseofulvin	*Penicillium griseofulvum*	Many (not Oomycota)	Fungal tubulins
Polyene macrolides	*Streptomyces* spp.	Many (not Oomycota)	Cell membrane
Polyoxins	*Strep. cacaoi*	Many (not Oomycota)	Chitin synthesis
Validamycin A	*Strep. hygroscopicus*	Some	Morphogen
Blasticidin-S	*Strep. griseochromogenes*	Some	Protein synthesis
Kasugamycin	*Strep. kasugaensis*	Some	Protein synthesis
Streptomycin	*Strep. griseus*	Oomycota	Calcium?
Pyrrolnitrin	*Pseudomonas* spp.		
Pyoluteorin	*Pseudomonas* spp.		
Gliotoxin	*Trichoderma virens*		
Gliovirin	*T. virens*		Implicated in biocontrol
Viridin	*T. virens*	Various plant	by nutrient competition,
Viridiol	*T. virens*	pathogens	antibiosis, parasitism of
Heptelidic acid	*T. virens*		other fungi, etc.
Trichodermin	*Trichoderma* spp.		
6-pentyl-α-pyrone	*Trichoderma* spp.		
Suzukacillin	*Trichoderma* spp.		
Alamethicine	*Trichoderma* spp.		

commercially for plant disease control were developed in Japan, mainly for the control of rice diseases (rice blast caused by *Magnaporthe grisea*, and rice sheath blight, caused by *Rhizoctonia solani*). In addition to these fermenter-produced antibiotics, several biological control agents are known to produce antifungal antibiotics. But these compounds are exploited indirectly by marketing the biocontrol agents as microbial inoculants – a strategy that can avoid the need to undertake detailed and expensive toxicological testing.

The Japanese agricultural antibiotics

Five novel types of antifungal antibiotic have been discovered and commercialized in Japan. They are discussed briefly below, and their structures are shown in Fig. 17.10.

Polyoxins and nikkomycin

The polyoxins, including **polyoxin D**, are specific inhibitors of chitin synthesis and therefore represent a unique mode of action. Their discovery in 1965 raised hopes that they could have significant roles in controlling fungi and insects – the two major groups of chitin-containing organisms – while having no effect on higher animals or plants. Nikkomycin is a similar compound that acts in the same way. Both compounds bind strongly to the active site of chitin

synthase, and compete with the normal substrate, UDP-*N*-acetylglucosamine from which chitin is synthesized (see Fig. 7.11). They have been used in practice to control some plant diseases, especially sheath blight of rice (*Rhizoctonia*) and black spot of pear (*Alternaria*) in Japan. However, these pathogens rapidly develop tolerance/resistance to the antibiotics in field conditions, because mutant strains of the fungi show reduced antibiotic uptake. Attempts to use these antibiotics to control diseases of human and animal hosts have also been unsuccessful. *In vitro* studies in laboratory media show that the polyoxins are powerful inhibitors of chitin synthesis. However, these compounds are taken into fungal cells through membrane proteins that are normally used for the uptake of dipeptides, and the abundance of small peptides in human or animal tissues competitively blocks the uptake of these antibiotics.

Blasticidin-S, kasugamycin, and validamycin

These three antibiotics have found limited roles in Japanese agriculture. **Blasticidin-S** inhibits protein synthesis in bacteria and a few fungi, by binding to the large subunit of 70S and 80S ribosomes. It was developed for control of rice blast (*M. grisea*) but later it was replaced by another protein synthesis inhibitor, **kasugamycin**, which has lower mammalian toxicity. Resistance has developed to both of these compounds in field conditions. It is based on reduced uptake of

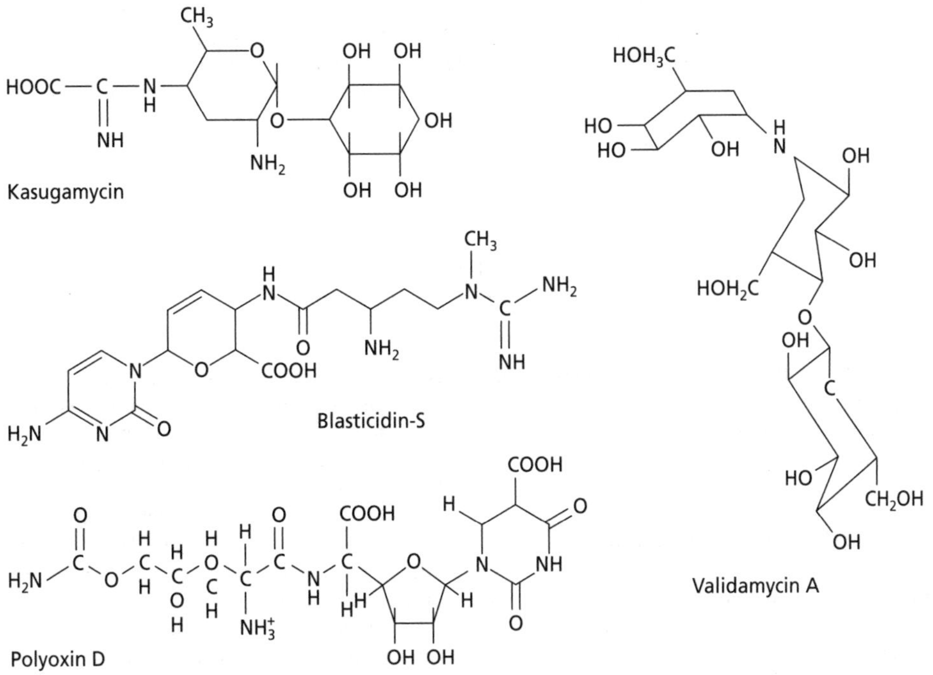

Fig. 17.10 Japanese agricultural antibiotics.

blasticidin-S, and reduced binding of kasugamycin to the ribosome.

Validamycin A is used to control several diseases caused by *Rhizoctonia*, including seedling diseases, black scurf of potato, and sheath blight of rice, but it is inactive against most fungi and bacteria. It has no effect on *Rhizoctonia* in rich culture media, but it causes abnormal branching and cessation of growth in weak media. Validamycin has been found to be converted to validoxylamine in fungi, and this compound is a strong inhibitor of trehalase, the enzyme that cleaves the fungal disaccharide trehalose to glucose (see Fig. 7.6).

Antibiotics implicated in biocontrol

The roles of antibiotics in interfungal interactions were discussed in Chapter 12, so here we consider only the practical applications of these compounds in disease control. Several new biocontrol products are undergoing field trials or have been marketed in the last few years. One of these, marketed as GlioGard™ but now as SoilGard, consists of small, air-dried alginate beads containing fermenter-grown biomass of *Trichoderma virens* and wheat bran as a food base. This product is thought to act mainly by producing the antibiotic gliotoxin. The dried granules are mixed with soil-less rooting media in glasshouses and left for several days after the medium is moistened, before seedlings or rooted cuttings are transplanted into the beds. The timing of these operations is crucial, because the fungus must grow on the wheat bran substrate and produce gliotoxin which diffuses into the rooting medium, protecting the plants from pathogenic attack. If there is a delay in planting then the antibiotic levels start to decline and the degree of protection is reduced.

Rhizosphere bacteria also produce many antifungal compounds. Table 17.4 shows the results of "non-selective" screening (on Trypticase Soy agar) of the dominant bacteria from a range of plants by a commercial company (Leyns *et al.* 1990). The most common bacteria with antifungal properties included *Pseudomonas* spp. (some of which produce the antibiotics pyrrolnitrin, pyoluteorin, phenazines, and diacetylphloroglucinol), *Xanthomonas maltophilia* (unknown antibiotics), *Bacillus subtilis* and other *Bacillus* spp. (some of which produce iturins, mycosubtilins, bacillomycin, fengymycin, mycobacillin, and mycocerein), and *Erwinia herbicola* (which produces herbicolins). Some of these bacteria are used in commercial biocontrol formulations. For example, *Bacillus subtilis* is marketed for seed treatment of cotton and soybeans, to protect against damping-off fungi. Others have been implicated in

Crop	No. of plants examined	No. of dominant bacterial isolates	% of isolates with antifungal properties
Sugarbeet	1550	6780	38
Maize	503	1508	27
Soybean	450	1139	21
Sunflower	450	1119	22
Barley	36	175	92
Grape	36	231	100

Table 17.4 Bacteria with antifungal properties isolated by "nonselective" methods from the root zone of crop plants. (Adapted from Leyns *et al.* 1990.)

natural biocontrol in field conditions, a classic example being the role of fluorescent pseudomonads (*P. fluorescens* and *P. putida*) in suppressing the take-all disease of cereals.

Control of fungal infections of humans

The control of fungal infections of humans presents considerable difficulties, because often the patient is severely compromised or immunosuppressed (Chapter 16). It is worth noting that even the common antibacterial antibiotics such as penicillins and cephalosporins do not work alone; they act in conjunction with the host's normal defenses. There are, however, several antimycotic agents that can be used successfully to treat human mycoses (Odds *et al.* 2003). We consider these below.

Griseofulvin

Griseofulvin (Fig. 17.11) is a naturally occurring antifungal antibiotic, produced by the fungus *Penicillium griseofulvum*. It is insoluble in water but can be taken orally in a microcrystalline form and it then accumulates in the keratin precursor cells of the skin, nails, and hair. It is used almost exclusively to control the dermatophyte infections of the keratinized tissues, but it

is fungistatic rather than fungicidal, so for serious infections the treatment must be prolonged until all the infected tissues have been shed.

As noted earlier, griseofulvin causes growth distortions of fungal hyphae in spore germination tests, including thickening of the hyphal walls, associated with heavy deposition of glucans. However, this is now known to be a secondary effect, and the primary role of griseofulvin is to interfere with the assembly of microtubules or of the tubulin-associated proteins. It causes abnormal nuclear division by disrupting the spindle microtubules, and abnormal growth, presumably by disrupting the microtubule-associated transport of materials to the hyphal tip. In these respects, the role of griseofulvin is similar to that of the benzimidazole fungicides used for plant disease control, because these also block microtubule assembly. But the benzimidazoles evidently act at a different site on microtubules because mutants that develop resistance to the benzimidazoles are not resistant to griseofulvin, and vice versa.

Griseofulvin has been the main drug used to treat dermatophyte infections for many years, and although resistance to the antibiotic can occur quite frequently *in vitro*, it has not been a major problem in clinical practice. However, griseofulvin is now being progressively replaced by newer synthetic drugs, including terbinafine and itraconazole, which are less toxic and require shorter treatment times.

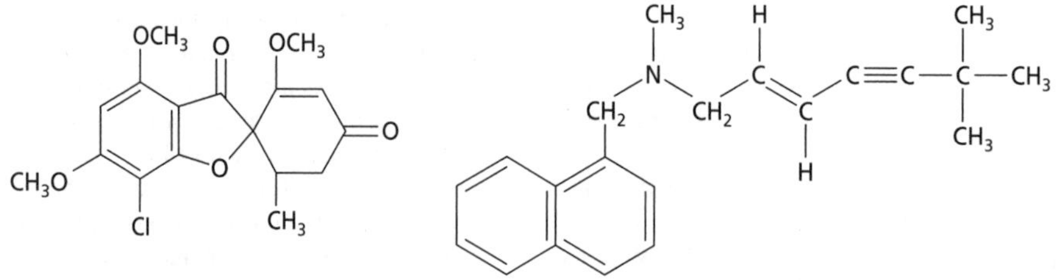

Fig. 17.11 Structures of griseofulvin and terbinafine, two compounds used to treat dermatophyte infections.

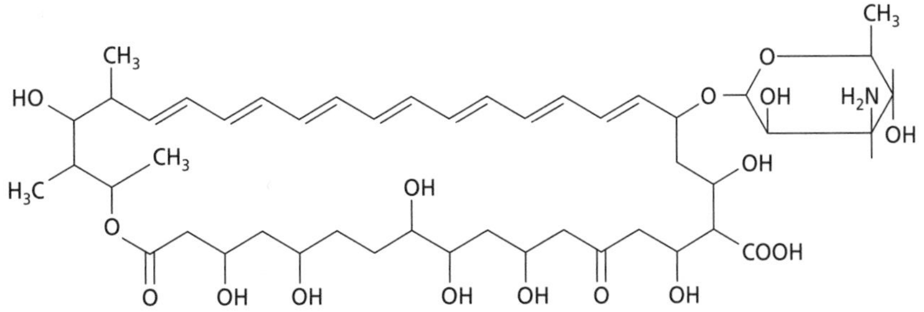

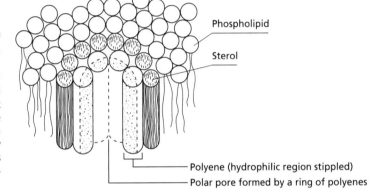

Fig. 17.12 Amphotericin B, a member of the polyene macrolide class of antifungal antibiotics.

Fig. 17.13 Proposed mode of action of amphotericin B and similar large polyene antibiotics. The antibiotic shows strong affinity for ergosterol in the fungal cell membrane. It binds to ergosterol molecules and repositions them so that the polyene–sterol complexes assemble to form a polar pore through which small ions diffuse freely. Note that only one layer of the phospholipid bilayer is shown in this diagram. (Based on a diagram in Gale *et al.* 1981.)

Phospholipid

Sterol

Polyene (hydrophilic region stippled)

Polar pore formed by a ring of polyenes

Terbinafine

Like griseofulvin, terbinafine (Fig. 17.11) can be taken orally (or topically) and it tends to accumulate in the skin, nails, hair, and fatty tissues. It inhibits ergosterol synthesis by interfering with an early step in sterol biosynthesis, where the enzyme squalene epoxidase starts the process of cyclization leading to lanosterol.

Polyene macrolide antibiotics

The polyene macrolide antibiotics are produced by several *Streptomyces* spp. Their typical structure is exemplified by **amphotericin B** (Fig. 17.12). An aminosugar group is attached to a large lactone ring with a series of double bonds on one face (hence the term polyene) and a series of hydroxyl groups on the other face. So the molecule has both a hydrophobic and a hydrophilic face. By X-ray crystallography the molecule is seen to have a rod-like structure, about 2.1 nm long in the case of amphotericin B, similar to a membrane phospholipid. The polyenes insert in the cell membrane by associating with sterols (the hydrophobic face) and are thought to cause rearrangement

of the sterols so that a group of eight polyene molecules forms a ring with the hydrophilic faces in the center (Fig. 17.13). Thus they form a polar pore through which small ions (K^+, H^+) can pass freely, disrupting the cell's ionic control.

Almost all eukaryotic cells have sterols in the membranes, although some Oomycota do not produce them or even need them for growth. So the polyene antibiotics would not, at first sight, seem to be selective antifungal agents. However, there are many different polyenes that differ in the ring size and presence or nature of the aminosugar group, and they show preferential binding to different sterols. Three of these compounds – amphotericin B, nystatin, and pimaricin – have a much higher affinity for ergosterol than for cholesterol, the mammalian membrane sterol, so they are selectively antifungal. Nystatin and pimaricin can be applied topically as creams to control *Candida* infection of the vagina, or as powder formulations to control athlete's foot.

Amphotericin B has lower mammalian toxicity than nystatin or pimaricin, and it also has some (low) degree of water solubility. Initial development work in the 1950s and 1960s showed that amphotericin B could be administered orally, with some success. But all the polyene antibiotics have adverse side effects,

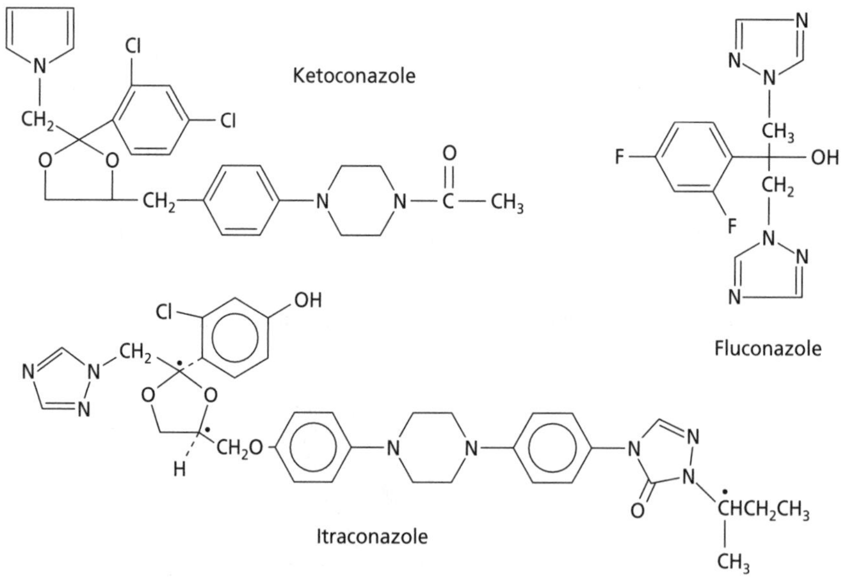

Fig. 17.14 Imidazole and triazole drugs.

so the main thrust of research has been to develop formulations that minimize these effects, such as chronic renal toxicity. There are now several lipid formulations of amphotericin B, either approved for human use or in stages of development. For example, the commercial formulation named AmBisome consists of small (<100 nm diameter) liposomes consisting of a single layer of phospholipids that incorporate amphotericin B. Single dose intravenous toxicity tests on mice showed that AmBisome had an LD_{50} value (the dose resulting in 50% mortality) of more that 175 mg/kg, compared with an LD_{50} of 2–3 mg/kg for a standard colloidal formulation of amphotericin B mixed with sodium deoxycholate. Amphotericin in the new formulations has the potential to control several systemic fungal pathogens, including *Candida*, *Cryptococcus*, *Aspergillus*, *Blastomyces*, *Coccidioides*, and *Histoplasma*. Further details of drug delivery systems for antifungal agents can be found in Adler-Moore & Proffitt (2004).

Azole drugs

The azole drugs are ergosterol biosynthesis inhibitors similar to those used for plant disease control, but with structures designed for the treatment of systemic mycoses of humans. They act by blocking sterol demethylation at the C-14 position, during the pathway from lanosterol to ergosterol (see Fig. 7.15). Depending on whether these drugs have two or three nitrogens in the five-membered heterocyclic ring, they are termed imidazoles or triazoles.

Ketoconazole (an imidazole; Fig. 17.14) was the first of these drugs to be developed for treatment of humans. It has strong activity against *Candida* and *Cryptococcus neoformans*, and also can be used against other systemic pathogens such as *Blastomyces dermatitidis* and *Coccidioides immitis*, but it has little activity against *Aspergillus*. However, ketoconazole has several adverse side effects, and it has been largely replaced by the more recent and less toxic triazole drugs such as **fluconazole** (Fig. 17.14) and **itraconazole** (Fig. 17.14). Although ketoconazole is still available, it is used mainly as a second-line drug for infections that do not respond to the triazoles.

Fluconazole is mainly used against *Candida* and *Cryptococcus* species, and is available as both oral and intravenous formulations. It also has limited activity against *Histoplasma* and *Blastomyces*, but almost no activity against other human pathogens. It is a widely used front-line drug against many *Candida albicans* infections, but is not effective against some of the less common *Candida* spp. such as *C. krusei* and *C. glabrata*.

Itraconazole is available in oral or intravenous forms and is active against many fungi including yeasts, dimorphic fungi, and mycelial fungi. Its spectrum includes *Candida* spp., *Cryptococcus*, *Aspergillus*, *Histoplasma*, *Blastomyces*, *Coccidioides*, and dermatophytes. However, resistance to itraconazole has been reported among many fungi, and this sometimes leads to cross-resistance to drugs with a similar mode of action. Three new triazoles with different spectra of activity have recently been developed. In one case,

voriconazole, the compound appeared on the market in 2002, and the other two compounds, **ravuconazole** and **posaconazole**, are in the final stages of clinical trials. Thus, compared with the situation only a decade ago, when ketoconazole was the drug of choice, there is now a wide range of azoles available for the treatment of specific mycoses.

For all these azole drugs, the consequence of blocking the demethylation step of the ergosterol biosynthetic pathway is to deplete the concentration of ergosterol so that fungi incorporate other sterols in the membrane. The effect of this can be studied in a model system, by exploiting the fact that *Saccharomyces cerevisiae* can grow both anaerobically and aerobically. When grown aerobically, *Saccharomyces* synthesizes ergosterol; but when grown anaerobically *Saccharomyces* needs to be supplied with sterols, because these are products of aerobic biosynthetic pathways (Chapter 7). If anaerobic cells of *Saccharomyces* are supplied with a different sterol such as lanosterol, they become leaky and cannot assemble the wall in the normal way. In this respect it will be recalled that the major wall-synthetic enzymes, chitin synthase and glucan synthase, are integral membrane proteins (Chapter 4), so presumably any change in the membrane composition could affect these enzyme functions. In *Candida*, one of the effects of the azole drugs is to prevent the phase transition from yeasts to hyphae. As we noted in Chapter 16, this phase transition is important for invasion of tissues and for allowing *Candida* to break out of phagocytes.

5-Flucytosine (5-FC)

5-FC is a fluorine-substituted nucleoside, developed originally as an antitumour agent but it has an additional role as an antimycotic agent for the treatment of systemic *Candida* and *Cryptococcus*. Its mode of action is shown in Fig. 17.15. It is taken into a fungal cell by the **cytosine permease** of the cell membrane, then deaminated to **5-fluorouracil** which, through a series of steps, is incorporated into RNA in place of uracil, causing impaired RNA function. It also impairs DNA synthesis because one of the products of 5-fluorouracil inhibits the enzyme **thymidylate synthase** in the pathway that generates thymidine nucleotides.

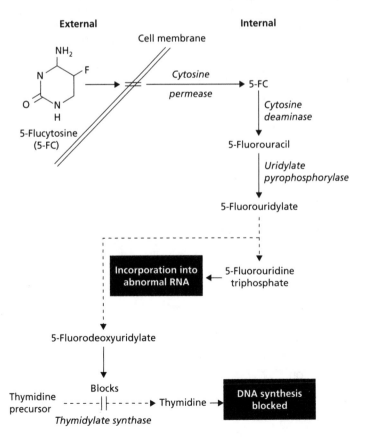

Fig. 17.15 Mode of action of the synthetic pyrimidine 5-flucytosine.

5-FC is selectively toxic to fungi because mammalian cells have a low rate of uptake of the molecule and have little or no ability to convert it to 5-fluorouracil because they lack the enzyme cytosine deaminase (Fig. 17.15). However, 5-FC affects only a small number of fungi, notably *Candida*, *Cryptococcus*, and *Aspergillus fumigatus*, and even these fungi easily develop resistance to it in clinical practice. The cause of this resistance can be studied *in vitro* by supplying cells with the various intermediates along the pathway shown in Fig. 17.15. Most resistant strains from clinical specimens are found to be deficient in uridylate pyrophosphorylase activity, but a few lack a cytosine permease or deaminase. Because of the rapid development of resistance, 5-FC is not used alone but in conjunction with other drugs, especially amphotericin B, with which it has a significant synergistic effect.

Echinocandins

Echinocandins are naturally occurring secondary metabolites of fungi, which are now in an advanced stage of development as novel antifungal agents because they inhibit the synthesis of β1-3 glucan, one of the major wall polymers of fungi. **Caspofungin** (Fig. 17.16), a polypeptide antifungal, was licensed for clinical use in 2002. It is active against some *Candida* spp. and *Aspergillus fumigatus*. Two other compounds of the same structural class, **anidulafungin** and **mycafungin**, are expected to be licensed soon.

Control of *Pneumocystis jirovici*

We noted in Chapter 16 that *P. jirovici* is an unusual fungus which for many years was classified as a protozoan. The principal means of controlling it is with the drug **pentamidine** which is thought to act on DNA. It is also sensitive to the antibacterial agents **trimethoprim** and **sulfomethoxazole**, which disrupt the folate pathway involved in transfer of groups from the amino acid serine to a range of other compounds. *P. jirovici* is not affected by the antifungal polyenes or azoles because it has cholesterol, not ergosterol, in its cell membrane. However, all three of the echinocandins mentioned above are active against *P. jirovici*.

The future . . .

Despite the impressive advances that have been made in controlling fungal infections of humans, and the recent introduction of the echinocandins, there is still concern that the pace of development of drugs with novel modes of action is slow. To quote Odds *et al.* (2003), ". . . the spectre of emergence of resistance is a real one . . . and new inhibitors will continue to be required for effective antifungal therapy in the future."

In parallel with the development of new antifungal agents, there is increasing interest in the development of vaccines to protect against some of the endemic human-pathogenic fungi. These studies are at a rel-

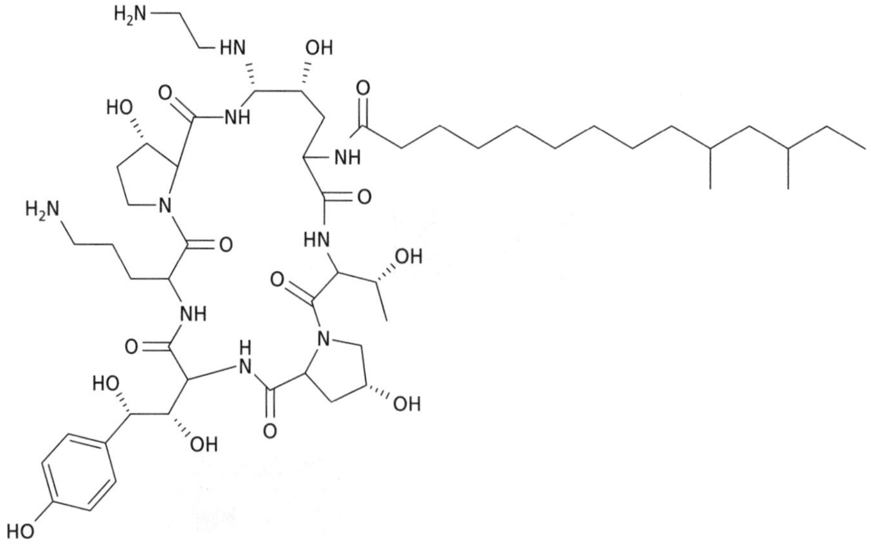

Fig. 17.16 Caspofungin, one of the new echinocandin antifungal agents.

atively early stage, but potential antigens have been identified for both *Coccidioides immitis* (and the related fungus *C. posadasii*) (Magee & Cox 2004) and *Paracoccidioides brasiliensis* (Travassos *et al.* 2004).

Online resources

Doctor Fungus website. http://www.doctorfungus.org [An excellent, comprehensive source of up-to-date information on all human mycoses.]

General texts

Deacon, J.W. (1991) Significance of ecology in the development of biocontrol agents against soil-borne plant pathogens. *Biocontrol Science and Technology* **1**, 5–20.

Georgopapadakou, N.H. & Walsh, T.J. (1994) Human mycoses: drugs and targets for emerging pathogens. *Science* **264**, 371–373.

Mintz, A.S. & Walter, J.R. (1993) A private industry approach: development of GlioGard™ for disease control in horticulture. In: *Pest Management: biologically based technologies* (Lumsden, R.D. & Vaughn, J.L., eds), pp. 398–404. American Chemical Society, Washington.

Shepherd, M.C. (1987) Screening for fungicides. *Annual Review of Phytopathology* **25**, 189–206.

Trinci, A.P.J. & Ryley, J.F. (1984) *Mode of Action of Antifungal Agents*. Cambridge University Press, Cambridge.

Cited references

Adler-Moore, J.P. & Proffitt, R.T. (2004) Novel drug delivery systems for antifungal agents. In: *The Mycota XII. Human Fungal Pathogens* (Domer, J.E. & Kobayashi, G.S., eds), pp. 339–362. Springer-Verlag Berlin.

Davidse, J.C. (1986) Benzimidazole fungicides: mechanisms of action and biological impact. *Annual Review of Phytopathology* **24**, 43–65.

Gale, E.F., Cundcliffe, E., Reynolds, P.E., Richmond, M.H. & Waring, M.J. (1981) *The Molecular Basis of Antibiotic Action*, 2nd edn. Wiley, London.

Godfrey, S.A.C., Monds, R.D., Lash, D.T. & Marshall, J.W. (2003) Identification of *Pythium oligandrum* using species-specific ITS rDNA PCR oligonucleotides. *Mycological Research* **107**, 790–796.

Katan, J. (1987) Soil solarization. In: *Innovative Approaches to Plant Disease Control* (Chet, I., ed.), pp. 77–105. Wiley, New York.

Leyns, F., Lambert, B., Joos, H. & Swings, J. (1990) Antifungal bacteria from different crops. In: *Biological Control of Soil-borne Plant Pathogens* (Hornby, D., ed.), pp. 437–444. CAB International, Wallingford.

Magee, M. & Cox, R.A. (2004) Vaccine development for Coccidioidomycosis. In: *The Mycota XII. Human Fungal Pathogens* (Domer, J.E. & Kobayashi, G.S., eds), pp. 243–257. Springer-Verlag, Berlin.

Odds, F.C., Brown, A.J.P. & Gow, N.A.R. (2003) Antifungal agents: mechanisms of action. *Trends in Microbiology* **11**, 272–279.

Olsen, C.M. & Baker, K.F. (1968) Selective heat treatment of soil and its effect on inhibition of *Rhizoctonia solani* by *Bacillus subtilis*. *Phytopathology* **58**, 79–87.

Stammler, G., Seemuller, E. & Duncan, J.M. (1993) Analysis of RFLPs in nuclear and mitochondrial DNA and the taxonomy of *Phytophthora fragariae*. *Mycological Research* **97**, 150–156.

Travassos, L.R., Taborda, C.P., Iwai, L.K., Cunha-Neto, E.C. & Puccia, R. (2004) The gp43 from *Paracoccidioides brasiliensis*: a major diagnostic antigen and vaccine candidate. In: *The Mycota XII. Human Fungal Pathogens* (Domer, J.E. & Kobayashi, G.S., eds), pp. 279–296. Springer-Verlag, Berlin.

Vanneste, J.L., Hill, R.A., Kay, S.J., Farrell, R.L. & Holland, P.T. (2002) Biological control of sapstain fungi with natural products and biological control agents: a review of the work carried out in New Zealand. *Mycological Research* **106**, 228–232.

Wolkow, P.M, Sisler, H.D. & Vigil, E.L. (1983) Effect of inhibitors of melanin biosynthesis on structure and function of appressoria of *Colletotrichum lindemuthianum*. *Physiological Plant Pathology* **22**, 55–71.

Sources

Chapter 1

Fig. 1.1 (Based on a diagram in Woese (2000) but showing only a few of the major groups of organisms.) Woese, C.R. (2000) Interpreting the universal phylogenetic tree. *Proceedings of the National Academy of Sciences, USA* **97**, 8392–8396.

Fig. 1.3 (Based on a drawing from: http://microscope.mbl.edu/scripts/microscope.php?func=imgDetail&imageID=4575)

Fig. 1.7(a,b) (Courtesy of Robert L. Anderson (photographer) and USDA Forest Service; www.forestryimages.org)

Chapter 2

Fig. 2.2 (From Deacon & Saxena 1997.) Deacon, J.W. & Saxena, G. (1997) Orientated attachment and cyst germination in *Catenaria anguillulae*, a facultative endoparasite of nematodes. *Mycological Research* **101**, 513–522.

Fig. 2.4 (Reproduced by courtesy of Schuessler *et al.* 2001, and the British Mycological Society.) Schuessler, A., Schwarzott, D. & Walker, C. (2001) A new fungal phylum, the Glomeromycota: phylogeny and evolution. *Mycological Research* **105**, 1413–1421.

Fig. 2.6 (Courtesy of Dirk Redecker and the AAAS; see Redecker *et al.* 2000.) Redecker, D., Kodner, R. & Graham, L.E. (2000) Glomalean fungi from the Ordovician. *Science* **289**, 1920–1921.

Fig. 2.7 (Courtesy of Redecker *et al.* 2000 and the AAAS.) Redecker, D., Kodner, R. & Graham, L.E. (2000) Glomalean fungi from the Ordovician. *Science* **289**, 1920–1921.

Fig. 2.13 (Courtesy of N. Read; from Read, N.D. & Lord, K.M. 1991 *Experimental Mycology* **15**, 132–139.)

Fig. 2.35 (Courtesy of Florian Siegert ©) (From: http://www.zi.biologie.unimuenchen.de/zoologie/dicty/dicty.html)

Fig. 2.37 ((a,b) Courtesy of J.P. Braselton, Ohio University; see http://oak.cats.ohiou.edu/~braselto/plasmos/)

Chapter 3

Fig. 3.2 (Courtesy of C. Bracker; from Grove & Bracker 1970.) Grove, S.N. & Bracker, C.E. (1970) Protoplasmic organization of hyphal tips among fungi: vesicles and Spitzenkörper. *Journal of Bacteriology* **104**, 989–1009.

Fig. 3.3 (Courtesy of C. Bracker; from Grove & Bracker 1970.) Grove, S.N. & Bracker, C.E. (1970) Protoplasmic organization of hyphal tips among fungi: vesicles and Spitzenkörper. *Journal of Bacteriology* **104**, 989–1009.

Fig. 3.4 (Courtesy of R. Roberson; see Roberson & Fuller 1988, 1990.) Roberson, R.W. & Fuller, M.S. (1988) Ultrastructural aspects of the hyphal tip of *Sclerotium rolfsii* preserved by freeze substitution. *Protoplasma* **146**, 143–149. Roberson, R.W. & Fuller, M.S. (1990) Effects of the demethylase inhibitor, Cyproconazole, on hyphal tip cells of *Sclerotium rolfsii*. *Experimental Mycology* **14**, 124–135.

Fig. 3.6 (From McCabe *et al.* 1999.) McCabe, P.M., Gallagher, M.P. & Deacon, J.W. (1999) Microscopic observation of perfect hyphal fusion in *Rhizoctonia solani*. *Mycological Research* **103**, 487–490.

Fig. 3.9 (Based on Hunsley & Burnett 1970.) Hunsley, D. & Burnett, J.H. (1970) The ultrastructural architecture of the walls of some hyphal fungi. *Journal of General Microbiology* **62**, 203–218.

Fig. 3.10 (Based on Trinci 1978.) Trinci A.P.J. (1978) *Science Progress* **65**, 75–99.

Fig. 3.13 (Courtesy of C.E. Bracker; from Bartnicki-Garcia *et al.* 1978.) Bartnicki-Garcia, S., Bracker, C.E., Reyes, E. & Ruiz-Herrera, J. (1978) Isolation of chitosomes from taxonomically diverse fungi and synthesis of chitin microfibrils *in vitro*. *Experimental Mycology* **2**, 173–192.

Fig. 3.14 (Photographs courtesy of B. Rees, V. Shepherd and A. Ashford; from Rees *et al.* 1994.) Rees, B, Shepherd, V.A. & Ashford, A.E. (1994) Presence of a motile tubular vacuole system in different phyla of fungi. *Mycological Research* **98**, 985–992.

Fig. 3.15 (Courtesy of N.D. Read; from Fisher-Parton *et al.* 2000.) Fisher-Parton, S., Parton, R.M., Hickey, P., Dijksterhuis, J., Atkinson, H.A. & Read, N.D. (2000) Confocal microscopy of FM4-64 as a tool for analysing endocytosis and vesicle trafficking in living fungal hyphae. *Journal of Microscopy* **198**, 246–259.

Fig. 3.17 (Courtesy of N.D. Read; from Fisher-Parton *et al.* 2000.) Fisher-Parton, S., Parton, R.M., Hickey, P., Dijksterhuis, J., Atkinson, H.A. & Read, N.D. (2000) Confocal microscopy of FM4-64 as a tool for analysing endocytosis and vesicle trafficking in living fungal hyphae. *Journal of Microscopy* **198**, 246–259.

Fig. 3.18 (From Fisher-Parton *et al.* 2000.) Fisher-Parton, S., Parton, R.M., Hickey, P., Dijksterhuis, J., Atkinson, H.A. & Read, N.D. (2000) Confocal microscopy of FM4-64 as a tool for analysing endocytosis and vesicle trafficking in living fungal hyphae. *Journal of Microscopy* **198**, 246–259.

Fig. 3.19 (Courtesy of H.C. Hoch; from Hoch & Staples 1985.) Hoch, H.C. & Staples, R.C. (1985) The microtubule cytoskeleton in hyphae of *Uromyces phaseoli* germlings: its relationship to the region of nucleation and to the F-actin cytoskeleton. *Protoplasma* **124**, 112–122.

Fig. 3.20 (Courtesy of R. Roberson; from Roberson & Fuller 1988.) Roberson, R.W. & Fuller, M.S. (1988) Ultrastructural aspects of the hyphal tip of *Sclerotium rolfsii* preserved by freeze substitution. *Protoplasma* **146**, 143–149.

Chapter 4

Fig. 4.6 (Based on a diagram in Wessels 1990.) Wessels, J.G.H. (1990) Role of cell wall architecture in fungal tip growth generation. In: *Tip Growth in Plant and Fungal Cells* (Heath, I.B., ed.), pp. 1–29. Academic Press, New York.

Fig. 4.7 (Based on Anderson & Smith 1971.) Anderson, J.G. & Smith, J.E. (1971) The production of conidiophores and conidia by newly germinated conidia of *Aspergillus niger* (microcycle conidiation). *Journal of General Microbiology* **69**, 185–197.

Fig. 4.8 (Redrawn from Robinson 1973.) Robinson, P.M. (1973) Oxygen-positive chemotropic factor for fungi? *New Phytologist* **72**, 1349–1356.

Fig. 4.9 (From Allan *et al.* 1992.) Allan, R.H., Thorpe, C.J. & Deacon, J.W. (1992) Differential tropism to living and dead cereal root hairs by the biocontrol fungus *Idriella bolleyi*. *Physiological and Molecular Plant Pathology* **41**, 217–226.

Fig. 4.10 (Mitchell & Deacon 1986.) Mitchell, R.T. & Deacon, J.W. (1986) Chemotropism of germ-tubes from zoospore cysts of *Pythium* spp. *Transactions of the British Mycological Society* **86**, 233–237.

Fig. 4.11 (Based on a drawing by Hartwell 1974.) Hartwell, L.L. (1974) *Saccharomyces cerevisiae* cell cycle. *Bacteriological Reviews* **38**, 164–198.

Fig. 4.12 (Reprinted from Mata, J. & Nurse, P. 1998, copyright 1998, with permission from Elsevier.) Mata, J. & Nurse, P. (1998) Discovering the poles in yeast. *Trends in Cell Biology* **8**, 163–167.

Fig. 4.14 (Courtesy of D.J. Read; from Read 1991.) Read, D.J. (1991) Mycorrhizas in ecosystems – nature's response to the "Law of the Minimum". In: *Frontiers in Mycology* (Hawksworth, D.L., ed.), pp. 29–55. CAB International, Wallingford, Oxon, pp. 101–130.

Fig. 4.17 (From Trinci 1992.) Trinci, A.P.J. (1992) Mycoprotein: a twenty-year overnight success story. *Mycological Research* **96**, 1–13.

Chapter 5

Fig. 5.1 (Based on Bartnicki-Garcia & Gierz 1993.) Bartnicki-Garcia, S. & Gierz, G. (1993) Mathematical analysis of the cellular basis of dimorphism. In: *Dimorphic Fungi in Biology and Medicine* (Van den Bossche, H., Odds, F.C. & Kerridge, D., eds), pp. 133–144. Plenum Press, New York.

Fig. 5.5 (Courtesy of N.D. Read; from Read *et al.* 1992.) Read, N.D., Kellock, L.J., Knight, H. & Trewavas, A.J. (1992) Contact sensing during infection by fungal pathogens. In: *Perspectives in Plant Cell Recognition* (Callow, J.A. & Green, J. R., eds), pp. 137–172. Cambridge University Press, Cambridge.

Fig. 5.6 (Courtesy of N.D. Read; from Read *et al.* 1992.) Read, N.D., Kellock, L.J., Knight, H. & Trewavas, A.J. (1992) Contact sensing during infection by fungal pathogens. In: *Perspectives in Plant Cell Recognition* (Callow, J.A. & Green, J.R., eds), pp. 137–172. Cambridge University Press, Cambridge.

Fig. 5.7 (Based on a drawing by H.C. Hoch & R.C. Staples; see Hoch *et al.* 1987.) Hoch, H.C., Staples, R.C., Whitehead, B., Comeau, J. & Wolfe, E.D. (1987) Signaling for growth orientation and differentiation by surface topography in *Uromyces*. *Science* **235**, 1659–1662.

Fig. 5.8 (Based on Allen *et al.* 1991.) Allen, E.A., Hazen, B.A., Hoch, H.C., *et al.* (1991) Appressorium formation in response to topographical signals in 27 rust species. *Phytopathology* **81**, 323–331.

Fig. 5.9 ((c,d) Courtesy of F.M. Fox; see Fox 1986.) Fox, F.M. (1986) Ultrastructure and infectivity of sclerotia of the ectomycorrhizal fungus *Paxillus involutus* on birch (*Betula* spp.) *Transactions of the British Mycological Society* **87**, 627–631.

Fig. 5.10 (From Christias & Lockwood 1973.) Christias, C. & Lockwood, J.L. (1973) Conservation of mycelial constituents in four sclerotium-forming fungi in nutrient-deprived conditions. *Phytopathology* **63**, 602–605.

Fig. 5.12 (Courtesy of F.M. Fox; from Fox 1987.) Fox, F.M. (1987) Ultrastructure of mycelial strands of *Leccinum scabrum*, ectomycorrhizal on birch (*Betula* spp.). *Transactions of the British Mycological Society* **89**, 551–560.

Fig. 5.13 (Courtesy of F.M. Fox; from Fox 1987.) Fox, F.M. (1987) Ultrastructure of mycelial strands of *Leccinum scabrum*, ectomycorrhizal on birch (*Betula* spp.). *Transactions of the British Mycological Society* **89**, 551–560.

Fig. 5.16 (Courtesy of C.E. Bracker; from Bracker 1968.) Bracker, C.E. (1968) The ultrastructure and development of sporangia in *Gilbertella persicaria*. *Mycologia* **60**, 1016–1067.

Fig. 5.19 (Based on Suzuki *et al.* 1977.) Suzuki, Y., Kumagai, T. & Oka, Y. (1977) Locus of blue and near ultraviolet reversible photoreaction in the stages of conidial development in *Botrytis cinerea*. *Journal of General Microbiology* **98**, 199–204.

Fig. 5.21 (Based on a diagram in Wessels 1996.) Wessels, J.G.H. (1996) Fungal hydrophobins: proteins that function at an interface. *Trends in Plant Science* **1**, 9–15.

Chapter 6

Fig. 6.3 ((a) From Gow 1984. (b) Based on Kropf *et al.* 1984.) Gow, N.A.R. (1984) Transhyphal electric currents in fungi. *Journal of General Microbiology* **130**, 3313–3318. Kropf, D.L., Caldwell, J.H., Gow, N.A.R. & Harold, F.M. (1984) Transhyphal ion currents in the water mould *Achlya*. Amino acid proton symport as a mechanism of current entry. *Journal of Cell Biology* **99**, 486–496.

Chapter 7

Fig. 7.7 (Based on Brownlee & Jennings 1981.) Brownlee, C. & Jennings, D.H. (1981) The content of carbohydrates and their translocation in mycelium of *Serpula lacrymans*. *Transactions of the British Mycological Society* **77**, 615–619.

Chapter 8

Fig. 8.1 (From Henry 1932.) Henry, A.W. (1932) Influence of soil temperature and soil sterilization on the reaction of wheat seedlings to *Ophiobolus graminis* Sacc. *Canadian Journal of Research* **7**, 198–203.

Fig. 8.6 (Based on Edwards & Bowling 1986.) Edwards, M.C. & Bowling, D.J.F. (1986) The growth of rust germ tubes towards stomata in relation to pH gradients. *Physiological and Molecular Plant Pathology* **29**, 185–196.

Fig. 8.8 (After Trinci *et al.* 1994.) Trinci, A.P.J., Davies, D.R., Gull, K., *et al.* (1994) Anaerobic fungi in herbivorous animals. *Mycological Research* **98**, 129–152.

Fig. 8.9 (Reproduced from Akhmanova *et al.* 1998, with permission; copyright Macmillan Publishing.) Akhmanova, A., Voncken, F., Van Alen, T., *et al.* (1998) A hydrogenosome with a genome. *Nature* **396**, 527–528.

Fig. 8.10 (Data from Ayerst 1969, for *Aspergillus* spp., and from Magan & Lacey 1984, for *F. culmorum*.) Ayerst, G. (1969) The effects of moisture and temperature on growth and spore germination in some fungi. *Journal of Stored Products Research* **5**, 127–141. Magan, N. & Lacey, J. (1984) Effect of temperature and pH on water relations of field and storage fungi. *Transactions of the British Mycological Society* **82**, 71–81.

Fig. 8.11 (From Northolt & Bullerman 1982.) Northolt, M.D. & Bullerman, L.B. (1982) Prevention of mould growth and toxin production through control of environmental conditions. *Journal of Food Production* **45**, 519–526.

Fig. 8.12 (Adapted from Hallsworth & Magan 1994.) Hallsworth, J.E. & Magan, N. (1994) Effect of carbohydrate type and concentration on polyhydric alcohol and trehalose content of conidia of three entomopathogenic fungi. *Microbiology* **140**, 2705–2713.

Chapter 9

Fig. 9.3 (Based on Oliver 1987.) Oliver, S.G. (1987) Chromosome organization and genome evolution in yeast. In: *Evolutionary Biology of the Fungi* (Rayner, A.D.M., Brasier, C.M. & Moore, D., eds), pp. 33–52. Cambridge University Press, Cambridge.

Fig. 9.4 (Courtesy of Rawlinson *et al.* 1975.) Rawlinson, C.J., Carpenter, J.M. & Muthyalu, G. (1975) Double-stranded RNA virus in *Colletotrichum lindemuthianum*. *Transactions of the British Mycological Society* **65**, 305–308.

Fig. 9.8 (Courtesy of M. Sweetingham; from MacNish & Sweetingham 1993.) MacNish, G.C. & Sweetingham, M.W. (1993) Evidence of stability of pectic zymogram groups within *Rhizoctonia solani* AG-8. *Mycological Research* **97**, 1056–1058.

Fig. 9.9 (After MacNish *et al.* 1993.) MacNish, G.C., McLernon, C.K. & Wood, D.A. (1993) The use of zymogram and anastomosis techniques to follow the expansion and demise of two coalescing bare patches caused by *Rhizoctonia solani* AG8. *Australian Journal of Agricultural Research* **44**, 1161–1173.

Fig. 9.17 (From Peter *et al.* 2003, with permission from the New Phytologist Trust.) Peter, M., Courty, P.-E., Kobler, A., *et al.* (2003) Analysis of expressed sequence tags from the ectomycorrhizal basidiomycetes *Laccaria bicolor* and *Pisolithus microcarpus*. *New Phytologist* **159**, 117–129.

Fig. 9.18 (From Freimoser *et al.* 2003; with permission from the Society for General Microbiology.) Freimoser, F.M., Screen, S., Bagga, S., Hu, G. & St Leger, R.J. (2003) Expressed sequence tag (EST) analysis of two subspecies of *Metarhizium anisopliae* reveals a plethora of secreted proteins with potential activity in insect hosts. *Microbiology* **149**, 239–247.

Chapter 10

Fig. 10.4 (Data from Deacon *et al.* 1983.) Deacon, J.W., Donaldson, S.J. & Last, F.T. (1983) Sequences and interactions of mycorrhizal fungi on birch. *Plant and Soil* **71**, 257–262.

Fig. 10.12 (Courtesy of M.S. Fuller; from Reichle & Fuller 1967.) Reichle, R.E. & Fuller, M.F. (1967).

Fig. 10.14 (Courtesy of M.S. Fuller; from Cho & Fuller 1989.) Cho, C.W. & Fuller, M.F. (1989) Ultrastructural organization of freeze-substituted zoospores of *Phytophthora palmivora*. *Canadian Journal of Botany* 67, 1493–1499.

Fig. 10.16 (Photographs courtesy of F. Gubler & A. Hardham; (a) from Hardham 1995; (b–d) from Gubler & Hardham 1988.) Hardham, A.R. (1995) Polarity of vesicle distribution in oomycete zoospores: development of polarity and importance for infection. *Canadian Journal of Botany* 73(suppl.) S400–407. Gubler, F. & Hardham, A.R. (1988) Secretion of adhesive material during encystment of *Phytophthora cinnamomi* zoospores, characterized by immunogold labeling with monoclonal antibodies to components of peripheral vesicles. *Journal of Cell Science* 90, 225–235.

Fig. 10.17 (From Deacon & Mitchell 1985.) Deacon, J.W. & Mitchell, R.T. (1985) Toxicity of oat roots, oat root extracts, and saponins to zoospores of *Pythium* spp. and other fungi. *Transactions of the British Mycological Society* 84, 479–487.

Fig 10.21 (All data from Warburton & Deacon 1998.) Warburton, A.J. & Deacon, J.W. (1998) Transmembrane Ca^{2+} fluxes associated with zoospore encystment and cyst germination by the phytopathogen *Phytophthora parasitica*. *Fungal Genetics and Biology* 25, 54–62.

Fig. 10.25 (From Lacey 1988; based on the work of P.H. Gregory and J.L. Monteith.) Lacey, J. (1988) Aerial dispersal and the development of microbial communities. In: *Micro-organisms in Action: concepts and applications in microbial ecology* (Lynch, J.M. & Hobbie, J.E., eds), pp. 207–237. Blackwell Scientific, Oxford.

Fig. 10.27 (From Carter 1965.) Carter, M.V. (1965) Ascospore deposition of *Eutypa armeniacae*. *Australian Journal of Agricultural Research* 16, 825–836.

Chapter 11

Fig. 11.11 (Based on Chang & Hudson 1967) Chang, Y. & Hudson, H.J. (1967) The fungi of wheat straw compost: paper I. *Transactions of the British Mycological Society* 50, 649–666.

Fig. 11.14 (From Krauss & Deacon 1994.) Krauss, U. & Deacon, J.W. (1994) Root turnover of groundnut (*Arachis hypogea* L.) in soil tubes. *Plant & Soil* 166, 259–270.

Fig. 11.15 (From Krauss & Deacon 1994.) Krauss, U. & Deacon, J.W. (1994) Root turnover of groundnut (*Arachis hypogea* L.) in soil tubes. *Plant & Soil* 166, 259–270.

Fig. 11.16 (From Lascaris & Deacon 1991.) Lascaris, D. & Deacon, J.W. (1991) Comparison of methods to assess senescence of the cortex of wheat and tomato roots. *Soil Biology and Biochemistry* 23, 979–986.

Fig. 11.17 (Based on Waid (1957) but with additional information and interpretation.) Waid, J.S. (1957) Distribution of fungi within the decomposing tissues of ryegrass roots. *Transactions of the British Mycological Society* 40, 391–406.

Chapter 12

Fig. 12.3 (Reproduced from Raaijmakers & Weller 1998.) Raaijmakers, J.M. & Weller, D.M. (1998) Natural plant protection by 2,4-diacetylphloroglucinol-producing *Pseudomonas* spp. in take-all decline soils. *Molecular Plant–Microbe Interactions* 11, 144–152.

Fig. 12.4 (From Wood *et al.* 1997.) Wood, D.W., Gong, F., Daykin, M.M., Williams, P. & Pierson, L.S. (1997) N-acyl-homoserine lactone-mediated regulation of gene expression by *Pseudomonas aureofaciens* 30–84 in the wheat rhizosphere. *Journal of Bacteriology* 179, 7663–7670.

Fig. 12.5 ((a) Courtesy of Samuels, G.J, Chaverri, P., Farr, D.F. & McCray, E.B. Trichoderma Online, Systematic Botany and Mycology Laboratory, ARS, USDA; from http://nt.ars-grin.gov/taxadescriptions/keys/TrichodermaIndex.cfm)

Fig. 12.10 (Courtesy of P. Jeffries; from Jeffries & Young 1976.) Jeffries, P. & Young, T.W.K. (1976) Ultrastructure of infection of *Cokeromyces recurvatus* by *Piptocephalis unispora* (Mucorales). *Archives of Microbiology* 109, 277–288.

Fig. 12.11 (From van den Boogert & Deacon 1994.) van den Boogert, P.H.J.F & Deacon, J.W. (1994) Biotrophic mycoparasitism by *Verticillium biguttatum* on *Rhizoctonia solani*. *European Journal of Plant Pathology* 100, 137–156.

Fig. 12.19 (Data from Deacon 1985.) Deacon, J.W. (1985) Decomposition of filter paper cellulose by thermophilic fungi acting singly, in combination, and in sequence. *Transactions of the British Mycological Society* 85, 663–669.

Chapter 13

Fig. 13.4 (From van der Heijden *et al.* 1998, with permission from the publisher.) van der Heijden, M.G.A., Wiemken, A. & Sanders, I.R. (1998) Mycorrhizal fungal diversity determines plant diversity, ecosystem variability and productivity. *Nature* 396, 69–72.

Fig. 13.5 (Data from van der Heijden *et al.* 1998, with permission from the publisher, but only some of the plant species are shown in this figure.) van der Heijden, M.G.A., Wiemken, A. & Sanders, I.R. (1998) Mycorrhizal fungal diversity determines plant diversity, ecosystem variability and productivity. *Nature* 396, 69–72.

Fig. 13.25 (Image courtesy of A. Scheussler & M. Kluge; from Schuessler & Kluge 2001.) Scheussler, A. & Kluge, M. (2001) *Geosiphon pyriforme*, an endocytosymbiosis between fungus and cyanobacteria, and its meaning as a model system for AM research. In: *The Mycota*, vol. IX. *Fungal Associations* (Hoch, B., ed.), pp. 151–161. Springer-Verlag, Berlin.

Fig. 13.26 (Courtesy of M.P. Coutts, J.E. Dolezal and the University of Tasmania; see Madden & Coutts 1979.) Madden, J.L. & Coutts, M.P. (1979) The role of fungi in the biology and ecology of woodwasps (Hymenoptera Siricidae). In: *Insect–Fungus Symbiosis* (Batra, L.R., ed.), p. 165. Allanheld, Osmun & Co., New Jersey.

Chapter 14

Fig. 14.9 (Based on Wastie 1960.) Wastie, R.L. (1960) Mechanism of action of an infective dose of *Botrytis* spores on bean leaves. *Transactions of the British Mycological Society* **45**, 465–473.

Fig. 14.14 (Based on Beckman & Talboys 1981.) Beckman, C.H. & Talboys, P.W. (1981) Anatomy of resistance. In: *Fungal Wilt Diseases of Plants* (Mace, C.E., Bell, A.A. & Beckman, C.H., eds), pp. 487–521. Academic Press, New York.

Fig. 14.18 (Source: http://www.aphis.usda.gov/vs/ceah/Equine/eq98endoph.htm)

Fig. 14.20 (Images courtesy of Joseph O'Brien, USDA Forest Service, www.invasive.org; accessed 22 March 2004.)

Fig. 14.21 (Based on a diagram in Voegele & Mendgen 2003.) Voegele, R.T. & Mendgen, K. (2003) Rust haustoria: nutrient uptake and beyond. *New Phytologist* **159**, 93–100.

Chapter 15

Fig. 15.2 ((a),(b) Courtesy of G.L. Barron; (c) courtesy of Shirley Kerr; http://www.kaimaibush.co.nz/; (d),(e) courtesy of Louis Tedders (photographer) and USDA, Agricultural Research Service, www.invasive.org)

Fig. 15.3 (Based on Charnley 1989.) Charnley, A.K. (1989) Mechanisms of fungal pathogenesis in insects. In: *Biotechnology of Fungi for Improving Plant Growth* (Whipps, J.M. & Lumsden, R.D., eds), pp. 85–125. Cambridge University Press, Cambridge.

Fig. 15.5 (Adapted from Whisler *et al.* 1975.) Whisler, H.C., Zebold, S.L. & Shemanchuk, J.A. (1975) Life history of *Coelomomyces psorophorae*. *Proceedings of the National Academy of Sciences, USA* **72**, 693–696.

Fig. 15.6 (Based on data in Wilding & Perry 1980.) Wilding, N. & Perry, J.N. (1980) Studies on *Entomophthora* in populations of *Aphis fabae* on field beans. *Annals of Applied Biology* **94**, 367–378.

Fig. 15.7 (Courtesy of Marlin E. Rice, University of Ohio.)

Fig. 15.8 (Courtesy of B.A. Jeffee; from Jaffee 1992.) Jaffee, B.A. (1992) Population biology and biological control of nematodes. *Canadian Journal of Microbiology* **38**, 359–364.

Fig. 15.9 (Courtesy of B.A. Jeffee; from Jaffee 1992.) Jaffee, B.A. (1992) Population biology and biological control of nematodes. *Canadian Journal of Microbiology* **38**, 359–364.

Fig. 15.10 (Courtesy of B.A. Jeffee; from Jaffee 1992.) Jaffee, B.A. (1992) Population biology and biological control of nematodes. *Canadian Journal of Microbiology* **38**, 359–364.

Fig. 15.11 (Drawn from photographs in Kerry & Crump 1980.) Kerry, B.R. & Crump, D.H. (1980) Two fungi parasitic on females of cyst-nematodes (*Heterodera* spp.). *Transactions of the British Mycological Society* **74**, 119–125.

Chapter 16

Fig. 16.2 (Reproduced by courtesy of the Canadian National Centre for Mycology; http://www2.provlab.ab.ca/bugs/webbug/mycology/dermhome.htm)

Fig. 16.6 (Reproduced by courtesy of the DoctorFungus website; http://www.doctorfungus.org/)

Fig. 16.10 (Reproduced with permission from Centers for Disease Control and Prevention, Image Library; http://www.dpd.cdc.gov/dpdx/HTML/ImageLibrary/Pneumocystis_il.htm)

Chapter 17

Fig. 17.8 (Courtesy of Marek Snowarski, Fungi of Poland; www.grzyby.pl)

Fig. 17.13 (Based on a diagram in Gale *et al.* 1981.) Gale, E.F., Cundcliffe, E., Reynolds, P.E., Richmond, M.H. & Waring, M.J. (1981) *The Molecular Basis of Antibiotic Action*, 2nd edn. Wiley, London.

Systematic index

General index